Biology and Utilization
of the Cucurbitaceae

Biology and Utilization of the Curcurbitaceae

EDITED BY

DAVID M. BATES

L. H. Bailey Hortorium, Cornell University

RICHARD W. ROBINSON

Department of Horticultural Sciences
New York State Agricultural Experiment Station, Geneva
Cornell University

CHARLES JEFFREY

Royal Botanic Gardens, Kew

Comstock Publishing Associates, a division of

Cornell University Press Ithaca and London

First published 1990 by Cornell University Press.

Library of Congress Cataloging-in-Publication Data

Biology and utilization of the Cucurbitaceae / edited by David M. Bates, Richard
W. Robinson, Charles Jeffrey.
    p. cm.
  Bibliography: p.
  Includes index.
  ISBN 0-8014-1670-1 (alk. paper)
    1. Cucurbitaceae.   2. Cucurbitaceae—Utilization.   I. Bates, David M.
II. Robinson, Richard W. (Richard Warren), 1930– .   III. Jeffrey, Charles.
SB351.C8B56   1989
583′.46—dc20   89-42885

Printed in the United States of America

Contents

Preface

The continuing humanization of the earth, and the consequent loss of natural habitats and the germplasm in them, gives urgency to studies concerned with the diversity, workings, preservation, and utilization of the plants and animals that compose our living world. It is imperative that we continue to increase and refine our total knowledge of the biota and that we apply this knowledge, in relevant ways, to better the state of humanity and the earth's environment. In the Cucurbitaceae, popularly known as the gourd family or simply as the cucurbits, the nature and extent of problems and opportunities facing humankind are exemplified, and the ultimate interrelatedness of the pure and applied aspects of science is demonstrated.

The cucurbits are widely distributed through tropical and warm-temperate regions of the world, being represented by some 118 extant genera and 825 species. Most species are components of native ecosystems and often are little known beyond basic descriptive taxonomic accounts. A few species are cultigens of major economic importance. They serve as sources of human nutrition or useful products and have been studied from many perspectives.

For much of humanity, the fruits of cultivated cucurbits characterize the family. Their remarkable variation in size, shape, and color patterning is a notable familial feature, although equally impressive is the prodigious growth of the herbaceous, vining stems of some species. The flesh of the fruits of many species, most notably those of the genera *Cucurbita* (squash, pumpkin), *Cucumis* (cucumber, melon), and *Citrullus* (watermelon), is eaten. Seeds, rich in oils and proteins, are important foods as well. Also sometimes eaten are leaves or shoots (for instance, those of the bitter melon, *Momordica charantia* L.), roots (for example, those of chayote, *Sechium edule* (Jacq.) Swartz), and flowers (commonly of *Cucurbita* species). Edible and industrial oils can be extracted from the seeds. The mature fruits of the calabash, *Lagenaria siceraria* (Mol.) Standl., which ranks among the world's oldest cultigens, provide humans with an enormous array of goods, ranging from containers to musical instruments, while the gourds of *Cucurbita pepo* L. are strikingly ornamental. The fibers of the fruits of *Luffa aegyptiaca* Mill. are vegetable sponges, which even today remain popular bath items. Many cucurbits are important elements of traditional medicine and enter the pharmacopoeias of Old and New World societies.

A striking feature of cultivated cucurbits is their adapatation to a wide variety of agricultural environments. They can be grown in extensive monocultures, typical of crop production in technologcal societies, but are equally at home in kitchen gardens and in traditional systems characterized by low energy requirements. Many cucurbits are adapted to environments considered marginal for agriculture. Some are gathered in the wild as food sources, for example, the fruits of *Acanthosycyos horridus* Welw. ex Hook. f. or the roots of *Coccinia rehmannii* Cogn., both of southern Africa; others are collected for medicinal uses.

From both biological and agricultural perspectives, the cucurbits present an array of intriguing problems. The widespread occurrence of bitter tetracyclic triterpenoids, known as cucurbitacins, provides a basis for studies of coevolution with insects. An ability to control the presence or absence of these compounds has implications for the biological control of insect pests, and their elimination or reduction in harvested parts is a critical step in the development of edible crops. Understanding the underlying controls of unisexuality, ubiquitious among cucurbits and expressed through both monoecy and dioecy, is another intriguing field for investigation. Its relevance to agriculture is seen in the potential to increase fruit production by eliminating dioecy, improving the ratio of female to male flowers, and facilitating the production of F_1 hybrid seed. Questions of broad concern and application can be addressed in studies of reproductive biology, of the genetic basis of character trait expression, and of the physiological nature of adaptations. Systematic investigations of taxa at all hierarchial levels remain intellectually stimulating, and many present added complexity and fascination because they deal with the origin of domesticated species and synergistic relationships with early agriculturists. Such studies draw on a wide range of data and a variety of analytical methods, thereby providing baseline information about phylogenetic relationships and germplasm resources—a necessary prerequisite for the identification of sources of resistance to pathogens, agronomically desirable characteristics, or even new species of economic potential.

As is so often the case with interdisciplinary publications, the genesis for this book was a conference, which in this instance dealt with various aspects of the biology, chemistry, and utilization of the Cucurbitaceae. The conference was held on the campuses of Cornell University, Ithaca, New York, and the New York State Agricultural Experiment Station, Geneva, New York, on August 3–6, 1980. During the conference, plans were made for publishing the proceedings of what was a productive and stimulating meeting of cucurbit specialists. Manuscripts were solicited and submitted, yet for unanticipated and unavoidable reasons, the project did not go forward until late 1985 when plans for the book were revived.

In its rebirth, the book has taken on a somewhat different character from that originally proposed. Because of the time that had elapsed since the conference, we recognized that most of the papers submitted would require revision. Furthermore, we thought that papers sampling recent work should be included in the volume if it was to represent the current state of knowledge and activity in the family. Toward these ends, most individuals who had participated in the conference graciously agreed to revise and update their manuscripts; others, nonparticipants, were invited to prepare papers outlining their research. The result is a book whose contents have remained timely yet have matured through the revision of manuscripts and the addition of new materials.

The contents of the book progress through five loosely defined parts, the first dealing with systematics and evolution, the last with approaches relevant to understanding and improving agricultural productivity. The purpose of this arrangement is to cluster related subjects. In whatever sequence they are read, the chapters demonstrate the interrelatedness of studies on the biology and utilization of cucurbits. They provide a sample of the kinds of problems existing in the family, a survey of the techniques and approaches used to seek solutions, and a sense of the extent of research still to be undertaken.

Part I begins with an overview of the state of systematics in the family. This is followed by surveys of cytology, amino acids, isozymes, and seed oils, each of which has implications for classification. Otherwise, focus is primarily on *Cucurbita* and *Cucumis*, although studies of *Luffa* and *Sechium* are also included. Two chapters, cast in a coevolutionary vein, complete the section. One deals with reproductive biology and natural history of the New World genera *Gurania* and *Psiguria*; the other with cucurbitacins and rootworm beetles, as viewed from both evolutionary and agricultural perspectives.

The second part includes comparative, familywide morphological and anatomical studies of embryos, pollen, trichomes, stomata, and seed coats, as well as a study concerned with adaptive anatomical features of xerophytic Madagascan genera. Although the systematic implications at the generic level and above are stressed, the chapters have a morphological theme that brings them together as a unit.

Four chapters concerned with sex expression in the family constitute Part

III. Genetic and hormonal determinants are discussed, and the ontogeny of male and female flowers in the cucumber is described and interpreted.

The fourth part is concerned with utilization of cucurbits, primarily those species with potential to be used more widely or even brought into cultivation. The focus of the first chapter is on the use of biodynamic compounds in traditional medicine in the American tropics and is illustrative of medicinal uses in other regions of the world. Succeeding chapters emphasize the use of cucurbits as sources of food or energy in low-energetic agricultural settings. Among food plants are the !nara (*Acanthosicyos horridus*) of the Namib Desert, the oyster nut (*Telfaria pedata* (Sims) Hook.) of Africa, and the Chinese lard fruit *(Hodgsonia macrocarpa* (Blume) Cogn.) of Asia. The buffalo gourd *(Cucurbita foetidissima* HBK) of North American deserts merits attention as a source of oil, starch, and, ultimately, energy in xeric regions of the world.

Chapters of the closing part are concerned with improvement and protection of cucurbits as crop plants. They introduce relevant propagation, agronomic, and breeding techniques and effectively illustrate the extent to which basic research on genetics, chemistry, physiology, ecology, and systematics, among other subject areas, not only enhances crop productivity, but also provides fundamental data about the biology of cucurbit crops and their relatives.

In books drawn from the contributions of authors of different specialities and perspectives, consistent application of scientific names throughout the text is seldom possible. In this volume, however, generic names and those at the subfamilial, tribal, and subtribal levels are in accord with Jeffrey's classification of the family, presented in the Appendix. Unfortunately, scientific names at the specific and infraspecific levels have no such universal guide.

Difficulties at the species level and below largely reflect an absence of modern revisions for many genera or honest disagreement among authors concerning species boundaries or the rules governing the application of names. For example, both *Cucurbita* and *Cucumis* are subjects of active systematic investigation, yet there is no modern comprehensive monograph of either to which research workers may refer. Hence, in this book different names are sometimes applied to the same taxon of a genus—for instance, the cushaw appears either as *Cucurbita mixta* Pang. or *C. argyrosperma* subsp. *callicarpa* Merrick & Bates. The application of some names appearing in this book is uncertain. Laura Merrick, for instance, points out that *C. radicans* L. H. Bailey (syn. *C. gracilior* L. H. Bailey) is a distinct species, but in the literature both it and the synonym are widely applied to what is now recognized as *C. argyrosperma* subsp. *sororia* (L. H. Bailey) Merrick & Bates (syn. *C. sororia* L. H. Bailey). When problems with names could be resolved with certainty, we have done so by indicating the accepted name. Such resolution was not always possible, however, nor was it always editorially acceptable to alter the taxonomic concepts and nomenclature of individual contributors. Eventually, these problems will be resolved with the preparation of systematic accounts.

Bringing together the work of scientists with wide-ranging interests serves to heighten both historical awareness and anticipation of the future. While each presentation owes its being to the author or authors who prepared it and who provide new data about and insights into long-standing or newly emerging questions, it also reflects the labor and ideas of the individuals whose work has gone before. Through the collective nature and historical continuity of its contents, the pages of this book summarize what we have learned about cucurbits. But, as neither science nor society as a whole can remains static, these pages are equally a foundation on which future studies can be built.

With a base of knowledge in place and significant new methods available for gathering and analyzing data, the stage is set for an expansion of studies in the pure and applied aspects of cucurbit biology. Just as these approaches are now complementary and interrelated, we suggest that in the future they will become less and less distinguishable. Expanded efforts in systematics and evolution will draw on increasingly refined knowledge of the biochemistry, genetics, ecology, physiology, morphogenesis, and other attributes of the family, much of which will be derived from applied research programs. Similarly, research designed to increase the use and productivity of cucurbits in a broad range of agroecosystems will depend not only on the application of findings from traditional agricultural and horticultural approaches and the rapidly developing field of biotechnology, but also on a detailed understanding of the nature and extent of the cucurbit germplasm, which owes its characterization to genetics and systematics. The prospects for future research on the Cucurbitaceae remain exciting and stimulating and promise both intellectual challenges and benefit to humankind.

Throughout the preparation of the manuscript, generous support was provided by the state of New York and Cornell University. One avenue of support came through the New York State College of Agriculture and Life Sciences and the L. H. Bailey Hortorium, which is a unit of both the college and the Division of Biological Sciences; the other avenue was through the New York State Agricultural Experiment Station, Geneva, and its Department of Horticultural Sciences. We are deeply indebted to all the individuals associated with these organizations who contributed their time and talents to various aspects of the production of the manuscript. Futhermore, since the book originated with a conference dealing with cucurbits, held on the university's campuses in Ithaca and Geneva, our thanks extend to those individuals whose efforts made the conference such an enjoyable and rewarding event.

During early phases of editing, Mary E. Porterfield and Francis VanKirk were responsible for computerizing manuscripts, and, during the last stages, Dorothy L. Johnson shouldered much responsibility. The bulk of computer entry tasks, however, were carried out admirably and with good cheer by Lucille S. Herbert. Preliminary editing of revised manuscripts was done by Tammy Markowitz. Bente King provided original art and graphics and readied plates for publication.

Last, two notes of special thanks: first, to the contributors, whose understanding, patience, and timely cooperation made the task of assembling this volume possible and rewarding, and, second, to Gerald A. Marx, whose friendship and perceptions ultimately led to the book's publication.

DAVID M. BATES
RICHARD W. ROBINSON
CHARLES JEFFREY

Ithaca, New York

Contributors

Thomas C. Andres, Department of Horticultural Sciences, New York State Agricultural Experiment Station, Cornell University, Geneva, New York 14456, USA

David M. Bates, L. H. Bailey Hortorium, Cornell University, Ithaca, New York 14853–4301, USA

W. P. Bemis,[1] Department of Plant Sciences, University of Arizona, Tucson, Arizona 85717, USA

Cyril E. Broderick, University of Monrovia, Monrovia, Liberia

Rebecca M. Cade, Department of Horticultural Science, North Carolina State University, Raleigh, North Carolina 27695–7609, USA

H. L. Chakravarty, 15/4 N. N. Ghosh Lane, Calcutta 700040, West Bengal, India

S. V. S. Chauhan, Department of Botany, R. B. S. College, Agra 282002, Uttar Pradesh, India

M. A. Condon, Herpetology Department, National Zoological Park, Smithsonian Institution, Washington, D.C. 20008, USA

J. B. M. Custers, Institute for Horticultural Plant Breeding (IVT), P.O. Box 16, 6700 AA Wageningen, The Netherlands

A. S. R. Dathan, Department of Botany, University of Rajasthan, Jaipur 302004, Rajasthan, India

Deena S. Decker-Walters, Department of Botany, University of Guelph, Guelph, Ontario, Canada N1G 2W1

Bithi Dutt, Department of Botany, Patna University, Patna 800005, Bihar, India

L. Fowden, Rothamsted Experimental Station, Harpenden, Herts AL5 2JQ, England

---

[1]Deceased.

M. Gangadhara,[1] Department of Biosciences, Sardar Patel University, Vallabh Vidyanagar 388120, Gujarat, India

A. C. Gathman, Department of Biology, Southeast Missouri State University, Cape Girardeau, Missouri 63701, USA

L. E. Gilbert, Department of Zoology, University of Texas, Austin, Texas 78712, USA

Martin C. Goffinet, Department of Horticultural Sciences, New York State Agricultural Experiment Station, Cornell University, Geneva, New York 14456, USA

J. D. Graham, Graham Horticultural Service, 834 Green Cove Drive, Middlesex, New York 14507, USA

Charles B. Heiser, Jr., Department of Biology, Indiana University, Bloomington, Indiana 47405, USA

C. Y. Hopkins, Chemistry Division, National Research Council of Canada, Ottawa, Ontario, Canada K1A OR6

J. A. Inamdar, Department of Biosciences, Sardar Patel University, Vallabh Vidyanagar 388120, Gujarat, India

T. J. Jacks, Southern Regional Research Center, ARS-USDA, P.O. Box 19687, New Orleans, Louisiana 70179, USA

Charles Jeffrey, Royal Botanic Gardens, Kew, Richmond, Surrey TW9 3AB, England

Monique Keraudren-Aymonin,[1] Laboratoire de Phanérogamie, Muséum National de'Histoire Naturelle, 75005 Paris, France

Robert D. Locy, Department of Horticultural Science, North Carolina State University, Raleigh, North Carolina 27695–7609, USA

R. L. Lower, Department of Horticulture, University of Wisconsin, Madison, Wisconsin 53706, USA

J. Brent Loy, Department of Plant Science, University of New Hampshire, Durham, New Hampshire 03824–3597, USA

Laura C. Merrick, Department of Plant and Soil Sciences, University of Maine, Orono, Maine 04469–0118, USA

Robert L. Metcalf, Department of Entomology, University of Illinois, Urbana, Illinois 61801, USA

L. E. Newstrom, Department of Botany, University of California, Berkeley, California 94720, USA

J. Nienhuis, Department of Horticulture, University of Wisconsin, Madison, Wisconsin 53706, USA

A. P. M. den Nijs, Foundation for Agricultural Plant Breeding (SVP), P.O. Box 117, 6700 AC Wageningen, The Netherlands

R. Provvidenti, Department of Plant Pathology, New York State Agricultural Experiment Station, Cornell University, Geneva, New York 14456, USA

Jerzy T. Puchalski, Botanical Garden of the Polish Academy of Sciences, 02–973 Warsaw, Poland

A. M. Rhodes, Department of Horticulture, University of Illinois, Urbana, Illinois 61801, USA

Richard W. Robinson, Department of Horticultural Sciences, New York State Agricultural Experiment Station, Cornell University, Geneva, New York 14456, USA

R. P. Roy, Department of Botany, Patna University, Patna 800005, Bihar, India

Jehoshua Rudich,[1] Faculty of Agriculture, The Hebrew University of Jerusalem, P.O. Box 12, Rehovot, Israel

B. H. Sandelowsky, Tucsin (The University Centre for Studies in Namibia), P.O. Box 11174, Windhoek 9000, Southwest Africa/Namibia

Sunil Saran, Department of Botany, Patna University, Patna 800005, Bihar, India

Edward E. Schilling, Department of Botany, University of Tennessee, Knoxville, Tennessee 37916, USA

Richard Evans Schultes, Botanical Museum of Harvard University, Cambridge, Massachusetts 02138, USA

K. N. Shenoy,[1] Department of Biosciences, Sardar Patel University, Vallabh Vidyanagar 388120, Gujarat, India

Oved Shifriss, Department of Horticulture and Forestry, Cook College, Rutgers—The State University, New Brunswick, New Jersey 08903, USA

Shridhar, Department of Botany, Government Postgraduate College, Sri Ganganagar 335001, Rajasthan, India

A. K. Singh, ICRISAT Center, Patancheru 502324, Andhra Pradesh, India

Dalbir Singh, Department of Botany, University of Rajasthan, Jaipur 302004, Rajasthan, India

Norman F. Weeden, Department of Horticultural Sciences, New York State Agricultural Experiment Station, Cornell University, Geneva, New York 14456, USA

Todd C. Wehner, Department of Horticultural Science, North Carolina State University, Raleigh, North Carolina 27695–7609, USA

Thomas W. Whitaker, 2534 Ellentown Road, La Jolla, California 92037, USA

PART I

Systematics and Evolution

1 | Systematics of the Cucurbitaceae: An Overview

Charles Jeffrey

ABSTRACT.   A brief discussion of the aims of systematics and the philosophical and methodological approaches employed in its practice is provided, in the light of which the present state of cucurbit systematics is reviewed, areas in which further systematic research is needed are identified, and references to recent and current systematic work are given.

Taxonomy is both the queen and servant of biology—queen in that the results of all other fields of biological research contribute to its data base, servant in that it provides a naming service for all practitioners of biology, without which repeatability and therefore the use of the scientific method in biology would be impossible. Most, though not all, taxonomists are staunch monarchists; many other biologists, especially those in applied and practical fields such as horticulture, might be said to wish for a palace revolution. This dichotomy in taxonomy reflects not only different views on the prime function of the discipline but also a profound difference in philosophical approach. One side is well illustrated by the dictum of Holttum (12): "If the diversity among existing plants is due to a process of evolution, an inbuilt system of classification must exist, and our aim must be to discover it." The other side is put succinctly by Erzinclioglu and Unwin (5): "Many taxonomists believe, almost as an act of faith, that a natural classification is there waiting to be discovered. We believe that classification is imposed by man." They argue further that taxonomy has "the main practical function of providing animals [and plants] with unequivocal labels." Philosophically, these two viewpoints can be termed the *empiricist* and the *instrumentalist*, respectively, and they reflect the uncertainty long felt about whether taxonomy should be regarded as a science or an art. Two decades ago, the instrumentalist approach was in the ascendancy (3) in conjunction with the advocacy of phenetic methods (42). Now the wheel has turned full circle and the empiricists prevail, riding the bandwagon of cladistics (11, 30, 32, 46).

The practical consequences of this shift of emphasis in terms of changes in classification, however, have been less than might have been envisaged. This is a result simply of the paucity of our past and present knowledge of the living world—a paucity that has precluded the adoption overall of any methodologically rigorous approach, phenetic or cladistic, yet one that permits the construction of classification by use of the traditional, intuitive, syncretistic approach. The validity of this approach has been discussed by Johnson (22) in his perceptive review of systematics and supported, from an entirely different viewpoint, by Reidl (37, 38).

Taxonomic reality therefore stands somewhat in between the ideals of the philosophical and methodological schools. Given that organisms carry information about the patterns in which the living world is ordered, it is surely the primary task of systematics to utilize that information to produce classifications that are hypotheses of those patterns. The better such hypotheses are, the more useful they will be in answering the questions asked of them by biological science. But there is no certainty in taxonomy—no difference in essence between the wildest speculation and the most well-informed and rigorous study. We can choose between them only on the basis of how well our observations of the properties of the organisms they consider are explained by them. Though undeniably important, nomenclatural stability in classification must always be a consideration secondary to explanatory power. Fortunately, it can often be maximized without loss of the latter by judicious choice of taxonomic ranking. Adherence to the provisions of the International Code of Botanical Nomenclature (8) will ensure that with any given taxonomic position, circumscription, and rank, a taxon will have only one name by which it should properly be known.

A classification of the Cucurbitaceae is presented in the Appendix. It represents the latest stage of a historical development of the classification of the family, which has been reviewed by Jeffrey (13–16) and is a product almost entirely of the intuitive, syncretistic approach. Modern floristic treatments additional to those mentioned by Jeffrey (16) include Porto (35) for Rio Grande do Sul, Brazil; Scholz (40) for central Europe; Telford (44) for Australia; Reekmans (36) for Rwanda; Nazimuddin and Naqvi (29) for Pakistan; Jeffrey (21) for Suriname; Lu An-Ming for China (26); and Chakravarty (1) for India, the last not always in accordance with the conclusions reached by Jeffrey (17–20) in his checklists for eastern Asia and the Indian subcontinent. An account of the cultivated *Citrullus* and *Cucurbita* of the USSR is given by Fursa and Filov (7), and of the genetic resources of Cucurbitaceae by Esquinas-Alcazar and Gulick (6). Treatments of the family for Ethiopia and Nicaragua by the present author are in press. However, the lack of modern monographic treatments of the major genera still persists and represents a serious problem for humanity. Systematic studies are concerned with the basic documentation of plant resources, the provision of names, descriptions, and means of identification, the establishment of geographical distributions, and the construction of systems of classification for data storage and retrieval, which make it possible to predict which plants are likely to have useful or interesting properties. They are therefore essential preliminaries to all programs of rational resource utilization, such as conservation, land use and reclamation, and crop introduction and improvement.

Descriptions of the family and discussions of its taxonomic affinity are given by Takhtajan (43), Jeffrey (13) and Cronquist (2); further discussion is given by Jeffrey (16). While the family is well defined, it is taxonomically isolated and best referred to a monotypic order, Cucurbitales. Its only identified relatives, to date, are the Begoniaceae and Datiscaceae of the order Begoniales, but the wider affinities of these three families as a whole are obscure. The Cucurbitaceae are predominantly tropical, with 90% of the species found in three main areas: Africa and Madagascar, Central and South America, and Southeast Asia and Malesia. There are about 118 extant genera and 825 species.

The two subfamilies are well characterized: Zanonioideae by the small striate pollen grains and Cucurbitoideae by the styles being united into a single column. The variation of the Zanonioideae is sufficiently restricted for its members all to be accommodated in a single tribe, the Zanonieae.

Of the subtribes of Zanonieae, the Gomphogyninae have a distinctive leaf architecture; the Sicydiinae a unilocular, uniovulate ovary; and the Fevilleinae very large, unwinged seeds. The Zanoniinae and Actinostemmatinae are less well characterized, being really no more than groupings of genera that cannot be placed elsewhere and requiring further study. While the genera seem reasonably well established, their relationships within these subtribes require further elucidation, a remark applicable to the subtribes of the Cucurbitoideae as well. Of the genera of Zanonioideae, *Fevillea* and *Siolmatra* are under revision by Robinson (University of South Florida). *Neoalsomitra* and *Hemsleya*, under study in China (24, 48), *Gynostemma*, revised for China (47), and *Sicydium* are most in need of overall revision. The rest of the genera of Zanonioideae are small, with four or fewer species each.

The much larger subfamily Cucurbitoideae is composed of seven tribes, the relationships of which have yet to be clarified. The Sicyeae, as here established on the basis of possession of a unique type of trichomatous nectary (45) and 4- to 10-colporate pollen grains, have two well-marked subtribes: the Cyclantherinae with ascending ovules and punctitegillate pollen, and the Sicyinae with solitary pendulous ovules and spinulose pollen.

Few genera in the Cyclantherinae, probably only *Hanburia, Marah,* and *Rytidostylis,* are satisfactorily characterized. Relationships in the subtribe are obscure and most genera require taxonomic revisions. The entire subtribe Sicyinae is in need of an overall cladistic study. Telford (National Botanic Gardens, Canberra) is unravelling the confusion caused by over-enthusiastic species description in Hawaiian *Sicyos,* and Newstrom (this volume) is studying *Sechium* and its allies.

The tribe Cucurbiteae is well characterized by its pantoporate, spiny pollen. Generic taxonomy and relationships again require clarification, and most genera need revision, especially the economically important *Cucurbita* and the large genus *Cayaponia.* Decker (4, this volume), Merrick (this volume), Andres (this volume), and Nee (New York Botanical Garden) are studying *Cucurbita* and its allies, and Kearns (University of Texas, Austin) is revising the genera *Polyclathra* and *Schizocarpum.*

The Benincaseae are as yet not well characterized as a tribe, although the genera on the whole appear to be well established; their relationships, and especially those of the endemic Madagascan genera with the rest, are still obscure, however. Most lack any overall revision of modern date, but numerous recent floristic treatments have resulted in the species of the majority being fairly well understood. *Luffa* is under study by Heiser and Schilling (9, 10, 39, this volume) and Dutt and Roy (this volume).

The tribe Trichosantheae and its subtribes (except Hodgsoniinae) are characterized by elongation of the hypanthium in the female flowers and by unique pollen morphologies. *Peponium* and the genera of the Trichosanthinae need further taxonomic study, especially *Trichosanthes,* which, while under study in China, where species of the genus have long been important in traditional medicine, requires overall monographic treatment (49).

The Joliffieae, as a tribe, are not very satisfactorily defined, and the relationships of the genera require clarification. The largest genus, *Momordica,* has been fairly well covered in recent floristic treatments. *Siraitia,* another genus important in Chinese traditional medicine, has been revised for China (28), and *Thladiantha* has been studied recently (27). The only genus of the Schizopeponeae, *Schizopepon,* has been revised recently, but its affinities are obscure (25).

The large tribe Melothrieae is another of the lesser known of the family. Of its subtribes, only the Cucumerinae is reasonably well characterized by its fringed anther thecae and seed coat anatomy, yet its relationships to the Guraniinae and the Trochomeriinae are uncertain. Generic relationships and limits in the Melothrieae need study in several areas, especially in the group of genera from *Kedrostis* to *Seyrigia* and those from *Cucurbitella* to *Wilbrandia* (see Appendix). Generic revisions are required, especially for *Ibervillea, Kedrostis, Ceratosanthes, Apodanthera, Psiguria, Gurania, Melothria, Zehneria,* and *Cucumis-Cucumella* treated as a unit. As an economically important genus, *Cucumis* has attracted attention (23, 31, 33, 34, 41) and is currently the subject of a taxonomic revision by Kirkbride (United States Department of Agriculture, Beltsville).

Other workers engaged in a study of Cucurbitaceae include Gomez (Wilson Botanical Garden at Las Cruces, Costa Rica), Lott (Universidad Nacional Autonoma de Mexico), and Sutton (British Museum [Natural History] London), who is writing an account of the family for the *Flora Mesoamericana*.

Literature Cited

1. Chakravarty, H. L. 1982. Cucurbitaceae. *In* K. Thothathri, ed., Fascicles of Flora of India. Fasc. 11. Botanical Survey of India, Howrah.
2. Cronquist, A. 1981. An Integrated System of Classification of Flowering Plants. Columbia Univ. Press, New York.
3. Davis, P. H., and V. H. Heywood. 1963. Principles of Angiosperm Taxonomy. Oliver and Boyd, Edinburgh.
4. Decker, D. S. 1988. Origin(s), evolution, and systematics of *Cucurbita pepo*. Econ. Bot. 42: 4–15.
5. Erzinclioglu, Y. Z., and D. M. Unwin. 1986. The stability of zoological nomenclature. Nature 320: 687.
6. Esquinas-Alcazar, J. T., and P. J. Gulick. 1983. Genetic Resources of Cucurbitaceae. International Board for Plant Genetic Resources Secretariat, Rome.
7. Fursa, T. V., and A. I. Filov. 1982. Tykhvennye. *In* D. D. Brezhnev, ed., Kul'turnaya Flora SSSR. Vol. XXI. Kolos, Moscow.
8. Greuter, W., et al. 1988. International Code of Botanical Nomenclature. Regnum Vegetabile 118. Koelz Scientific Books, Konigstein.
9. Heiser, C. B., and E. E. Schilling. 1988. Phylogeny and distribution of *Luffa* (Cucurbitaceae). Biotropica 20: 185–191.
10. Heiser, C. B., E. E. Schilling, and B. Dutt. 1988. The American species of *Luffa* (Cucurbitaceae). Syst. Bot. 13: 135–145.
11. Hennig, W. 1966. Phylogenetic Systematics. Univ. Illinois Press, Urbana.
12. Holttum, R. E. 1986. Retrospect on a 90th birthday. Kew Bull. 41: 485–489.
13. Jeffrey, C. 1966. Cucurbitaceae. *In* J. C. Willis [H. K. Airy-Shaw], A Dictionary of the Flowering Plants and Ferns, 7th ed. Cambridge Univ. Press, London.
14. Jeffrey, C. 1967. On the classification of the Cucurbitaceae. Kew Bull. 20: 417–426.
15. Jeffrey, C. 1978. Further notes on Cucurbitaceae. V. Some New World taxa. Kew Bull. 33: 347–380.
16. Jeffrey, C. 1980. A review of the Cucurbitaceae. J. Linn. Soc. Bot. 81: 233–247.
17. Jeffrey, C. 1980. Further notes on Cucurbitaceae. V. The Cucurbitaceae of the Indian subcontinent. Kew Bull. 34: 789–809.
18. Jeffrey, C. 1980. The Cucurbitaceae of Eastern Asia. Royal Botanic Gardens, Kew.

19. Jeffrey, C. 1981. The Cucurbitaceae of Eastern Asia. Supplement: Corrigenda and Determination List. Royal Botanic Gardens, Kew.
20. Jeffrey, C. 1982. Further notes on Cucurbitaceae. VI. Cucurbitaceae of the Indian Subcontinent. Corrigenda and addenda. Kew Bull. 36: 737–740.
21. Jeffrey, C. 1984. Cucurbitaceae. In A. L. Stoffers and J. C. Lindeman, eds., Flora of Suriname 6: 457–518. E. J. Brill, Leiden.
22. Johnson, L. A. 1968. Rainbow's end: the quest for an optimal taxonomy. Proc. Linn. Soc. New S. Wales 93: 8–45.
23. Leeuwen, L. van. 1979. Een bijdrage tot de identificatie van wilde Cucumis L. species. L. van Leeuwen, Wageningen.
24. Lu A.-M. 1982. Materials for the genus Hemsleya Cogn. Acta Phytotax. Sin. 20: 87–90.
25. Lu A.-M. 1985. Studies on the genus Schizopepon Max. (Cucurbitaceae). Acta Phytotax. Sin. 23: 106–120.
26. Lu A.-M, and Chen S.-K., eds. 1986. Cucurbitaceae. Flora Reipublicae Popularis Sinica 73: 84–301.
27. Lu A.-M., and Zhang Z.-Y. 1982. A revision of genus Thladiantha Bunge (Cucurbitaceae). Bull. Bot. Res. (Harbin) 1: 61–96.
28. Lu A.-M., and Zhang Z.-Y. 1984. The genus Siraitia Merr. in China. Guihaia 4: 27–33.
29. Nazimuddin, S., and S. S. H. Naqvi. 1984. Cucurbitaceae. In E. Masir and S. L. Ali, eds., Flora of Pakistan. No. 154. Pakistan Agricultural Research Council, Islamabad.
30. Nelson, G., and M. Platnick. 1982. Systematics and Biogeography. Columbia Univ. Press, New York.
31. Pas, H. N. ten, J. W. P. Schoenaker, E. H. Oost, and C. E. Jarvis. 1985. Relectotypification of Cucumis sativus L. Taxon 34: 288–293.
32. Patterson, C. 1982. Cladistics and classification. New Sci. 94: 303–306.
33. Perl-Treves, R., and E. Galun. 1985. The Cucumis plastome: physical map, intrageneric variation and phylogenetic relationships. Theor. Appl. Genet. 71: 417–429.
34. Perl-Treves, R., D. Zamir, N. Navot, and E. Galun. 1985. Phylogeny of Cucumis based on isozyme variability and its comparison with plastome phylogeny. Theor. Appl. Genet. 71: 430–436.
35. Porto, M. L. 1974. Cucurbitaceae. In A. R. Schultz, ed., Flora Illustrado do Rio Grande do Sul. Fasc. VIII. Univ. Fed. Rio Grande do Sul, Porto Alegre.
36. Reekmans, M. 1983. Cucurbitaceae. In G. Troupin, ed., Flora du Rwanda 2: 453–480. Musée Royale de l'Afrique Centrale, Tervuren.
37. Reidl, R. 1978. Order in Living Organisms. Wiley, New York.
38. Reidl, R. 1984. Biology of Knowledge. Wiley, New York.
39. Schilling, E. E., and C. B. Heiser. 1981. Flavonoids and the systematics of Luffa. Biochem. Syst. Ecol. 9: 263–285.
40. Scholz, H. 1979. Cucurbitaceae. In G. Hegi (H. J. Conert et al., eds.) Illustrierte Flora von Mitteleuropa, 2nd ed. Vol 6, part 2. Verlag Paul Parey, Berlin.
41. Singh, A. K., and K. S. Yadava. 1984. An analysis of interspecific hybrids and phylogenetic implications in Cucumis (Cucurbitaceae). Pl. Syst. Evol. 147: 237–252.

42. Sokal, R. R., and P. H. A. Sneath. 1963. Principles of Numerical Taxonomy. Freeman, San Francisco.
43. Takhtajan, A. L. 1966. Sistema i Filogeniya Tsvetkovykh Rastenii. Nauka, Moscow.
44. Telford, I. R. 1982. Cucurbitaceae. *In* A. S. George, ed., Flora of Australia. Vol. 8. Austral. Gov. Publ. Service, Canberra.
45. Vogel, S. 1981. Trichomatische Blutennektarien bei Cucurbitaceen. Beitr. Biol. Pflanzen 55: 325–353.
46. Wiley, E. O. 1981. Phylogenetics. Wiley, New York.
47. Wu C.-Y., and Chen S.-K. 1983. A study on the genus *Gynostemma* Bl. (Cucurbitaceae) from China. Acta Phytotax. Sin. 21: 355–369.
48. Wu C.-Y., and Chen Z.-L. 1985. Materia ad floram Cucurbitaceam Sinensium. *Hemsleya* Cogn. Acta Phytotax. Sin. 23: 121–143.
49. Yue C.-X, and Zhang Y.-L. 1986. Studies on the pollen morphology of Chinese *Trichosanthes*. Bull. Bot. Res. (Harbin) 6: 21–31.

2 | Cytogenetics and Evolution in the Cucurbitaceae

A. K. Singh

ABSTRACT. Cytogenetic investigations of the Cucurbitaceae, with particular reference to the role of cytogenetic mechanisms in species differentiation, are reviewed. On the basis of karyomorphological affinities, cross-compatibilities, chromosomal associations, and pollen fertility in F_1 hybrids, a number of evolutionary trends related to speciation are indicated. In obligate sexual species, trends are found in structural rearrangements, alterations of gene sequences, and differences in the morphology of the chromosome complement, chromosome number, or both. In facultative asexual or perennial species, hybridization and polyploidy are the additional factors involved in species differentiation.

The members of the Cucurbitaceae are widely distributed throughout the tropics and subtropics of the world, and cultivation brings the family into the warmer temperate regions. Systematic studies at the generic and species levels remain challenging (8, 9, 25, Jeffrey, this volume). Cytogenetically, the family has not been well investigated because of technical constraints; however, since the late 1960's improvements in cytogenetic and biochemical methods have provided valuable information on many taxa and have established a better understanding of species differentiation and phylogenetic relationships in several economically important genera, including *Cucumis*, *Cucurbita*, *Citrullus*, *Luffa*, *Trichosanthes*, *Momordica*, and *Coccinia*. These investigations have revealed the main cytogenetic mechanisms responsible for differentiation and evolution of species in the Cucurbitaceae. This chapter discusses the progress that such studies have made in revealing evolutionary trends in various cucurbits.

10

Cucumis

The genus *Cucumis* contains about 30 annual and perennial species. The cultivated species include *C. anguria* L., *C. melo* L., and *C. sativus* L. Among the wild species, 17 have been identified from southern Africa (31) and 13 from tropical eastern Africa (25). Of these, *C. prophetarum* L. f. reaches India, where *C. hardwickii* Royle—alternatively, *C. sativus* var. *hardwickii* (Royle) Alef.—is indigenous.

Cucumis has two base chromosome numbers: $x = 7$ for *C. sativus* and *C. hardwickii*, and $x = 12$ for *C. melo, C. anguria*, and the remaining wild species (11, 12, 50, 67, 83). From phytogeographic evidence and studies of host-parasite relationships, it has been suggested that eastern Africa is the primary gene center and southern Africa the secondary gene center for the species with $2n = 24$, 48, and 72 (30). For *C. melo*, India, Turkey, and Afghanistan have been regarded as secondary gene centers (80), except by Filov and Vilenskaya (18). *Cucumis sativus* apparently was domesticated in India (14). It shares the chromosome number of $2n = 14$ with *C. hardwickii*, which may have been its progenitor.

At one point $x = 7$ was considered primitive in *Cucumis* and was thought to have given rise to $x = 12$ through fragmentation of the chromosomes with secondary constrictions (2, 7). However, this hypothesis could not be confirmed cytologically, and $x = 12$ is now generally regarded as primitive (56, 67, 83). The base number of $x = 7$ probably evolved from $x = 12$ through fusion or unequal translocations between nonhomologous chromosomes, presumably involving subtelocentric chromosomes, with or without the loss of the centromere and heterochromatic regions (56, 67). This process is commonly regarded as a means for reducing chromosome numbers and may result in the evolution of aneuploid series (64). Further credence is given to a reductional hypothesis because $x = 12$ is the predominant base number for *Cucumis* and the Cucurbitaceae (60), *Cucumis* taxa with $x = 12$ have greater phenotypic variability and geographic distribution, and despite the higher chromosome number, they have almost the same half chiasma per chromosome value as taxa with $x = 7$ (84). These factors suggest that both groups of species have a common chromosome complement. The evolution of species with $x = 7$ may have been favored to restrict recombinations by reducing the chromosome number but retaining the same chiasma frequency. This could be a factor contributing to their comparatively restricted phenotypic variability and distribution.

On the basis of karyomorphological affinities, Yadava (83) placed the *Cucumis* species with $2n = 24$ into five major groups. The results suggested that the South African species *C. africanus*[1] L. f. and *C. leptodermis*

[1]Meeuse (31) pointed out that among the African species of *Cucumis* what has been known as *C. africanus* = *C. zeyheri*, that *C. leptodermis* = *C. myriocarpus* subsp. *leptodermis*, and that *C. hookeri* = the true *C. africanus*.

Schweick are the most primitive, thereby strengthening the same earlier conclusions based on the presence of genes conferring disease resistance in the South African species (30). The remaining four groups in the order of relative advancement were: 2) *C. ficifolius* A. Rich., *C. hookeri* Naud., and *C. dipsaceus* Ehrenb. ex. Spach; 3) *C. myriocarpus* Naud., *C. zeyheri* Sond., accession CUCU 44/74, *C. prophetarum*, and *C. anguria*; 4) *C. metuliferus* E. Mey ex. Schrad.; and 5) *C. sagittatus* Peyr. and *C. melo*.

Among the species with $2n = 14$, a collection of *C. hardwickii*, misidentified as *C. callosus* (Rottl.) Cogn., was considered the most primitive because of its symmetrical karyotype and absence of a pair of chromosomes with a secondary constriction. *Cucumis sativus* and other collections of *C. hardwickii*, each with three pairs of chromosomes with a secondary constriction, were considered advanced (56). *Cucumis sativus* was suggested to be the closest relative of *C. melo* based on a similarity in total chromatin length (67). Later investigations (83), however, revealed that the chromatin length of both species with $x = 7$ is similar and is significantly less than that of species with $x = 12$.

A series of interspecific hybridization studies involving *Cucumis* species provided valuable information on species relationships (10, 13, 15, 34, 35, 61). Deakin et al. (13) presented the first comprehensive data on cross-compatibility relationships of *Cucumis* species and on the pollen fertility of F_1 hybrids. As a result, they placed *Cucumis* species into four groups: 1) the Anguria Group, including all the spiny-fruited interfertile annual, perennial, and advanced polyploid species; 2) the Melo Group, comprising the wild, non-spiny-fruited species and the cultivated *C. melo*; 3) the Metuliferus Group, represented by a single species, *C. metuliferus*, having a distinct morphology and being cross-incompatible with other *Cucumis* species; and 4) the Sativus Group, containing the cultivated *C. sativus* and the wild *C. hardwickii* with $2n = 14$.

Recently, the extension of interspecific hybridization studies to the analysis of chromosome associations and pollen and seed fertility has provided a better understanding of species relationships (10, 34, 61). A high degree of chromosome pairing in hybrids between the spiny-fruited species (61) indicated strong genomic homology between them. The hybrids between the species belonging to the third karyomorphological group of Yadava (83), i.e., *C. zeyheri*, *C. prophetarum*, *C. myriocarpus*, accession CUCU 44/74, and *C. anguria*, predominantly showed bivalent associations with a few univalents. Only the hybrids involving *C. prophetarum* as one of the parents occasionally showed multivalent associations. This suggests that in the group, *C. prophetarum* has differentiated from the other species through structural alterations involving more than one chromosome, whereas the other species have differentiated because of intrachromosomal changes. The hybrids between *C. dipsaceus*, a member of the second karyomorphological group, and the species of the third group showed significantly more univalents than the intragroup hybrids, thereby confirming a slightly distinct genetic nature and the group status of *C. dipsaceus*. Other studies involving intergroup hybridizations supported the above conclusions on speciation (10).

Despite different degrees of morphological, cytological, and genetic divergence among the spiny-fruited species, on the whole they are best treated as members of a single large group, the Anguria Group, which can be further subdivided based on cross-compatibility and karyomorphological relationships. Furthermore, karyological specialization has not always proceeded from a symmetrical to an asymmetrical karyotype. Both progressive and retrogressive specialization of karyotypes has occurred through chromosomal alterations, such as unequal translocations and pericentric inversions.

Hybridization attempts between cultivated *C. melo* and spiny- and non-spiny-fruited wild species have not yielded entirely conclusive results (13, 33, 61). Norton and Granberry (35) reported fertile F_1 and F_2 progeny from crossing *C. melo* and *C. metuliferus*, but others have not been successful in making this cross, other than producing abortive embryos (den Nijs and Custers, this volume). When crossed with *C. melo*, non-spiny-fruited wild species may also yield seeds with abortive embryos (13). A collective evaluation of the cytogenetic results in species with $2n = 24$ suggests they fall into at least three broad groups: 1) wild, interfertile, spiny-fruited species, and 2) non-spiny-fruited species that are cross-incompatible with spiny-fruited species and weakly cross-compatible with *C. melo*, and 3) *C. melo*. The relationship of *C. metuliferus* and *C. melo* requires further investigation.

Polyploidy has also led to the evolution of new species in *Cucumis*. A total of seven polyploid species has been reported, of which *C. heptadactylus* Naud., *C. aculeatus* Naud., *C. zeyheri*, *C. pustulatus* Hook. f., *C. membranifolius* Naud., and *C. meeusei* C. Jeffr. are tetraploid, $2n = 48$, and *C. figarei* is a hexaploid, $2n = 72$ (12, 60). Because of the nature chromosomal associations or absence of them and a high percentage of tetrapartite pollen, it is suggested that *C. pustulatus*, *C. membranifolius*, *C. aculeatus*, and *C. zeyheri* are allotetraploids, *C. meeusei* and *C. heptadactylis* are autotetraploids, and *C. figarei* is an autoallopolyploid (12, 60). The autotetraploid species have exhibited a progression toward diplontic behavior in the subsequent generations (60).

Hybridization studies among polyploid and diploid species have revealed their genomic constitution and ancestry (10). The hybrids between tetraploid *C. aculeatus* and diploid *C. ficifolius* showed normal pairing between one set of chromosomes of *C. aculeatus* and the chromosomes of *C. ficifolius*, indicating that *C. ficifolius* is one of the probable parents. On the other hand, the hybrids between *C. aculeatus* and diploid *C. anguria* and diploid *C. zeyheri* showed high univalent frequency, indicating genomic nonhomology and the distant nature of these species. The hybrid between diploid and tetraploid *C. zeyheri* showed pairing between the one set of chromosomes, revealing a common genome between the two and indicating that the diploid *C. zeyheri* is one progenitor of tetraploid *C. zeyheri*. The F_1 hybrid between hexaploid *C. figarei* and tetraploid *C. aculeatus* showed almost normal pairing between the two common sets of chromosomes of the two species, with the remaining chromosomes appearing as univalents and multivalents. Such chromosome associations indicate that *C. aculeatus* is a step in the final evolution of hexaploid *C. figarei*. The multivalent associations observed in this hybrid may be the result of reciprocal trans-locations, allosyndetic pairing, or both.

The foregoing data suggest that in *Cucumis* geographic isolation, struc-tural and numerical alteration of chromosomes, and hybridization are the principal factors in species evolution and differentiation. South African spe-cies with $2n = 24$ are the most primitive. The genus can be divided into two subgenera, *Cucumis* and *Melo*, each with a different chromosome base number, and the subgenus *Melo* can be further subdivided into at least three groups.

Cucurbita

Cucurbita is a New World genus containing about 20 species, of which only five are cultivated (77, 80, 81). Recent evidence from phytogeographic and host-parasite relationships suggest south-central Mexico as the center of origin of this genus (80, 82). *Cucurbita* species can be divided into mesophytic and xerophytic species. The cultivated species apparently origi-nated from mesophytic species and are part of this group.

Weiling (73) was the first to examine this genus cytologically. He re-ported that all species contain 20 small rod-shaped chromosomes. He sug-gested that *Cucurbita* species are secondary polyploids with the base num-ber of $x = 10$, and concluded that cultivated annuals, *C. pepo* L., *C. mixta* Pang., *C. moschata* (Duch. ex Lam.) Duch. ex Poir., and *C. maxima* Duch. ex Lam. have identical genomes, but that *C. ficifolia* Bouché may be genom-ically distinct (see Andres, this volume). Species differentiation among the annuals was thought to have resulted from structural rearrangements through translocations. Allosyndetic pairing in some species hybrids was also observed in this work. Later studies of microsporogenesis in *C. lundelliana* L. H. Bailey, *C. foetidissima* HBK., *C. palmata* S. Wats., and *C. cylindrata* L. H. Bailey and some species hybrids supported Weiling's hypothesis (19).

Interspecific hybridization studies have provided greater information on species differentiation relationships in the genus (3, 4, 5, 38, 39, 75, 78, 79). Whitaker and Bemis (78) were the first to provide a comprehensive account of relationships and evolution of *Cucurbita* species on the basis of cross-compatibility and pollen fertility of the hybrids. The mesophytic species are the most primitive and form a homogeneous group. *Cucurbita lundelliana* has a wide spectrum of cross-compatibility within the mesophytic and cultivated species and on this basis has been considered primitive in the genus (78), although Merrick (this volume) provides a different interpretation. *Cucurbita lundelliana* crosses with *C. sororia* L. H. Bailey and *C. okeechobeensis* (Small) L. H. Bailey to produce fertile F_1 and F_2 progenies. Furthermore, *Cucurbita sororia* and *C. martinezii* L. H. Bailey hybridize with *C. okeechobeensis* to produce fertile offspring. This crossing behaviour suggests a close phylogenetic relationship between all the mesophytic species. The annual *C. sororia* forms a subgroup (6).

   Cucurbita lundelliana crosses with all the five cultivated species. It produces partially fertile hybrids with *C. moschata*, *C. maxima*, and *C. ficifolia*. With *C. pepo* and *C. mixta* it produces shrivelled seeds, which can be raised into plants through embryo culture. Compared with *C. lundelliana*, the other mesophytic species are weakly compatible with the cultivated *Cucurbita* species. *Cucurbita andreana* Naud. and *C. texana* A. Gray, two wild species, cross freely with *C. maxima* and *C. pepo*, respectively, to produce fertile hybrids. A close phylogenetic relationship apparently exists between these wild species and the two cultivated species (75, Decker, this volume).

   Xerophytic species form the other major group of the genus *Cucurbita*. They have been considered an advanced terminal group derived from mesophytic species. Based on cross-compatibility the group can be subdivided into two subgroups, one consisting of *C. palmata*, *C. digitata* A. Gray, and *C. cylindrata*, and the other only *C. foetidissima*, a highly polymorphic species.

   The xerophytic species cross only with the cultivated *C. moschata* to produce vigorous, sterile hybrids (3). Cytological studies of the hybrids suggest that among the xerophytic species *C. foetidissima* is closer to *C. moschata* than to *C. palmata* and *C. digitata*, although the F_1 hybrids of *C. foetidissima* and *C. moschata* show a high univalent frequency (5, 19, 79).

   The cultivated species are cross-compatible among themselves but generally produce partly or entirely sterile hybrids. The cultivated *C. moschata*, which crosses with both the mesophytic and xerophytic groups, has been suggested as a connecting link between wild and cultivated taxa (79). It has also been considered as the main axis of genetic diversity affecting the cultivated species. Yet, evidence suggests that the domesticated species probably came into cultivation independently in three different areas at different periods (79).

   Numerical taxonomic studies and those concerned with cross-compatibility and bee-pollinator/plant relationships broadly support the above grouping of *Cucurbita* species (6, 23, 40). Phenetic analyses of 21 *Cucurbita* species placed 16 species in five groups (40). Two groups included the xerophytic species, one containing the interfertile *C. digitata*, *C. palmata*, (including *C. californica* Torr. ex S. Wats.), *C. cylindrata*, and *C. cordata* S. Wats., and the other *C. okeechobeensis* and *C. lundelliana*. The third group encompassed the mesophytic species, *C. sororia*, *C. gracilior* L. H. Bailey, and *C. palmeri* L. H. Bailey, together with the cultivated *C. mixta*. *Cucurbita maxima* and *C. andreana* formed the fourth group, and *C. pepo* and *C. texana* the fifth group. *Cucurbita moschata*, *C. ficifolia*, *C. pedatifolia* L. H. Bailey, *C. foetidissima*, and *C. ecuadorensis* Cutler & Whitaker were independent and did not cluster. Later, Bemis et al. (6) used correlation values of 0.28 or higher to cluster *Cucurbita* species in ten groups. The species that cluster at these values are genetically cross-compatible, and their progenies do not show genetic breakdown in subsequent generations. Hurd et al. (23) using geographical distribution and the behavior of the bees *Peponapis* and *Xenoglossa*, placed *Cucurbita* species in nine groups (23). Seven groups were mesophytic and two were xerophytic. The five cultivated species fell into different groups. Phylogenetic analyses of *Cucurbita* are now awaited.

A collective evaluation of the preceding data, in conjunction with new archaeological and genetic evidence, indicates that *Cucurbita* is native both to South and North America. Among the cultivated species, *C. maxima* (79) and perhaps *C. ficifolia* (Andres, this volume) are native to South America, while *C. mixta* (Merrick, this volume, as *C. argyrosperma*), *C. pepo*, (Decker, this volume), and *C. moschata* (79) are native to North America. *Cucurbita moschata* is probably closest to the ancestral taxon.

Citrullus and *Praecitrullus*

Citrullus is a xerophytic genus of the Old World tropics containing three species, *C. colocynthis* (L.) Schrad., *C. lanatus* (Thunb.) Matsum. & Nakai, and *C. ecirrhosus* Cogn. A fourth species, *C. fistulosus* Stocks (*C. vulgaris* var. *fistulosus* (Stocks) Stewart), is now referred to the monotypic, Indian genus *Praecitrullus*, as *P. fistulosus* (Stocks) Pang. (36). The primary gene center for *Citrullus* is not known, although tropical Africa or India have been suggested. Cytogenetic investigations support the segregation of *Praecitrullus* from *Citrullus*. The base chromosome number of *C. colocynthis* and *C. lanatus* is $x = 11$, and that of *P. fistulosus* is $x = 12$ (27, 28, 59, 67). In addition to a different base numbers, *P. fistulosus* has chromatin 1.5 times longer than that of *Citrullus* species, has a distinct karyomorphology, and is cross-incompatible with *Citrullus* species.

Citrullus colocynthis and *C. lanatus*, each $2n = 22$, have strong karyomorphological affinities (59). These two species have been crossed successfully (28) and have also been observed to cross in nature to produce partially fertile hybrids (53). These observations confirm close phylogenetic relationships between the two taxa. The appearance of multivalents and univalents in their hybrids indicates that structural rearrangements between and within the chromosomes have probably accompanied or led to the differentiation of the two species. The domesticated forms of C. *lanatus* apparently resulted from selection for fruit size and quality from wild progenitors lacking bitterness in the fruits.

Luffa

The systematics and cytogenetics of *Luffa* have been widely considered (16, 17, 21, 22, 43, 62, 66, and by Heiser and Schilling, and Dutt and Roy, this volume.) The following account summarizes major points. *Luffa* consists of seven species. Four occur in the Old World, ranging from Africa to India and Australia; three occur in the Neotropics. The base and only gametic number is $n = 13$ (46).

Interspecific hybridization studies have provided evidence on species relationships (16, 17, 43, 66). Hybrids between the two cultivated species, L. *cylindrica* (L.) M. J. Roem. and L. *acutangula* (L.) Roxb., showed a high frequency of bivalents and moderate pollen fertility (66). Those between L. *graveolens* Roxb. and both L. *cylindrica* and L. *acutangula* also showed a high bivalent formation but complete pollen sterility, while hybrids between L. *echinata* Roxb. and the three other Old World species showed a higher frequency of univalents and complete pollen sterility (16, 17, 43). These results demonstrate progressively greater genomic divergence.

Results of hybridization studies involving the New World taxa also show variable results. Heiser and his coworkers (21, 22, Heiser and Schilling, this volume) reported that a single hybrid between L. *astorii* Svens. and L. *operculata* (L.) Cogn. was obtained only with difficulty, while hybrids between L. *quinquefida* (Hook. & Arn.) Seem. and L. *operculata* showed pollen fertility of up to 52% and even set some seeds. Finally, hybrids in several combinations were made between Old and New World species. Those of L. *graveolens* and L. *quinquefida* showed pollen fertility of 28% in one plant.

While not entirely consistent with the phylogenetic interpretation of Heiser and Schilling (this volume), cytogenetic data indicate varying degrees of genomic divergence among the species of *Luffa*, but at the same time retention of a relatively high degree of chromosomal homology throughout the genus. *Luffa graveolens* seems to have the most generalized chromosomal complement, as evidenced by its crossability with Old and New World species, and it may be closest to the ancestral type.

Momordica

The genus *Momordica* comprises about 45 species of tropical and subtropical distribution. The primary center of diversity is tropical Africa, where about 30 species occur, but the presence of several species in the Indo-Malayan region suggests that this area may be another gene center (72). *Momordica charantia* L. and *M. dioica* Roxb. ex Willd. are cultivated (8). They also occur in a wild state in eastern India. Recent cytogenetic studies have confirmed $2n = 22$ for *M. charantia* and *M. balsamina* L., and $2n = 28$ for *M. dioica* (7, 41, 47, 68). *Momordica charantia* and *M. balsamina* have fairly strong karyomorphological similarities, although the latter has smaller chromosomes. *Momordica dioica*, a perennial dioecious taxon, differs from the other two species in chromosome number and in having a markedly asymmetrical karyotype (47); however, contrary to an earlier report (41), it is without a heteromorphic chromosome pair related to dioecy.

Momordica dioica appears to be the most advanced of the three investigated species. Its evolution has been accompanied by an increased chromosome number and development of an asymmetrical karyotype, dioecy, and facultative asexual reproduction. The increase in chromosome number probably did not result from fragmentation, because there is an absence of small chromosomes. More likely it resulted from nondisjunction, or delayed separation, of certain pairs of chromosomes during meiotic anaphase. Populations of progenitors of *M. dioica*, carrying extra chromosomes, could have established themselves vegetatively, a strategy suggested in this species by the duplication of certain loci/chromosomes (54). The absence of multivalent associations is probably due to the early evolution of diplontic behavior.

Cytological analysis of *M. dioica* populations collected from the Khasi and the Jayanta Hills in India revealed two cytotypes, a tetraploid (47) and a triploid (1). Quadrivalent and trivalent associations in these two cytotypes suggest an autopolyploid origin. However, evidence is needed to establish a reasonable explanation for their evolution. Considering the facultative asexual nature of *M. dioica*, there are three possibilities: 1) polysomaty, resulting in the production of cells then shoots with a ploidy constitution different from that of the parental shoot; 2) formation of unreduced gametes through restitution nuclei, which following fertilization may result in the production of tetraploid and triploid seeds; and 3) formation of triploids following crosses between diploid and tetraploid cytotypes.

Attempts to cross the two cultivated species, *M. dioica* and *M. charantia*, have failed (47); however, we succeeded in crossing *M. charantia* and *M. balsamina* and obtained two fruits and four seeds after over 200 pollinations (Singh and Yadava unpubl.). The progeny had a high bivalent frequency with a normal meiotic cycle, indicating that although the two taxa are phylogenetically close, they probably have developed prefertilization barriers ensuring their reproductive isolation.

Trichosanthes

Trichosanthes includes about 40 species, of which 16 have been recorded from India. The geographic distribution of the genus indicates either an Indo-Malayan or Chinese center of origin. Recent cytogenetic analyses of four species have provided an account of their phylogenetic relationships (57, 58). The base chromosome number of the genus is apparently $x = 11$. Of the five taxa investigated, *T. cucumerina* L. vars. *cucumerina* and *anguina* (L.) Haines (syn. *T. anguina* L.), *T. lobata* Roxb., and *T. dioica* Roxb. are diploid with $2n = 22$, and *T. tricuspidata* Lour. (syn. *T. palmata* Roxb.) is a hexaploid with $2n = 66$.

The first three taxa are monoecious and show strong karyomorphological affinities to each other. They have identical chromosome morphology and almost similar chromatin length. Their affinities are further suggested by the free cross-compatibility, nearly normal meiotic cycles, and high pollen fertility in F_1 hybrids between them (58). In the main, they appear to have evolved through minor cryptic structural alterations, accompanied by geographic isolation and selection for fruit size. However, occasional multivalent associations also suggest differentiation resulting from reciprocal translocations involving nonhomologous chromosomes. It has been suggested that *T. cucumerina* var. *cucumerina* is the most primitive taxon of the three. It may have given rise to var. *anguina* and *T. lobata* directly. Alternatively var. *anguina* may have arisen from var. *cucumerina*, followed by the evolution of *T. lobata* from a cross between the varieties. The second possibility is unlikely since var. *anguina* is known only in cultivation.

Trichosanthes dioica is dioecious and in cultivation is vegetatively propagated. It differs from the other three diploid taxa in its karyomorphology and dioecy. An earlier report (37) of a heteromorphic pair of chromosomes associated with dioecy has not been confirmed by recent studies (57). An autotriploid has been recorded for *T. dioica* (42). The strong likelihood is that this cytotype has evolved through polysomaty, in which triploid nuclei give rise to triploid shoots asexually. Vegetative propagation of these shoots could then perpetuate the cytotype. The chance of fertilization between a spontaneous unreduced male gamete and a reduced female gamete seems less likely.

Aberrant sex forms with sterile hermaphroditic flowers found in *T. cucumerina* var. *anguina* and *T. dioica* (63) were suggested to represent a case of reversal to an ancestral condition, indicating evolution of dioecy in *T. dioica* from hermaphrodite parents through a monoecious stage (46).

Trichosanthes tricuspidata, which propagates by rhizomatous tubers, has a somatic complement of 66 chromosomes and is a natural hexaploid (55, 57). Counts of $2n = 22$ and $2n = 44$ have also been reported for this species (69, 70), in one investigation under the name *T. bracteata* (Lam.) Voigt. (65). It has been argued that either there are different cytotypes in *T. tricuspidata* or there has been confusion about the identification of the species, possibly with *T. cordata* Roxb. or *T. bracteata*. Detailed observations of the *T. tricuspidata* karyotype showed the total chromatin length to be less than three times that of the investigated diploid species. This reduction in chromatin length has been ascribed to tight coiling, loss of heterochromatin, or both. These features are common in polyploid forms and have often been considered as the main factor for diminution in chromosome size in polyploid taxa with subtelocentric chromosomes. The identification of a set of four homologous chromosomes in the karyogram and multivalent associations up to hexavalents during meiosis suggests that *T. tricuspidata* is either an autopolyploid or a segmental allopolyploid.

Trichosanthes is unusual because of its relatively large chromosomes, and the role of polyploidy in differentiation of populations, either independently or in combination with hybridization (54). It provides a presumed example of the evolution of genetic dioecy from plants with hermaphrodite flowers.

Lagenaria

Lagenaria was earlier thought to be monotypic, including a single cultivated species, but recently six species have been recognized (25, 26). One is the monoecious, cultivated *L. siceraria* (Mol.) Standl. and five are wild, perennial, dioecious species of Africa and Madagascar. Tropical Africa is the probable primary gene center. Recent work has confirmed the existence of two subspecies of *L. siceraria*, one in Asia, and the other in both Africa and the New World (20), presumably introduced in the latter.

The somatic complement of an Indian accession of *L. siceraria* has been studied (42). The base and gametic number of the genus is $x = 11$, although it has been suggested that the present forms of *Lagenaria* are secondary polyploids that evolved from an ancestor with base number $x = 5$, as $5 + 5 + 1 = 11$ (71). *Lagenaria siceraria* has medium-sized chromosomes with mostly median centromeres. Three pairs have a secondary constriction. Phylogenetic conclusions are not possible because of the lack of data for other species. Nevertheless, the wild perennial *L. sphaerica* (Sond.) Naud. has been successfully crossed with *L. siceraria*. In the hybrids, pistillate flowers abort and staminate flowers have poor pollen viability (76).

Coccinia

With the exception of *C. grandis* (L.) Voigt. (syn. *C. indica* Wight. & Arn.), which occurs wild from Africa to the Indo-Malayan region, *Coccinia*, with 30 species, is confined to tropical Africa. Cytological investigations in *C. grandis*, $2n = 24$, revealed the presence of a distinct heteromorphic pair of sex chromosomes in male plants. Diploid males are heterogametic with 2A + XY, and the females are homogametic with 2A + XX. The Y chromosome is conspicuously large and heterochromatic with a secondary constriction. The X chromosome can be identified because it stains darker and has a submedian centromere.

Detailed cytogenetic investigations suggested a *Melandrium* type of sex mechanism in this species, in which Y plays a decisive role in differentiation of male plants (45). The earlier record of sterile gynodioecious forms with 2A + XY has been taken as a reversal to the ancestral hermaphrodite form (29). It has been suggested that this represents an important step in the evolution of dioecious XY mechanism of *C. grandis*, which is more refined than the genetically controlled dioecy found in other cucurbits (46). A natural triploid in *C. grandis* (29) reveals the role of polyploidy in this species in evolving new cytotypes like that of *Trichosanthes dioica* and *Momordica dioica*. All these species are facultative apomicts. Comparative studies on other species of the genus are not available for phylogenetic conclusions. However, *C. grandis* has been successfully crossed as female with *Diplocyclos palmatus* (L.) C. Jeffr. (syn. *Bryonopsis laciniosa* Auct.) (44). All progeny of the cross were female, confirming the XY mechanism of sex differentiation in *C. grandis* and a close phylogenetic relationship between the two taxa.

Melothria and Related Genera

Melothria was once considered the largest genus of the Cucurbitaceae, with some 80 species in the Old and New World tropics (8). Re-evaluation of the genus (8, 24, 25, Appendix, this volume) restricts it to 12 New World species. The remaining species have been transferred to the paleotropical genera *Zehneria*, *Mukia*, and *Solena*, with about 35, 4, and a single species, respectively. All four genera are members of the tribe Melothrieae, but *Melothria*, *Zehneria*, and *Mukia* are members of the subtribe Cucumerinae, while *Solena* is placed in the subtribe Trochomeriinae. Cytological information is limited, and although it indicates a common base number of $x = 12$ in the four genera, that number is widespread in the family.

The majority of species investigated in this complex have $2n = 24$. *Zehneria indica* (Lour.) Keraudren (syn. *Melothria japonica* (Thunb.) Maxim. ex Cogn.) is the single diploid exception with $2n = 22$. Three species, *Solena amplexicaulis* (Lam.) Gandhi (syn. *M. heterophylla* (Lour.) Cogn.), *Zehneria scabra* (L. f.) Sond. (syn. *M. perpusilla* sensu Chakravarty), and *Mukia javanica* (Miq.) C. Jeffr. (syn. *Melothria assamica* Chakravarty), have been reported to have $2n = 48$ (51). Diploid *Mukia maderaspatana* (L.) M. J. Roem. (syn. *Melothria maderaspatana* (L.) Cogn.), the diploid cytotype of *S. amplexicaulis*, and the tetraploid *Mukia javanica* have been studied karyologically (51). The most symmetrical, and presumably the least specialized karyotype, with seven pairs of median chromosomes, occurs in *Mukia maderaspatana*.

Corallocarpus

Corallocarpus is a genus of 17 paleotropical species, of which five occur in India (8, 25). Among the Indian species, the chromosome number for *C. epigaeus* (Rottl.) C. B. Clarke (as *C. conocarpus* Dalz. & Gibs.) has been determined as $2n = 24$ (53), while among the African species, *C. welwitchii* (Naud.) Hook. f. ex Welw. has been reported as $2n = 72$ (32).

In some plants of *C. epigaeus*, 13 and 14 bivalents, rather than 12, were observed in some pollen mother cells. The production of such aneuploid nuclei in plants is not unusual, especially when asexual reproduction dominates (49). It may result from the nondisjunction of chromosomes in premeiotic mitotic cycles or somatic inconsistencies expressed at the gametic level.

Other Taxa of Cucurbitaceae

Besides the genera considered in this chapter, there are many others for which limited cytological information is available and further detailed studies are required. *Sechium*, for example, was once considered monotypic, but now includes as many as eight species (26, Newstrom, this volume). Karyotypic analysis of an *S. edule* (Jacq.) Sw. showed 22 fairly large somatic chromosomes, of which three pairs have a secondary constriction and the remaining pairs submedian constrictions (42). No data for the other species are available. Similar situations exist in *Marah* and *Gomphogyne*. In the former, only *M. macrocarpus* (Greene) Greene (syn. *Echinocystis macrocarpa* Greene), the first undoubted polyploid reported in the genus, is known

(74). In the monotypic *Gomphogyne*, *G. cissiformis* Griff., an auto-polyploid (2n =32) with exceptionally large chromosomes, has been reported (48), but neither of the other two genera of the subtribe Gomphogyninae are known cytologically. Last, even among the economically important taxa additional cytological work could be fruitful, *Benincasa hispida* (Thunb.) Cogn. being a case in point.

Overview of Cytogenetics in the Cucurbitaceae

Cytological analyses in various taxa of Cucurbitaceae have revealed wide variations in chromosome number, structure, and behavior, indicative of a cytological dynamism in the family. Chromosome size is not invariably small. Euploid and aneuploid changes have been involved in the evolution of species and genera. Isolating mechanisms and related aspects of speciation involve both genetic factors and structural changes in chromosomes.

Evolution in Sexual Species

Cytogenetically, speciation in sexual species has occurred principally through chromosomal alterations leading to repatterning of gene sequences. In a majority of cases these are minor structural rearrangements, which have accumulated in the course of evolution and provided better adaptive gene combinations. They may be cryptic or cytologically discernible. In some cases, e.g., *Cucumis sativus* and related taxa, fusion or unequal transloca-tion between nonhomologous chromosomes apparently has reduced the chromosome number to a lower base number. In other cases, nondisjunction of chromosomes, either at the premeiotic mitotic stage or during the meiotic cycle, has produced aneuploid gametes. Outcrossing of such gametes may result in increased chromosome number, which may stabilize at a higher base number, e.g., as is probable in *Momordica*.

Evolution in Facultative Asexual Species

Asexual reproduction in cucurbits is limited, and rather than playing an evolutionary role by itself, has contributed to it by maintaining genotypic variability produced either sexually or asexually. Because of their dual re-productive system, facultative asexual species have been found to be highly labile from the evolutionary point of view. These taxa provide ideal opportunities for introgression, leading to enormous variability, e.g., as is found in *Cucumis figarei*. Furthermore, asexual propagation has helped in establishing polyploid clones and populations.

Interspecific hybridization may lead to new populations that are sustained in nature asexually even though sterile sexually, e.g., in *Citrullus*. There is evidence that euploid or aneuploid nuclei that are produced in somatic tissue by endoduplication or nondisjunction of chromosomes in somatic cells may produce genotypically different shoots. Alternatively, the outcrossing of spontaneous, unreduced or aneuploid gametes has resulted in the sexual production of polyploid populations. Gametes of these kinds are produced as a result of premitotic or meiotic irregularity under adverse environmental conditions, e.g., in *Momordica dioica*, *Trichosanthes dioica*, and *Corallocarpus epigaeus*. Furthermore, polyploidy has permitted sterile hybrids to regain fertility through amphidiploidization and to establish sexual populations, e.g., in *Cucumis* and *Trichosanthes*.

Literature Cited

1. Agrawal, P. K., and R. P. Roy. 1976. Natural polyploids in Cucurbitaceae. I. Cytogenetical studies in triploid *Momordica dioica* Roxb. Caryologia 29: 7–13.
2. Ayyangar, K. R. 1967. Taxonomy of Cucurbitaceae. Bull. Natl. Inst. Sci. India 34: 380–396.
3. Bemis, W. P. 1964. Interspecific hybridization in *Cucurbita*. II. *C. moschata* Poir. × xerophytic species of *Cucurbita*. J. Heredity 54: 285–289.
4. Bemis, W. P. 1970. Polyploid hybrids from the cross *Cucurbita moschata* (Lam.) Poir. × *C. foetidissima* HBK. J. Amer. Soc. Hort. Sci. 95: 529–531.
5. Bemis, W. P., and J. M. Nelson. 1963. Interspecific hybridization within the genus *Cucurbita*. I. Fruit set, seed and embryo development. J. Ariz. Acad. Sci. 2: 104–107.
6. Bemis, W. P., A. M. Rhodes, T. W. Whitaker, and S. G. Carmer. 1970. Numerical taxonomy applied to *Cucurbita* relationships. Amer. J. Bot. 57: 404–412.
7. Bhaduri, P. N., and P. C. Bose. 1947. Cytogenetical investigations in some common cucurbits with special reference to fragmentation of chromosomes as physical basis of speciation. J. Genet. 48: 237–256.
8. Chakravarty, H. L. 1959. Monograph on Indian Cucurbitaceae. Rec. Bot. Survey India 17: 1–234.
9. Chakravarty, H. L. 1966. Monograph on the Cucurbitaceae of Iraq. Tech. Bull. No. 1. Ministry of Agriculture, Baghdad.
10. Dane, F., D. W. Denna, and T. Tsuchiya. 1980. Evolutionary studies of wild species in the genus *Cucumis*. Z. Pflanzenzucht. 85: 89–109.
11. Dane, F., and T. Tsuchiya. 1976. Chromosome studies in the genus *Cucumis*. Euphytica 25: 367–374.
12. Dane, F., and T. Tsuchiya. 1979. Meiotic chromosome and pollen morphological studies of polyploid *Cucumis* species. Euphytica 28: 563–567.
13. Deakin, J. R., G. W. Bohn, and T. W. Whitaker. 1971. Interspecific hybridization in *Cucumis*. Econ. Bot. 25: 195–211.
14. De Candolle, A. L. 1886. Origin of Cultivated Plants. 2nd ed. (reprinted 1959). Hafner, New York.

15. De Ruiter, A. C. 1973. Interspecific Cross Between Cucumber (*C. sativus*) and Muskmelon (*C. melo*). Eucarpia, Montfavet-Avignon.

16. Dutt, B., and R. P. Roy. 1969. Cytogenetical studies in the interspecific hybrids of *Luffa cylindrica* L. and *L. graveolens* Roxb. Genetica 40: 7–18.

17. Dutt, B., and R. P. Roy. 1971. Cytogenetic investigations in Cucurbitaceae. I. Interspecific hybridization in *Luffa*. Genetica 42: 139–156.

18. Filov, O. I., and G. M. Vilenskaya. 1973. Cultivated cucurbits in different languages of the world. Pl. Breed. Abstr. 43: 755.

19. Groff, D., and W. P. Bemis. 1967. Meiotic irregularities in *Cucurbita* species hybrids. J. Heredity 58: 109–111.

20. Heiser, C. B. 1973. Variation in bottle gourd. *In* B. J. Meggers, E. S. Ayensu, and W. D. Duckworth, eds., Tropical Forest Ecosystems in Africa and South America: A Comparative Review. Smithsonian Institute Press, Washington, DC.

21. Heiser, C. B., and E. E. Schilling. 1988. Phylogeny and distribution of *Luffa* (Cucurbitaceae). Biotropica 20: 185–191.

22. Heiser, C. B., E. E. Schilling, and B. Dutt. 1988. The American species of *Luffa* (Cucurbitaceae). Syst. Bot. 13: 138–145.

23. Hurd, P. D., Jr., E. G. Linsley, and T. W. Whitaker. 1971. Squash and gourd bees (*Peponapis*, *Xenoglossa*) and the origin of cultivated *Cucurbita*. Evolution 25: 218–234.

24. Jeffrey, C. 1962. Notes on Cucurbitaceae, including a proposed new classification of the family. Kew Bull. 15: 337–371.

25. Jeffrey, C. 1967. Cucurbitaceae. *In* E. Milne-Redlead and R. M. Polhill, eds., Flora of Tropical Africa. Crown Agents, London.

26. Jeffrey, C. 1980. A review of the Cucurbitaceae. J. Linn. Soc. Bot. 81: 233–247.

27. Khoshoo, T. M. 1955. Cytotaxonomy of Indian species of *Citrullus*. Curr. Sci. 24: 377–378.

28. Khoshoo, T. N., and S. P. Viz. 1963. Biosystematics of *Citrullus vulgaris* var. *fistulosus*. Caryologia 16: 341 352.

29. Kumar, L. S. S., and S. Vishveshwaraiah. 1952. Sex mechanism in *Coccinia indica* Wight & Arn. Nature 170: 330–331.

30. Leppik, E. E. 1966. Searching gene centers of the genus *Cucumis* through host-parasite relationship. Euphytica 15: 323–328.

31. Meeuse, A. D. J. 1962. The Cucurbitaceae of southern Africa. Bothalia 8: 1–111.

32. Meige, J. 1962. Quatrième liste de nombres chromosomiques d'espèces d'Afrique Occidentale. Rev. Cytol. Biol. Veg. 24: 149–164.

33. Nijs, A. P. M. den, and E. H. Oost. 1980. Effect of mentor pollen on pistil-pollen incongruities among species of *Cucumis* L. Euphytica 29: 267–271.

34. Nijs, A. P. M. den, and D. L. Visser. 1985. Relationships between African species of the genus *Cucumis* L. estimated by the production, vigour and fertility of F_1 hybrids. Euphytica 34: 279–290.

35. Norton, J. D., and D. M. Granberry. 1980. Characteristics of progeny from an interspecific cross of *Cucumis melo* with *C. metuliferus*. J. Amer. Soc. Hort. Sci. 105: 174–180.

36. Pangalo, K. I. 1944. A new genus of the Cucurbitaceae. *Praecitrullus*, an ancestor of the contemporary watermelon (*Citrullus* Forsk.). Bot. Zurr. SSSR 29: 200–204.

37. Patel, G. I. 1952. Chromosome basis of dioecism in *Trichosanthes dioica*. Curr. Sci. 21: 343–344.

38. Pearson, O. H., R. Hopp, and G. W. Bohn. 1951. Notes on species crosses in *Cucurbita*. Proc. Amer. Soc. Hort. Sci. 57: 310–322.

39. Rhodes, A. M. 1959. Species hybridization and interspecific gene transfer in the genus *Cucurbita*. Proc. Amer. Soc. Hort. Sci. 74: 546–551.

40. Rhodes, A. M., W. P. Bemis, T. W. Whitaker, and S. G. Carmer. 1968. A numerical taxonomic study of *Cucurbita*. Brittonia 20: 251–266.

41. Richharia, R. H., and P. N. Ghosh. 1953. Meiosis in *Momordica dioica* Roxb. Curr. Sci. 22: 17–18.

42. Roy, R. P. 1973. Cytogenetical Investigations in the Cucurbitaceae. PL-480 Research Project Final Technical Report, June 1967–May 1972. Dept. Bot., Patna Univ., Patna.

43. Roy, R. P., A. R. Mishra, R. Thakur, and A. K. Singh. 1970. Interspecific hybridization in the genus *Luffa*. J. Cytol. Genet. 2: 16–26.

44. Roy, R. P., and P. M. Roy. 1971. An intergeneric cross in the Cucurbitaceae (*Coccinia indica* W. & A. × *Bryonopsis laciniosa* Arn.). Curr. Sci. 40: 46–48.

45. Roy, R. P., and P. M. Roy. 1971. Mechanism of sex determination in *Coccinia indica*. J. Indian Bot. Soc. 50A: 391–400.

46. Roy, R. P., S. Saran, and B. Dutt. 1972. Speciation in relation to the breeding system in the Cucurbitaceae. *In* Y. S. Murty, ed., Advances in Plant Morphology, Sarita Prakashan, Meerut.

47. Roy, R. P., V. Thakur, and R. N. Trivedi. 1966. Cytogenetical studies in *Momordica* L. J. Cytol. Genet. 1: 30–40.

48. Roy, R. P., and R. N. Trivedi. 1966. Cytology of *Gomphogyne cissiformis*. Curr. Sci. 35: 420–421.

49. Sharma, A. K. 1974. Plant Cytogenetics. Vol 2. The Cell Nucleus. Academic Press, New York.

50. Shimotsuma, M. 1965. Chromosome studies of some *Cucumis* species. Rep. Kihara Inst. Biol. Res. 17: 11–16.

51. Singh, A. K. 1974. Cytological studies in *Melothria* L. Ann. Arid Zone Res. 13: 266–268.

52. Singh, A. K. 1975. Cytogenetics of semi-arid plants. II. Cytological studies in *Corallocarpus conocarpus* Dalz. & Gibs of Cucurbitaceae. Curr. Sci. 44: 511.

53. Singh, A. K. 1978. Cytogenetics of semi-arid plants. III. A natural interspecific hybrid of Cucurbitaceae (*Citrullus colocynthis* Schrad. × *C. vulgaris* Schrad.). Cytologia 43: 569–574.

54. Singh, A. K. 1979. Cucurbitaceae and polyploidy. Cytologia 44: 897–905.

55. Singh, A. K., and R. P. Roy. 1973. Cytological studies in *Trichosanthes palmata* Roxb. A natural hexaploid. Sci. Cult. 39: 505–506.

56. Singh, A. K., and R. P. Roy. 1974. Karyological studies in *Cucumis* L. Caryologia 27: 153–160.

57. Singh, A. K., and R. P. Roy. 1979. Cytological studies in *Trichosanthes* L. J. Cytol. Genet. 14: 50–51.

58. Singh, A. K., and R. P. Roy. 1979. An analysis of interspecific hybrids in *Trichosanthes* L. Caryologia 32: 329–334.

59. Singh, A. K., and K. S. Yadava. 1977. Cytomorphological studies in *Citrullus* L. Biol. Contemp. 4: 168–172.
60. Singh, A. K., and K. S. Yadava. 1984. Cytogenetics of *Cucumis* L. IV. Comparative study of natural and induced polyploids. Cytologia 49: 69–78.
61. Singh, A. K., and K. S. Yadava. 1984. An analysis of interspecific hybrids and phylogenetic implications in *Cucumis* (Cucurbitaceae). Pl. Syst. Evol. 147: 237–252.
62. Singh, D., and M. M. Bhandari. 1963. Identity of an imperfectly known hermaphrodite *Luffa*, with a note on related species. Baileya 11: 132–141.
63. Singh, R. N. 1953. Studies on the sex expression, sex ratio and floral abnormalities in the genus *Trichosanthes*. Indian J. Hort. 10: 98–106.
64. Stebbins, G. L. 1971. Chromosomal Evolution in Higher Plants. Edward Arnold, London.
65. Thakur, G. K. 1973. A natural tetraploid in genus *Trichosanthes* from Bihar. Proc. 55th Indian Sci. Congr. Vananasi, Part 3: 334 (Abstract).
66. Thakur, M. R., and B. Choudhury. 1967. Interspecific hybridization in the genus *Luffa*. Indian J. Hort. 1: 87–94.
67. Trivedi, R. N., and R. P. Roy. 1970. Cytological studies in *Cucumis* and *Citrullus*. Cytologia 35: 561–569.
68. Trivedi, R. N., and R. P. Roy. 1972. Cytological studies in the genus *Momordica*. Genetica 43: 282–291.
69. Vavilov, N. I. 1931. The role of central Asia in the origin of cultivated plants. Bull. Appl. Bot. Genet. Pl. Breed. 26: 1–24.
70. Verghese, B. M. 1971. Cytology of *Trichosanthes palmata* Roxb. Cytologia 36: 205–209.
71. Verghese, B. M. 1973. Cytology and origin of tetraploid *Trichosanthes palmata* Roxb. Genetica 42: 292–301.
72. Verghese, B. M. 1973. Studies on the cytology and evolution of south Indian Cucurbitaceae. Ph.D dissertation, Kerala Univ., India.
73. Weiling, F. 1959. Genomanalytische Untersuchungen bei Kürbis (*Cucurbita* L.). Zuchter 29: 161–179.
74. Whitaker, T. W. 1950. Polyploidy in *Echinocystis*. Madrono 10: 109–211.
75. Whitaker, T. W. 1951. A species cross in *Cucurbita*. J. Heredity 42: 65–69.
76. Whitaker, T. W. 1971. Endemism and pre-Colombian migration of the bottle gourd, *Lagenaria siceraria* (Mol.) Standl. *In* C. L. Riley, ed., Man Across the Sea: Problems of pre-Colombian contacts: Univ. Texas Press, Austin.
77. Whitaker, T. W. 1974. Cucurbitales. *In* Encyclopaedia Britannica, 15th ed. Encyclopaedia Britannica, Chicago.
78. Whitaker, T. W., and W. P. Bemis. 1965. Evolution in the genus *Cucurbita*. Evolution 18: 553–559.
79. Whitaker, T. W., and W. P. Bemis. 1975. Origin and evolution of the cultivated *Cucurbita*. Bull. Torrey Bot. Club. 102: 362–368.
80. Whitaker, T. W., and W. P. Bemis. 1976. Cucurbits: *Cucumis, Citrullus, Cucurbita, Lagenaria* (Cucurbitaceae). *In* N. W. Simmonds, ed., Evolution of Crop Plants. Longman, London.
81. Whitaker, T. W., and G. N. Davis. 1962. Cucurbits: Botany, Cultivation and Utilization. Leonard Hill, London.

82. Whitaker, T. W., and R. J. Knight, Jr. 1980. Collecting cultivated and wild cucurbits in Mexico. Econ. Bot. 34: 312–319.
83. Yadava, K. S. 1982. Cytogenetic investigation in Cucurbitaceae. Ph.D. dissertation, Univ. Jodhpur, India.
84. Yadava, K. S., A. K. Singh, and H. C. Arya. 1984. Cytogenetic investigation in *Cucumis* L. I. Meiotic analysis in twenty-four *Cucumis* species. Cytologia 49: 1–9.

3 | Amino Acids as Chemotaxonomic Indices

L. Fowden

ABSTRACT. Certain unusual nonprotein amino acids occur in members of the Cucurbitaceae. Information concerning their nature and distribution may be helpful in refining relationships at the generic, subtribal, or tribal levels of classification. A large measure of agreement is seen between groupings reached on such chemical criteria and those resulting from the use of more traditional characters.

A large proportion of the nitrogen taken up by plants invariably is incorporated into 20 amino acids that are present mainly in proteins, although smaller amounts exist free in the tissues of plants. Lesser quantities of nitrogen are found in other essential materials such as the nucleic acids, chlorophyll, and auxin. In addition, several groups of nitrogenous secondary products are recognized. These include the alkaloids, the cyanogenetic glucosides, the glucosinolates, and the nonprotein amino acids, a class of amino acids not normally encountered in protein molecules. Compounds representative of these groups do not occur universally in members of the plant kingdom and thus are not considered to play key roles in intermediary plant metabolism. Although by definition secondary products are not essential for the maintenance of life in an individual plant, they nevertheless may be important for the survival of the species as a whole in a particular natural habitat. The restricted distributions of individual compounds, which reflect evolutionary trends in the biosynthetic enzyme complements of plants, enable them to be used as chemotaxonomic markers in plant classification. While all groups of nitrogenous secondary products have proved to be useful indices for chemical taxonomy, the nonprotein amino acids undoubtedly have been used most widely and successfully.

Amino Acids Characterizing Members of the Cucurbitaceae

Several amino acids were first identified and characterized as constituents of cucurbit species, and a number of these compounds are still known to exist only in selected members of the family. Citrulline (α-amino-δ-ureido-pentanoic acid, Figure 1, I) acquired its trivial name following isolation in 1930 (12) from watermelon, *Citrullus lanatus* (Thunb.) Matsum. & Nakai. It is an intermediate in the biosynthetic pathway leading to arginine and, a priori, is then present in all plants, albeit at very different concentrations in different species. Watermelon was also the original source of an unusual pyrazole-containing amino acid, whose structure was established (8) as β-pyrazol-l-ylalanine (Figure 1, II). This heterocyclic β-substituted alanine occurs together with its γ-glutamyl peptide (Figure 1, III) in the seeds of certain cucurbit species (1). These compounds can constitute a significant fraction of the total free amino acids present in such seeds. It is rare to encounter natural products that contain linked nitrogen atoms, and the occurrence of the pyrazole ring system is very restricted. In addition to its presence in these cucurbit amino acids, the heterocycle is present in the structures of pyrazomycin, an antifungal and antiviral agent produced by *Streptomyces candidus* (10); and of withasomnine, an alkaloid present in the roots of the solanaceous drug plant of India, *Withania somnifera* (L.) Dunal (19). Occasionally, the common amino acid amides, glutamine and asparagine, are encountered in plants as substituted forms containing alkyl groups attached to their amide-N atoms (4, 5). The N^4-substituted as-paragines appear to be restricted to a very few cucurbit species, and the alkyl residues identified (2) have been methyl, ethyl, and 2-hydroxyethyl structures (Figure 1, IVa, b, and c, respectively). A further β-substituted alanine (*m*-carboxyphenylalanine, Figure 1, V) occurs in seeds of a variety of cucurbits. This amino acid has been identified as a constituent of plants assigned to a number of families, being initially isolated from the genus *Iris* of the Iridaceae (11). Chinese workers (3) isolated and characterized cucur-bitin (3-aminopyrrolidine-3-carboxylic acid, Figure 1, VI) from *Cucurbita moschata* (Duch. ex Lam.) Duch. ex Poir. This interesting structure, com-bining the features of an α-amino acid and a cyclic β-imino acid, is claimed to possess anthelmintic properties.

When chromatography is used to reveal the amino acids present in plant extracts, it is usual to record a number of unidentified constituents, and in this respect, the species forming the Cucurbitaceae are no exception. In our survey of some 50 cucurbits (2), about 20 unknown ninhydrin-positive compounds were recognized.

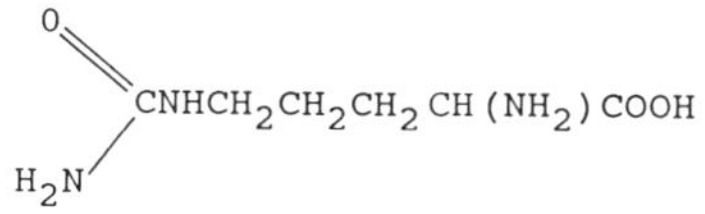

O
‖
H_2N–CNHCH$_2$CH$_2$CH$_2$CH(NH$_2$)COOH

I

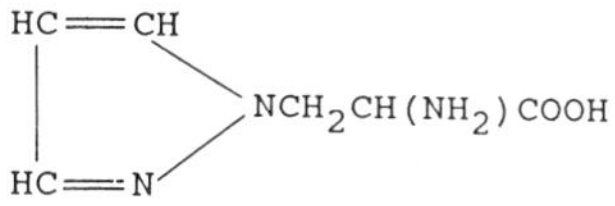

HC═CH
│
HC═N⟩NCH$_2$CH(NH$_2$)COOH

II

HC═CH
│
HC═N⟩NCH$_2$CHCOOH
                    │
                   NH
                    │
         OCCH$_2$CH$_2$CH(NH$_2$)COOH

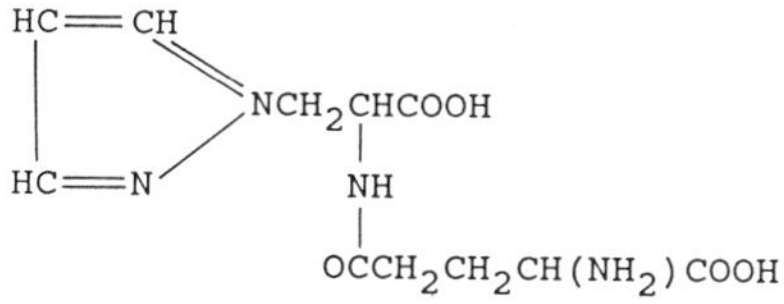

III

HOOCCH(NH$_2$)CH$_2$CONHR

a,   R = CH$_3$

b,   R = CH$_2$CH$_3$

c,   R = CH$_2$CH$_2$OH

IV

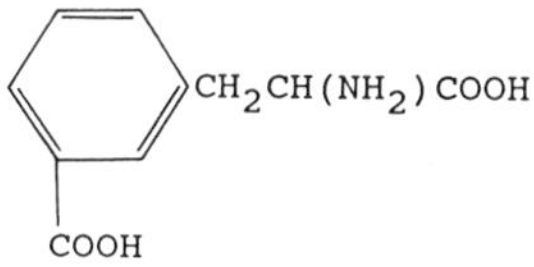

CH$_2$CH(NH$_2$)COOH

COOH

V

NH$_2$

COOH

N
H

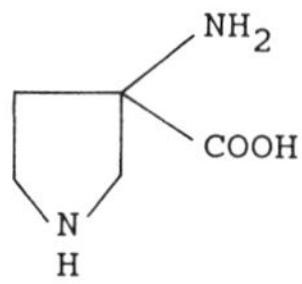

VI

Figure 1. Nonprotein amino acids in members of the Cucurbitaceae. I. Citrulline (α-amino-δ-ureidopentanoic acid). II. β-Pyrazol-l-ylalanine. III. γ-Glutamyl peptide. IV. N⁴-substituted asparagines: (a) methyl, (b) ethyl, (c) 2-hydroxyethyl residues. V. *m*-Carboxyphenylanine. VI. Cucurbitin (3-aminopyrrolidine-3-carboxylic acid).

Specificity of Amino Acid Synthesis

The presence of a particular nonprotein amino acid in a plant species requires that a complex of enzymes needed for the consecutive steps of its synthesis be present in the plant's cells and that their activities should be expressed. The absence of any enzyme operating early in the pathway will result in a failure to synthesize the final amino acid. Under these circumstances an intermediary compound may accumulate. Furthermore, if the product of the absent enzyme is supplied to the plant, then the later stages of the biosynthetic pathway may proceed. This last situation was encountered when β-pyrazol-l-ylalanine (βPA) biosynthesis was examined (4) in three cucurbit species: *Cucurbita maxima* Duch. ex Lam., *Luffa cylindrica* (L.) M. J. Roem., and *Sicyos angulatus* L. Analysis of plant extracts indicated that none of these species normally produced the amino acid, but when pyrazole was supplied to their seedlings, βPA accumulated after three days in concentrations comparable with those present in βPA-producing species. Other experiments established that a number of noncucurbit species possessed a similar ability to synthesize βPA if pyrazole was provided. The unique ability of cucurbits to produce βPA therefore apparently depends on their possession of an enzyme complex specifically capable of forming pyrazole.

Regrettably, the precursors of pyrazole and the mechanism whereby the heterocycle is produced in cucurbit species remain unclear. Biogenetic hypotheses have sought to involve 2,4-diaminobutyric acid and/or 1,3-diaminopropane as precursors, but these necessitate an unlikely step in which distal nitrogen atoms on a 3-carbon skeleton become joined. The unspecific nature of the enzymic reaction forming βPA from pyrazole became more evident with the demonstration that when several other heterocycles were supplied to cucumber seedlings, the corresponding β-substituted alanines were produced in a facile manner (4).

Different considerations of enzyme specificity underlie the biosynthesis of the N^4-substituted asparagines. The formation of compounds of this type requires the presence within the producer species of an enzyme capable of effecting the replacement (or exchange) of the amido-N grouping in asparagine by a residue deriving from the appropriate substituent amine. The properties of such an amidotransferase enzyme present in seedlings of *Ecballium elaterium* (L.) A. Rich. were studied in some detail, and it was shown that the specificity requirements were not very stringent (5). A range of primary alkyl amines, i.e., where R_2 possesses from one to five carbon atom skeletons, could act as substrates for the enzyme with formation of the corresponding N^4-alkylasparagine. While asparagine ($R_1 = H$) is regarded as the normal physiological substrate, the enzyme could also catalyze the interchange of N^4-alkyl substituents. For example, N^4-methylasparagine ($R_1 = CH_3$) underwent reaction with ^{14}C-ethylamine to yield labelled N-ethylasparagine, and this reaction proceeded more rapidly than that between asparagine itself and ethylamine. We may therefore conclude that the range of N^4-substituted asparagines encountered in species possessing the amidotransferase is determined by the amines available within the plant at the subcellular sites of the enzyme.

$$HOOCCH(NH_2)CH_2CONHR_1 + R_2NH_2 \xrightarrow{\text{transferase}}$$
$$HOOCCH(NH_2)CH_2CONHR_2 + R_1NH_2$$

There is no evidence that the amidotransferase activity in the reaction is

ATP-dependent, as is the case with asparagine synthetase, and its substrate affinity apparently is restricted to primary amines, i.e., dimethylamine is unreactive.

A similar transferase enzyme was demonstrated in seedlings of *Bryonia dioica* Jacq. *Bryonia* is one of only three other genera in which N^4-substituted asparagines have been detected, and experiments with extracts of other cucurbit species have failed to establish transferase activity. The enzyme was shown to be distinct from that responsible for the synthesis of N^5-substituted glutamines, of which the most familiar is N^5-ethylglutamine (theanine), present in the leaves of the tea plant.

Amino Acid Distribution and Cucurbit Classification

Studies concerning the distribution of amino acids within members of the Cucurbitaceae are few in number, and that of Dunnill and Fowden (2) remains the most extensive. The results of their study were published shortly after Jeffrey had completed a major revision of the classification of the family (6), and slightly modified it by recognizing the importance of pollen morphology (7). He concluded that the modification definitely delimited the natural limits of subfamilies, tribes, and subtribes. This version contained two subfamilies, Cucurbitoideae and Zanonioideae. The Cucurbitoideae was divided into eight tribes, of which five were further separated into subtribes. The Zanonioideae possessed only one tribe, divided into four subtribes. It is convenient to review information on amino acid distribution against the framework of Jeffrey's division of the family (see Appendix).

 Amino acid analyses are based on seed contents, partly because of the convenience of seed materials, but also because the composition of seeds tends to be less affected than that of vegetative tissues by nutritional and environmental factors to which the plant is subjected. The survey (2) of seed composition was limited; the main comparison involved just over 50 species, and two more detailed surveys of species in the genera *Cucumis* and *Cucurbita* were also undertaken. In making comparisons of composition, emphasis was placed on occurrence of the amino acids I, II, III, IVa, b, c, and V (Figure 1). In some cases, unidentified amino acids present as characteristic seed components have also proved useful in establishing chemotaxonomic affinity or distinction. Examples of seed compositions are included in Table 1, where concentrations of individual amino acids are indicated on a relative scale.

The genera assigned to individual subtribes generally had similar compositions, but some striking anomalies were encountered. Distinct differences of composition could often be identified between subtribes or tribes, but this was not invariably so, and some groups of genera clearly separated by morphological characters had very similar amino acid compositions.

Table 1. Nonprotein amino acids in the seeds of the Cucurbitaceae

Taxon	I	II	III	IVa,b,c,	V
Subfamily Cucurbitoideae					
Tribe Melothrieae					
Subtribe Dendrosicyinae					
Apodanthera undulata A. Gray	W	0	0	0	S[2]
Corallocarpus bainesii (Hook. f.) Meeuse	W	0	0	0	M
C. epigaeus (Rottl.) C. B. Clark	M	0	0	S (IVa)	M
C. grevei (Keraudren) Keraudren)	M	0	0	0	M
Kedrostis africana Cogn.	M	S	M	0	0
K. elongata Keraudren	T	0	0	W (IVa)	W
K. foetidissima (Jacq.) Cogn.	M	0	0	0	T
Subtribe Cucumerinae					
Cucumis melo L.	M	S	S	0	0
Melothria japonica (Thunb.) Maxim. ex Cogn.					
(= *Zehneria indica* (Lour.) Keraudren)	T	W	W	0	W
M. pendula L.	S	S[2]	S	0	0
M. scabra Naud.	W	S[2]	S	0	0
Mukia maderaspatana (L.) M. J. Roem.	M	M	W	0	0
Subtribe Trochomeriinae					
Ctenolepis cerasiformis C. B. Clarke	M	S	M	0	0
Dactyliandra welwitschii Hook. f.	W	S	M	0	0
Tribe Joliffieae					
Subtribe Thladianthinae					
Momordica cymbalaria Fenzl. ex Hook. f.	S[2]	0	0	0	0
Tribe Trichosantheae					
Subtribe Ampelosicyinae					
Peponium hirtellum Keraudren	M	S[2]	M	0	0
Subtribe Trichosanthinae					
Trichosanthes tricuspidata Lour.	W	0	0	0	S
Tribe Benincaseae					
Subtribe Benincasinae					
Acanthosicyos horridus Welw. ex Hook. f.	M	S[2]	M	0	0
Benincasa hispida (Thunb.) Cogn.	M	S[2]	S	0	0
Bryonia dioica Jacq.	S	W	W	S[2] (IVb) M (IVc)	M
Citrullus colocynthis (L.) Schrad.	S	S[2]	S	0	0
Cogniauxia podolaena Baill.	T	0	0	0	M
Ecballium elaterium (L.) A. Rich	M	W	W	W (IVb)	M
Lagenaria abyssinica (Hook. f.) C. Jeffr.	W	M	W	0	0
Subtribe Luffinae					
Luffa cylindrica (L.) M. J. Roem.	W	0	0	0	S
Tribe Cucurbiteae					
Cayaponia sp.	T	0	0	0	0
Cucurbita maxima Duch. ex Lam.	W	0	0	0	W
Sicana odorifera Naud.	W	0	0	0	S
Tribe Sicyeae					
Subtribe Cyclantherinae					
Cyclanthera brachystachya (Ser.) Cogn.	W	0	0	0	S
Echinopepon wrightii Cogn.	W	0	0	0	S[2]
Marah macrocarpus (Greene) Greene	T	0	0	0	S
Subtribe Sicyinae					
Sicyos angulatus L.	T	0	0	0	0

Table 1.   (Continued)

Taxon	Amino acid[1]				
	I	II	III	IVa,b,c,	V
Subfamily Zanoniodeae					
Tribe Zanonieae					
Subtribe Fevilleinae					
Fevillea peruviana (Huber) C. Jeffr.	T	0	0	0	0
Subtribe Zanoniinae					
Alsomitra macrocarpa (Blume) M. J. Roem.	S	0	0	0	0
Gerrardanthus grandiflorus Gilg. ex Cogn.	S	0	0	0	0
Subtribe Gomphogyninae					
Gynostemma pentaphyllum (Thunb.) Mak.	S	0	0	0	0

[1]I, citrulline; II, β-pyrazol-I-ylalanine; III, γ-glutamyl peptide; IV, N^4-substituted asparagines; (a) methyl, (b) ethyl, (c) 2-hydroxyethyl residues; V, *m*-carboxyphenylalanine. 0, absent; T, trace; W, weak concentration; M, medium concentration; S, strong concentration.
[2]Major amino acid constituent.

The largest single group of species included in the survey belonged to the subtribe Benincasinae. Table 1 shows the composition of approximately half the number of this group examined. The revised classification of Jeffrey (6) included *Acanthosicyos* and *Ecballium* within this subtribe, two genera earlier split off as a distinct subtribe, the Acanthosicyinae. Jeffrey's reason for placing these genera in the Benincasinae was that palynology indicated the Acanthosicyinae are not a natural group, *Acanthosicyos* being closer to *Citrullus* and *Ecballium* to *Bryonia* in pollen type than they are to each other. This conclusion is certainly supported by the chemical evidence. *Ecballium* and *Bryonia* differ from all other members of the subtribe, including *Acanthosicyos*, by their content of N^4-ethyl- and N^4-hydroxyethylasparagine. The chemical separation of these two genera aligns with their geographic distinctness as the only European members of the Benincasinae, which is formed mainly of species of African origin. *Acanthosicyos* and *Citrullus* have closely similar compositions; they possess a high content of βPA, a characteristic common to most species in the subtribe. On chemical grounds, *Cogniauxia* appears an exceptional member of the group, and more closely resembles *Luffa*, which is split off as the only genus in Luffinae (a related subtribe). The tendril and androecial characters of *Cogniauxia* are the least specialized within the members of the Benincasinae, and the genus may be regarded as forming a link between the rest of the subtribe and the Luffinae. Many years ago a species of *Cogniauxia* was referred to the genus *Luffa*, pending knowledge of its female flowers and fruit (13).

The structure of the fruits of *Luffa* species resemble, most closely within the family, those of members of the subtribe Cyclantherinae. Both groups showed similar amino acid compositions. Their seeds do not contain βPA or its γ-glutamyl peptide, but all species synthesize *m*-carboxyphenylalanine. The Cyclantherinae are more specialized metabolically than the Luffinae, as seen in their ability to synthesize unidentified amino acids.

The Melothrieae forms one of the most diverse tribes on chemical grounds. The *Cucumis* and *Mukia* species of Cucumerinae that were examined displayed compositions, including βPA and comparable unidentified constituents, almost identical to those of the African genera constituting the Benincasinae and the two members analyzed from the subtribe Trochomeriinae. In contrast, the genus *Apodanthera* (Dendrosicyinae) lacks βPA but, like members of the Cyclantherinae, possesses a high content of *m*-carboxyphenylalanine. Another chemical variant is encountered in certain species of *Corallocarpus* and *Kedrostis* (also Dendrosicyinae) that uniquely synthesize N⁴-methylasparagine. Morphologically, the Melothrieae have been insufficiently studied since they exhibit considerable diversity in many characters. Pollen morphology suggests a division of the subtribe into two differentiated groups. Among the genera analyzed, *Cucumis* and *Mukia* fall into the same group, and therefore their chemical likeness is compatible with the morphological information. Morphologically, *Kedrostis* is a heterogeneous genus, and we have noted chemical diversity to occur among the few species examined. The genus *Melothria* is also diverse, New World species being combined with others of Asiatic origin. Jeffrey (6) suggested that the genus properly might be confined to New World members, leaving the others to be reestablished as *Zehneria* species. Certainly the amino acid chromatograms prepared from *Melothria japonica* (Thunb.) Maxim. ex Cogn. (= *Zehneria indica* (Lour.) Keraudren) of Asia differed entirely from those derived from *M. pendula* L. and *M. scabra* Naud. of the New World, which contrasts with seed coat and palynological similarity (see Singh and Dathan,and Shridhar and Singh, this volume).

The tribe Trichosantheae includes both Asian and African species. *Peponium* species, like the African members of the Benincasinae, are marked by high contents of βPA and its peptide and by the absence of *m*-carboxyphenylalanine. *Trichosanthes* contrasts sharply; it is an Asiatic genus with a composition more reminiscent of *Luffa* or *Cyclanthera*, i.e., no βPA but a high level of *m*-carboxyphenylalanine. Species assigned to the tribe Cucurbiteae have rather unremarkable amino acid compositions, although *m*-carboxyphenylalanine is usually detected, and cucurbitin was a component of all *Cucurbita* species examined in the more detailed survey of this genus.

This survey was particularly incomplete in relation to members of the subfamily Zanonioideae; too few species were examined to permit useful comparisons. The limited evidence indicates that the main nonprotein amino acids characterizing other cucurbit species are not elaborated by members of this subfamily. Citrulline was always present, usually as a significant component of the free amino acids.

The overall picture emerging from the survey of the amino acid composition of seeds of the family is that the chemical information derived provided useful support for many features of the current system based on morphological characteristics. When the chemical evidence was clearly in conflict with recognized relationships, closer examination often revealed that a measure of uncertainty still existed concerning the morphological associations between species or genera. Amino acid data can therefore provide a valuable index for refining relationships, especially in situations where considerable morphological diversity is encountered within genera or subtribes.

Literature Cited

1. Dunnill, P. M., and L. Fowden. 1963. γ-L-glutamyl-β-pyrazol-l-yl-L-alanine, a peptide from cucumber seed. Biochem. J. 86: 388–391.
2. Dunnill, P. M., and L. Fowden. 1965. The amino acids of seeds of the Cucurbitaceae. Phytochemistry 4: 933–944.
3. Fang, S., L. Li, C. Niu, and K. Tseng. 1961. Chemical studies on *Cucurbita moschata* Duch. I. The isolation and structural studies of cucurbitine, a new amino acid. Sci. Sin. 10: 845–851.
4. Frisch, D. M., P. M. Dunnill, A. Smith, and L. Fowden. 1967. The specificity of amino acid biosynthesis in the Cucurbitaceae. Phytochemistry 6: 921–931.
5. Gray, D. O., and L. Fowden. 1961. N^4-ethyl-L-asparagine: a new amino acid amide from *Ecballium*. Nature 189: 401–402.
6. Jeffrey, C. 1961. Notes on Cucurbitaceae, including a proposed new classification of the family. Kew Bull. 15: 337–371.
7. Jeffrey, C. 1964. A note on pollen morphology in Cucurbitaceae. Kew Bull. 17: 473–477.
8. Noe, F. F., and L. Fowden. 1960. β-Pyrazol-l-ylalanine, a new amino acid from watermelon seed (*Citrullus vulgaris*). Biochem. J. 77: 543–546.
9. Schroter, H. B., D. Neumann, A. R. Katritsky, and F. J. Swinbourne. 1966. Withasomnine, a pyrazole alkaloid from *Withania somnifera*. Tetrahedron Letters 22: 2895–2897.
10. Sweeney, M. J., F. A. Davis, G. E. Gutowski, R. L. Hamill, D. H. Hoffman, and G. A. Poore. 1973. Experimental antitumor activity of pyrazomycin. Cancer Res. 33: 2619–2623.
11. Thompson, J. F., C. J. Morris, S. Asen, and F. Irreverre. 1961. *m*-Carboxyphenyl-L-alanine: a dicarboxylic aromatic amino acid from *Iris* bulbs. J. Biol. Chem. 236: 1183–1185.
12. Wada, M. 1930. Citrullin, a new amino acid in the press juice of the watermelon, *Citrullus vulgaris*. Biochem. Z. 224: 420–429.
13. Wright, C. H. 1896. Hooker's Icones Plantarum 25: plate 2490.

4 | Fatty Acids of Cucurbitaceae Seed Oils in Relation to Taxonomy

C. Y. Hopkins

ABSTRACT. The occurrence of specific long-chain fatty acids in the seed oils of 55 species of Cucurbitaceae is reviewed. The majority of the oils are composed of ordinary fatty acids, but a considerable number of species have unusual acids that may be useful markers in taxonomy. One of these is punicic acid, which was found in 15 species and is a characteristic feature of the family. It occurs at least once in all six of the tribes studied and is prominent in two of them. Eleostearic acid is a notable component in one tribe, the Joliffieae. There is substantial agreement between Jeffrey's classification of the family and the composition of the seed oils, with a number of exceptions.

Much information has accumulated on the nature of the long-chain fatty acids in the seed oils of the Cucurbitaceae. This information is utilized to examine the relationship between the occurrence of specific acids and the classification of plants in the family. The classification prepared by Jeffrey (28, 29, 30, Appendix, this volume) provides a logical basis for the botanical aspects of the study.

Occurrence of Fatty Acids

Long-chain fatty acids occur in seed oils mainly as glycerides, i.e., neutral esters of glycerol. Since the glycerol portion is common to all, it is not necessary to consider it further.

The most common fatty acids of seed oils are palmitic, stearic, oleic, and linoleic. However, the seed fats or oils of many plants also contain one or more fatty acids that differ from the common acids in chain length, degree of unsaturation, or other structural features. The uncommon acids tend to have taxonomic relationships and hence are of special interest to botanists.

A number of plant families have a less common acid that is typical of the family. For example, lauric acid is predominant in the seed oils of many species of Lauraceae, capric acid is typical of Ulmaceae, and erucic acid of Cruciferae (22). Another characteristic of families concerns classes of acids rather than specific acids. Thus, the Myristicaceae is noted for simple saturated acids, Gramineae for simple unsaturated acids (oleic, linoleic), and Santalaceae for very complex unsaturated acids (22, 23).

In Cucurbitaceae, two unusual acids, eleostearic and punicic, have been found in some genera.

Eleostearic acid, $C_{18}H_{30}O_2$
(Octadeca-*cis*-9, *trans*-11, *trans*-13-trienoic acid)
$$CH_3(CH_2)_3CH{=}CHCH{=}CHCH{=}CH(CH_2)_7COOH$$
$\quad\quad\quad\quad\quad\quad\; t\quad\quad\quad\; t\quad\quad\quad\; c$

Punicic acid, $C_{18}H_{30}O_2$
(Octadeca-*cis*-9, *trans*-11, *cis*-13-trienoic acid)
$$CH_3(CH_2)_3CH{=}CHCH{=}CHCH{=}CH(CH_2)_7COOH$$
$\quad\quad\quad\quad\quad\quad\; c\quad\quad\quad\; t\quad\quad\quad\; t$

Each of these acids has three centers of unsaturation, i.e., double bonds, and are conjugated, since the double bonds are separated by one single bond in each case. Eleostearic and punicic acids are referred to as conjugated trienoic acids or conjugated trienes.

Eleostearic and punicic acids are isomers. They differ in their geometric form, designated respectively as *cis*, *trans*, *trans* for eleostearic acid and *cis*, *trans*, *cis* for punicic acid. They have different solubilities, melting points, and optical properties (23, 25). Eleostearic acid is sometimes called α-eleostearic acid, but it seems unnecessary to use the prefix α for the natural acid. The all-*trans* isomer is β-eleostearic acid, which is prepared artificially.

Eleostearic acid occurs in the seed oils of several species of Euphorbiaceae, Rosaceae, and Valerianaceae, as well as in the Cucurbitaceae (22). It is the major component of the commercially important tung oil, which is obtained from *Aleurites* species (Euphorbiaceae). Tung oil is used in making varnish and in other industries.

Punicic acid is less widely distributed in nature. Its occurrence is limited to the Cucurbitaceae, except for its original discovery (37) in the seed oil of the pomegranate, *Punica granatum* L. (Punicaceae). Before the structure of punicic acid was established, it was sometimes called "trichosanic acid," but this name is no longer used.

Punicic acid has been found in 15 species of Cucurbitaceae (Tables 1–6) and can be regarded as a key substance in the taxonomy of the family. Eleostearic and punicic acids are valuable markers because they are readily detected in oils by their ultraviolet and infrared absorption spectra and can be determined quantitatively by the same means (23). They can be identified unequivocally by isolation from an oil if an appreciable amount is present (23). Recent work has demonstrated that these and related acids also can be detected by means of the ^{1}H and ^{13}C nuclear magnetic resonance spectra of oils (40). Determination of two different triene acids in the same oil is possible by this method (36, 39), which should prove useful for rapid analysis of large numbers of samples.

Acids with conjugated unsaturation are much more reactive than ordinary acids and are usually considered to be nonedible. Thus, seeds with an appreciable amount of conjugated acid in their oils would not ordinarily be regarded as a food source (27).

Fatty Acids of Cucurbitaceae Oils

Many of the cucurbit seed oils have ordinary component acids, and the amounts of each are in the normal range for seed oils (3, 22). Thus, a sample of *Cucurbita pepo* oil was composed of 15% palmitic plus stearic acids, 30% oleic acid, 51% linoleic acid, and 4% all others (15). Hilditch and Williams (22) listed results of analyses for some 25 species, of which 16 were more or less similar to *C. pepo*, although with variation in the amounts of each acid.

Nine species were reported as having a conjugated triene acid in quantities from 10 to 55% of the total acids. Other acids found in quite small amounts were lauric, myristic, palmitoleic, arachidic, and linolenic. Some of the earlier analyses may require confirmation by modern methods. Linolenic acid was reported in only seven species and in amounts of 1 to 3%.

Occurrence of Fatty Acids by Tribes

Data on the various fatty acids of the seed oils are presented in Tables 1 to 6. The species listed as having "ordinary" acids have mainly the usual saturated, oleic, and linoleic acids. When the species has an unusual acid or an exceptional quantity the acid is named, and the amount is given as a percentage of the total acids of the oil.

The tables reflect Jeffrey's classifications of the family (28, 29, Appendix, this volume), but the order of the tribes is modified. All species in these tables are from one subfamily, Cucurbitoideae. However, the subfamily Zanonioideae was sampled by Tulloch and Bergter, who examined *Fevillea trilobata* L. (40). Its seed oil had punicic acid as a major acid and eleostearic acid in a minor amount.

Joliffieae

In the subtribe Telfairiinae (Table 1), *Telfairia occidentalis* has eleostearic acid in moderate amounts but *T. pedata* has none. The latter, however, has an unusually large proportion of saturated acids, amounting to a total of 46%.

In the subtribe Thladianthinae, four species of *Momordica* have eleostearic acid, but one, *M. balsamina*, has punicic acid. *Momordica* species have been studied in a number of laboratories, and the results are in agreement (see Table 1 for references). The content of eleostearic acid in four *Momordica* species is quite large. It is the major acid in these seed oils. The presence of 58% punicic acid in *M. balsamina* is a striking exception and indicates the extreme specificity of the enzymes involved in the biosynthetic process.

Joliffieae appears to be the only tribe in the subfamily in which eleostearic acid has been found to date. On the other hand, punicic acid is known in several other tribes.

Benincaseae

Recent data for 13 species of the tribe Benincaseae are given in Table 2. Most of these have only the common acids. The two exceptions, *Diplocylos palmatus* and *Ecballium elaterium*, have moderately large amounts of punicic acid.

The genus *Ecballium* was originally placed by Jeffrey (28) in the tribe Cucurbiteae, as was *Acanthosicyos*. Later he transferred both genera and others to the Benincaseae subtribe Benincasinae (29). Work of Dunnill and Fowden (12) on the amino acids of the seeds showed that *Ecballium* and *Bryonia* differed markedly from other genera of Benincasinae. This point was noted by Gibbs (16). On the chemical evidence, therefore, *Ecballium* may not be happily positioned in Benincasinae. Barclay and Earle

Table 1. Fatty acids of seed oils of Joliffieae

Species	Acid	Amount (%)	Reference
Subtribe Thladianthinae			
Momordica balsamina L.	Punicic	58, 68	9, 15, 24
M. charantia L.	Eleostearic	50, 61	7, 9, 42
M. cochinchinensis (Lour.) Spreng.	Eleostearic	65	26
M. cymbalaria Hook f.	Eleostearic	49	26
M. dioica Roxb. ex Willd.	Eleostearic	55, 57	5, 8, 9
Subtribe Telfairiinae			
Telfairia occidentalis Hook f.	Eleostearic	19	21
T. pedata (Sims) Hook.	Ordinary	See text	7, 21

Table 2. Fatty acids of seed oils of Benincaseae

Species	Acid	Amount (%)	Reference
Subtribe Benincasinae			
Acanthosicyos horridus Welw. ex Hook. f.	Ordinary		8
Benincasa hispida (Thunb.) Cogn. (as *Cucurbita cerifera* Sav.)	Ordinary		8
Bryonia alba L.	Conjugated diene	3	8
B. dioica Jacq.	Conjugated diene	4	Kleiman, pers. comm.
Citrullus colocynthis (L.) Schrad.	Conjugated diene	2	2
C. lanatus (Thunb.) Matsum. & Nakai (as *C. vulgaris* Schrad.)	Linoleic	67	13, 18
Diplocyclos palmatus (L.) C. Jeffr.	Punicic	38	17
Ecballium elaterium (L.) A. Rich.	Punicic	22	2, 7
Lagenaria siceraria (Mol.) Standl. (as *L. vulgaris* Ser.)	Linoleic	76	7
L. sphaerica (Sond.) Naud. (as *L. mascarena* Naud.)	Linoleic	78	18
Subtribe Luffinae			
Luffa acutangula (L.) Roxb.	Ordinary		15
L. cylindrica (L.) M. J. Roem. (as *L. aegyptica* Mill.)	Ordinary		6, 7
L. operculata (L.) Cogn.	Ordinary		2

(2) confirmed the presence of large amounts of punicic acid in *Ecballium elaterium*.

Although *Bryonia* species have mainly the common acids (with much linoleic), a small amount of conjugated diene acid has been observed in *B. alba* (8). Barclay and Earle (2) found similar proportions, up to 5%. As discussed later, conjugated diene acids in small quantities may be artifacts.

Three species in the Benincasinae have unusually high contents of linoleic acid (67, 76, and 78%). This feature also occurs in two species of Melothrieae. It suggests, although we have no proof of this, that biosynthesis in the seed leads either to the nonconjugated linoleic acid; or to the conjugated triene, punicic acid, and that these two pathways may be mutually exclusive.

Melothrieae

A few examples of the Melothrieae have been studied (Table 3). The four species of *Cucumis* have only the ordinary acids, although two have an unusually large content of linoleic acid, and one has a small amount of conjugated diene acid. On the other hand, *Apodanthera undulata* has 30% punicic acid (4), and *Ibervillea sonorae* gives strong evidence of a conjugated triene acid, possibly punicic (14, Kleiman, pers. comm.).

Table 3. Fatty acids of seed oils of Melothrieae

Species	Acid	Amount (%)	Reference
Subtribe Dendrosicyinae			
Apodanthera undulata A. Gray	Punicic	30,33	4, 9, 14, Kleiman, pers. comm.
Ibervillea sonorae (S. Wats.) Greene (as *Maximowiczia sonorae* S. Wats.)	Conjugated triene	n.a.	31
Subtribe Cucumerinae			
Cucumis anguria L.	Conjugated diene	3	2
C. dipsaceus Ehrenb. ex Spach	Linoleic	76	7
C. melo L.	Linoleic	70	7
C. sativus L.	Ordinary		6, 31

Sicyeae

Except for *Cyclanthera brachystachya*, the subtribe Cyclantherinae (Table 4) is characterized by species with only the ordinary acids (13, 15). *Cyclanthera brachystachya* has punicic acid in the seed oil (8). Rather surprisingly, punicic acid was not present in *C. pedata*, at least not in our sample. The one example of the subtribe Sicyinae, *Sicyos angulatus*, has only the common acids (8, 31).

Trichosantheae

Seed oils of *Trichosanthes* species consistently have major proportions of punicic acid (Table 5). The seeds constitute a readily available source of punicic acid. *Hodgsonia macrocarpa*, placed by Jeffrey in the subtribe

Table 4. Fatty acids of seed oils of Sicyeae

Species	Acid	Amount (%)	Reference
Subtribe Cyclantherinae			
Cyclanthera brachystachya (Ser.) Cogn. (as *C. explodens* Naud.)	Punicic	25	8
C. pedata (L.) Schrad.	Ordinary		8
Echinopepon racemosus (Steud.) C. Jeffr. (as *E. horridus* Naud.)	Ordinary		2
Marah fabaceus (Naud.) Greene (as *Echinocystis fabaceae* Naud.)	Ordinary		13
M. gilensis (Greene) Greene	Ordinary		15
M. macrocarpus (Greene) Greene	Ordinary		15
M. oreganus (Torr. & A. Gray) Howell (as *Echinocystis oregana* Torr. & A. Gray)	Ordinary		13
Subtribe Sicyinae			
Sicyos angulatus L.	Ordinary		8, 31

Table 5. Fatty acids of seed oils of Trichosantheae

Species	Acid	Amount (%)	Reference
Subtribe Hodgsoniinae			
Hodgsonia macrocarpa (Blume) Cogn. (as *H.*			
capniocarpa (Ridl.) Cogn.)	Ordinary	see text	19
Subtribe Trichosanthinae			
Trichosanthes cordata Roxb.	Punicic	32	26
T. cucumerina var. *anguina* (L.) Haines	Punicic	43, 56	6, 7, 9
T. cucumerina L. var. *cucumerina*	Punicic	47, 56	2, 26, 34
T. dioica Roxb.	Punicic	28	32
T. ovigera Blume (as *T. cucumeroides* (Ser.)			
Maxim.)	Punicic	29	38

Hodgsoniinae (29), has only the common acids, represented by an unusually high content of saturated acids, mainly palmitic and stearic, which totalled 46% (19).

Cucurbiteae

Data for the tribe Cucurbiteae are given in Table 6. The genus *Cucurbita* has mainly ordinary acids, but there are three exceptions, *C. cordata*, *C. digitata*, and *C. palmata*. They have punicic acid in small to moderate amounts. The ability of *C. palmata* and *C. digitata* to produce punicic acid was reported by Ault (1) and has been amply confirmed (3, 8).

The genus *Cayaponia* has a substantial content of punicic acid, which was

Table 6. Fatty acids of seed oils of Cucurbiteae

Species	Acid	Amount (%)	Reference
Cayaponia africana (Hook. f.) Exell	Punicic	38	8
C. grandifolia (Torr. & A. Gray) Small	Punicic	39	26
Cayaponia sp.	Conjugated triene	34	2
Cucurbita cordata S. Wats.	Punicic	9	3
C. digitata A. Gray	Punicic	10, 18	1, 3
C. ficifolia Bouché	Ordinary		3, 7, 14
C. foetidissima H. B. K.	Conjugated diene	2	14, 35, 42
C. lundelliana L. H. Bailey	Ordinary		3, 31
C. maxima Duch. ex Lam.	Ordinary		3, 14
C. moschata (Duch. ex Lam.) Duch. ex.			
Poir.	Ordinary		3, 14
C. okeechobeensis (Small) L. H. Bailey	Ordinary		3, 14
C. palmata S. Wats.	Punicic	11–29	1, 3, 9
C. palmeri L. H. Bailey	Ordinary		2, 3
C. pepo L.	Ordinary		3, 14, 15
Peponopsis adhaerens Naud. (as			
Cayaponia maximowiczii Cogn.)	Conjugated triene	37	2

identified in two species. Similar amounts of conjugated triene acid, possibly punicic acid, have been observed (2, 14) in two other *Cayaponia* species.

Discussion

It is convenient to examine the data by considering the various component acids as markers in the classification process.

Eleostearic acid

Five of the seven species studied in the Joliffieae had eleostearic acid as an important component of their seed oils. In the four *Momordica* species it was the major component. It was absent or in insignificant amounts in the other 50 species of the subfamily Cucurbitoideae that were examined. One should not conclude that it was completely absent in all these, but the methods were such that even relatively small amounts would have been detected. Since eleostearic acid was not found in the other tribes, it appears to be a possible marker for Joliffieae, subject to confirmation by testing a much larger number of species. The two exceptions already found have been noted in Table 1.

Punicic acid

Punicic acid occurred in moderate to large proportions in 15 of the 55 species examined. This relatively wide distribution in the family has not been recognized until now. It may be a key substance in the classification of Cucurbitaceae. Although punicic acid occurs in six tribes, as presently classified, it was prominent only in two. It was found in Trichosantheae in five of the six species studied (Table 5). In the Cucurbiteae, it occurred in three species of *Cucurbita* and in two species of *Cayaponia* and was indicated in two additional species of the latter (Table 6).

Scattered appearances of punicic acid in the other tribes were as follows: once in 7 species of Joliffieae, twice in 13 species of Benincaseae, once in 6 species of Melothrieae, and once in 7 species of Sicyeae. These isolated cases are worthy of further study concerning classification, since the amount of punicic acid was substantial in each one.

Saturated Acids

Two species have been shown to have a high content of saturated acids. These were *Telfairia pedata* of the Joliffieae (7, 21) and *Hodgsonia macrocarpa* of the Trichosantheae (19). Both had 46% saturated acids (palmitic plus stearic). The usual amount in the family is 15–20%, with a few examples ranging up to 30% (3, 22). *Hodgsonia macrocarpa* was placed by Jeffrey (28) in a separate subtribe, removed from the *Trichosanthes* species, which have punicic acid. Thus, the content of saturated acids in the oil could be a useful criterion.

Linoleic Acid

Four instances of an unusually high content of linoleic acid (70% or more) have been noted. The amount is commonly 45–65% in this family. Linoleic acid production is subject to appreciable variation as a result of climatic conditions. Hence, it is of relatively minor value for purposes of taxonomy. However, the high values of 76% and 78% in *Lagenaria* (Table 2, two species) and 70% and 76% in *Cucumis* (Table 3, two species) are worthy of note. Linoleic acid is an essential factor in human nutrition.

Conjugated Diene Acid

Evidence of small amounts (up to 4%) of conjugated diene acid was obtained in five of the 55 species listed (Tables 2, 3, 6). The acid itself was not identified further. Although this could be a distinctive characteristic, the amounts are small and do not warrant consideration at this time without further study. The acid may be an artifact arising from oxidative degradation of linoleic or other unsaturated acids during storage of the seed. A search for this diene acid would require freshly harvested seed from mature plants to determine whether or not it is a normal component of the oil.

Theory of Biosynthesis of Punicic Acid

Little was known of the origin of the conjugated trienoic acids until 1984–85, when Crombie and Holloway (10, 11) reported a careful study of the biosynthesis of calendic acid, octadeca-*trans*-8, *trans*-10, *cis*-12-trienoic acid. This acid occurs in the seed oil of *Calendula officinalis* L. and is an isomer of punicic acid. The study showed that oleic acid, labelled with the isotope carbon-14, was converted in the seed tissue to calendic acid. Administration of labelled linoleic acid gave a similar result, but no calendic acid was produced from labelled linolenic acid.

It is well known that oleic acid is readily dehydrogenated to linoleic acid in plants. The hypothesis that linoleic acid is the progenitor of conjugated trienoic acids was proposed first by Morris and Marshall (33). The work of Crombie provided the necessary experimental support for the mechanism with respect to calendic acid. The theory was extended to include punicic acid and others.

On the basis of the isotopic experiments, Crombie postulated that enzymic action converts linoleic acid to a radical with double bonds at C-10 and C-12, which is then dehydrogenated enzymically at C-8 to calendic acid, the 8, 10, 12-triene. He suggested further that a different enzyme could produce an isomeric radical with double bonds at C-9 and C-11, followed by dehydrogenation to punicic acid, the 9, 11, 13-triene. The presumed route for conversion of linoleic and to punicic acid is the following:

$$CH_3(CH_2)_4CH \underset{c}{=\!=} CHCH_2CH \underset{c}{=\!=} CH(CH_2)_7CO_2H$$
linoleic acid (9, 12-diene)
↓

$$CH_3(CH_2)_4\overset{.}{C}H\text{-}CH \underset{t}{=\!=} CHCH \underset{c}{=\!=} CH(CH_2)_7CO_2H$$
radical (9, 11-diene)
↓

$$CH_3(CH_2)_3CH \underset{c}{=\!=} CHCH \underset{t}{=\!=} CHCH \underset{c}{=\!=} CH(CH_2)_7CO_2H$$
punicic acid (9, 11, 13-triene)

The enzymes involved are doubtless characteristic of the respective plant species. The stereospecific nature of the enzyme determines whether or not it produces a *cis* or *trans* double bond at the point of action. α-Eleostearic acid is also presumed to originate from linoleic acid but by the action of a different enzyme in the final step.

Conclusions

Although it would be desirable to examine many more species, some conclusions can be drawn from the present group. First, punicic acid is characteristic of the subfamily Cucurbitoideae, occurring in 15 species in six tribes, most of them in Trichosantheae and Cucurbiteae. Second, eleostearic acid was found only in Joliffieae of this subfamily and may be a useful marker for the tribe. Third, some of the other fatty acids of the seed oils may have taxonomic importance but the present data are insufficient to demonstrate this. Fourth, species within some genera, especially *Momordica*, *Cayaponia*, and *Trichosanthes*, are generally homogeneous in fatty acid composition, although there is one exception in *Momordica*. Last, the results tend to substantiate Jeffrey's classification in certain respects.

Literature Cited

1. Ault, W. C., M. L. Swain, and L. C. Curtis. 1947. Oils from perennial gourds. J. Amer. Oil Chem. Soc. 24: 289–290.
2. Barclay, A. S., and F. R. Earle. 1974. Chemical analysis of seeds. III. Oil and protein content of 1,253 species. Econ. Bot. 28: 178–236.
3. Bemis, W. P., J. W. Berry, M. J. Kennedy, D. Woods, M. Moran, and A. J. Deutschman, Jr. 1967. Oil composition of *Cucurbita*. J. Amer. Oil Chem. Soc. 44: 429–430.
4. Bemis, W. P., M. Moran, J. W. Berry, and A. J. Deutschman, Jr. 1967. Composition of *Apodanthera undulata* oil. Canad. J. Chem. 45: 2637.
5. Chakrabarty, M. M., S. Bhattacharya, M. J. Desai, and S. A. Patel. 1956. Studies on the seed fats of Cucurbitaceae family. Naturwis. 43: 523–525.
6. Chakrabarty, M. M., D. K. Chowdhury, and B. K. Mukherji. 1955. Studies on some seed fats of Cucurbitaceae family. Naturwis. 42: 344–345.
7. Chisholm, M. J., and C. Y. Hopkins. 1964. Fatty acid composition of some Cucurbitaceae seed oils. Canad. J. Chem. 42: 560–564.
8. Chisholm, M. J., and C. Y. Hopkins. 1967. Conjugated fatty acids in some Cucurbitaceae seed oils. Canad. J. Biochem. 45: 1081–1086.
9. Conacher, H. B. S., F. D. Gunstone, G. M. Hornby, and F. B. Padley. 1970. Glyceride studies. Part IX. Intraglyceride distribution of vernolic acid and of five conjugated octadecatrienoic acids. Lipids 5: 434–441.
10. Crombie, L., and S. J. Holloway. 1984. Origins of conjugated triene fatty acids. The biosynthesis of calendic acid by *Calendula officinalis*. J. Chem. Soc., Chem. Commun. No. 15: 953–955.
11. Crombie, L., and S. J. Holloway. 1985. Biosynthesis of calendic acid, octadeca-(8E, 10E, 12Z)-trienoic acid by developing marigold seeds. Origins of (E.E.Z) and (Z.E.Z) conjugated triene acids in higher plants. J. Chem. Soc. Perkin Trans. 1: 2425–2434.
12. Dunnill, P. M., and L. Fowden. 1965. The amino acids of seeds of the Cucurbitaceae. Phytochemistry 4: 933–944.
13. Earle, F. R., C. A. Glass, G. C. Geisinger, I. A. Wolff, and Q. Jones. 1960. Search for new industrial oils. IV. J. Amer. Oil Chem. Soc. 37: 440–447.
14. Earle, F. R., and Q. Jones. 1962. Analysis of seed samples from 113 plant families. Econ. Bot. 16: 221–250.
15. Earle, F. R., E. H. Melvin, L. H. Mason, C. H. VanEtten, and I. A. Wolff. 1959. Search for new industrial oils. I. Selected oils from 24 plant families. J. Amer. Oil Chem. Soc. 36: 304–307.
16. Gibbs, R. 1974. Chemotaxonomy of Flowering Plants. Vol. 2. McGill-Queen's University Press, Montreal.
17. Gowrikumar, G., V. V. S. Mani, T. Chandrasekhara Rao, T. N. B. Kamal, and G. Lakshminarayana. 1981. *Diplocyclos palmatus* L. A new seed source of punicic acid. Lipids 16: 558–559.
18. Gunstone, F. G., G. M. Taylor, J. A. Cornelius, and T. W. Hammonds. 1968. New tropical seed oils. II. Component acids of leguminous and other seed oils. J. Sci. Food Agric. 19: 706–709.
19. Hilditch, T. P., M. L. Meara, and W. H. Pedelty. 1939. Fatty acids and glycerides

of solid seed fats. VIII. The seed fat of *Hodgsonia capniocarpa.* J. Soc. Chem. Ind. 58: 27–29.

20. Hilditch, T. P., and J. P. Riley. 1946. The use of low-temperature crystallization in the determination of component acids of liquid fats. III. Fats which contain eleostearic acid. J. Soc. Chem. Ind. 65: 74–81.

21. Hilditch, T. P., I. C. Sime, Y. A. H. Zaky, and M. L. Meara. 1944. The component acids of various vegetable fats. J. Soc. Chem. Ind. 63: 112–114.

22. Hilditch, T. P., and P. N. Williams. 1964. The Chemical Constitution of Natural Fats, 4th ed. Chapman and Hall, London.

23. Hopkins, C. Y. 1972. Fatty acids with conjugated unsaturation. *In* F. D. Gunstone, ed., Topics in Lipid Chemistry. Vol. 3. Logos, London.

24. Hopkins, C. Y., and M. J. Chisholm. 1962. Identification of conjugated triene fatty acids in certain seed oils. Canad. J. Chem. 40: 2078–2082.

25. Hopkins, C. Y., and M. J. Chisholm. 1968. A survey of the conjugated fatty acids of seed oils. J. Amer. Oil Chem. Soc. 45: 176–182.

26. Hopkins, C. Y., M. J. Chisholm, and J. A. Ogrodnik. 1969. Identity and configuration of conjugated fatty acids in certain seed oils. Lipids 4: 89–92.

27. Jacks, T. J., T. P. Hensarling, and L. Y. Yatsu. 1972. Cucurbit seeds. I. Characterization and uses of oils and proteins. A review. Econ. Bot. 26: 135–141.

28. Jeffrey, C. 1962. Notes on Cucurbitaceae, including a proposed new classification of the family. Kew Bull. 15: 337–371.

29. Jeffrey, C. 1964. A note on pollen morphology in Cucurbitaceae. Kew Bull. 17: 473–477.

30. Jeffrey, C. 1966. On the classification of the Cucurbitaceae. Kew Bull. 20: 417–426.

31. Jones, Q., and F. R. Earle. 1966. Chemical analysis of seeds. II. Oil and protein content of 759 species. Econ. Bot. 20: 127–155.

32. Mathur, H. H., and J. S. Aggarwal. 1953. Component fatty acids of the drying oil from the seeds of *Trichosanthes dioica.* J. Sci. Ind. Res. India 12B: 60–62.

33. Morris, L. H., and M. O. Marshall. 1966. Occurrence of *cis*, *t*rans linoleic acid in seed oils. Chem. Ind. (London) 1493–1494.

34. Patel, S. A., S. Battacharya, and M. M. Chakrabarty. 1958. Seed fats of Cucurbitaceae. II. Component fatty acids of *Trichosanthes cucumerina.* J. India Chem. Soc. 35: 67–71.

35. Shahani, H. S., F. G. Dollear, K. S. Markley, and J. R. Quinby. 1951. The buffalo gourd, a potential oilseed crop of the southwestern drylands. J. Amer. Oil Chem. Soc. 28: 90–95.

36. Takagi, T., and Y. Itabashi. 1981. Occurrence of mixtures of geometrical isomers of conjugated octadecatrienoic acids in some seed oils. Lipids 16: 546–551.

37. Toyama, Y. and T. Tsuchiya. 1935. A new stereoisomer of eleostearic acid in pomegranate seed oil. J. Soc. Chem. Ind. Japan 38 (suppl. binding): 182–185.

38. Toyama, Y., and T. Tsuchiya. 1935. Another new stereoisomer of eleostearic acid in the seed oil of Karasu-uri (*Trichosanthes cucumeroides* Maxim.). J. Soc. Chem. Ind. Japan 38 (suppl. binding): 185–187.

39. Tulloch, A. P. 1982. [13]C nuclear magnetic resonance spectroscopic analysis of seed oils containing conjugated unsaturated acids. Lipids 17: 544–550.

40. Tulloch, A. P., and L. Bergter. 1979. Analysis of the conjugated trienoic acid-

containing oil from *Fevillea trilobata* by ^{13}C nuclear magnetic resonance. Lipids 14: 996–1002.
41. Vasconcellos, J. A., J. W. Berry, C. W. Weber, J. C. Scheerens, and W. P. Bemis. 1980. The properties of *Cucurbita foetidissima* oil. J. Amer. Oil Chem. Soc. 57: 310–313.
42. Verma, J. P., and J. S. Aggarwal. 1956. A note on component fatty acids of the oil from the seed of *Momordica charantia* L. J. Indian Chem. Soc. 33: 357–358.

5 | Isozyme Studies in *Cucurbita*

Norman F. Weeden and Richard W. Robinson

ABSTRACT.   Considerable allozyme polymorphism has been identified in the genus *Cucurbita*. This polymorphism has been used to help reveal phylogenetic relationships in the genus and to distinguish cultivars of *C. pepo*. One isozyme locus appears to be useful as a genetic marker for resistance to watermelon mosaic virus 2. Four multilocus linkage groups are identified, at least one of which appears to be conserved throughout much of the genus. The high proportion of duplicated isozyme loci provides additional evidence for the polyploid nature of the genome. Finally, the nature of genetic change during speciation is examined by observing the segregation at isozyme loci in the interspecific cross *C. maxima* × *C. ecuadorensis*.

Why study isozymes in *Cucurbita*? More specifically, what questions currently being asked by scientists investigating *Cucurbita* can be approached by using isozyme techniques? In order to address these questions we must understand the nature of the information that is available from banding patterns on starch gels. Isozymes are proteins and therefore generally carry a net charge determined primarily by their amino acid composition. To a great extent this charge determines the rate at which the protein will migrate in an electric field. Alterations in the DNA sequences of the coding gene often produce changes in the amino acid sequence of the protein, thereby altering its net charge and, concomitantly, its electrophoretic mobility. Hence, a direct, albeit imperfect, correlation exists between genetic diversity in DNA sequences and variability in its isozyme phenotypes. Thus, isozyme analysis often exposes simple genetic variation not otherwise detectable in the taxon being studied.

The initial isozyme analyses for most plants are usually simple descriptive surveys of easily scored systems. The first applications of isozyme techniques in *Cucurbita* were a brief report of esterase isozymes (29) and two

51

surveys of peroxidase variation (6, 18). These results were soon extended by the excellent investigations of Wall and Whitaker. Not only did Wall (32) survey a large number of *Cucurbita* species for variation of both esterase and leucine aminopeptidase, but he and Whitaker (33) also examined the genetic basis of the variation and used the allozyme polymorphism identified to investigate the nature of hybrid breakdown in the interspecific cross *C. maxima* Duch. ex Lam. × *C. ecuadorensis* Cutler & Whitaker. Despite this early and auspicious start, isozyme analysis in *Cucurbita* has lagged behind that in many other crops. In her 1983 review, Dane (2) identified only five reviewed papers on *Cucurbita* isozymes, involving only four isozyme systems, published during the 1970's.

More recently, the number of isozyme studies has increased. Mulcahy et al. (20) utilized a microelectrophoretic technique with individual pollen grains of *Cucurbita* interspecific hybrids to demonstrate segregation for isozymes in pollen. Segregation was noted for 30% of the 37 zones seen on zymograms stained for acid phosphatase. They concluded that postmeiotic, gametophytically transcribed loci are common. Investigations in H. D. Wilson's laboratory have greatly expanded genetic and systematic studies in *C. pepo* L. and the closely related wild species *C. texana* (Scheele) A. Gray (3, 4, 16, 17). Robinson and Puchalski used isozyme phenotypes to help establish the synonymy of *C. martinezii* L. H. Bailey and *C. okeechobeensis* (Small) L. H. Bailey (25) and to examine systematic relationships (23, 26). Laura Merrick (pers. comm., 1988) investigated phylogenetic relationships in the Sororia Group by comparing allele frequencies in the different species, and Decker (4) also obtained isozyme evidence for a close affinity of *C. mixta* Pang. and *C. moschata* (Duch. ex Lam.) Duch. ex Poir. Isozyme analyses of species relationships in *Cucurbita* have also been reported by Ferguson (9). Many of these recent studies have used extensive surveys of isozyme polymorphism in populations in order to determine genetic similarity among the taxa being compared. In the remainder of this chapter we will take a different approach, focusing on applications of isozyme analysis that involve dealing with the loci themselves rather than with allele frequencies.

Mendelian Genetics and Marker Loci

One of the most important features of allozyme variation is the simple inheritance it exhibits. Codominance of alleles, freedom from epistatic or pleiotropic interactions, freedom from environmental influence, expression in many tissues, and selective neutrality are all characteristics that are generally associated with isozymes (10). Thus, an allozyme variant provides an excellent genetic marker with numerous applications.

Cucurbita pepo is one of the most variable vegetables. It includes all summer squashes and vegetable marrows, 'Table Queen' winter squash, the traditional pumpkin of Halloween, the unique 'Vegetable Spaghetti', and ornamental gourds of various shapes and colors. This highly polymorphic species also displays considerable allozymic diversity (3–5, 8, 9, 14, 17, 19), which can be used for cultivar identification and to determine the percentage of seeds derived from self-pollination in a hybrid seed lot.

Hybrid seed of summer squash is usually produced by treating the maternal parent with the growth regulator ethephon, thereby preventing the development of staminate flowers for an extended period (27). Bees are relied on to bring pollen from the male parent in an adjacent row in the field. The effect of the treatment eventually subsides, and self- or sib-pollinations are possible when the maternal parent produces staminate flowers. Hybrid seed can be easily identified if the maternal and paternal parents are homozygous for different allozymes. A true hybrid must display the allozymes from both parents, whereas seed resulting from self-fertilization exhibits only the allozyme from the maternal parent. We have surveyed over 20 different hybrids from several seed companies and all possess at least one allozyme phenotype that can be used to distinguish true hybrids from self-fertilized seed.

Beyond economic usefulness, isozyme analysis provides important information on the structure and evolution of the genome. Our work on three segregating F_2 populations provided linkage information on 11 isozyme loci and three morphological characters (36). Normal Mendelian segregation was obtained at all loci except *Bu*, which determines plant habit (vine vs. bush), in the cross of *C. pepo* cultivars 'Senator' × 'Table Queen'. Joint segregation analysis identified two multilocus linkage groups: one with *Gpi-c2* and *Aat-p2*, the other with *Est*, *Aldo-p*, and *Skdh*, in that order. Kirkpatrick (16) and Kirkpatrick et al. (17) found linkage between *Idh-3* and *Pgm-2* in the F_2 of *C. pepo* × *C. texana*.

The locus designations used in our work are slightly different from the conventional terms adopted by Kirkpatrick et al. (17). Our reasons for departing from convention were primarily to identify homologous loci in other species. Most of the enzyme systems we investigated exhibited more than one isozyme, and it is risky to assign interspecific homologies solely on the basis of mobility. However, many enzyme systems display subcellular compartmentation of the particular isozymes, such compartmentation being highly conserved during the diversification of land plants (11). Compartmentation of isozymes in *Cucurbita* was first reported by Dvorak and Cernohorska (7), who found that some peroxidases of *C. pepo* were in the mitochondria and others were not. Determining the subcellular location of the enzyme product makes it much easier to identify homologous loci, especially those coding isozymes of aspartate aminotransferase, glucose-

phosphate isomerase, malate dehydrogenase, phosphoglucomutase, 6-phos-
phogluconate dehydrogenase, and triosephosphate isomerase. We have included a one- or two-letter suffix on the locus designation as a means of
indicating the compartmentation of the coded enzyme (c = cytosol, m =
mitochondria, p = plastid, mb = microbody). The actual names we use for
the enzymes (and the respective acronyms) are those specified by the International Union of Biochemistry (15).

When both isozyme loci and other characters segregate in the same F_2 or
backcross population, it may be possible to identify an isozyme marker for
the particular morphological or physiological character. We analyzed C.
pepo crosses for isozymes and several morphological characters, including
plant habit, fruit shape and fruit color (36). Unfortunately, no clear correlations were found between allozyme genotype and morphological phenotype. Puchalski et al. (23) also failed to find linkage between isozyme loci
and morphological characters in the cross C. moschata × C. martinezii. A
third cross, C. maxima × C. ecuadorensis, gave more interesting results.
Although no correlations were observed between allozyme segregation and
that for the leaf mottle gene (M), resistance to zucchini yellow mosaic virus
(ZYMV), or resistance to squash mosaic virus (SqMV), a consistent pattern
was seen between the Aldo-2 phenotype and resistance to watermelon
mosaic virus 2 (WMV). Genetic studies on the resistance of C. ecuadorensis
to WMV indicated that the resistance was polygenic. The locus coding the
aldolase band appeared to be linked to a gene that was required for expression of resistance but not sufficient to produce the resistant phenotype by
itself. This conclusion was based on the presence of the aldolase band in
nearly all resistant plants as well as a significant number of the susceptible
plants (Table 1). Although the use of aldolase as a marker does not conform
to the simple one-to-one correlation normally associated with a marker
gene, the results demonstrate an even more important application of genetic

Table 1.   Joint segregation analysis for resistance to watermelon mosaic virus 2 and Aldo-2
phenotypes in (C. maxima × C. ecuadorensis) × C. maxima backcross populations

Population	WMV-susceptible		WMV-resistant		Chi square[2]
	Aldo-2 slow	Aldo-2 het.[1]	Aldo-2 slow	Aldo-2 het.[1]	
83-978	10	7	1	7	4.73*
83-979	10	2	0	7	12.31**
Total	20	9	1	14	15.37**

[1]het., heterozygous phenotype.
[2]Chi-square values were calculated assuming an expected ratio of 1 : 1 : 1 : 1.
*Statistically significant deviation at P < 0.05.
**Statistically significant deviation at P < 0.01.

markers: the dissection of quantitative characters into their major genetic components.

Conserved Linkage Groups

By comparing the linkage relationships of homologous loci, it is possible to identify linkage groups that have been conserved among species of *Cucurbita*. Two linkage groups in *C. palmata* S. Wats. were identified by means of monosomic alien addition lines developed from a cross with *C. moschata* (Graham and Bemis, this volume). One linkage group contained the locus coding fumarase and a gene for the hard rind character; while the second contained the loci *Gpi-p*, *Gpi-c2*, and *Aat-p2* (34). Apparently the *Gpi-c2–Aat-p2* linkage has been conserved in much of *Cucurbita*, for it has also been found in *C. maxima* and *C. ecuadorensis* (35) and in *C. pepo* (36). Interestingly, the relative mobility of the isozymes and the degree of polymorphism have been maintained in these systems, for the other cytosolic GPI isozyme is faster and monomorphic in all the species mentioned, as is the other plastid aspartate aminotransferase. If linkage groups are conserved within *Cucurbita*, then *Pgm-p*, *Pgm-c2*, and *Idh-3* probably are on the same chromosome, since the first two were found to be linked in *C. maxima* (35), whereas *Pgm-2* (= *Pgm-c2*) and *Idh-3* were linked in *C. pepo* (17). It remains to be determined whether the *Skdh–Aldo-p–Est* linkage group in *C. pepo* (35) or the linkage of *Fum* and the hard rind gene in *C. palmata* (34) are conserved in other cucurbit species.

Polyploidy

The high chromosome number ($2n = 40$) of *Cucurbita* suggests that this genus is of polyploid origin. Evidence for allotetraploidy in several *Cucurbita* species has been provided by the cytogenetic investigations of Weiling (37). However, remarkably few examples of characters controlled by paired loci have been reported. Genetic analysis of isozyme phenotypes provides an excellent method for identifying duplicated genes, particularly in systems for which the number of genes expressed in diploid plants has been shown to be highly conserved (11). The additive expression of diploid genomes in a tetraploid has been demonstrated most clearly in the recent allotetraploid, *Tragopogon miscellus* M. Ownb. (28); however, even an ancient tetraploid can still express a significant number of duplicated isozyme loci (1).

Eight isozyme systems were selected for a study of gene expression in *Cucurbita*. In diploid plants each system is specified by a predictable number of genes (Table 2), usually one for each subcellular compartment in

Table 2. Comparison of number of isozyme loci expressed in
Cucumis sativus ($n = 7$) and *Cucurbita* species ($n = 20$)

Enzyme system	Loci expressed in *Cucumis*[1]	Loci expressed in *Cucurbita*[1]
Aspartate aminotransferase	4 (4)	7 (8)
Glucosephosphate isomerase	2 (2)	4 (4)
Isocitrate dehydrogenase	1 (1)	3 (2)
Malate dehydrogenase[2]	2 (2)	4 (4)
Phosphoglucomutase	2 (2)	3 (4)
6-Phosphogluconate dehydrogenase	2 (2)	3 (4)
Shikimate dehydrogenase	1 (1)	1 (2)
Triosephosphate isomerase	3 (2)	4 (4)
Total	17 (16)	29 (32)

[1] Numbers in parentheses are the expected number of loci expressed in a diploid (under *Cucumis*) and a tetraploid (under *Cucurbita*).

[2] Cytosolic and mitochondrial forms only; the microbody-specific form was not resolved.

which the enzyme is found. The total number of gene products predicted in a diploid plant for the eight systems is 16. Preliminary analysis of *Cucumis sativus* L. ($2n = 14$) indicated that 17 loci were being expressed. In contrast, genetic studies of the isozyme phenotypes of *Cucurbita maxima, C. ecuadorensis, C. pepo, C. moschata,* and *C. palmata* demonstrated that at least 29 loci were contributing isozymic forms, approximating the number expected for a tetraploid. Similar studies of isozyme expression in *C. pepo* and *C. texana* produced additional evidence for duplication in these species (17).

The fact that many duplicate isozymes are expressed in *Cucurbita* may provide additional phylogenetic information on the genus. Just as gene duplications have proved useful in identifying phylogenetic relationships (12, 21), loss of duplicate gene expression could also be used as a phylogenetic marker (38). It is probable that such loss would occur more frequently than duplication events, for there appears to be a natural tendency for polyploids to revert back to a diploid level of gene expression (22). In *Cucurbita*, several examples of diminished expression of one of a pair of homoeologous loci have been observed. Loss of duplicate loci coding for malate dehydrogenase in several *Cucurbita* species was suggested by Ferguson (9). The *Tpi-c2* locus in *C. ecuadorensis* appears to produce much less polypeptide than *Tpi-c1* (35). Similarly, in several accessions of *C. maxima* the isozyme GPI-3 shows a lower level of activity than its counterpart, GPI-2. The latter divergence appears to have arisen within *C. maxima*, for other accessions show equal expression of both isozymes (T. C. Andres, pers. comm., 1988). It is probable that such changes in the level of isozyme expression are rare events that are not easily reversed. If so, loss of expression could be used to identify monophyletic groups, just as gene duplications have been.

Speciation

The analysis of allozyme segregation in F_2 and backcross progeny from the interspecific cross *C. maxima* × *C. ecuadorensis* has permitted important conclusions regarding the genetic changes that occurred during speciation in *Cucurbita*. The hybrid produced from this cross and those generated from several other interspecific crosses in *Cucurbita* are fertile, but considerable sterility appears in the F_2 and the backcrosses to either parent. The phenomenon has been referred to as "hybrid breakdown" and has appeared in several genera, being best studied in *Gossypium* (31) and *Lycopersicon* (24). Historically, this phenomenon has been attributed to small chromosomal rearrangements that occurred since the divergence of the two species (13, 30, 31). If such were the case, recombination during meiosis in the hybrid would often produce gametes with duplications or deletions in their chromosomal complement. The recombinant gametes would tend to make less of a contribution to the succeeding generation than nonrecombinant gametes because the former would often lack or contain an imbalance of critical genetic material. Thus, the nonrecombinant genotypes would be expected to predominate in F_2 and backcross generations (30). Wall and Whitaker (33) examined esterase and leucine aminopeptidase allozyme segregation in the backcross *C. ecuadorensis* × (*C. maxima* × *C. ecuadorensis*) and observed the predicted favoring of nonrecombinant genotypes at the *Est* locus. However, our studies on the alternative backcross and F_2 populations (35) did not confirm these results. Although skewed segregation ratios were observed at many loci, the skewing did not favor nonrecombinant genotypes, nor did any single locus display skewing in a consistent direction when results from different populations were compared. We were forced to conclude that cryptic chromosomal rearrangements are not the cause of hybrid breakdown in this cross.

Conclusions

The considerable allozyme polymorphism of *Cucurbita* provides an extremely useful tool for analyzing problems of biological importance. The study of *Cucurbita* allozymes has already provided significant insights on the phylogenetic relationships among the species, on the nature of the genome, and on genetic changes accompanying speciation. Many more applications undoubtedly will emerge as new questions are asked. It is probable that as our knowledge of *Cucurbita* expands, the genus will become an excellent model for the study of evolution in polyploids.

Literature Cited

1. Bailey, G. S., A. C. Wilson, J. E. Halver and C. L. Johnson. 1970. Multiple forms of supernatant malate dehydrogenase in salmonid fishes. J. Biol. Chem. 245: 5927–5940.
2. Dane, F. 1983. Cucurbits. *In* S. D. Tanksley and T. J. Orton, eds., Isozymes in Plant Genetics and Breeding. Part B. Elsevier, Amsterdam.
3. Decker, D. S. 1985. Numerical analysis of allozyme variation in *Cucurbita pepo*. Econ. Bot. 39: 300–309.
4. Decker, D. S. 1986. A biosystematic study of *Cucurbita pepo*. Ph.D. dissertation, Texas A & M Univ., College Station.
5. Denna, D. W., and M. B. Alexander. 1975. The isoperoxidases of *Cucurbita pepo* L. *In* C. L. Markert, ed., Isozymes. Vol. 2. Academic Press, New York.
6. Dvorak, M., and J. Cernohorska. 1967. Peroxidases of different parts of the pumpkin plant (*Cucurbita pepo* L.). Biol. Pl. 9: 308–316.
7. Dvorak, M., and J. Cernohorska. 1972. Comparison of effects of calcium deficiency and IAA on the pumpkin plant (*Cucurbita pepo* L.). Biol. Pl. 14: 26–38.
8. Eguchi, H., and T. Matsui. 1969. Phylogenetic study of *Cucurbita* species by means of esterase zymogram. J. Fac. Ag. Kyushu Univ. 15: 345–352.
9. Ferguson, J. E. 1985. Disposition of cucurbitacin as a factor in host plant selection by diabroticite beetles and electrophoretic analysis of the coevolved genus *Cucurbita*. Ph.D. dissertation, Univ. Illinois, Urbana.
10. Gottlieb, L. D. 1981. Electrophoretic evidence and plant populations. *In* L. Reinholdt et al., eds., Progress in Phytochemistry. Vol. 7. Pergamon, New York.
11. Gottlieb, L. D. 1982. Conservation and duplication of isozymes in plants. Science 216: 373–380.
12. Gottlieb, L. D., and N. F. Weeden. 1979. Gene duplication and phylogeny in *Clarkia*. Evolution 33: 1024–1039.
13. Grant, V. 1975. Genetics of Flowering Plants. Columbia Univ. Press, New York.
14. Ignart, F., and N. F. Weeden. 1984. Allozyme variation in cultivars of *Cucurbita pepo* L. Euphytica 33: 779–785.
15. International Union of Biochemistry. 1984. Enzyme Nomenclature. Academic Press, New York.
16. Kirkpatrick, D. J. 1984. The relationship between isozyme phenotype and morphological variation in *Cucurbita*. M.S. thesis, Texas A & M Univ., College Station.
17. Kirkpatrick, D. J., D. S. Decker, and H. D. Wilson. 1985. Allozyme differentiation in the *Cucurbita pepo* complex: *C. pepo* var. *medullosa* vs. *C. texana*. Econ. Bot. 39: 289–299.
18. Loy, J. B. 1967. Peroxidases and differentiation in cucurbits. Ph.D. dissertation, Colorado State Univ., Fort Collins.
19. Loy, J. B. 1972. A comparison of stem peroxidases of squash (*Cucurbita maxima* Duch. and *C. pepo* L.). J. Exptl. Bot. 23: 450–457.

20. Mulcahy, D. L., R. W. Robinson, M. Ihara, and R. Kesseli. 1981. Gametophytic transcription for acid phosphatases in pollen of *Cucurbita* species hybrids. J. Heredity 72: 353–354.

21. Odrzykoski, I. J., and L. D. Gottlieb. 1984. Duplications of genes coding 6–phosphogluconate dehydrogenase in *Clarkia* (Onagraceae) and their phylogenetic implications. Syst. Bot. 9: 479–489.

22. Ohno, S. 1970. Evolution by Gene Duplication. Springer-Verlag, New York.

23. Puchalski, J. T., R. W. Robinson, and J. W. Shail. 1978. A comparative analysis of isozymes of *Cucurbita* species. Cucurbit Genet. Coop. Rep. 1: 20.

24. Rick, C. M. 1969. Controlled introgression of chromosomes of *Solanum pennellii* into *Lycopersicon esculentum*: segregation and recombination. Genetics 62: 753–768.

25. Robinson, R. W., and J. T. Puchalski. 1980. Synonomy of *Cucurbita martinezii* and *C. okeechobeensis*. Cucurbit Genet. Coop. Rep. 3: 45–46.

26. Robinson, R. W., and J. T. Puchalski. 1980. Systematics of the melonsquash. Cucurbit Genet. Coop. Rep. 3: 47.

27. Robinson, R. W., T. W. Whitaker, and G. W. Bohn. 1970. Promotion of pistillate flowering in *Cucurbita* by 2–chloroethylphosphonic acid. Euphytica 19: 180–183.

28. Roose, M. L., and L. D. Gottlieb. 1976. Genetic and biochemical consequences of polyploidy in *Tragopogon*. Evolution 30: 818–830.

29. Schwarz, H. M., S. I. Biedron, M. M. von Holdt, and S. Rehm. 1964. A study of some plant esterases. Phytochemistry 3: 189–200.

30. Stebbins, G. L. 1950. Variation and Evolution in Plants. Columbia Univ. Press, New York.

31. Stephens, S. G. 1949. The cytogenetics of speciation in *Gossypium*. I. Selective elimination of the donor parent genotype in interspecific backcrosses. Genetics 34: 627–637.

32. Wall, J. R. 1969. A partial survey of the genus *Cucurbita* for electrophoretic variants of esterase and leucine aminopeptidase. Southwestern Naturalist 14: 141–148.

33. Wall, J. R., and T. W. Whitaker. 1971. Genetic control of leucine aminopeptidase and esterase isozymes in the interspecific cross *Cucurbita ecuadorensis* × *C. maxima*. Biochem. Genet. 5: 223–229.

34. Weeden, N. F., J. D. Graham, and R. W. Robinson. 1986. Identification of two linkage groups in *Cucurbita palmata* using alien addition lines. HortScience 21: 1431–1433.

35. Weeden, N. F., and R. W. Robinson. 1986. Allozyme segregation ratios in the interspecific cross *Cucurbita maxima* × *C. ecuadorensis* suggest that hybrid breakdown is not caused by minor alterations in chromosome structure. Genetics 114: 593–609.

36. Weeden, N. F., R. W. Robinson, and J. W. Shail. 1986. Genetic analysis of isozyme variants in *Cucurbita pepo*. Cucurbit Genet. Coop. Rep. 9: 104–106.

37. Weiling F. 1959. Genomanalytische Untersuchungen bei Kürbis (*Cucurbita* L.). Zuchter 29: 161–179.

38. Wilson, H. D., S. C. Barber, and T. Walters. 1983. Loss of duplicate gene expression in tetraploid *Chenopodium*. Biochem. Syst. Ecol. 11: 7–13.

6 | Electrophoretic Analysis of Isozymes
in *Cucurbita* and *Cucumis* and Its
Application for Phylogenetic Studies

Jerzy T. Puchalski and Richard W. Robinson

ABSTRACT. Electrophoretic patterns of esterase and peroxidase were compared in species of *Cucurbita* and *Cucumis*. The greatest interspecific differences in both genera were for fast migrating anodic bands of esterases. On the basis of zymogram patterns, *Cucurbita* species were placed in seven groups, the composition of which were similar to those resulting from phenetic studies or hybridization tests. *Cucumis* species were placed in six major groups, two of which were subdivided into subgroups. The isozyme classification of *Cucumis* is in general agreement with relationships suggested by hybridization studies.

Isozymes are widely recognized as valuable biochemical markers in genetic and systematic studies. They are analyzed by means of the zymogram technique, in which multiple enzyme forms are electrophoretically separated in different gels and then histochemically stained to reveal specific enzyme activities. The technique is simple, rapid, precise, and applicable for mass screening of plant materials. Since isozyme analysis offers some advantages over that of complex morphological or cytological characters, it is a very useful tool for plant systematics and for evolutionary and phylogenetic studies (27).

The thrust of our research was to study intraspecific and interspecific isozyme variation in cultivated and wild species of *Cucurbita* and *Cucumis* and to use the results in evaluating species classifications in the respective genera. Very few such applications of isozyme data had been made in the Cucurbitaceae when these investigations began a decade ago (61, 62), but in recent years considerably more isozyme research with cucurbits has been undertaken (14, 21, 42, 60, 73, 78, Weeden and Robinson, this volume).

60

Materials and Methods

These studies involved 19 species of *Cucurbita* and 28 of *Cucumis* from the Cucurbitaceae collection of the New York State Agricultural Experiment Station, Geneva, New York. Most of the species were represented by more than one accession. Included were wild species as well as different cultivars of squash and pumpkin (*Cucurbita moschata* (Duch. ex Lam.) Duch. ex Poir., *C. mixta* Pang., *C. pepo* L., and *C. maxima* Duch.), cultivars and botanical varieties of muskmelon (*Cucumis melo* L.), and interspecific hybrids within *Cucumis* and *Cucurbita*.

Each sample consisted of a young leaf from five field-grown plants. Enzymes were extracted from the fresh leaves by crushing them in a chilled mortar with polyvinyl polypyrrolidone and 0.1 M (2 mg/g) TRIS-hydrochloride buffer (pH 7.8) with 0.5 M sucrose, 6 mM L-cysteine hydrochloride and 6 mM ascorbic acid. The crude extract was centrifuged for 20 minutes at 4°C and 20,000 g, and the clear supernatant was used directly for electrophoresis or was frozen and stored for later analysis. Multiple enzyme forms were separated by means of starch gel electrophoresis, according to the technique of Kristjansson (45), in a horizontal system. The gel buffer was 0.014 M TRIS + 0.004 M citric acid (pH 7.6), and the electrode buffer was 0.3 M boric acid + 0.1 M NaOH (pH 8.6). After electrophoretic separation, the specific activities of esterase (EC 3.1.1) and peroxidase (EC 1.11.1.7) were stained histochemically by the procedures of Shaw and Prasad (70), with the following modifications. Nonspecific esterases were detected with α-naphthyl acetate and Fast Garnet GBC in 0.1 M phosphate buffer (pH 6.0). Peroxidase activity was stained in the gel by means of 3-amino-9-ethylcarbazole (dissolved in N,N-dimethyl formamide) and 0.01% H_2O_2 in 0.2 M acetate buffer (pH 5.0), excluding $CaCl_2$. Analyses were also made for leucine amino peptidase and acid phosphatase, but are not reported here since there was less interspecific variation than for esterase and peroxidase.

Cucurbita

Both the esterase and peroxidase systems were very heterogenous; 32 multiple electrophoretic forms of nonspecific esterases and 23 electrophoretic peroxidase isoforms were found in *Cucurbita* species. Isozymes moved during electrophoresis in both directions, to the cathode and the anode, but the most significant interspecific and intervarietal differences were detected for esterase in the anodal part of the gel, and for peroxidase in the cathodal part. *Cucurbita* is highly variable for patterns of both enzymes, and distinct interspecific differences as well as similarities can be seen (Figure 1). The most interesting variation occurred in the zone of esterase bands showing the greatest electrophoretic migration mobility.

The comparison of esterase and peroxidase electrophoretic patterns gave us an opportunity to characterize and classify *Cucurbita* species. Their reciprocal similarity was estimated on the basis of the number of common and different electrophoretic bands. As a result, *Cucurbita* was divided into seven species groups (Figure 1).

Group I

Group I contains four very similar xerophytic species: *Cucurbita cylindrata* L. H. Bailey, *C. digitata* A. Gray, *C. palmata* S. Wats., and *C. cordata*

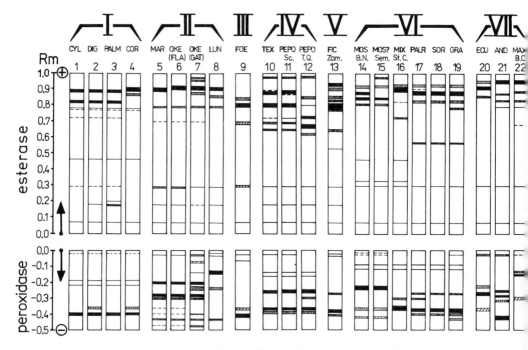

Figure 1. Esterase (above) and peroxidase (below) patterns of *Cucurbita* species, arranged according to their similarity. 1. *C. cylindrata.* 2. *C. digitata.* 3. *C. palmata.* 4. *C. cordata.* 5. *C. martinezii.* 6. *C. okeechobeensis* (from Florida). 7. Accession misidentified as *C. okeechobeensis* (from Gatersleben, GDR). 8. *C. lundelliana.* 9. *C. foetidissima.* 10. *C. texana.* 11. *C. pepo* 'Scallop'. 12. *C. pepo* 'Table Queen'. 13. *C. ficifolia* 'Zamba'. 14. *C. moschata* 'Butternut'. 15. *C. moschata* 'Seminole'. 16. *C. mixta* 'Striped Cushaw'. 17. *C. palmeri.* 18. *C. sororia.* 19. *C. gracilior.* 20. *C. ecuadorensis.* 21. *C. andreana.* 22. *C. maxima* 'Buttercup'.

S. Wats. *Cucurbita cordata*, however, lacked one esterase band characteristic of the other species.

Interspecific hybridization data support the conclusion that species of isozyme group I are very closely related and may be conspecific. Bemis and Nelson (6) obtained good fruit set and seed production when *C. digitata*, *C. cylindrata*, and *C. palmata* were crossed in all combinations, and Bemis and Whitaker (9) obtained fertile hybrids in reciprocal crosses of each of these species with *C. cordata*. Bemis and Whitaker (8) reported that natural hybridization of *C. digitata* and *C. palmata* occurs where the two species are sympatric.

Group II

Included in group II are three mesophytic species: *C. okeechobeensis* L. H. Bailey, *C. martinezii* L. H. Bailey, and *C. lundelliana* L. H. Bailey. The esterase and peroxidase patterns of *C. okeechobeensis* and *C. martinezii* were quite similar. The taxa are also identical in morphology and disease resistance, and they produce fertile hybrids (66). We conclude they are the same species.

There was some morphological variation, particularly for leaf lobing, among different accessions of *C. martinezii*, but all had the same esterase and peroxidase banding patterns as accessions of *C. okeechobeensis* from Florida. However, an accession received as *C. okeechobeensis* from Gatersleben, East Germany differed significantly from the other accessions in isozymes and morphology. This accession had some isozyme bands in common with the Florida accessions of *C. okeechobeensis*, but differed for other esterase and peroxidase allozymes. The German accession may have been derived from an interspecific cross but is not *C. okeechobeensis*.

We have obtained fertile hybrids between *C. okeechobeensis* and *C. lundelliana* and *C. martinezii* and *C. lundelliana* (66). There are distinctive electrophoretic and morphological differences between *C. lundelliana* and *C. okeechobeensis*, but they are both included in isozyme group II because they have common esterase bands.

Group III

The xerophytic species *C. foetidissima* HBK had a unique electrophoretic pattern and is placed in its own group. Crosses have been made between *C. foetidissima* and *C. moschata* (5, 30), but the hybrids are sterile, and *C. foetidissima* stands apart from other species of the genus.

Group IV

Group IV includes two very closely related and apparently conspecific taxa: *Cucurbita pepo* and *C. texana* A. Gray (20, Decker, this volume). The affinity of these two species has long been recognized (4, 22, 28). They cross readily and form fully fertile hybrids (29). The isozyme evidence is in accord with other data. The esterase and peroxidase patterns were the same for *C. texana* as for *C. pepo* 'Scallop' (Figure 1). Very thorough investigations in the laboratory of H. D. Wilson (21, 42) have provided additional electrophoretic evidence for the closeness of the relationship.

Distinctive intraspecific isozyme variation occurred within *C. pepo*. Four types of zymograms, which we classified as the 'Scallop', 'Kousa', 'Table Queen', and 'Zucchini' types, were observed. More recent and complete isozyme investigations by Decker (19) and Ignart and Weeden (33) have confirmed that considerable electrophoretic variation occurs among *C. pepo* cultivars. *Cucurbita pepo* is the most highly polymorphic species of *Cucurbita*, in isozymes as well as morphology.

Group V

Cucurbita ficifolia Bouché had a distinct zymogram and constitutes a separate group. It does, however, share some electrophoretic bands with *C. pepo*. Both *C. pepo* and *C. maxima* can be crossed with *C. ficifolia*, but the hybrids are highly sterile (31). Partly fertile hybrids have been made with *C. lundelliana* (80). Although *C. ficifolia* can be crossed with species of three different isozyme groups, sterility barriers and isozyme patterns indicate that it is not closely related to other *Cucurbita* species.

Group VI

Cucurbita mixta had similarities in its isozyme banding pattern to *C. palmeri* L. H. Bailey, *C. sororia* L. H. Bailey, and *C. gracilior* L. H. Bailey. (Plants grown as the latter species generally have proved to be *C. sororia*, Merrick, this volume.) They cross readily, and natural hybrids of *C. mixta* and taxa of the *C. sororia* group have been reported (50). Merrick (this volume) considers the species of this complex and proposes a new classification for them.

Cucurbita moschata is included in this isozyme group because it had isozyme bands in common with the other species of the group, although its electrophoretic pattern was distinctive. *Cucurbita mixta* and *C. moschata* are somewhat similar in morphology and were once considered to be the same species, but Pangalo (57) distinguished them on the basis of crossing relationships. They can be crossed, but with difficulty, and the hybrids are mostly self-sterile (81, but see Merrick, this volume). The isozyme evidence supports continued recognition of *C. moschata* and *C. mixta*.

A squash that has been cultivated for many years by the Seminole Indians in Florida was considered by Bailey to be *C. moschata* (3), but Morton (51) suggested it might be derived from a cross between *C. moschata* and *C. okeechobeensis*. 'Seminole Pumpkin' had isozymes similar to *C. moschata*, although it differed in some esterases from the five tested cultivars of *C. moschata*. There was no evidence of gene introgression to 'Seminole Pumpkin' from *C. okeechobeensis* (67).

Group VII

The last group of *Cucurbita* species, based on isozyme similarities, includes *C. maxima*, *C. andreana* Naud. and *C. ecuadorensis* Cutler & Whitaker. *Cucurbita maxima* and *C. andreana* are very closely related, having many isozymes in common. They can be crossed in all combinations (6, 79).

Cucurbita ecuadorensis is included in this group because of its esterase and peroxidase similarities to *C. maxima*, but it also showed some similarities to species of group II. The thorough investigations of Weeden (78), however, substantiated the distinctness of the species. He found that *C. ecuadorensis* and *C. maxima* had different isozyme phenotypes for each of the 12 enzyme systems studied, and segregation for 20 isozyme loci was detected in progeny of the interspecific hybrid.

Again, the results of interspecific hybridizations are in general agreement with the isozyme tests. *Cucurbita ecuadorensis* can be crossed with *C. maxima* and *C. andreana*. It also crosses with *C. okeechobeensis* and *C. lundelliana* of group II (12), although some sterility occurs in the hybrids and their progeny. We obtained seed for each of these crosses without having to use embryo culture.

Our proposed groupings of species based on isozyme characters are very similar to those derived from phenetic correlations of morphological characters (7). One difference is in the placement of *C. ecuadorensis* in isozyme group VII, while in the phenetic classification of Bemis et al. (7) it was placed it a separate group, close to *C. lundelliana* and *C. okeechobeensis*. Also, Bemis et al. placed *C. moschata* in a separate group, but one close to *C. mixta* and its allies, while we included them all in the same isozyme group. Although *C. moschata* and *C. ecuadorensis* have affinities for other species in their respective isozyme groups, other members of those groups are more closely related to each other than they are to either *C. moschata* or *C. ecuadorensis*.

Cucumis

Investigations of *Cucumis* species have been hampered by the absence of a modern, comprehensive, systematic treatment of the genus. Many species designations in the literature are incorrect. A number of United States Department of Agriculture plant introductions of *Cucumis* were originally misclassified as to species. This problem was compounded by the use of open-pollination to increase stocks of USDA plant introductions of *Cucumis*, often resulting in natural hybridization when compatible species were grown together. Fortunately, Dutch researchers (47, 75, den Nijs and Custers, this volume) determined the correct identity for many previously misidentified accessions of *Cucumis* obtained from the USDA and from botanical gardens. Their corrected species designations are used for this investigation and, wherever possible, for references to the literature.

Cucumis is an Old World genus, composed of two subgenera. Subgenus *Cucumis*, with a base chromosome number of $x = 7$ is indigenous to the Indian subcontinent, and here is considered as isozyme group VI. The wild species of subgenus *Melo* (L.) C. Jeffr., with a base number of $x = 12$, are indigenous to Africa and are placed in isozyme groups I–V.

Group I

Group I is the largest and most complex species group in the subgenus *Melo*. It is often referred to as the 'anguria' group. *Cucumis anguria* L. of the West Indies was thought by Naudin (52) to be of American origin, although Hooker (18) believed it was derived from an African species. Meeuse (48) suggested that African slaves introduced the species to the Caribbean Islands. He considered *C. longipes* Hook. f. to be the most likely progenitor of *C. anguria*, since it is similar to it morphologically, shares with it identical disease and insect resistance, and produces fertile hybrids when crossed with it. F_2 segregation of three plants with bitter fruit to one with nonbitter fruit indicated a single gene distinguishes the bitter-fruited *C. longipes* from nonbitter *C. anguria*. The normal Mendelian segregation, indicative of good chromosome pairing and segregation in the interspecific hybrid, established that the two taxa are closely related. Meeuse (48) treated *C. longipes* as a variety of *C. anguria* (as *C. anguria* var. *longipes* (Hook. f.) Meeuse).

Our isozyme results (Figure 2) fully support this conclusion. The two taxa differed only in the mobility of one fast esterase band. Dane (13) also found similarity in peroxidases between *C. anguria* var. *anguria* and var. *longipes*. Esquinas-Alcazar (23) reported no significant electrophoretic differences between the varieties for six enzyme systems. His results indicate that *C. anguria* is more closely related to *C. longipes* than to any other species in its compatibility group. Perl-Treves et al. (60) also presented isozyme evidence suggesting that *C. anguria* and *C. longipes* are conspecific. Although closely related, morphological, biochemical, and some isozyme differences (72, 73) indicate varietal distinction is appropriate.

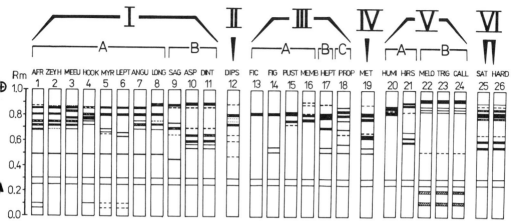

Figure 2. Species-specific esterase patterns of *Cucumis* species arranged according to their reciprocal similarity. 1. *C. africanus*. 2. *C. zeyheri*. 3. *C. meeusei*. 4. *C. hookeri*. 5. *C. myriocarpus*. 6. *C. leptodermis*. 7. *C. anguria*. 8. *C. longipes*. 9. *C. sagittatus*. 10. *C. asper*. 11. *C. dinteri*. 12. *C. dipsaceus*. 13. *C. ficifolius*. 14. *C. figarei*. 15. *C. pustulatus*. 16. Accession received as *C. membranifolius*, later identified (16) as *C. aculeatus*. 17. *C. heptadactylus*. 18. *C. prophetarum*. 19. *C. metuliferus*. 20. *C. humifructus*. 21. *C. hirsutus*. 22. *C. melo*. 23. *C. trigonus*. 24. *C. callosus*. 25. *C. sativus*. 26. *C. hardwickii*.

Cucumis anguria can be crossed with a number of species, including *C. africanus* L. f., *C. zeyheri* Sond., *C. myriocarpus* Naud., *C. leptodermis* Schweick., *C. ficifolius* A. Rich., and *C. prophetarum* (2, 17), but none of the hybrids are as fertile as those resulting from crosses of *C. anguria* with *C. longipes* (17, 54). This is added reason to believe that *C. longipes* or a close relative is the progenitor of *C. anguria*. Wild forms of *C. anguria* do not exist in Africa today, so domestication, involving a reduction in fruit spines and cucurbitacins, must have occurred long ago.

Only recently has the name *Cucumis africanus* been correctly applied (49). Many reports referring to *C. africanus* actually pertain to *C. zeyheri* (49, 75). Part of the problem may be attributed to the morphological variation between the diploid and tetraploid forms of *C. zeyheri* (16), the latter of which may be allotetraploids (Singh, this volume). *Cucumis africanus* and diploid *C. zeyheri* are similar morphologically and form hybrids when crossed (17). At one point, Jeffrey (34) considered *C. zeyheri* to be a subspecies of *C. prophetarum*, but later reestablished *C. zeyheri* as a species in its own right on the basis of the reduced fertility of the interspecific hybrid (37).

Diploid *C. zeyheri* and *C. africanus* were similar in their isozymes (Figure 2) and, in fact, were more alike in isozymes than either was to *C. anguria* or *C.*

longipes (Figure 2), although each of the four taxa shared common esterase bands. This is in agreement with their crossing relationships (55, 76).

Cucumis meeusei C. Jeffr. (35) had isozymes similar to those of *C. africanus* but differed for specific esterases. *Cucumis hookeri* Naud., considered by Jeffrey (34) and Meeuse (49) to be the same as *C. africanus*, had isozymes similar to those of that species (Figure 2).

The close relationship of *C. myriocarpus* Naud. and *C. leptodermis* was recognized by Meeuse (49), who reported they are morphologically and biochemically similar and can be crossed easily. He suggested they perhaps should be considered as varieties or subspecies, while Jeffrey (34) considered *C. leptodermis* a synonym of *C. myriocarpus*. Their almost identical isozymes support the conclusion that the taxa are conspecific (Figure 2).

Based on esterase allozyme similarities, *C. asper* Cogn., *C. dinteri* Cogn., *C. sagittatus* Wawra & Peyr., and *C. angolensis* Hook. f. ex Cogn. are related. Jeffrey (34) and Fernandes and Fernandes (26) considered *C. dinteri* and *C. angolensis* to be synonyms of *C. sagittatus*. Dane et al. (15) argued that *C. dinteri* and *C. sagittatus* are conspecific and Perl-Treves et al. (60) found the two species to be identical in chloroplast DNA, but we found the peroxidase patterns for *C. dinteri* to be different than those for *C. sagittatus*. Despite their morphological differences, *C. asper* and *C. dinteri* were very similar in esterase electromorphs (Figure 2).

Group II

Cucumis dipsaceus Ehrenb., the only species of group II, hybridizes with many species of group I. Its fruits are morphologically distinct (17), as was its esterase pattern (Figure 2). Staub and Frederick (73) reported it also has a unique banding pattern for shikimic dehydrogenase.

Group III

Species forming group III fall into three subgroups. One subgroup comprises *C. ficifolius* and *C. figarei*, which were very similar in isozymes (Figure 2). These findings agree with Meeuse's conclusion (49) that these species are .conspecific. Jeffrey (36), however, considers *C. figarei* to be a distinct species.

Pollen tubes grow to the ovary within 24 hours after *C. figarei* is pollinated with *C. africanus*, *C. prophetarum*, *C. zeyheri*, *C. anguria* var. *longipes*, *C. angolensis* Hook. f. ex Cogn., or *C. heptadactylus* Naud., but fruit and seed do not develop (43). *Cucumis ficifolius* is also separated by partial sterility barriers from *C. dipsaceus* (17) and the species of group I (17, 41).

Cucumis pustulatus Hook. f., which Jeffrey (36) considered to be synonymous with *C. figarei*, was similar but not identical to that species in isozymes (Figure 2). A plant introduction (PI273650) obtained from the USDA as *C. membranifolius* Hook. f., was considered by Dane et al. (15) to be a tetraploid form of *C. aculeatus* Cogn., and Leeuwen and den Nijs (47) suggested it might be *C. ficifolius*. We found its isozymes were similar to the other members of this subgroup.

Cucumis heptadactylus Naud. is very distinctive, with its elongate, narrow leaves and perennial, tuberous roots. It is also a dioecious tetraploid. Despite these characteristics, it had esterase bands in common with those of group III species (Figure 2). It also had unique esterases, however, and thus was placed in its own subgroup. Similarly, *C. prophetarum* shared isozyme similarities with group III species, but was sufficiently different to warrant its own subgroup.

Group IV

The African horned cucumber, *C. metuliferus* Naud., had two iso-esterases in common with group III species (Figure 2), but constitutes group IV because of its unique banding pattern. Hybrids of *C. metuliferus* have been reported with *C. anguria* (25), *C. melo* (50, 56), and accessions considered to be *C. africanus* (11) but later classified as *C. zeyheri* (75). However, these crosses are difficult to make (24), and *C. metuliferus* is not closely related to any other species.

Group V

Group V includes taxa considered closest to *C. melo* L. *Cucumis trigonus* Roxb. and *C. callosus* (Rottl.) Cogn. had the same isozymes characteristic of *C. melo* (Figure 2). Naudin (53) produced fertile hybrids of *C. melo* and *C. trigonus* but nevertheless maintained each as a species. Jeffrey (38) considered *C. callosus* to be a synonym of *C. melo*, and Leeuwen and den Nijs (47) considered both *C. callosus* and *C. trigonus* to be synonyms of *C. melo*. Our isozyme results support the contention that the three taxa represent the same species.

Abo-Baker et al. (1) reported that botanical varieties of *C. melo* were almost identical in leaf amino acids. Esquinas-Alcazar (23) made a very thorough electrophoretic study of *C. melo* cultivars, land races, and botanical varieties, and concluded that there was no genetic justification for considering *C. callosus* as a distinct species. He found allozyme variation within *C. melo*, but the zymograms of *C. callosus* were among the most standard for *C. melo*. *Cucumis callosus* crosses readily with *C. melo* (58, 64) and meiosis is normal in the hybrid (58). Thus, the evidence is convincing that *C. callosus* is conspecific with *C. melo*.

Cucumis humifructus Stent. and *C. hirsutus* Sond. were described by

Jeffrey (36) as showing some similarity to *C. melo*. We include them in this group but set apart in a different subgroup because of isozyme differences (Figure 2).

Group VI

The cucumber, *C. sativus* L., is so different from other *Cucumis* species in chromosome number, center of origin, and other respects that Jeffrey (39) placed it in a separate subgenus. Our isozyme data are in agreement with this classification. *Cucumis sativus* can be distinguished by its enzymes from all other species of the genus except *C. hardwickii* Royle, the only other taxon with seven pairs of chromosomes. Furthermore, since the isozyme patterns of *C. sativus* and *C. hardwickii* were the same and they cross readily, forming a fully fertile hybrid (17, 65), we accept that they are conspecific (65), distinguishable as var. *cucumis* and var. *hardwickii* (Royle) Alef.

Esquinas-Alcazar (23) reported that PI271337, identified as *C. trigonus* by USDA Plant Introductions, had isozymes typical of *C. sativus*. He recognized, however, that it had morphological characteristics of *C. sativus*. Leeuwen and den Nijs (47) confirmed that it was misidentified, and our isozyme results indicate it is *C. sativus*.

Despite many attempts, hybrid plants have never been obtained in crosses between the cucumber and any species in subgenus *Melo*. Darwin, Naudin, Bailey, Whitaker, and many others (46) have tried to cross the cucumber and muskmelon, but without success. Fruits can be obtained from this cross, but they are parthenocarpic or produce nonhybrid progeny resembling the maternal parent (32). Thus, it was surprising when it was reported that the cross had been made by using the mentor pollination technique, in which cucumber was pollinated with a mixture of cucumber and muskmelon pollen (44). The majority of the putative hybrid plants, known as megurk, were reported to have 19 chromosomes. Despite presumably having an unbalanced chromosome number and being derived from very distinct parents, the plants were fertile and bred true for type. Electrophoretic analyses of isozymes were used to authenticate the parentage of the plants (68). The isozymes were identical to the *C. sativus* parent (Figure 3), proving the plants were not interspecific hybrids. It was found later that the plants were homozygous for a single recessive gene, 'rosette', that produces a phenotype in cucumber plants somewhat like that expected for a cucumber-muskmelon hybrid (69). The report that the plants had 19 chromosomes, the expected number for a cross between $n = 7$ and $n = 12$ species, presumably was erroneous.

Our isozyme classification of *Cucumis* species should be considered as preliminary. Classification of the genus is very difficult because of large

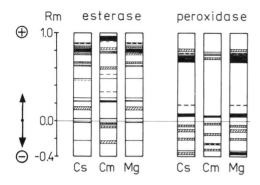

Figure 3. Isozyme patterns of megurk (Mg), *Cucumis sativus* (Cs), and *C. melo* (Cm). Electrophoretic patterns of both enzyme systems indicate that the megurk is *C. sativus*, but not a hybrid of *C. sativus* × *C. melo*.

intraspecific and interspecific variation, misidentification of species, and disagreement about species definitions. There is reasonable agreement, however, between our isozyme classification of *Cucumis* species and the classification based on crossing relationships of Deakin et al. (17). Our groups II and III, the Dipsaceus and Ficifolius Groups, were not segregated from group I, the Anguria Group, by Deakin et al. Also, they included *C. sagittatus* and *C. dinteri* with the Melo Group (group IV), while we made them a subdivision of the Anguria Group, but otherwise the isozyme and hybridization classifications are similar.

Dane et al. (15) made morphological, hybridization, and cytological studies of *Cucumis* species. They included *C. ficifolius* and *C. figarei* in the cross-compatible Anguria Group, while we include them in a separate group. In many other respects, however, their results are in agreement with our classification.

Esquinas-Alcazar (23) analyzed 21 *Cucumis* species for isozymes and classified them into four groups. His groupings were in general agreement with ours, although they differ in some respects, such as his including *C. heptadactylus* with the Anguria Group and *C. sagittatus*, *C. dinteri*, and *C. asper* with the Melo Group.

Singh and Yadava (71) used morphology, cytology, and crossing relationships to place *Cucumis* species into four groups. This classification is in general agreement with ours, although we differed by including *C. dipsaceus* and *C. metuliferus* in separate groups, while they included them with the Anguria and Melo Groups, respectively. The classifications of Jeffrey (39) and den Nijs and Visser (54), on the basis of crossability of *Cucumis* species and fertility of interspecific hybrids, also placed *C. metuliferus* in a separate group.

Perl-Treves et al. (59, 60) investigated the chloroplast DNA and isozymes of *Cucumis* species. Their dendogram includes *C. metuliferus* in a distinct group, and in many other respects agrees with our classification.

Brown et al. (10) investigated chromatographic flavonoid patterns of

Cucumis species, but the specific flavonoid compounds involved were not identified. Phylogenetic interpretation of their findings and its relationship to our isozyme classification is difficult because some of the plant introductions they used probably were misidentified. For example, the introductions they considered to be *C. trigonus* and *C. leptodermis* were probably actually *C. sativus* var. *hardwickii* and *C. myriocarpus*, respectively. Nevertheless, it is of interest that they found *C. humifructus, C. metuliferus, C. dinteri,* and *C. hirsutus* to have clearly different chromatographic patterns from each other and from the other species tested.

Cucurbitacins have not been of major systematic value in the study of *Cucumis* but provide support for the separation of the subgenera. *Cucumis sativus* is distinguished from other species of *Cucumis* by having only cucurbitacin C (63). All other *Cucumis* species tested lacked this compound, and most had cucurbitacins B, D, and G. Interestingly, *C. hookeri, C. leptodermis,* and *C. myriocarpus* differed from the others by having a high content of cucurbitacin A.

Progress had been rapid in recent years in the application of electrophoresis for genetic and phylogenetic research. Just how much advancement has been accomplished can be gauged by the additional genetic information that has been obtained for the first interspecific cross in the Cucurbitaceae investigated for isozyme variation. Wall and Whitaker (77) reported segregation for two unlinked genes for two enzymes in progeny of *Cucurbita maxima* × *C. ecuadorensis*, but recently Weeden and Robinson (78) obtained evidence for 20 isozyme loci and five linkage groups by analyzing F_2 and backcross populations of the same interspecific cross for 12 enzyme systems.

The electrophoretic investigation of species relationships in *Cucumis* and *Cucurbita* reported here was limited to only two enzyme systems. A better understanding of phylogenetic relationships can be expected in the future as additional enzyme systems, newer techniques of electrophoresis, restriction length fragment polymorphisms, and other biochemical markers are used to investigate species relationships in the Cucurbitaceae. To some extent (59, 60, 73, 78), these advanced techniques have already been applied to investigations of the systematics of the cucurbits.

Literature Cited

1. Abo-Bakr, M. A., M. M. Hussein, and Y. Hanna. 1978. Cytological features in six botanical varieties of *Cucumis melo* and different fifteen hybrids. Egypt. J. Genet. Cytol. 7: 266–276.
2. Andrus, C. F., and G. Fassuliotis. 1965. Crosses among *Cucumis* species. Veg. Improv. Newslett. 7: 3.

3. Bailey, L. H. 1929. Addenda in volume II, particularly in relation to nomenclature. Gentes Herb. 2: 427–430.

4. Bailey, L. H. 1943. Species of *Cucurbita*. Gentes Herb. 6: 267–316.

5. Bemis, W. P. 1963. Interspecific hybridization in *Cucurbita*. II. *C. moschata* Poir. × xerophytic species of *Cucurbita*. J. Heredity 54: 285–289.

6. Bemis, W. P., and J. M. Nelson. 1963. Interspecific hybridization within the genus *Cucurbita* I. Fruit set, seed and embryo development. J. Ariz. Acad. Sci. 2: 104–107.

7. Bemis, W. P., A. M. Rhodes, T. W. Whitaker, and S. G. Carmer. 1970. Numerical taxonomy applied to *Cucurbita* relationships. Amer. J. Bot. 57: 404–412.

8. Bemis, W. P., and T. W. Whitaker. 1965. Natural hybridization between *Cucurbita digitata* and *C. palmata*. Madrono 18: 39–47.

9. Bemis, W. P., and T. W. Whitaker. 1969. The xerophytic *Cucurbita* of northwestern Mexico and southwestern United States. Madrono 20: 33–41.

10. Brown, G. B., J. R. Deakin, and M. B. Wood. 1969. Identification of *Cucumis* species by paper chromatography of flavonoids. J. Amer. Soc. Hort. Sci. 94: 231–234.

11. Custers, J. B. M., and G. van Ee. 1980. Reciprocal crosses between *Cucumis africanus* L. f. and *C. metuliferus* Naud. Cucurbit Genet. Coop. Rep. 4: 53–55.

12. Cutler, H. C., and T. W. Whitaker. 1968. A new species of *Cucurbita* from Ecuador. Ann. Missouri Bot. Gard. 55: 392–396.

13. Dane, F. 1976. Evolutionary studies on the genus *Cucumis*. Ph.D. dissertation, Colorado State Univ., Fort Collins.

14. Dane, F. 1983. Cucurbits. *In* S. D. Tanksley and T. J. Orton, eds., Isozymes in Plant Genetics and Breeding. Part B. Elsevier, Amsterdam.

15. Dane, F., D. W. Denna, and T. Tsuchiya. 1980. Evolutionary studies of wild species in the genus *Cucumis*. Z. Pflanzenzucht. 85: 89–109.

16. Dane, F., and T. Tsuchiya. 1976. Chromosome studies in the genus *Cucumis*. Euphytica 25: 367–374.

17. Deakin, J. R., G. W. Bohn, and T. W. Whitaker. 1971. Interspecific hybridization in *Cucumis*. Econ. Bot. 25: 195–211.

18. De Candolle, A. L. 1885. Origin of Cultivated Plants. Appleton, New York.

19. Decker, D. S. 1985. Numerical analysis of allozyme variation in *Cucurbita pepo*. Econ. Bot. 39: 300–309.

20. Decker, D. S. 1988. Origin(s), evolution, and systematics of *Cucurbita pepo*. Econ. Bot. 42: 4–15.

21. Decker, D. S., and H. D. Wilson. 1987. Allozyme variation in the *Cucurbita pepo* complex: *C. pepo* var. *ovifera* vs. *C. texana*. Syst. Bot. 12: 263–273.

22. Erwin, A. T. 1938. An interesting Texas cucurbit. *Cucurbita pepo* L. var. *ovifera* Alef. (*C. texana* Gray). Iowa State College J. Sci. 12: 253–261.

23. Esquinas-Alcazar, J. T. 1977. Allozyme variation and relationships in the genus *Cucumis*. Ph.D. dissertation, Univ. of Calif., Davis.

24. Fassuliotis, G. 1977. Self-fertilization of *Cucumis metuliferus* Naud. and its cross-compatibility with *C. melo*. J. Amer. Soc. Hort. Sci. 102: 336–339.

25. Fassuliotis, G. and B. V. Nelson. 1986. Conversion in vitro of *Cucumis metuliferus* × *C. anguria* embryos. *In* A. Walters and P. G. Alderson, eds., Plant Tissue Culture and its Agricultural Applications. Butterworths, London.

26. Fernandes, R., and A. Fernandes. 1970. Cucurbitaceae. Conspectus Florae Angolensis 4: 232–289.
27. Gottlieb, L. D. 1977. Electrophoretic evidence and plant systematics. Ann. Missouri Bot. Gard. 64:161–180.
28. Gray, A. 1868. Field, Forest, and Garden Botany. American Book, New York.
29. Grebenscikov, I. 1955. Notulae cucurbitologicae. II. Über Cucurbita texana A. Gray und ihre Kreuzung mit einer hochgezuchteten C. pepo-Form. Kulfurpflanze 3: 50–59.
30. Grebenscikov, I. 1958. Über zwei Cucurbita-Artkreuzungen. Zuchter 28: 233–237.
31. Grebenscikov, I. 1965. Notulae cucurbitologicae. VI. Über einige Artkreuzungen in der Gattung Cucurbita. Kulturpflanze 13: 145–161.
32. Hagiwara, T., and K. Kamimura. 1937. Genetic studies in Cucumis. Jap. J. Genet. 13: 71–79.
33. Ignart, F., and N. F. Weeden. 1984. Allozyme variation in cultivars of Cucurbita pepo L. Euphytica 33: 779–785.
34. Jeffrey, C. 1962. Notes on Cucurbitaceae including a proposed new classification of the family. Kew Bull. 16: 337–371.
35. Jeffrey, C. 1965. Further notes on Cucurbitaceae. Kew Bull. 19: 215–223.
36. Jeffrey, C. 1967. Cucurbitaceae. In E. Milne-Redhead and R. M. Polhill, eds., Flora of Tropical East Africa. Crown Agents, London.
37. Jeffrey, C. 1975. Further notes on classification of Cucurbitaceae. III. Some southern African taxa. Kew Bull. 30: 475–493.
38. Jeffrey, C. 1979. Further notes on Cucurbitaceae. V. The Cucurbitaceae of the Indian subcontinent. Kew Bull. 34: 789–809.
39. Jeffrey, C. 1980. A review of the Cucurbitaceae. J. Linn. Soc. Bot. 81: 233–247.
40. Kho, Y. D., J. Franken, and A. P. M. den Nijs. 1981. Species crosses under controlled temperature conditions. Cucurbit Genet. Coop. Rep. 4: 56–57.
41. Kho, Y. D., A. P. M. den Nijs, and J. Franken. 1980. In vivo pollen tube growth as a measure of interspecific incongruity in Cucumis L. Cucurbit Genet. Coop. Rep. 3: 52–54.
42. Kirkpatrick, K. J., D. S. Decker, and H. D. Wilson. 1985. Allozyme differentiation in the Cucurbita pepo complex: C. pepo var. medullosa vs. C. texana. Econ. Bot. 39: 289–293.
43. Kishi, Y., and N. Fujishita. 1970. Studies on the interspecific hybridization in the genus Cucumis. II. Pollen tube growth fertilization and embryogenesis of post-fertilization stage in incompatible crossing. J. Jap. Soc. Hort. Sci. 39: 149–156.
44. Knaap, B. J. van der, and A. C. de Ruiter. 1978. An interspecific cross between cucumber (Cucumis sativus) and muskmelon (Cucumis melo). Cucurbit Genet. Coop. Rep. 1: 6–8.
45. Kristjansson, F. K. 1963. Genetic control of two pre-albumins in pigs. Genetics 18: 1059–1063.
46. Kroon, G. H., J. B. M. Custers, Y. O. Kho, A. P. M. den Nijs, and H. Q. Varekamp. 1979. Interspecific hybridization in Cucumis (L.). I. Need for genetic variation, biosystematic relations, and possibilities to overcome crossability barriers. Euphytica 28: 723–728.

47. Leeuwen, L. van, and A. P. M. den Nijs. 1980. Problems with the identification of *Cucumis* L. taxa. Cucurbit Genet. Coop. Rep. 3: 55–60.
48. Meeuse, A. D. J. 1958. The possible origin of *Cucumis anguria* L. Blumea (Suppl.) 4: 196–205.
49. Meeuse, A. D. J. 1962. The Cucurbitaceae of Southern Africa. Bothalia 8: 1–111.
50. Merrick, L. C., and G. P. Nabhan. 1984. Natural hybridization of wild *Cucurbita sororia* group and domesticated *C. mixta* in southern Sonora, Mexico. Cucurbit Genet. Coop. Rep. 7: 73–75.
51. Morton, J. F. 1975. The sturdy Seminole pumpkin provides much food for thought. Proc. Florida Hort. Soc. 88: 137–142.
52. Naudin, C. 1859. Essais d'une monographie des espèces et des variétés du genre *Cucumis*. Ann. Sci. Nat. Bot., Ser. 4, 11: 5–87.
53. Naudin, C. 1865. *Cucumis meloni-trigonus*. Nouv. Arch. Mus. Hist. Nat. 1: 118–132.
54. Nijs, A. P. M. den, and D. L. Visser. 1985. Relationships between African species of the genus *Cucumis* L. estimated by the production, vigour and fertility of F_1 hybrids. Euphytica 34: 279–290.
55. Nijs, A. P. M. den, D. L. Visser, and J. B. M. Custers. 1981. Seedling death in interspecific crosses with *Cucumis africanus* L. f. Cucurbit Genet. Coop. Rep. 4: 58–60.
56. Norton, J. D., and D. M. Granberry. 1980. Characteristics of progeny from an interspecific cross of *Cucumis melo* with *Cucumis metuliferus*. J. Amer. Soc. Hort. Sci. 105: 174–180.
57. Pangalo, K. I. 1930. A new species of cultivated pumpkin (in Russian, English summary). Bull. Appl. Bot. Pl. Breed. 23: 253–265.
58. Parthasarathy, V. A. 1980. Taxonomy of *Cucumis callosus* (Rottl.) Cogn.—the wild melon of India. Cucurbit Genet. Coop. Rep. 3: 66–67.
59. Perl-Trev, R., and E. Galun. 1985. The *Cucumis* plastome: physical map, intragenic variation and phylogenetic relationships. Theor. Appl. Genet. 71: 417–429.
60. Perl-Trev, R., D. Zamir, N. Navat, and E. Galun. 1985. Phylogeny of *Cucumis* based on isozyme variability and its comparison with plastome phylogeny. Theor. Appl. Genet. 71: 430–436.
61. Puchalski, J. T., R. W. Robinson, and J. W. Shail. 1978. Comparative electrophoretic analysis of isozymes of *Cucumis* species. Cucurbit Genet. Coop. Rep. 1: 39.
62. Puchalski, J. T., R. W. Robinson, and J. W. Shail. 1978. Isozyme variation in the genus *Cucurbita*. HortScience 13: 387 (Abstract).
63. Rehm, S., R. P. Enslin, A. D. J. Meeuse, and J. H. Wessells. 1957. Bitter principles of the Cucurbitaceae. VII. The distribution of bitter principles in the plant family. J. Sci. Food Agric. 8: 679–686.
64. Robinson, R. W. 1976. *Cucumis* species: Phylogenetic relationships and breeding potential. Pickling Cucumber Improvement Committee Rep.
65. Robinson, R. W., and E. Kowalewski. 1978. Interspecific hybridization of *Cucumis*. Cucurbit Genet. Coop. Rep. 1: 40.
66. Robinson, R. W., and J. T. Puchalski. 1980. Synonymy of *Cucurbita martinezii* and *C. okeechobeensis*. Cucurbit Genet. Coop. Rep. 3: 45–46.

67. Robinson, R. W., and J. T. Puchalski. 1987. Electrophoretic classification of *Cucurbita* cultivars. Cucurbit Genet. Coop. Rep. 10: 83–84.

68. Robinson, R. W., J. T. Puchalski, and A. C. de Ruiter. 1979. Isozyme analysis of the megurk. Cucurbit Genet. Coop. Rep. 2: 17–18.

69. Ruiter, A. C. de, B. J. van der Knap, and R. W. Robinson. 1980. Rosette, a spontaneous cucumber mutant arising from cucumber-muskmelon mentor pollination. Cucurbit Genet. Coop. Rep. 3: 4.

70. Shaw, C. R., and R. Prasad. 1970. Starch gel electrophoresis of enzymes—a compilation of recipes. Biochem. Genet. 4: 297–320.

71. Singh, A. K., and K. S. Yadava. 1984. An analysis of interspecific hybrids and phylogenetic implications in *Cucumis* (Cucurbitaceae). Pl. Syst. Evol. 147: 237–252.

72. Staub, J. E. 1986. Malate dehydrogenase variation in African *Cucumis* species: a testable genetic hypothesis. Cucurbit Genet. Coop. Rep. 9: 18–21.

73. Staub, J., and L. Frederick. 1985. Electrophoretic variation among wild species in the genus *Cucumis*. Cucurbit Genet. Coop. Rep. 8: 22–25.

74. Staub, J., and R. S. Kupper. 1984. Electrophoretic comparison of six species of *Cucumis*. Cucurbit Genet. Coop. Rep. 7: 27–30.

75. Varekamp, H. Q., D. L. Visser, and A. P. M. den Nijs. 1982. Rectification of the names of certain accessions of the IVT-Cucumis collection. Cucurbit Genet. Coop. Rep. 5: 59–60.

76. Visser, D. L., and A. P. M. den Nijs. 1983. Variation for interspecific crossability of *Cucumis anguria* L. and *C. zeyheri* Sond. Cucurbit Genet. Coop. Rep. 6: 100–101.

77. Wall, J. R., and T. W. Whitaker. 1971. Genetic control of leucine aminopeptidase and esterase isozymes in the interspecific cross *Cucurbita ecuadorensis* × *C. maxima*. Biochem. Genet. 5: 223–229.

78. Weeden, N. F., and R. W. Robinson. 1986. Allozyme segregation ratios in the interspecific cross *Cucurbita maxima* × *C. ecuadorensis* suggest that hybrid breakdown is not caused by minor alterations in chromosome structure. Genetics 114: 593–609.

79. Whitaker, T. W. 1951. A species cross in *Cucurbita*. J. Heredity 42: 65–69.

80. Whitaker, T. W., and W. P. Bemis. 1964. Evolution in the genus *Cucurbita*. Evolution 18: 553–559.

81. Whitaker, T. W., and G. N. Davis. 1962. Cucurbits: Botany, Cultivation, and Utilization. Interscience, New York.

7 | Systematics and Evolution of a Domesticated Squash, *Cucurbita argyrosperma*, and Its Wild and Weedy Relatives

Laura C. Merrick

ABSTRACT. Systematic studies of *Cucurbita argyrosperma* (previously known as *C. mixta*) and related Mexican and Central American species are reported. Data are derived from studies of extant populations, preserved materials, controlled hybridizations, and isozymes. The latter two approaches reveal that little genetic differentiation has taken place in *C. argyrosperma*, although the species exhibits a relatively high degree of geographically correlated morphological diversity. Interpretation of the results indicates that *C. arygyrosperma* is a complex of domesticated and feral forms (subsp. *argyrosperma*) and wild populations (subsp. *sororia*). Domestication, from subsp. *sororia*, probably took place in southern Mexico and is represented in subsp. *argyrosperma* by four evolutionary lines, three of which are cultivated, vars. *argyrosperma*, *stenosperma*, and *callicarpa*, and one of which is feral, var. *palmeri*. *Cucurbita argyrosperma* is probably most closely related to *C. moschata*. Other species previously allied to *C. argyrosperma* are shown to be distant from it. The systematic and evolutionary implications of this study are discussed.

Studies concerned with the origin of domesticated plants are challenging intellectually and rewarding because of their potential to improve the crops on which humankind depends. In *Cucurbita argyrosperma* Huber, which generally has been known under the name *C. mixta* Pang., such potential is realized. This species, native to Mexico and Central America, includes domesticates, such as the 'Silverseed Gourd' and the 'Green Striped Cushaw', feral forms, and truly wild populations. In this chapter the results of a variety of different approaches to the study of plant evolution are brought together to provide a reasoned understanding of the origin and diversification of *C. argyrosperma*.

77

Nomenclature

Systematic studies of *Cucurbita argyrosperma* indicate that it is a complex of domesticated, feral, and wild forms; that it is most closely related to *C. moschata* (Duch. ex Lam.) Duch. ex Poir.; and that some taxa previously related to it are, in fact, quite distant. Morphological variation, especially as related to geographic distribution, and data from isozyme analyses and controlled hybridization studies, each of which is discussed in succeeding sections of this paper and in greater detail by Merrick (29), outline an evolutionary pattern in *C. argyrosperma* that is expressed in the recognition of two subspecies, *argyrosperma* and *sororia* (L. H. Bailey) Merrick & Bates (*C. sororia* L. H. Bailey, *C. kellyana* L. H. Bailey). The latter is composed of wild populations from which the domesticated and weedy forms of subsp. *argyrosperma* ultimately were derived. The domesticates are represented in subsp. *argyrosperma* by three varieties—var. *argyrosperma* (*C. mixta* var. *cyanoperizona* Pang.), var. *stenosperma* (Pang.) Merrick & Bates (*C. mixta* var. *stenosperma* Pang.), and var. *callicarpa* Merrick & Bates (*C. mixta* auct. non Pang.)—and the weedy taxon is represented by var. *palmeri* (L. H. Bailey) Merrick & Bates (*C. palmeri* L. H. Bailey). A full discussion of the nomenclatural aspects of these changes is presented elsewhere (30), but some highlights follow. Throughout, the intraspecific taxa are referred to under their respective subspecific or varietal epithets.

Until nearly the mid-twentieth century, plants of subsp. *argyrosperma* were widely recognized as *C. moschata* (12, 13, 43, 54). In 1930, on the basis of several distinct morphological traits and cross-incompatibility with three other domesticated *Cucurbita* species, the Russian geneticist, Pangalo, described *C. mixta* in two varieties, var. *cyanoperizona* and var. *stenosperma* (36). The first, from Guatemala, was characterized by large, broad, white-bodied seeds with blue-gray margins; the second, from Mexico, was characterized by long, narrow, white-bodied seeds. Not included in Pangalo's description or concepts were forms typical of northwestern Mexico, such as the 'Green Striped Cushaw', which Cutler and Whitaker (13) placed in *C. mixta*. The publications of Whitaker and Bohn (54) and Cutler and Whitaker (13) served to establish the broad concept of *C. mixta* in the literature.

Pangalo did not designate the typical variety for the species, but since his concept was limited to the two varieties, one must be considered typical, in this instance var. *stenosperma* (30). The designation of 'Green Striped Cushaw' as the typical element (13) is inappropriate, since it was not included in *C. mixta* by Pangalo. Nomenclaturally, the effect of this typification is to deprive the 'Green Striped Cushaw' of a botanical name, if it is considered distinct from the originally described elements of *C. mixta*.

In 1867 the name *C. argyrosperma* was published in the nursery catalogue of Charles Huber Bros. and Co., yet the origin of the name remained unknown until 1985 when Mabberly (26) mentioned the catalogue and the new species described in it. In the intervening years, *C. argyrosperma* persisted as a horticultural name, and Bailey (4), unable to determine its origin, gave it status by redescribing it. He assumed it was the same as *C. mixta* var. *cyanoperizona*, a point that has bearing on typification and nomenclature (30). The original description of *C. argyrosperma* by Huber was brief, but it accurately described the nature of the seeds, which is the crucial character. Recognition that *C. argyrosperma* and *C. mixta* var. *cyanoperizona* represent the same taxon leads, in relation to current understanding of evolutionary patterns, to the classification and nomenclature of this paper.

Two previous taxonomic treatments of *Cucurbita*, in which the genus was divided into nine or ten groups of presumably allied species (7, 24), depicted affinities of some of the taxa now recognized as *C. argyrosperma* with other species—namely *C. gracilior* L. H. Bailey, *C. moorei* L. H. Bailey, and *C. radicans* Naud. *Cucurbita radicans*, which actually includes *C. gracilior*, and *C. pedatifolia* L. H. Bailey, which includes *C. moorei*, are perennial species adapted to high elevations (3, 4). They are more closely related to each other and to other perennial *Cucurbita* species, i.e., *C. digitata* A. Gray and *C. foetidissima* HBK, than they are to the annual species of *Cucurbita*, such as *C. argyrosperma* (2, 29). The contradiction between this interpretation of the affinities of *C. gracilior* and *C. radicans* and that of Bemis et al.(7) and Hurd et al. (24) is found in the early misidentification of some accessions of subsp. *sororia* (as *C. sororia*) as these two species (27). Subsequently, the misidentified samples were included in numerous experimental studies and reviews concerned with the genus (6, 7, 19, 24, 33, 37, 38, 40, 52, 53), giving a misleading impression of species relationships.

Geographic Distribution

Almost all of the genetic diversity of domesticated subsp. *argyrosperma* is found within traditional cultivars or landraces, which continue to be grown in Mexico, Central America, and the southwestern United States (29). Plants of subsp. *argyrosperma* are adapted to warm climates and are principally grown at low elevations (49, 58), mostly well below an altitudinal limit of about 1800 m (29). Variety *argyrosperma* is cultivated in eastern and southern Mexico, including the Yucatán Peninsula, and in Central America. Variety *stenosperma* is endemic to Mexico and is grown in the region south of Mexico City. Landraces of var. *callicarpa* are cultivated in central and northwestern Mexico and the southwestern United States (29). Varieties *argyrosperma* and *callicarpa* are essentially allopatric, while there is a degree of overlap in distribution and integration of characters of vars. *stenosperma* and *argyrosperma* in south-central Mexico (29).

In contrast to the situation of *C. pepo* L. and *C. maxima* Duch. ex Lam., only a few commercial cultivars of subsp. *argyrosperma* exist (12, 45, 57). Those grown in the United States are virtually all var. *callicarpa* and consist of some of the cushaw type squashes such as 'Green Striped Cushaw', 'White Cushaw', 'Japanese Pie', and 'Tennessee Sweet Potato' (12, 13, 43). Notable exceptions are the 'Golden Cushaw' type cultivars, which are *C. moschata*, and several other cultivars of that species, e. g., 'Chirimin', 'Kikuza', 'Saikyo', and 'Yokohama', erroneously placed in *C. argyrosperma* (13). A few cultivars of var. *callicarpa* are grown in Argentina and probably are relatively recent introductions from the northern part of the range of subsp. *argyrosperma*. A single commercial cultivar of var. *argyrosperma*, 'Silverseed Gourd', has occasionally been grown as a horticultural curiosity in the United States (22).

The wild and feral taxa of *C. argyrosperma* are locally abundant throughout a widespread geographic range in Mexico and Central America (29). There is considerable overlap in the distribution of both subsp. *sororia* and var. *palmeri* with that of domesticated subsp. *argyrosperma*, although the domesticates are also cultivated in areas where wild taxa of *C. argyrosperma* are absent, e.g., the Yucatán Peninsula, the Sonoran Desert, or higher elevations on the Mexican Central Plateau. The range of altitude for both subsp. *sororia* and var. *palmeri* is sea level to perhaps 1300 m. Both taxa are weedy in terms of ecological adaptation, and populations tend to occur in disturbed areas—along roadsides or waterways, near agricultural fields or human habitations, or in garbage dumps. They thrive in high light environments and often can be found climbing by means of tendrils on neighboring vegetation or man-made structures.

In Mexico, subsp. *sororia* ranges from Jalisco south through Chiapas and from Tamaulipas south through Veracruz in the coastal areas and lower elevation slopes of the Sierras. It also occurs in two deep valleys in southern Mexico, i.e., along the Río Balsas and the central depression of Chiapas, and across the Isthmus of Tehuantepec. In Central America, subsp. *sororia* occurs in coastal zones and lower elevation mountain slopes of the Pacific side of Guatemala, El Salvador, Honduras, and Nicaragua (29).

Variety *palmeri* is distributed to the northwest of the geographic range of subsp. *sororia*. It ranges from the vicinity of the border between Nayarit and Jalisco northward into Sonora. It is absent from the extremely arid coastal desert zone of northern Sonora, but otherwise can be found in the coastal lowlands, western canyons, and lower elevations of the escarpment of the Sierra Madre Occidental. The northernmost extension of geographic range for var. *palmeri* coincides with the northern limit of the latter mountain range, and thus native populations do not occur in the United States (29).

Prior to this account, *C. lundelliana* L. H. Bailey was thought to be the only wild *Cucurbita* occurring in Central America (3, 14, 19, 24, 52). Found in Yucatán, Belize, and northern Guatemala, it does not, or rarely, overlaps the range of subsp. *sororia* (29). On the east coast of Mexico, north of the Yucatán Peninsula, the range of subsp. *sororia* somewhat overlaps *C. martinezii* L. H. Bailey and *C. fraterna* L. H. Bailey (1, 29). Populations of wild, feral, and domesticated taxa of *C. argyrosperma* frequently co-occur with *C. moschata* throughout lower elevations of Mexico and Central America (28, 29), but infrequently grow in the vicinity of either *C. pepo* or *C. ficifolia* Bouché, which typically are cultivated at higher elevations in Latin America (28, 29, 49, 58). With respect to the perennial species that had been placed in the Sororia Group, i.e., *C. radicans* and *C. pedatifolia*, and their currently presumed relatives, i.e., *C. digitata* and *C. foetidissima*, *C. argyrosperma* is largely ecologically and geographically distinct (8, 29).

Morphological Variation

Phenetic studies reveal that intraspecific taxa of *C. argyrosperma* are more similar in morphology to each other than they are to other *Cucurbita* species and that, of the latter, *C. moschata* is the species most similar to *C. argyrosperma* (7, 16, 29). Morphological traits characterizing *Cucurbita* are described in detail elsewhere (3, 4, 29, 54, 57); however, a selected number of fruit and seed traits that can be used to distinguish intraspecific taxa of *C. argyrosperma* are given next.

Fruit shapes of cultivated *C. argyrosperma* range along a continuum expressed in the gradual definition and lengthening of a thick neck, i.e., from depressed-globose to globose to pyriform, and finally to elongate with crooked to straight necks. In Mexico, fruit shapes of both var. *stenosperma* and var. *argyrosperma* typically range from globose to stoutly pyriform and average about 18–19 cm in length and diameter, yet in Central America fruits of var. *argyrosperma* tend to be long-necked and up to about 50 cm long. In contrast, fruits of var. *callicarpa* from a given field or region are often highly variable in shape, ranging from the one extreme to the other. They tend to be relatively long, i.e., 33 cm on average, and may reach 50 cm in length. The diameter is generally similar to that of the two other domesticated varieties.

Fruits of the wild subsp. *sororia* typically are ovate and relatively small (about 8 × 7 cm in length and diameter, on average). Those of var. *palmeri* are variable in shape, i.e., depressed-globose to flask-shaped, and usually are somewhat larger (10 × 9 cm) than those of subsp. *sororia*, although a wide range of sizes are found within and between populations.

The two subspecies share the same typical green and white rind color, but the patterning differs. Fruits of var. *stenosperma* and var. *argyrosperma* are predominantly white (rarely cream-colored) with green-mottled stripes occurring in bands, whereas the fruits of subsp. *sororia* and var. *callicarpa* and var. *palmeri* tend to be dominanted by green. The distribution of green on the rinds of var. *callicarpa* is usually much more netted than that found on fruits of the other two domesticated varieties. Fruits that are solid white or solid green in color are common in var. *callicarpa* and occasional in var. *palmeri*, but rarely occur in the other three taxa. Irregular blotches or mottled stripes of bright orange or yellow are sometimes present on the rinds of var. *callicarpa*; however, with age the green stripes on fruits of all *C. argyrosperma* taxa may fade to yellow. Unlike domesticated *C. pepo* and *C. moschata*, which are often characterized by shallow to deep longitudinal furrows and ridges on the surface of the fruit (12, 43, 54), *C. argyrosperma* fruits, with rare exceptions, have an unfurrowed surface (29).

All fruits of domesticated subsp. *argyrosperma* possess pale yellow to orange, nonbitter flesh, but the dark green tint to the placental tissue of var. *stenosperma* is unique to that taxon. Variety *argyrosperma* tends to have paler flesh than var. *callicarpa*. Variety *palmeri* typically has pale yellow to orange flesh like the domesticates, and although the flesh usually is bitter, nonbitter flesh is common. In contrast, fruits of subsp. *sororia* are characterized by bitter white flesh.

There is no difference in average length of the peduncles among the intraspecific taxa of *C. argyrosperma*. However, maximum peduncle width and texture distinguish var. *argyrosperma* from var. *stenosperma* and var. *callicarpa*, and var. *palmeri* from subsp. *sororia*. Varieties *stenosperma* and *callicarpa* typically have greatly enlarged corky peduncles (in some cases, as large as the fruits of subsp. *sororia*), whereas those of var. *argyrosperma* are only slightly enlarged. Warts or ridges of corky tissue, which radiate downward from the peduncle, are common on the surface of the fruits of var. *callicarpa*. The peduncles of var. *palmeri* are typically enlarged but slightly narrower and more variable than those of var. *argyrosperma*, ranging from thin and hard to thickened and corky. Peduncles of subsp. *sororia* are thin and hard, lacking any corky enlargement during development.

A sample of the diversity of seed types within each subspecific taxon is illustrated in Figure 1. Although both var. *argyrosperma* and var. *stenosperma* have smooth-surfaced white seeds, their shapes and margins differ. Variety *argyrosperma* has the broader seeds (ca. 23 × 12 mm long and broad), often with quite wide gray to bluish gray margins; var. *stenosperma* typically has more slender seeds (ca. 22 × 9 mm long and broad) with thin grayish or cream-colored margins. Together these seed variants constituted the "silverseed" type of Cutler and Whitaker (13).

Several major seed variants occur within var. *callicarpa*. The most common is white with a tan margin, i.e., the "green striped cushaw" type (13),

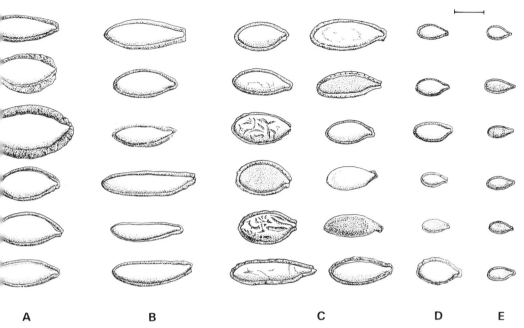

A   B   C   D   E

Figure 1. Diversity of seed types in *C. argyrosperma* from the United States, Mexico, and Central America. The seed sources are listed from top to bottom in each column. *Column A:* var. *argyrosperma.* 1. PI442231, San Luis Potosí. 2. PI442238, Veracruz. 3. PI512121, Veracruz. 4. PI438706, Campeche. 5. CATIE12041, Nicaragua. 6. PI512116 'Silverseed Gourd', United States. *Column B:* var. *stenosperma.* 1. PI512123, Veracruz. 2. BGCU82, Oaxaca. 3. PI511935, Guerrero. 4. PI512131, Veracruz. 5. PI438710, Yucatán. 6. BGCU167, Morelia. *Column C, left:* var. *callicarpa.* 1. PI512128 'Green Striped Cushaw', United States. 2. E01-028, Sonora. 3. E01-002, Arizona. 4. E01-023, Sonora, "taos" type. 5. E01-023, Sonora. 6. E01-029, Chihuahua. *Column C, right:* var. *callicarpa.* 1. PI512135, Guanajuato. 2. *Kelly 59,* Jalisco, "taos" type. 3. PI511907, Sonora. 4. PI511907, Sonora, marginless type. 5. PI442241, Zacatecas, marginless "taos" type. 6. PI442241, Zacatecas. *Column D:* var. *palmeri.* 1. PI512210, Nayarit. 2. E02-001, Sonora. 3. PI511900, Sinaloa. 4. PI512211, Sinaloa. 5. PI512228, Sonora, marginless type. 6. PI512230, Sonora. *Column E:* subsp. *sororia.* 1. PI512209, Jalisco. 2. PI512217, Oaxaca. 3. PI512216, Guerrero. 4. PI512212, Veracruz. 5. TWW21, El Salvador. 6. TWW22, Nicaragua.

in which the seed surface ranges from smooth to slightly or deeply "etched." A second *callicarpa* seed variant includes the "taos" type (13), which is brownish gold with a tan margin and smooth, thick seed coat. Two other seed variants are marginless counterparts of both the brownish gold and white color morphs (10, 29, 34). Seeds of var. *calliparpa* are generally about 19 × 11 mm long and broad (29).

Subspecies *sororia* produces small tan seeds (ca. 9 × 5 mm long and broad) with tan margins (29). Seeds of var. *palmeri* are somewhat larger (ca. 11 × 6 long and broad) than those of subsp. *sororia*, and typically are white with a tan margin. A marginless seed form of var. *palmeri* also exists, but is not widespread (29, 32, 34).

Genetic Relationships

Genetic affinities within *C. argyrosperma* and between it and other species of the genus have been studied using isozymes and hybridization data (29). Isozyme banding patterns of the intraspecific taxa of *C. argyrosperma* are indistinguishable. In their unity they distinguish *C. argyrosperma* from other *Cucurbita* species (16, 29, 38, 39, Puchalski and Robinson, this volume). The latter studies indicate that among the species of the genus, *C. moschata* is closest to *C. argyrosperma*, since the two species have more alleles in common than either does with other species of *Cucurbita*.

The intraspecific taxa of *C. argyrosperma* are interfertile (29). Table 1 summarizes cross-compatibility relationships among four of the five taxa of *C. argyrosperma* (var. *stenosperma* is not included). No significant differences in seed set were observed among the intraspecific crosses. The average number of seeds per fruit exceeds 250, which is comparable to the number of seeds per fruit resulting from *C. argyrosperma* sib- or self-pollinations (29).

Unlike seed set, fruit set response distinguished nondomesticated from domesticated taxa. The percentages of successful fruit set depended on the state of the maternal parent, whether it was a domesticated or a wild/weedy taxon. The identity of the intraspecific pollen donor did not affect fruit set (Table 1). When the percentage of fruit set is averaged for subsp. *sororia*, var. *palmeri*, var. *callicarpa*, and var. *argyrosperma*, as the maternal parent, values of 96%, 90%, 28%, and 44%, respectively, are obtained. These data indicate high fruit set for pollinations with wild taxa as female parents, and low fruit set for those with domesticated taxa as female parents. The same trend was seen in the results from sib- and self-pollinations of plants from a total of 86 different populations of wild and domesticated *C. argyrosperma* (29). In these pollinations, nondomesticates averaged 91% fruit set, whereas domesticates averaged 36% fruit set.

Possible explanations for the differences in fruit set behavior of domesticates vs. nondomesticates are found in a hormone response or source/sink relationship, which is triggered by the presence of prior fruit set on domesticated plants (25, 42), a mechanism that does not operate in the wild or feral plants. An analagous situation was described for another pair of cucurbits: the cucumber (*Cucumis sativus* L.) and its wild relative (*C. sativus* var. *hardwickii* (Royle) Alef.) (25). Breeding programs are underway to incorporate the multiple fruiting ability of the wild variety into cucumbers, since the number of fruit set per cultivated plant is often extremely low (25).

Table 1. Cross-compatibility relationships among intraspecific taxa of
Cucurbita argyrosperma

Parentage (female × male)	No. seeds per fruit (avg. ± SD)[1]	Fruit set percentage[2]	N[3]
sororia × *sororia*	298 ± 89	97	38
sororia × *palmeri*	244 ± 96	95	43
sororia × *callicarpa*	229 ± 104	92	12
sororia × var. *argyrosperma*	219 ± 68	94	18
palmeri × *sororia*	283 ± 130	92	24
palmeri × *palmeri*	294 ± 131	91	44
palmeri × *callicarpa*	251 ± 129	93	14
palmeri × var. *argyrosperma*	280 ± 160	82	12
callicarpa × *sororia*	399	50	2
callicarpa × *palmeri*	273 ± 158	19	16
callicarpa × *callicarpa*	345 ± 98	67	3
callicarpa × var. *argyrosperma*	439	25	4
var. *argyrosperma* × *sororia*	341 ± 69	22	9
var. *argyrosperma* × *palmeri*	325 ± 66	67	6
var. *argyrosperma* × *callicarpa*	264 ± 102	100	2
var. *argyrosperma* × *argyrosperma*	No fruit set	0	1

Note: A total of 156 different accessions of *Cucurbita* were included in the experimental hybridization study summarized, in part, in Tables 1 and 2 (29). *Cucurbita argyrosperma* comprised 92 of the accessions (var. *stenosperma* is included in the crosses summarized in Table 2, but not in Table 1); *C. moschata*, 20; *C. pepo*, 19; and *C. maxima*, *C. martinezii*, *C. foetidissima*, *C. radicans*, *C. pedatifolia*, *C. digitata*, and *C. lundelliana*, from 2 to 6 accessions each.
[1] Average number and standard deviation of seeds with viable embryos per fruit.
[2] Percentage of fruits per total number of pollinations.
[3] Total number of pollinations.

High levels of fertility (over 90% viable pollen) for experimentally and spontaneously produced intraspecific *C. argyrosperma* hybrids provides additional evidence for lack of reproductive isolation among these taxa (29). Bemis and Nelson (6) and Whitaker and Bemis (52) also found complete fertility in artificial crosses between populations of subsp. *sororia*, but Whitaker and Bemis reported that only partially fertile hybrids resulted from matings between subsp. *argyrosperma* and subsp. *sororia*, in contrast to the results reported here.

Interspecific crosses between *C. argyrosperma* and other species of *Cucurbita*, with the exception of *C. moschata*, seldom produce fertile hybrids, and if produced, the hybrids exhibit reduced fertilities (6, 11, 13, 29, 44, 47, 52, 54, 57). The results of an interspecific hybridization study among *Cucurbita* species, in part summarized in Table 2, indicate that the species can be placed into four general groups on the basis of their hybridization behavior with *C. argyrosperma* (29).

The first group consists of species that are capable of producing numerous, highly fertile interspecific progeny in crosses with *C. argyrosperma*. It is

Table 2. Cross-compatibility relationships between *Cucurbita argyrosperma* and other *Cucurbita* species

Parentage (female × male)	No. seeds per fruit (avg. ± SD)[1]	Fruit set percentage[2]	N[3]
Crosses with annual species			
C. argyrosperma × C. moschata	181 ± 172	66	64
C. moschata × C. argyrosperma	0	17	42
C. argyrosperma × C. pepo[4]	14 ± 16	57	44
C. pepo × C. argyrosperma	no fruit	0	95
C. argyrosperma × C. maxima	4 ± 5	61	13
C. maxima × C. argyrosperma	0	25	4
C. argyrosperma × C. lundelliana	0	55	11
C. lundelliana × C. argyrosperma	0	68	47
C. argyrosperma × C. martinezii	0	60	5
C. martinezii × C. argyrosperma	0	68	31
Crosses with perennial species			
C. argyrosperma × C. foetidissima	8 ± 11	80	5
C. foetidissima × C. argyrosperma	no fruit	0	1
C. argyrosperma × C. pedatifolia	0	50	2
C. pedatifolia × C. argyrosperma	no fruit	0	6
C. argyropserma × C. radicans	0	6	18
C. radicans × C. argyrosperma	no fruit	0	10
C. argyrosperma × C. digitata[5]	0	37	8
C. digitata × C. argyrosperma	no fruit	0	4

Note: See the note in Table 1 for a summary of the taxa used.
[1] Average number and standard deviation of seeds with viable embryos per fruit.
[2] Percentage of fruits per total number of pollinations.
[3] Total number of pollinations.
[4] Includes domesticated taxa, C. pepo subsp. pepo and subsp. ovifera (L.) Decker and the wild C. pepo subsp. ovifera var. texana (Scheele) Decker (17).
[5] Includes both C. digitata and C. cylindrata, in the sense of Gathman et al. (20).

composed solely of *C. moschata*. Seed set was fairly high on average when *C. argyrosperma* was the female parent, although embryos typically were somewhat reduced developmentally and the number of seeds per fruit was more variable and generally slightly less than in intraspecific crosses (Tables 1 and 2). Both F_1 and F_2 hybrids of *C. argyrosperma* × *C. moschata* were found to be highly fertile with about 89% viable pollen (29). The reciprocal cross, *C. moschata* × *C. argyrosperma*, failed to produce viable embryos.

The second hybridization group consists of species that produce few sterile to partially fertile hybrid progeny in crosses with *C. argyrosperma*. It includes two annual species, *C. pepo* (both wild and domesticated taxa) and *C. maxima*, and a perennial species, *C. foetidissima*. The third group consists of species that tend to produce fruits in reciprocal crosses with *C.*

argyrosperma, but the fruits are devoid of viable embryos. *Cucurbita lundelliana* and *C. martinezii*, which are completely interfertile with one another (29, 52), belong to this group. Finally, the fourth group consists of species that set fruit only when used as the male parent in crosses with *C. argyrosperma*. Viable embryos were not produced. The group comprises three perennial species, *C. pedatifolia*, *C. radicans*, and *C. digitata*, which are cross-compatible among themselves although not fully interfertile (2, 29).

In an evolutionary sense, the second group is unnatural in that the three species included in it are not closely related. Although not described here, their respective hybridization behavior with *C. argyrosperma* differs qualitatively (29). The species composition of the third and fourth groups is in accord with recognized intrageneric alliances (7, 29, 52).

In all instances of fruit set in *C. argyrosperma* when the male parent in an interspecific cross was one of the species of groups 3 or 4, the female parent was a nondomesticated plant, i.e., either subsp. *sororia* or var. *palmeri* (29). In fact, subsp. *sororia* and var. *palmeri*, and also the wild annual species *C. lundelliana* and *C. martinezii*, set fruit in approximately two-thirds of the crosses in which they were the female parent, irrespective of the genetic compatibility of the male parent. Embryo development, however, occurred in those fruits only if the parents were compatible (29). Bemis and Nelson (6) made a similar observation for *C. lundelliana*, but such hybridization behavior has not previously been reported for subsp. *sororia*, var. *palmeri*, or *C. martinezii*.

Ethnobotany

The pattern of variation in traits of the fruits and seeds of subsp. *argyrosperma* can be explained, in part, by the geographic patterns of human use. In southern Mexico and Guatemala, human selection in subsp. *argyrosperma* has emphasized seeds rather than fruits; both var. *stenosperma* and var. *argyrosperma* are cultivated principally for seed production (29, 35). The seeds are often extremely large, particularly those of var. *argyrosperma* landraces from the states of Veracruz and Puebla in Mexico, and, in comparison to the diversity of fruit shapes, colors, and sizes observed for var. *callicarpa* in the north, the southern domesticated varieties exhibit less diversity in fruit types. The fruit flesh tends to be stringy and is usually discarded or used as animal feed (29, 35, 49). The seeds are roasted and consumed, whole or peeled, as a snack (*pepitas*) or are ground and used in sauces, such as *pipián* or *mole verde*. In addition to production for home consumption, seeds of both varieties are produced on a commercial scale and sold in large volume in markets. The most common terms for these varieties in Mexico and Guatemala, *pipiana* and *pepitória*, are a reflection of their selection for seed consumption. Further south in Central America, selection of subsp. *argyrosperma* has been predominantly for consumption of the immature fruits as a vegetable. There, landraces usually produce long-necked fruits, which is the shape generally preferred when fruits are harvested at an immature stage (29).

In the northern part of its range, subsp. *argyrosperma* has been selected simultaneously for a wide variety of purposes. Variety *callicarpa* is usually consumed in three main forms: seeds, mature fruits, and immature fruits. The consumption of flowers and stem tips, or use of the thick, hard fruit rinds as receptacles (14) is only occasional (28, 29). Seeds are both consumed locally and sold in markets. As a probable result of selection for human consumption, the flesh of mature fruits of var. *callicarpa* is generally of higher quality, i.e., much smoother in texture and deeper orange, than that of var. *stenosperma* and var. *argyrosperma*. In northwestern Mexico, long-necked fruits of var. *callicarpa* are typically preferred for consumption as an immature vegetable; globose, flattened forms are said to be preferentially used for animal feed, indicating that several different fruit shapes may be consciously selected in the region (28).

The great variety of fruit shapes of var. *callicarpa* is analogous to that found in the landraces of *C. moschata*; however, the range of shapes for *C. moschata* far exceeds that of *C. argyrosperma*, since the latter fruits basically vary in terms of relative extension of a thickened neck (28, 29). Two of the common names of var. *callicarpa* in Mexico, *tamalyota* (a reference to smooth-textured flesh) and *calabaza pinta* (striped squash) refer to fruit characteristics. Other common terms for var. *callicarpa* in Mexico include *arota, tecomata, calabaza caliente* (squash of the hot lowlands), *guayeca, sopoma,* and *temprana* (early). In a few localities, different terms are used to discriminate different forms within var. *callicarpa*, e.g., forms with thick rinds vs. forms with thin rinds. Yet, for the most part, throughout the geographic range of subsp. *argyrosperma* as a whole, terms for landraces serve to distinguish *C. argyrosperma* from other cultivated species of *Cucurbita* (28, 29).

The flesh of the fruits of subsp. *sororia* and var. *palmeri* is not consumed because of the presence of high concentrations of the intensely bitter cucurbitacins (33). However, interviews with rural Mexicans revealed a number of folk uses, including 1) the consumption of seeds locally as snacks or, at times, sold in markets, or use as a vermifuge for intestinal worms or as a biocide to purify water; 2) use of the saponin-containing, suds-producing fruit pulp for washing animals or for washing and removing stains from clothing; 3) use of the intact rinds as receptacles; and 4) use of the bitter pulp to wean babies or young animals from breastfeeding (29, 31). Informants generally describe the folk uses of the wild taxa as being more prevalent in the past, but no geographic pattern for uses was discerned (29).

The most prevalent names for both var. *palmeri* and subsp. *sororia* in western Mexico are *chichicayota*, or the related terms, *chicayota* and *cala-bacilla de cayota* (or *coyote*). In eastern Mexico, the terms *morchete* or *calabacilla* are used for subsp. *sororia*. In the southern part of the range, subsp. *sororia* is commonly known as *calabacilla de caballo* (of the horse), or *ayote de caballo* (*ayote* is the generic term for cultivated squash in Central America). *Calabacilla* is a diminutive of the Mexican term for culti-vated squash, *calabaza*, and is used widely throughout Mexico to refer to various wild *Cucurbita* species (29).

Origin and Evolution

Although of limited extent, archaeological evidence supports the theory that *C. argyrosperma* was domesticated in southern Mexico. The earliest date for subsp. *argyrosperma* is about 5200 B.C., which was determined from specimens excavated from the Tehuacán Valley in Puebla and Oaxaca (15). The next dates for subsp. *argyrosperma* are some 5000 to 6000 years later and are from sites in northeastern and northwestern Mexico and southwestern United States: 200–900 A.D. at the Ocampo site in Tam-aulipas (56); 700 A.D. at the Río Zape site in Durango (9); and 380–1340 A.D. in caves in Arizona and New Mexico (14). These studies indicate that in the northern Mexican sites subsp. *argyrosperma* either was introduced thousands of years later or, if contemporary with *C. moschata* and *C. pepo*, was substantially less abundant. According to Whitaker (51), *C. argyrosper-ma* is absent from all other major archaeological sites in Mexico, United States, and South America where specimens of other *Cucurbita* species, dating to over 8000 B.C., have been recovered.

The taxonomic identity of the early archaeological specimens of subsp. *argyrosperma*, with one exception, corresponds to that of the varieties grown in the vicinity of those sites today: var. *stenosperma* near the Tehua-cán Valley, var. *argyrosperma* in Tamaulipas; var. *callicarpa* in north-western Mexico and the southwestern United States (29). However, from a single cave in Tehuacán Valley, Cutler and Whitaker (15) identified some samples of slender, brown, smooth-coated seeds as "taos" type. Since brown-seeded subsp. *argyrosperma* occurs today only as a segregate of var. *callicarpa* with a distribution hundreds of miles farther to the north and west of Tehuacán, this finding indicates that in the past either 1) brown-seeded forms of var. *stenosperma* may have occurred, or 2) "taos" type var. *callicarpa* may have been present.

Domestication in *C. argyrosperma* appears to have involved such changes as a loss of seed dormancy; a reduction of toxic secondary compounds, i.e., bitter cucurbitacins; a decrease in length of trichomes; an increase in overall size of plants, but in particular of the fruits and seeds, which are most commonly utilized by humans; and changes in color, shape, and texture of these parts (29). Such changes have accompanied the domestication of many other crops (5, 21, 41).

Subspecies *sororia* has been proposed as the ancestor of the domesticated varieties of *C. argyrosperma* (29). It possesses traits characteristic of other wild *Cucurbita* (3, 4), but no other wild *Cucurbita* exhibits such a close genetic, phenetic, ecological, and geographic similarity to subsp. *argyrosperma* (29). Morphological studies indicate that var. *argyrosperma* includes the least specialized forms of subsp. *argyrosperma*, and the northernmost landraces of var. *callicarpa* are the most derived (29). While it is not inconceivable that domesticated varieties were derived independently from subsp. *sororia*, the explanation favored here suggests that var. *argyrosperma* was the first to be selected. Derived from it in southern regions of Mexico was the narrowly seeded var. *stenosperma*, and, presumably later, the more variable and northerly var. *callicarpa*. This is in keeping with the suggestion that seeds of wild *Cucurbita* were initially utilized by humans for nutrition and only later were the plants selected for nonbitter flesh and otherwise domesticated (53, 55). Furthermore, there is morphological convergence of the vars. *stenosperma* and the more southerly landraces of var. *callicarpa* in south-central Mexico. Here populations of var. *callicarpa* tend to vary morphologically toward var. *argyrosperma*. Although different interpretations are again possible, the one favored here suggests these intermediate populations were selected less intensively for fruit characteristics than those from northwestern Mexico and the southwestern United States and tend to retain some character expressions more typical of var. *stenosperma*.

Aside from its often bitter flesh, var. *palmeri* essentially appears to be a smaller version of var. *callicarpa* (29, 31, 50). While var. *palmeri* could have diverged from subsp. *sororia*, all of the traits that distinguish it from subsp. *sororia* are characteristics of the domesticates rather than expressions that one might expect of two wild taxa diverging from each other. Such an example is found in the evolution of the large, fleshy, perennial roots of wild *Cucurbita* in arid northern Mexico, whose progenitors were fibrous-rooted, mesic annuals from more southerly regions (8, 52).

Hybridization between wild and domesticated plants of *C. argyrosperma* is one mechanism (18) that could explain the presence of domesticate traits in populations of var. *palmeri*. Indeed, the variety tends to be highly variable both within and between populations. Many traits—e.g., fruit and seed color, size, and shape; peduncle size and consistency; and bitterness—segregate in populations, strongly suggesting that hybridization is taking place (29). However, hybridization and introgression between domesticated and wild/weedy plants apparently occurs periodically not only in the range of var. *palmeri* (28, 29, 31, 32, 34, 40, 50) but also in the more southern and eastern range of subsp. *sororia* (15, 16, 24, 29, 52). Since gene exchange between co-occurring wild/feral and cultivated C. *argyrosperma* apparently occurs on occasion throughout the range in Mexico and perhaps in Central America, periodic hybridization alone does not explain why wild traits continue to be the norm in populations of subsp. *sororia*, whereas a mixture of wild and domesticate traits characterizes populations of var. *palmeri*.

A more likely explanation for the presence of domesticate traits in var. *palmeri* is in its evolution from escaped populations of var. *callicarpa*, although periodic occurrences of cultivated/escaped/weedy hybridizations probably also were involved. It is not uncommon to find domesticated *Cucurbita* growing along roadsides, in trash dumps or near cultivated fields (29). However, the adaptive syndrome for persistence of *Cucurbita* outside of cultivation seems to involve traits such as high cucurbitacin content, relatively small fruit size, hard fruit rinds, and multiple fruits per plant— characteristics that either serve to protect the plants against herbivores or to aid in seed dispersal. Natural selection for such traits, following mutation rather than hybridization in spontaneous populations of var. *callicarpa*, could have given rise to var. *palmeri*. It is also likely that var. *palmeri* has evolved repeatedly from populations of var. *callicarpa*, hence its pattern of variability (29). The hypothesized mode of origin of var. *palmeri* may also exist between the interfertile, cultivated and wild species pair of C. *maxima* and C. *andreana* Naud. (46).

Relationship of *Cucurbita argyrosperma* and *C. moschata*

The evidence presented here challenges the oft-repeated hypotheses that C. *lundelliana* is the ancestor to the cultivated species of *Cucurbita* in general (47, 52, 53, 55, 57) or C. *moschata* in particular (23, 24, 48, 50, 59). Aside from a distribution in the area thought to be the center of origin for the genus as a whole (52, 53), the cross-compatibility of C. *lundelliana* with other *Cucurbita* taxa and with C. *moschata* in particular has been the basis for arguing the progenitor status of C. *lundelliana*. Whitaker and Bemis (52) suggested that the ability of C. *lundelliana* to set fruit when pollinated by other *Cucurbita* species (6, 47) was a good indication of close evolutionary relationship. However, Merrick (29) found that propensity in *Cucurbita* to set fruit regardless of the species identity of the pollen parent is not unique to C. *lundelliana*. It is a characteristic shared by the nondomesticated taxa of C. *argyrosperma* and C. *martinezii*, in which embryo development and hybrid fertility, rather than fruit set, are more reliable indicators of genetic compatibility. The level of fertility reported by Whitaker (47, 48) for interspecific hybrids of C. *lundelliana* and C. *moschata* (on average, about 42% pollen viability) is about one half the level of fertility found by Merrick (29) in interspecific hybrids of var. *palmeri* or subsp. *sororia* and *C. moschata*, which showed, on average, about 89% pollen viability. Finally, the results of studies of morphological (7, 29) and isozymatic (29, 38, Puchalski and Robinson, this volume) variation suggest that *C. argyrosperma* is much more closely related to *C. moschata* than is *C. lundelliana*.

Cucurbita moschata and the domesticated and wild forms of *C. argyrosperma* are frequently sympatric in Mexico, yet apparently the two species continue to maintain their genetic integrity (29, 32). *Moschata*-like traits that are dominant in experimentally produced *C. argyrosperma* × *C. moschata* hybrids, i.e., tan fruit rind, darker hue to seed color, hard peduncle that is flared at the base, soft elongate trichomes, and pronounced foliaceous sepals, have not been reported as traits combined in plants from native populations of *C. argyrosperma* (29).

Merrick and Nabhan (32) and Merrick (29) have identified a series of factors, including temporal and spatial separation, pollinator behavior, genetic relationships, and differences in physiology, that are likely to be among the primary mechanisms controlling the incidence and direction of gene flow in *Cucurbita*. In consort, these factors appear to limit the amount of interspecific hybridization between *C. argyrosperma* and *C. moschata* despite their relatively high degree of genetic compatibility. A series of ethnobotanical interviews with farmers in northwestern Mexico indicated that gene exchange between wild and domesticated forms of *C. argyrosperma* is occasional, whereas, in their opinion, *C. moschata* and *C. argyrosperma* rarely, if ever, hybridize (28, 29, 31). However, Zizumbo (59) found that Mayan farmers in Yucatán prefer *C. argyrosperma*/*C. moschata* hybrids for seed production because the hybrids are said to combine the *moschata*-like trait of high seed numbers with the *argyrosperma*-like trait of large seed size.

Literature Cited

1. Andres, T. C. 1987. *Cucurbita fraterna*, the closest wild relative and progenitor of *C. pepo*. Cucurbit Genet. Coop. Rep. 10: 69–71.
2. Andres, T. C. 1987. Hybridization of *Cucurbita foetidissima* with *C. pedatifolia*, *C. radicans*, and *C. ficifolia*. Cucurbit Genet. Coop. Rep. 10: 72–73.
3. Bailey, L. H. 1943. Species of *Cucurbita*. Gentes Herb. 6: 1–321.
4. Bailey, L. H. 1948. Jottings in the cucurbitas. Gentes Herb. 7: 449–477.
5. Baker, H. G. 1972. Human influences on plant evolution. Econ. Bot. 26: 32–43.
6. Bemis, W. P., and J. M. Nelson. 1963. Interspecific hybridization in *Cucurbita*. I. Fruit set, seed and embryo development. J. Ariz. Acad. Sci. 2: 104–107.
7. Bemis, W. P., A. M. Rhodes, T. W. Whitaker, and S. G. Carmer. 1970. Numerical taxonomy applied to *Cucurbita* relationships. Amer. J. Bot. 57: 404–412.
8. Bemis, W. P., and T. W. Whitaker. 1969. The xerophytic *Cucurbita* of northwestern Mexico and southwestern United states. Madrono 20: 33–41.

9. Brooks, R. H., L. Kaplan, H. C. Cutler, and T. W. Whitaker. 1962. Plant material from a cave on the Rio Zape, Durango, Mexico. Amer. Antiq. 27: 356–369.

10. Bye, R. A., Jr. 1976. Ethnoecology of the Tarahumara of Chihuahua, Mexico. Ph.D. dissertation, Harvard Univ., Cambridge.

11. Castetter, E. F. 1930. Species crosses in the genus *Cucurbita*. Amer. J. Bot. 17: 41–57.

12. Castetter, E. F., and A. T. Erwin. 1927. A systematic study of the squashes and pumpkins. Iowa Agric. Exp. Sta. Bull. 244: 107–135.

13. Cutler, H. C., and T. W. Whitaker. 1956. *Cucurbita mixta* Pang.: its classification and relationships. Bull. Torrey Bot. Club 83: 253–260.

14. Cutler, H. C., and T. W. Whitaker. 1961. History and distribution of cucurbits in the Americas. Amer. Antiq. 26: 469–485.

15. Cutler, H. C., and T. W. Whitaker. 1967. Cucurbits from the Tehuacan caves of Mexico. *In* D. S. Byers, ed., The Prehistory of the Tehuacan Valley. Vol. 1. Environment and Subsistence. Univ. Texas Press, Austin.

16. Decker, D. S. 1986. A biosystematic study of *Cucurbita pepo*. Ph.D. dissertation, Texas A & M Univ., College Station.

17. Decker, D. S. 1988. Origin(s), evolution, and systematics of *Cucurbita pepo* (Cucurbitaceae). Econ. Bot. 42: 4–15.

18. DeWet, J. M. J., and J. R. Harlan. 1975. Weeds and domesticates: evolution in the man-made habitat. Econ. Bot. 29: 99–107.

19. Esquinas-Alcazar, J. T., and P. J. Gulick. 1983. Genetic Resources of Cucurbitaceae: A Global Report. IBPGR Secretariat, Rome.

20. Gathman, A. C., C. Young, J. C. Scheerens, J. M. Nelson, and G. P. Nabhan. 1989. Phenetic variation and systematic revision of the xerophytic gourd group, *Cucurbita digitata* Gray. Amer. J. Bot. (submitted).

21. Harlan, J. R. 1975. Crops and Man. Amer. Soc. Agron. and Crop Sci. Soc. Amer., Madison, WI.

22. Heiser, C. B., Jr. 1979. The Gourd Book. Univ. Oklahoma Press, Norman.

23. Heiser, C. B., Jr. 1979. Origins of some new world plants. Ann. Rev. Ecol. Syst. 10: 309–326.

24. Hurd, P. D., Jr., E. G. Linsley, and T. W. Whitaker. 1971. Squash and gourd bees (*Peponapis, Xenoglossa*) and the origin of the cultivated *Cucurbita*. Evolution 25: 218–234.

25. Lower, R. L., J. Nienhuis, and C. H. Miller. 1982. Gene action and heterosis for yield and vegetative characteristics in a cross between a gynoecious pickling cucumber inbred and a *Cucumis sativus* var. *hardwickii* line. J. Amer. Soc. Hort. Sci. 107: 75–78.

26. Mabberly, D. J. 1985. Die neuen Pflanzen von Ch. Huber frères & Co. in Hyères. Taxon 34: 448–456.

27. Merrick, L. C. 1987. Imperiled collections of cucurbit germplasm. Report to the California Genetic Resources Conservation Program.

28. Merrick, L. C. 1987. Wild and cultivated cucurbits from the Sierra Madre Occidental of Northwest Mexico and the Rio Balsas Valley of Southwest Mexico. Report to the International Board for Plant Genetic Resources.

29. Merrick, L. C. 1989. Systematics, evolution, and ethnobotany of a domesticated squash, its wild relatives and allied species in the genus *Cucurbita*. Ph.D. dissertation, Cornell University, Ithaca, NY (in prep.).

30. Merrick, L. C., and D. M. Bates. 1989. Classification and nomenclatural of *Cucurbita argyrosperma* Huber. Baileya 23: 94–102.
31. Merrick, L. C., and G. P. Nabhan. 1984. Natural hybridization of wild *Cucurbita* group and domesticated *C. mixta* in southern Sonora, Mexico. Cucurbit Genet. Coop. Rep. 7: 73–75.
32. Merrick, L. C., and G. P. Nabhan. 1985. Natural hybridization of wild and cultivated *Cucurbita* in Mexico: implications for crop evolution. Ecol. Soc. Amer. Bull. 66: 231 (Abstract).
33. Metcalf, R. L., A. M. Rhodes, R. A. Metcalf, J. Ferguson, E. R. Metcalf, and P.-Y. Lu. 1982. Cucurbitacin contents and Diabroticite (Coleoptera:Chrysomelidae) feeding upon *Cucurbita* spp. Environ. Ent. 11: 931–937.
34. Nabhan, G. P. 1984. Evidence of gene flow between cultivated *Cucurbita mixta* and a field edge population of wild *Cucurbita* at Onavas, Sonora. Cucurbit Genet. Coop. Rep. 7: 76.
35. Otzoy, M., M. Gonzalez, C. Azurdia, and M. Holle. 1988. Why were *Cucurbita* spp. domesticated? Unpublished mss.
36. Pangalo, K. I. 1930. A new species of cultivated pumpkin (in Russian, English summary). Bull. Appl. Bot. Gen. Pl. Breed. 23: 253–265.
37. Provvidenti, R., R. W. Robinson, and H. M. Munger. 1978. Resistance in feral species to six viruses infecting *Cucurbita*. Pl. Dis. Reporter 62: 326–329.
38. Puchalski, J. T., and R. W. Robinson. 1978. Comparative electrophoretic analysis of isozymes in *Cucurbita* species. Cucurbit Genet. Coop. Rep. 1: 28.
39. Puchalski, J. T., and R. W. Robinson. 1987. Electrophoretic classification of *Cucurbita* cultivars. Cucurbit Genet. Coop. Rep. 10: 84.
40. Rhodes, A. M., W. P. Bemis, T. W. Whitaker, and S. G. Carmer. 1968. A numerical taxonomic study of *Cucurbita*. Brittonia 20: 251–266.
41. Schwanitz, F. 1966. The Origin of Cultivated Plants. Harvard Univ. Press, Cambridge.
42. Stephenson, A. G. 1981. Flower and fruit abortion: proximate causes and ultimate functions. Ann. Rev. Ecol. Syst. 12: 253–279.
43. Tapley, W. T., W. D. Enzie, and G. P. van Eseltine. 1937. The Vegetables of New York, Vol. 1, part 4: The Cucurbits. N.Y. Agric. Exp. Sta., Geneva.
44. Van Eseltine, G. P. 1937. *Cucurbita* hybrids. Amer. Soc. Hort. Sci. 34: 577–581.
45. Whealy, K. 1985. The Garden Seed Inventory. Seed Savers Exchange Publ., Decorah, IA.
46. Whitaker, T. W. 1951. A species cross in *Cucurbita*. J. Heredity 42: 65–69.
47. Whitaker, T. W. 1956. The origin of the cultivated *Cucurbita*. Amer. Nat. 90: 171–176.
48. Whitaker, T. W. 1959. An interspecific cross in *Cucurbita* (*C. lundelliana* Bailey × *C. moschata* Duch.) Madrono 15: 4–13.
49. Whitaker, T. W. 1968. Ecological aspects of the cultivated *Cucurbita*. Hort Science 3: 9–11.
50. Whitaker, T. W. 1980. Cucurbitáceas Americanas Utiles al Hombre. Comisión de Investigaciones Científicas, La Plata.
51. Whitaker, T. W. 1981. Archeological cucurbits. Econ. Bot. 35: 460–466.
52. Whitaker, T. W., and W. P. Bemis. 1964. Evolution in the genus *Cucurbita*. Evolution 18: 553–559.

53. Whitaker, T. W., and W. P. Bemis. 1975. Origin and evolution of the cultivated *Cucurbita*. Bull. Torrey Bot. Club. 102: 362–368.

54. Whitaker, T. W., and G. W. Bohn. 1950. The taxonomy, genetics, production, and uses of the cultivated species of *Cucurbita*. Econ. Bot. 4: 52–81.

55. Whitaker, T. W., and H. C. Cutler. 1965. Cucurbits and cultures in the Americas. Econ. Bot. 19: 344–349.

56. Whitaker, T. W., H. C. Cutler, and R. S. MacNeish. 1957. Cucurbit materials from three caves near Ocampo. Amer. Antiq. 22: 352–358.

57. Whitaker, T. W., and G. N. Davis. 1962. Cucurbits: Botany, Cultivation, and Utilization. Interscience, New York.

58. Whitaker, T. W., and R. J. Knight, Jr. 1980. Collecting wild and cultivated cucurbits in Mexico. Econ. Bot. 34: 312–319.

59. Zizumbo-V., D. 1986. Aspectos etnobotánicos de las calabazas silvestres y cultivadas (*Cucurbita* spp.) de la peninsula de Yucatán. Bol. Esc. Cien. Antro. Univ. Yucatan 13: 15–29.

8 | Evidence for Multiple Domestications of *Cucurbita pepo*

Deena S. Decker-Walters

ABSTRACT. *Cucurbita pepo*, which comprises pumpkins, squashes, and ornamental gourds, has a relatively well-documented history and prehistory in North America. Nevertheless, the past 50 years have witnessed much debate concerning the origins and subsequent evolution of this domesticate. Recent evidence from new sources, e.g., allozyme analysis, provides a fresh perspective on these unresolved issues. In particular, the evidence suggests independent domestications of *C. pepo* in the eastern United States and in Mexico, with *C. texana* and *C. fraterna* as the respective wild progenitors.

The archaeological record indicates that *Cucurbita pepo* L., which consists of pumpkins, marrows, acorn squashes, crooknecks, fordhooks, scallop squashes, and ornamental gourds, has been cultivated over a large portion of North America. The earliest remains are seeds more than 9000 years old from the Valley of Oaxaca in southern Mexico (35). Seeds from Ocampo Caves, Tamaulipas, in northeastern Mexico date from 5000 to 7000 B.C. (36). In the eastern United States, there is good evidence of *C. pepo* as early as 2700 B.C. (8, 24), and 7000 year old rind from a site in Illinois may also represent this species (3, 6). *Cucurbita pepo* seeds, peduncles, and rind from archaeological sites in the southwestern United States and in northwestern Mexico are much younger (9, 24).

Modern and archaeological *C. pepo* material from the southwestern United States reveals little morphological diversity in the fruits (9, 34). Today, most fruits from this region and Mexico are medium-sized pumpkins with relatively large peduncles and narrow seeds (11, 34, 37). In contrast to this homogeneity, a variety of cultivated forms, e.g., scallop squashes and ornamental gourds, have been associated with the eastern United States both historically (31, 32, 34) and prehistorically (13, 19, 24).

96

Morphological diversity in archaeological and modern *C. pepo* material has led to several hypotheses concerning the origin or origins of this species. Some of these hypotheses accept a single domestication; others assert that there were multiple domestications. A recent advocate for the former, Ford (17, 18), proposed that *C. pepo* was selected from an undiscovered or now extinct wild species in Mexico. He hypothesized that different forms of *C. pepo* were carried to the United States along two corridors, one leading from northeastern Mexico to the eastern United States and the other connecting western Mexico to the southwestern United States.

Hypotheses supporting multiple domestications of *C. pepo* have their roots in the early works of Carter and Whitaker (5, 34). Although the idea of multiple domestications lost momentum shortly after it was proposed, Heiser (21, 22, 23) has recently revived it. He suggested that the progenitor of *C. pepo* was a wild species that once had a more extensive distribution in the eastern United States and Mexico. After the independent domestication of *C. pepo* in these two regions, the wild species was eliminated throughout most of its range by extensive harvesting, habitat destruction, or introgression with cultivars (22).

Heiser (21) postulated that the wild progenitor of *C. pepo* was *C. texana* (Scheele) A. Gray, a native and endemic inhabitant of river banks in southern and central Texas (4, 7). While his suggestion was not new, e.g., see Erwin (16), others have pointed out the lack of evidence favoring the ancestral position of *C. texana* over the possibility that it is a naturalized escape of *C. pepo* (20, 33). That *C. texana* is closely related to *C. pepo* is clear; the overall morphology of the wild species is nearly identical to that of some ornamental gourd cultivars (4) and the two taxa can hybridize without loss of fertility (33). The earliest scientific collection of *C. texana* dates to 1835 (16). More recently, spontaneous (self-sustaining in the wild) populations of *texana*-like plants have been documented for Alabama, Arkansas, Illinois, and Missouri (14, 26, 28, 29). Generally, these disjunct populations are considered ornamental gourd escapes. However, in accordance with Heiser's (22) hypothesis of a more widespread distribution for the progenitor of *C. pepo*, it has been suggested that these populations are relicts of *C. texana* in an area northeast of its currently recognized distribution in Texas (14, 15).

Recently, data have been generated that provide additional support for the theory of multiple origins of *C. pepo*. Much of this new evidence comes from the field of allozyme analysis, the comparison of variant enzyme forms. Numerical analyses of large morphological data sets have also been informative.

An exploratory study (10) of allozymes in *C. pepo* indicated substantial divergence between the Mexican landraces, pumpkins, marrows, and a few ornamental gourd cultivars on the one hand and acorn squashes, crooknecks, fordhooks, scallop squashes, and the bulk of ornamental gourd cultivars on the other hand. This dichotomy, which was firmly supported by more in-depth studies (11, 15), is comparable to differences that exist between species and even genera in other plant families. The association of bitter, small-fruited gourd cultivars with each lineage seems reasonable if one assumes that these forms represent the early selections from wild gourds and preceded larger, edible fruits. None of the cultivars, gourd or otherwise, expressed intermediacy in the allozyme analyses; in other words, there was no evidence of a common domesticated stock from which the two lineages evolved. The observed divergence within *C. pepo* was supported, albeit to a lesser extent, by recent numerical analyses of quantitative data taken from seeds (14), fruits, seedlings, young plants, and flowers (11).

The close morphological and allozyme association of pumpkin and marrow cultivars with modern Mexican landraces suggests that these cultivars are rooted in Mexican stock (11, 37). In contrast, development of diversity in the bulk of ornamental gourd cultivars and evolution of acorn squash, crookneck, fordhook, and scallop squash cultivars, appear to have a separate and dominant history in the eastern United States (31, 32).

Cucurbita texana and populations of *texana*-like plants from Alabama, Arkansas, and Illinois were included in some of the allozyme analyses (11, 15, 25) as well as in the morphological studies (11, 14). Although plants of *C. texana* do have an overall morphology very similar to that of some ornamental gourd cultivars, the wild species has a few distinct features, including the shape of the seed near the seed scar and the possession of unique allozyme alleles (14, 15, 25). The extent of the differentiation and the inability to correlate these characters with any selective pressure argue against the possibility that *C. texana* is a recent escape of *C. pepo*. In these analyses, the wild species almost always exhibited a closer relationship to the *C. pepo* lineage dominated by crooknecks, scallop squashes, and relatives than the two cultivated lineages did to each other.

The spontaneous populations of *texana*-like plants from Alabama, Arkansas, and Illinois generally expressed intermediacy with respect to *C. texana* and *C. pepo* in the allozyme and morphological analyses. Since large distances between *C. texana* in Texas and similar populations beyond Texas would preclude genetic exchange, genetic similarity among all these populations has been interpreted as evidence that they descended from a wild species that once had a widespread distribution in the eastern United States (15). Affinity of the non-Texas populations for *C. pepo* may be the result of subsequent introgression with cultivated forms. Introgression apparently did not occur in eastern and central Texas, perhaps because *C. pepo* was not cultivated in these areas prehistorically (30).

The discovery of characters that were useful in differentiating *C. texana* seeds from those of ornamental gourd cultivars provided a new methodology for examining archaeological remains. Numerous whole, uncharred seeds, identified initially as *C. pepo*, were recovered from water-saturated deposits at Hontoon Island in northeastern Florida (27). The earliest radiocarbon date associated with deposits containing cucurbit seeds was about 800 A.D. (13). In a discriminant analysis that included over 250 of these seeds, most were classified as *C. texana*, others were classified as belonging to the eastern United States lineage of *C. pepo*, and only a few seeds in Spanish period strata exhibited affinities with the Mexican lineage (13). These data are evidence that *C. texana* has an archaeological past and that it once inhabited Florida.

If *C. texana* is ancestral to the domesticated lineage whose origins presumably lie in the eastern United States, then what ancestral populations gave rise to the lineage that is of Mexican origin? The answer may lie in a little known wild species described as the "brother" of *C. texana* (4), *C. fraterna* L. H. Bailey. Recent collections of this species from northeastern Mexico have provided the opportunity to compare it to *C. pepo* and *C. texana*. In the preliminary morphological and allozyme analyses of Andres et al. (2), *C. fraterna* exhibited the highest affinity for 'Miniature Ball', an ornamental gourd cultivar of the Mexican lineage. Its location in Mexico also makes *C. fraterna* a likely candidate for the progenitor of the Mexican lineage of *C. pepo*.

In summary, evidence from various sources indicates a significant divergence within *C. pepo*. Furthermore, this divergence appears to separate a group of cultivars rooted in Mexico from another group whose origins are apparently in the eastern United States. Spontaneous populations of *texana*-like plants from beyond Texas and archaeological seeds from a site in Florida suggest that *C. texana* is an old established taxon that once was more widely distributed to the north and east of Texas than it is today. Reduced range and population size may be the result of extensive harvesting or habitat destruction by aborigines or introgression with cultivars as described by Heiser (22). Introgression with the domesticate is supported by the intermediate nature of relictual populations in Alabama, Arkansas, and Illinois. This and other evidence, including recent morphological and allozyme analyses, strongly suggest that *C. texana* is ancestral to the eastern United States lineage of the domesticate. Exploratory studies on *C. fraterna* (1, 2) indicate that it may be the Mexican counterpart to *C. texana*, having given rise to another lineage of *C. pepo*.

If both *C. texana* and *C. fraterna* descended from wild populations that gave rise to *C. pepo*, then new questions emerge. For example, what factors were involved in the divergence of *C. texana* and *C. fraterna* from a presumed common ancestor? At what point in the divergence of the wild taxa did the domestications of *C. pepo* take place? Were *C. texana* and *C. fraterna* separate species during the domestication process? In fact, do they represent good species today? Given that *C. pepo*, *C. texana*, and *C. fraterna* are fully cross-compatible (1) as well as very similar with respect to their allozymes and overall morphology, it has been suggested that they be treated as conspecific (1, 11, 12). While these proposed changes in nomenclature do not provide definitive answers to all of the above questions, they do provide a new point of reference for future studies.

Literature Cited

1. Andres, T. C. 1987. *Cucurbita fraterna*, the closest wild relative and progenitor of *C. pepo*. Cucurbit Genet. Coop. Rep. 10: 69–71.
2. Andres, T. C., M. Nee, N. F. Weeden, and J. Wyland. 1986. Rediscovery of *Cucurbita fraterna* Bailey, the alleged "brother" to *C. texana*. 27th Annual Meeting Soc. Econ. Bot., New York Botanical Garden, Bronx, NY.
3. Asch, D. L., and N. E. Asch. 1985. Prehistoric plant cultivation in west-central Illinois. *In* R. I. Ford, ed., Prehistoric Food Production in North America. Pap. Mus. Anthropol., Univ. Michigan 75, Ann Arbor, MI.
4. Bailey, L. H. 1943. Species of *Cucurbita*. Gentes Herb. 6: 267–322.
5. Carter, G. F. 1945. Plant Geography and Culture History in the American Southwest. Viking Fund Publ. Anthropol. 5, New York.
6. Conrad, N., D. L. Asch, N. B. Asch, D. Elmore, H. Gove, M. Rubin, J. A. Brown, M. D. Wiant, K. B. Farnsworth, and T. G. Cook. 1984. Accelerator radiocarbon dating of evidence for prehistoric horticulture in Illinois. Nature 308: 443–446.
7. Correll, D. S., and M. C. Johnston. 1979. Manual of the Vascular Plants of Texas. (Reprint of 1970 ed.). Univ. Texas Press, Dallas.
8. Cowan, C. W. 1985. Understanding the evolution of plant husbandry in eastern North America: Lessons from botany, ethnography and archaeology. *In* R. I. Ford, ed., Prehistoric Food Production in North America. Pap. Mus. Anthropol., Univ. Michigan 75, Ann Arbor, MI.
9. Cutler, H. C., and T. W. Whitaker. 1961. History and distribution of the cultivated cucurbits in the Americas. Amer. Antiq. 26: 469–485.
10. Decker, D. S. 1985. Numerical analysis of allozyme variation in *Cucurbita pepo*. Econ. Bot. 39: 300–309.
11. Decker, D. S. 1986. A biosystematic study of *Cucurbita pepo*. Ph.D. dissertation, Texas A & M Univ., College Station.
12. Decker, D. S. 1988. Origin(s), evolution, and systematics of *Cucurbita pepo*. Econ. Bot. 42: 4–15.
13. Decker, D. S., and L. A. Newsom. 1988. Numerical analysis of archaeological *Cucurbita* seeds from Hontoon Island, Florida. J. Ethnobiol. 8: 35–44.
14. Decker, D. S., and H. D. Wilson. 1986. Numerical analyses of seed morphology in *Cucurbita pepo*. Syst. Bot. 11: 595–607.
15. Decker, D. S., and H. D. Wilson. 1987. Allozyme variation in the *Cucurbita pepo* complex: *C. pepo* var. *ovifera* vs. *C. texana*. Syst. Bot. 12: 263–273.
16. Erwin, A. T. 1938. An interesting Texas cucurbit. Iowa State Coll. J. Sci. 12: 253–255.

17. Ford, R. I. 1980. 'Artifacts' that grew: their roots in Mexico. Early Man 2: 19–23.

18. Ford, R. I. 1985. Patterns of prehistoric food production in North America. *In* R. I. Ford, ed., Prehistoric Food Production in North America. Pap. Mus. Anthropol., Univ. Michigan 75, Ann Arbor, MI.

19. Gilmore, M. R. 1931. Vegetable remains of the Ozark Bluff-Dweller culture. Pap. Michigan Acad. Sci. 14: 83–102.

20. Gray, A. 1868. Field, Forest and Garden Botany. American Book, New York.

21. Heiser, C. B., Jr. 1979. The Gourd Book. Univ. Oklahoma Press, Norman.

22. Heiser, C. B., Jr. 1985. Of Plants and People. Univ. Oklahoma Press, Norman.

23. Heiser, C. B., Jr. 1985. Some botanical considerations of the early domesticated plants north of Mexico. *In* R. I. Ford, ed., Prehistoric Food Production in North America. Pap. Mus. Anthropol., Univ. Michigan 75, Ann Arbor, MI.

24. King, F. B. 1985. Early cultivated cucurbits in eastern North America. *In* R. I. Ford, ed., Prehistoric Food Production in North America. Pap. Mus. Anthropol., Univ. Michigan 75, Ann Arbor, MI.

25. Kirkpatrick, K. J., D. S. Decker, and H. D. Wilson. 1985. Allozyme differentiation in the *Cucurbita pepo* complex: *C. pepo* var. *medullosa* vs. *C. texana*. Econ. Bot. 39: 289–299.

26. Mohlenbrock, R. H., and D. M. Ladd. 1978. Distribution of Illinois Vascular Plants. Southern Illinois Univ. Press, Carbondale.

27. Newsom, L. A. 1986. Plants, human subsistence, and environment: a case study from Hontoon Island (8–VO-202), Florida. M.S. thesis, Univ. Florida, Gainesville.

28. Smith, E. B. 1978. An Atlas and Annotated Checklist of the Vascular Plants of Arkansas. Univ. Arkansas Press, Fayetteville.

29. Steyermark, J. A. 1963. Flora of Missouri. Iowa State Univ. Press, Ames.

30. Story, D. A. 1985. Adaptive strategies of archaic cultures of the West Gulf Coastal Plain. *In* R. I. Ford, ed., Prehistoric Food Production in North America. Pap. Mus. Anthropol., Univ. Michigan 75, Ann Arbor, MI.

31. Sturtevant, E. L. 1890. The history of garden vegetables. Amer. Naturalist 24: 719–744.

32. Tapley, W. T., W. D. Enzie, and G. P. van Eseltine. 1937. The Vegetables of New York. Vol. 1, part 4: The Cucurbits. N. Y. Agric. Exp. Sta., Geneva.

33. Whitaker, T. W., and W. P. Bemis. 1964. Evolution in the genus *Cucurbita*. Evolution 18: 553–559.

34. Whitaker, T. W., and G. F. Carter. 1946. Critical notes on the origin and domestication of the cultivated species of *Cucurbita*. Amer. J. Bot. 33: 10–15.

35. Whitaker, T. W., and H. C. Cutler. 1971. Prehistoric cucurbits from the Valley of Oaxaca. Econ. Bot. 25: 123–127.

36. Whitaker, T. W., H. C. Cutler, and R. S. MacNeish. 1957. Cucurbit materials from three caves near Ocampo, Tamaulipas. Amer. Antiq. 22: 352–358.

37. Whitaker, T. W., and R. J. Knight, Jr. 1980. Collecting cultivated and wild cucurbits in Mexico. Econ. Bot. 34: 312–319.

9 | Biosystematics, Theories on the Origin, and Breeding Potential of *Cucurbita ficifolia*

Thomas C. Andres

ABSTRACT. The domesticate *Cucurbita ficifolia* merits the attention of the squash breeder; the vines are cold-tolerant with some disease resistance and produce long-keeping fruits. Biosystematic evidence indicates that the species is not as highly distinguished morphologically and genetically from the other domesticated squashes as was once believed; it does not differ from these other species in having perennial rather than annual growth habit as has been widely reported, and it may be hybridized to them by the use of embryo culture. In contrast to the other *Cucurbita* cultigens, *C. ficifolia* is relatively uniform in morphology and genotype. Variation, however, does occur in the fruit and seed color, some isozymes, and sensitivity to photoperiod. Although *C. ficifolia* has no known wild progenitor, it appears to have been domesticated in the highlands of northern South America and was introduced into Mexico and the southern Andes Mountains during pre-Columbian times.

Cucurbita ficifolia Bouché, commonly called the fig-leaf gourd or black-seeded squash, is the least known and perhaps most misunderstood species among the five domesticated species of squash. Based on phenetic and compatibility relationships, it is generally recognized to be an evolutionarily aberrant member of the genus. Reports vary as to whether *C. ficifolia* is perennial or annual, whether it originated in Mexico, South America, or elsewhere, and whether it is even a truly domesticated species. It is infrequently used in developed countries, except as a rootstock for grafted cucumbers, but is a significant secondary food in some cool highland regions of tropical America where few other vegetables are able to grow. There have been no breeding programs for the improvement of *C. ficifolia*, apparently because of its unrecognized potential economic value in developed countries and because of interspecific sterility barriers. However, the species does contain potentially valuable genes for disease resistance, cold

102

hardiness, and the long storage life of fruits. The present biosystematic study was undertaken to investigate questions concerning the origin and domestication of *C. ficifolia* and to explore its potential use in squash breeding.

Nomenclature

Cucurbita ficifolia was described by Bouché in 1837 (14); however, the species was first reported in Europe and named *C. melanosperma* by Braun in 1824 in a seed catalog from Carlsruhe Garden, Germany, but without a description or indication of place of origin (11, 65). Gasparrini (30), apparently unaware of either earlier publication, also used the name *C. melanosperma* in his 1848 description of the species. Braun et al. (15) later described *C. melanosperma* in detail, and this name was subsequently used by Naudin (45–48) and others. However, the name *C. ficifolia* has priority. In addition to *C. melanosperma,* another name preceded Bouché's. *Pepo malabaricus* was proposed by Sageret in 1826 (54, 55) for the "melon du Malabar," a vernacular name used in France for the fig-leaf gourd. Because there was no accompanying description or reference to a publication or a specimen, the name was not validly published (6). In the late 1800's *C. mexicana* Hort., originally described in the Dammann Catalogue (23), was cited (3) as a unique species from Mazatlan, Mexico "very similar to *C. melanosperma,* but with the leaf of a different shape, and flower of a different hue, the seeds are large and black." There is nothing in the description to indicate this name to be anything but a synonym of *C. ficifolia,* the only known species of *Cucurbita* with black seeds. Two species have been incorrectly ranked as varieties or cultivars of *C. ficifolia* in the past, including *C. argyrosperma* Huber (7, 36) and *C. moorei* L. H. Bailey (28), which is probably a synonym for *C. pedatifolia* L. H. Bailey (T. C. Andres, unpublished observation).

Biosystematics

In general, *Cucurbita* is a well-defined genus composed of closely related species that are often highly polymorphic and poorly differentiated. For example, all *Cucurbita* species are monoecious or rarely gynoecious with 20 pairs of chromosomes and can be crossed in numerous interspecific combinations (see Singh, this volume). Among the cultigens *C. pepo* L., *C. moschata* (Duch. ex Lam.) Duch. ex Poir., *C. argyrosperma* Huber, and *C. maxima* Duch. ex Lam., parallelisms in fruit morphology are common; four different species are sometimes recognized as winter squash or pumpkins. In order to identify the species confidently, complete specimens are often needed, including leaves, stems, roots, flowers, fruits (or at least their photograph), pedicels, and seeds along with detailed field notes. Such evidence has rarely been assembled and exists in only a few herbaria.

In contrast, *C. ficifolia* is relatively homogeneous in morphology and genetic composition. This facilitates recognition of the species but has given a misleading impression of its uniqueness. The fruits, which are the part of the plant most likely to have undergone diversification as a result of human selection, are relatively uniform in shape, rind and flesh characteristics. This is in marked contrast to the other species of squash, in which numerous cultivars of striking variation exist.

Several characters have been incorrectly used to separate *C. ficifolia* from the other four domesticated species of squash, the most notable being perennial versus annual habit. Taxonomic keys to the squashes invariably delineate *C. ficifolia* as the only perennial cultigen in the genus (74, 77); however, no morphological features are ever given by which the duration of the plant can be recognized. To further confuse matters, it is sometimes stated that *C. ficifolia* is perennial but is often cultivated as an annual (8, 25).

It is now evident that *C. ficifolia* does not differ in longevity from the other squash species. All domesticated species of *Cucurbita* have extensive fibrous root systems and indeterminate growth. Under suitable growing conditions, each of these species, with the exception of bush varieties, will continue to grow indefinitely when the stems are permitted to root at the nodes. *Cucurbita ficifolia* is tolerant of low temperatures, although it and all species of *Cucurbita* are frost-tender (59). *Cucurbita ficifolia*, however, is grown in the high-altitude tropics, where it will often maintain vigor through the cool winters while the other, less cold-tolerant species perish, thereby appearing to differ from the other cultivated cucurbits grown in the same region. In warmer climates, landraces of *C. moschata* have been reported to last for several years (44), although this species is generally reported to be annual. All of the domesticated squash species are usually grown as annuals, but in their native habitats, i.e., the presumed habitat of their wild progenitors, they may grow spontaneously as short-lived herbaceous perennials.

One character used to delineate *C. ficifolia* is the color of its seeds. It is the only species of *Cucurbita* with black seeds, but not all *C. ficifolia* seeds are black; some may be dark brown, while others are more or less pale buff-colored like those of most other species of *Cucurbita* (Figure 1). Bukasov (16) listed two forms of *C. ficifolia* based on the seed colors, *C. ficifolia* f. *melanosperma* and *C. ficifolia* f. *leucosperma*. There are landraces of *C. moschata* grown in northern South America that also have dark brown seeds.

A final character that is often mistakenly used to distinguish *C. ficifolia*

Figure 1. Range of seed variation within cultivated *Cucurbita ficifolia.*

from the other species of *Cucurbita* are the leaves shaped like those of *Ficus carica* L., hence the specific epithet. Other species, such as *C. lundelliana* L. H. Bailey, *C. ecuadorensis* Cutler & Whitaker, and some landraces of *C. moschata* and *C. pepo* may also have figlike leaf shapes. Furthermore, most species of *Cucurbita* and many other related genera are heterophyllous, with late-developing leaves generally more deeply lobed than those produced early in the growth cycle. Unfortunately, most herbarium specimens of cucurbits include only one stage of leaf development. This variation in leaf shape has perhaps contributed to an excessive splitting of the number of described species of *Cucurbita* and led Bailey (8) to believe there was another undescribed species of black-seeded gourd.[1]

In summary, of the three characters most often used in taxonomic keys to identify *C. ficifolia*, one is misinterpreted (perennial growth habit), one is not unique to *C. ficifolia* (leaf shape), and the third is not always true for *C. ficifolia* (black seeds).

There are several characters that may be used, at least in concert, to distinguish *C. ficifolia* from all other species. As already mentioned, *C. ficifolia* is the only species of *Cucurbita* with black seeds, but some forms and immature seeds are light buff-colored as in other cucurbits. The shape

[1]Bailey grew the unidentified accession with black seeds in Ithaca, New York, but the plants were evidently sensitive to a short-day photoperiod and never produced fruits. The voucher specimens indicate that they are *C. ficifolia.*

of the seeds is fairly diagnostic, however. They are large, 15–25 mm in length, and oblong-ellipsoidal with a width-to-length ratio of 3:2, which is generally broader than that in the other cucurbit cultigens. They are flat in cross section and hard, without a thick, spongy epidermis that is characteristic of the seeds of *C. maxima* (58) and some forms of *C. argyrosperma*. Furthermore, the surface of the seed appears minutely pitted or pebbled and not polished or crazed like the surface of the seeds of *C. maxima*. There is a uniformly thin margin around the edge of *C. ficifolia* seeds with the same color and texture as the rest of the seed.

A more reliable distinguishing feature of *C. ficifolia*, which has not been reported previously, is the presence of trichomes on the filaments in the male flowers. Most other species of *Cucurbita* have glabrous filaments or filaments with just a few scattered trichomes at their base. The only exceptions to this are *C. foetidissima* HBK, *C. pedatifolia* Bailey, and *C. radicans* Naud., which also have these trichomes, albeit shorter (< 1 mm in length) than those of *C. ficifolia* (> 1 mm in length and clearly visible without a hand-lens). Trichomes on the axis of *C. ficifolia*, particularly along the growing tips and young leaves, bear a brown glandular resin that will stain the fingers.

Other morphological features of *C. ficifolia* have been adequately described elsewhere (7, 8, 16, 45, 75). The stem trichomes of *C. ficifolia* are setaceous, like those of *C. maxima*. The peduncle is hard, smoothly angled, and slightly flaring at the attachment to the fruit, like that of *C. ecuadorensis* and some landraces of *C. moschata*. The exterior color and shape of the fruit are very similar to some landraces of *C. maxima*; and the fruit flesh is white, somewhat dry, and coarsely fibrous, like some forms of *C. argyrosperma*. The fruits of *C. ficifolia*, which range from 15 to 50 cm long, with reports by roadside vendors of up to 100 cm long, are sometimes even confused with watermelon, *Citrullus lanatus* (Thunb.) Matsum. & Nakai, a case of convergent rather than parallel evolution.

Whitaker and Bohn (75) report that *C. ficifolia* seems to require a short photoperiod for flowering, whereas the other domesticated species are insensitive to day length. While this is generally true, not all accessions of *C. ficifolia* require short-days for flowering, and many landraces of other *Cucurbita* species are short-day plants. There are accessions of *C. ficifolia* that produce fruits at the extreme latitude of Norway (38). On the other hand, there are accessions of squash, primarily in the species *C. ficifolia* and *C. moschata*, that do not flower north or south of the torrid zone.

What little variation exists in *C. ficifolia* appears to occur throughout its range. The buff-seeded form, although less common than the black-seeded form, occurs throughout the range, usually in the same field with black-seeded plants. The fruits have basically three different color patterns that occur throughout Mexico, Central America, and South America (Figure 2):

Figure 2. The three basic fruit types of *Cucurbita ficifolia*.

all white; a distinct reticulated pattern of green on white, sometimes with ten radial stripes spreading from the apex to the base; and dark green, without reticulations but sometimes also with white longitudinal stripes. Slight variations of the reticulations and stripes occur, but nothing like the wide assortment of color variants, warts, corky outgrowths, and shapes and sizes that occur in the fruits of the other domesticated species. There is no association between seed color and fruit color. Uses of these various forms of *C. ficifolia* are generally the same and no named cultivars have been recognized.

Even at the molecular level, there appears to be little genetic diversity throughout the range of *C. ficifolia*. In an isozyme survey using starch-gel electrophoresis, 60 accessions of *C. ficifolia* were assayed for 20 enzyme systems coded by a putative 50 scorable loci (T. C. Andres, unpublished data). This study included six individuals each of 20 accessions procured from seed companies and botanical gardens throughout the United States, Europe, and Japan; 28 accessions from Mexico; nine accessions from Central America; and three accessions from South America. Five of the 50 loci had variant alleles or allozymes, an average of only 1.1 allozymes per enzyme locus, contrasted to around 1.6 allozymes per locus in *C. pepo* (Puchalski and Robinson, this volume; T. C. Andres, unpublished data). The few polymorphic loci detected in *C. ficifolia* were in the enzyme systems isocitrate dehydrogenase, malic dehydrogenase, acid phosphatase, and peroxidase. A larger sample from South America would be desirable before concluding whether or not the entire species is genetically uniform.

Weiling (69) proposed that species in the genus *Cucurbita* are secondary tetraploids, with *C. ficifolia* having a different genome set from the other cultivated species. He proposed that *C. pepo*, *C. argyrosperma*, *C. moschata*, and *C. maxima* have the identical genome pairs AABB, while the genome pairs of *C. ficifolia* are AACC. This conclusion was based on a study of the chromosome complement at meiosis of the cultivated species and some F_1 species hybrids, including *C. maxima* × *C. ficifolia*. According to Weiling (69), however, the genome BB has a weak mutual affinity for CC. Whitaker and Davis (77) stated that these conclusions are largely theoretical and deserve further analysis.

Several investigators have attempted to cross *C. ficifolia* with the other cultivated species. Successful hybrids have been produced in a very low percentage of the pollinations between *C. ficifolia* and *C. maxima*, *C. moschata*, and *C. pepo* (26, 32, 47, 62, 68, 70, 77, T. C. Andres, unpublished data). Only partially developed embryos form, and embryo culture must be employed to obtain subsequent generations. Unfortunately, sterility barriers have thus far prevented obtaining progeny beyond the F_1 and first backcross generations. Variable results, however, have been obtained when different cultivars of a species are used. It is hoped that more compatible cultivar combinations may exist. Whitaker and Bemis (74) state that a fig-leafed form of *C. ficifolia* is probably allied with *C. pepo* or *C. maxima*, whereas a form with *moschata*-like leaves is probably closely related to *C. moschata*. Evidence for these two forms was not documented nor have they been observed by the author.

Since there is relatively little variation found within *C. ficifolia*, the question arises whether it is a truly domesticated species. Although it is usually cultivated, it sometimes spreads and becomes semi-feral, giving the impression of a nondomesticated plant. As with the other domesticated species of squash, *C. ficifolia* fruits contain no detectable bitterness (42). Based on this evidence and the allometric growth of those plant parts desired by humans (wild *Cucurbita* species have much smaller, extremely bitter fruit and smaller seeds), *C. ficifolia* is a domesticated species.

The reason(s) why *C. ficifolia* has remained morphologically and genetically monomorphic is not known, but possible explanations include 1) initial genetic variation in the species was low; 2) a single domestication event occurred that resulted in a taxon biologically isolated from the wild progenitor; 3) only one genotype was extensively introduced throughout its present range; 4) intensity of selection for variant types has remained low; 5) domestication has taken place in only one type of habitat; 6) it is not the oldest domesticated squash and therefore has had less time than the other squash species to diversify.

Origin and Distribution

Like all other species of *Cucurbita*, *C. ficifolia* originated in the Americas. It is presently grown in many tropical highland regions of the world, including south-central Asia and the Philippines (33). However, it is primarily used under sustainable agriculture systems in Latin America, where personal observations and herbarium records show that it is generally grown between 1000 and 2800 m above sea level. The Latin American distribution of *C. ficifolia* exceeds that of all the other cultivated *Cucurbita* species, extending from the highlands of northern Mexico, through Central America, and throughout Andean America from Colombia and Venezuela to central Chile and northwestern Argentina. The climate in these areas is generally very moist and too cool for any other species of squash to grow, except for some short-season landraces of *C. pepo* and *C. maxima*.

The first fruits of *C. ficifolia* to reach Europe apparently took a circuitous route from South America to the Malabar coast of India along the much traveled Portuguese and Dutch trade routes in the 16th and 17th centuries, before finally reaching Europe (41). Bouché (14) did not know where his cultivated material came from, but vernacular names such as *melon de Malabar* (Malabar melon), *courge de Siam* (Siam squash), and *Angurian-kurbis* (Angora squash) suggested a southern Asian origin.

In 1854 a well-documented shipment of yaks was introduced to France from the border of Tibet with India. In Shanghai a large quantity of *C. ficifolia* fruits were loaded to serve as fodder for these animals during the long ocean voyage. More than a year after the voyage ended many of these fruits were left over and still intact at the Muséum d'Histoire Naturelle, Paris. They became a curiosity in European botanical gardens. This convinced the botanists of the time and for many years afterward that *C. ficifolia* was native to the Orient (7, 16, 19, 46). However, De Candolle (24) doubted this because all known perennial *Cucurbita* species were native to North America. He assumed that only the "annual" species were of Old World origin.

The etymological, ethnobotanical, and archaeological evidence prove that *C. ficifolia* is of American origin. In Asia, *C. ficifolia* is primarily used only for livestock feed, and no Sanskrit name exists, whereas in Latin America there are archaeological records and many local names and uses of the plant. The most frequently used vernacular name for *C. ficifolia* in Mexico and parts of Central America is *chilacayote* (also spelled *chilicayota*, *chilacayoti*, *chilacayotl*, and other variations), probably a corruption of *tzilacayotli*, but other names include *alcayota*, *pachayota*, *silacayote*, *cuicuilticayotli*, *cidracayote*, *cidra*, *mail*, and *chiberre*, *chibesse*, or *chiverre* (the latter three names used only in Costa Rica) (16, 52, 53, 67). Martínez (40) lists several additional Mexican names of generally Nahuatl origin for *C. ficifolia*. In Colombia, *C. ficifolia* is usually referred to as *anjama*, *vic-*

toria, or *vitoriera*; in Venezuela, the Quechuan name *zapallo* is generally used for all *Cucurbita*; and in Ecuador, *zambo* (*zambu*, *sambo*, or *tambo*) is used. The names used in Bolivia, Peru, and northern Argentina, in the Quechua and Aymara language areas, are surprisingly similar if not identical to the Mexican names: *lacayote* or *lacahuiti*, *silacoyote*, *alcayote* or *alcallota*, *cayote*, and *tintimoro*.

While *C. ficifolia* was clearly cultivated in Mexico prior to the Spanish conquest, it is uncertain where it was first domesticated in its extensive Latin American range. The etymological evidence is inconclusive at best in this regard. The Nahuatl names could be derived from or precede the similar sounding Quechuan and Aymara names. For example, Francisco Hernández in the 16th century was the first to clearly describe the *chilacayote* in Mexico (see 61). He speculated that the Nahuatl name, *tzilacayotli*, is derived from either the words *tzilac-ayotli*, meaning "smooth calabaza or squash," or *tzilictic-ayotli*, meaning "squash that resonates or sounds" when struck. The name *cuicuilticayotli* literally means the "painted squash." Even if Hernández is correct, these names may still be derivations from South America that have taken on slightly new meanings in Nahuatl by folk etymology.

Sahagun (56), a Franciscan friar, recorded in the 16th century that the hollowed-out fruit of *tzilacayotli* was used by the Aztecs as a bowl to contain offerings, such as pulque, for their deities. The bowl was placed in front of the image or images, and with their marbled appearance, were said to be made of precious stones which the Aztecs called *chalchihuitl*. This religious practice suggests an ancient use, if not origin, in Mexico and subsequent spread to the south.

The archaeological record tells a different story. Numerous seed remnants and pedicels of *C. ficifolia* have been recovered from archaeological sites along the central to northern coastal desert of Peru. These have been recovered from several horizons spanning several thousand years, beginning from a pre-ceramic, pre-maize horizon dated at 3000 B.C. (12, 60). In contrast, no definitive archaeological specimen of *C. ficifolia* has been found north of South America. A single seed, recovered from the Valley of Oaxaca, dated at 700 A.D., and tentatively identified as *C. ficifolia* (72, 76), appears instead to be *C. pepo*.

Despite the archaeological record, *C. ficifolia* is generally considered to have originated in Mesoamerica, specifically southern Mexico and Guatemala (16, 35, 57, 63, 71, 73). This conclusion is based on several lines of circumstantial evidence in addition to that already mentioned and suggests a pattern of dispersal evident in the often closely associated crops maize, beans, and the squash species, *C. pepo*. On the basis of the estimated time taken for *C. ficifolia* to have diffused from Mesoamerica to Peru by 3000 B.C., Cutler and Whitaker (21) concluded that it must be the oldest cultivated plant in the Americas.

Sauer (57) noted that *C. ficifolia* is absent today from the hot, dry low-land region where the archaeological specimens were found. He concluded from this that *C. ficifolia* must be an ancient yet still primitive cultigen that was dropped from cultivation except in the cool high country when better squashes were developed. He did not consider the possibility that these archaeological squashes were not autochthonous but were instead brought down from farms at higher altitudes. Cohen (20) presented evidence that the early Peruvian coastal cultures depended primarily on resources from the ocean and trading with the Andean cultures. Agriculture was initially of lesser importance in the lowlands than at the higher elevations. The earliest archaeological seeds of *C. ficifolia* are the same size as those today, which suggests that it was an imported domesticate.

Hurd et al. (35) used two genera of solitary bees (*Peponapis* and *Xenoglossa*), commonly known as squash and gourd bees, to help discern the species relationships in the genus *Cucurbita*. These bees depend almost exclusively on squash flowers for pollen and nectar for themselves and their larvae, and therefore appear to have coevolved with the cucurbits. The center of diversity of these bees occurs in southern Mexico, where the greatest concentration of *Cucurbita* species is located.

There is one species of squash bee, *P. atrata* (Smith), which is reported to be restricted to the pollen of *C. ficifolia* (35). It is found only in the high-lands of Mexico and adjacent Central America and not in South America. There are other more generalized species of squash and gourd bees, which occur throughout the Americas and use the pollen of all the cultivated species of squash, including *C. ficifolia*. The absence of *P. atrata* in South America is used as evidence to support the hypothesis that *C. ficifolia* was introduced by humans into South America from Mexico (35). This is based on the assumption that *P. atrata* did not concordantly extend its range into South America with *C. ficifolia* because it was unable to cross the interven-ing lowland areas. Before reaching a definitive conclusion about human dispersal of *C. ficifolia* based on the distribution of its pollinators, a thor-ough sampling needs to be conducted on the species of pollinators visiting *C. ficifolia* flowers throughout its range, including South America. *Peponapis atrata* may not be as monolectic as reported in the literature; a female *P. atrata* was collected from a flower of *C. argyrosperma* in Chiapas, Mexico in December, 1985 (specimen collected by T. C. Andres and species determined by W. E. LaBerge; deposited in the Cornell University Insect Collection, lot number 1166).

The hypotheses resulting from the squash and gourd bee study of Hurd et al. (35) are not conclusive. While they are derived from extensive field observations, they have not been tested experimentally. The study did not demonstrate whether the various species of bees coevolved with specific species of *Cucurbita* or with the genus in general. The latter is the rule among oligolectic bees (G. C. Eickwort, pers. comm., 1988). The fact that these bees and *Cucurbita* species share centers of diversity may not be due to their coevolution but rather to other causes such as their independent adaptation to the extreme habitat diversity of the region. Experimental tests need to be conducted to determine if species such as *P. atrata*, given the choice between *C. ficifolia* and other species of *Cucurbita*, are able to differentiate between them.

Hurd et al. (35) stated that the pollen collecting devices of the bees are species-specific, being adapted to carry the pollen grains of different *Cucurbita* species, which vary in size and structure. However, Andres (1) showed by light and scanning electron microscopy that the pollen exine characteristics of *Cucurbita* do not show significant variation between species and therefore probably do not play an important role in the evolution of the pollinators.

There are also few consistent morphological differences in the flowers of *Cucurbita* species. For example, the ultraviolet patterns, i.e. "nectar guides", are indistinguishable among *Cucurbita* species (T. C. Andres, unpublished observation). Whether the bees are able to cue in on other interspecific differences in the cucurbit flowers has yet to be determined.

Whitaker (71, 73) and Whitaker and Davis (77) also assume that *C. ficifolia* was domesticated in Mesoamerica rather than South America. This conclusion is based on compatibility data and the distribution of the possible wild progenitors. *Cucurbita ficifolia* and *C. moschata* are the only domesticated species of squash with no known logical wild ancestor. Whitaker (71) suggests that *C. ficifolia* may be derived from the wild mesophytic species, *C. lundelliana* or *C. martinezii* L. H. Bailey. Evidence from comparative morphology, geographic distribution, ecology, and genetic relationships do not, however, indicate a very close relationship between these wild species and *C. ficifolia*. *Cucurbita lundelliana* is the only wild mesophytic cucurbit that has been reported to cross with *C. ficifolia*, but attempts to obtain subsequent generations have failed because of sterility factors (71). Furthermore, pollen fertility was reported to be higher in the F_1 hybrids between *C. lundelliana* and the other domesticates than between *C. lundelliana* and *C. ficifolia* (70). Neither wild species occurs in the cool highland areas where *C. ficifolia* is grown; *C. lundelliana* is confined to the Yucatan peninsula, which includes a part of the countries of Mexico, Belize, and northern Guatemala, whereas *C. martinezii* extends from southern Tamaulipas, through Veracruz to northern Chiapas, Mexico. These two species are closely related and both have small gray colored (green when wet) seeds that bear little resemblance to those of *C. ficifolia*.

A recent extensive crossing program showed that when *C. ficifolia* is used as the pistillate parent it may be crossed with the interfertile xerophytic species, *C. foetidissima* and *C. pedatifolia*, without the need for embryo culture (2). Sterility barriers prevented obtaining subsequent generations.

These two xerophytic species and *C. radicans* share with *C. ficifolia* pubescent filaments but otherwise bear little resemblance to *C. ficifolia* in morphology and ecology. In the study on the relationships of *Cucurbita* species with the squash and gourd bees (35), however, *C. pedatifolia* was grouped with *C. ficifolia*.

In South America there are less likely candidates for the wild progenitor of *C. ficifolia*. Only two wild species of *Cucurbita* are endemic to South America. One, *C. andreana* Naud., is fairly certain to be the wild progenitor of *C. maxima* (43). The other, *C. ecuadorensis*, has some phenetic similarity to *C. ficifolia*, such as the similar leaf shape already mentioned and an unusually large fruit for a wild cucurbit, but crossing studies indicate it is more closely aligned with *C. maxima* and the *C. lundelliana* group (22).

There is perhaps more persuasive evidence suggesting that *C. ficifolia* originated in northern South America rather than Mesoamerica. Intensive searches for a compatible wild progenitor of *C. ficifolia* in Mexico have produced no evidence of any such populations. Although less intensive searches have been conducted south of Mexico, there are sketchy reports of possible wild or at least spontaneous populations. Dieterle (25) reports populations of *C. ficifolia* in some highland localities of Guatemala to "have become thoroughly naturalized and look like native plants". The difference between wild and naturalized populations of cucurbits is often ambiguous or at least difficult to determine. In Colombian markets Cardenas (18) reported seeing "victoria" squashes with "necks like a bottle." This shape has not been found elsewhere and, if this report is correct, may represent a primitive character or a derived morph from an ancestral population. In Bolivia, Cardenas (18) reported that *C. ficifolia* plants with viviparous seeds appear wild because they are spontaneous and hanging off the mountain sides. No description or documentation is provided on the morphology of the plants. Granado (31) stated that *C. ficifolia* originated in Bolivia, but he did not provide any evidence. Agronomists at Inquisivi, Departamento La Paz, reported a wild form of *C. ficifolia*, but this has not yet been confirmed (M. Nee, pers. comm., 1988).

The hypothesis of an Andean origin and domestication of *C. ficifolia* rather than a Mexican origin explains the archaeological record and does not necessitate a change in habitat as Sauer proposed (57). *Cucurbita ficifolia* is most likely uniquely adapted to cool, moist conditions because it evolved in a habitat characterized by such a climate. The preponderance of short-day flowering accessions points to an area of origin within the torrid zone. If domestication occurred in the northern Andes around or before 3000 B.C., there was ample time for a gradual spread of the crop to the north and use by the Aztecs in Mexico beginning in the 12th century. A similar situation occurred in *C. pepo,* which spread northward during pre-Columbian times from central and northern Mexico to southwestern and eastern United States to as far north as adjacent Canada (75, 77). However, in the case of C. *pepo*, this dispersal may have occurred several millennia before C. *ficifolia* was even domesticated, thus allowing sufficient time for secondary diversification of cultivars to occur within the new range (see Decker, this volume). Other examples of New World crops involved in very early long distance cultural exchanges include peanuts and guava, which spread from Peru to Mexico by 200 b.c., and maize, which spread from Mexico to Peru by around 3000 b.c. (29).

It is conceivable that there are still extant wild conspecific populations of C. *ficifolia* in some isolated cool moist cloud forests on the eastern flank of the Andes in Venezuela, Colombia, Ecuador, Peru, or Bolivia. Furthermore, in such an area this species may contain greater genetic diversity than has been found elsewhere. Germplasm collecting and evaluations need to be conducted in this region. There are few dry caves in this region where archaeological preservation could have occurred, but those fruits that were traded with cultures along the dry Peruvian coast may have been preserved. This type of trading of the highland grown C. *ficifolia* for lowland crops, such as C. *moschata*, still takes place today in much of Latin America.

Economic Uses

Cucurbita ficifolia is usually cultivated in the traditional maize-beans-squash milpa (garden). However, the feasibility of large-scale cultivation of C. *ficifolia* is presently being investigated in Chile (37). It is consumed in a multitude of ways throughout the year, in basically the same manner as the other domesticated species of squash (57). The young leaves and vine tips may be prepared as a green vegetable; the carotene-rich male flowers and buds are used in soups, stews, salads, or as *rellenos* (stuffed). The tender immature fruits, like a summer squash or cucumber, are popular in many dishes such as a potato soup in Ecuador called *locro de zambo* (27); and the raw or roasted seeds are often eaten as a nutritious snack food, ground to make a *pipián* sauce for use in a variety of dishes, and sometimes used as a vermifuge (16). The mature fruit is primarily used to make various types of sweets or dulce. This is perhaps the most popular use of C. *ficifolia*, and is prepared by first boiling the peeled, cut-up fruit with the seeds in alkali or lime water and then in syrup until it becomes translucent and impregnated with the sugar. A similar dessert type food, known from the time of the Aztecs, called in Spanish *cabello de angel* (angel's hair), is prepared by boiling the pulp of well-ripened fruits in plain water until the fibers separate into long tender fleshy strings, which are strained and then boiled in syrup and flavored with vanilla or other ingredients (4, 16). Younger fruits or seeds are sometimes used in Bolivia (18) and southern Mexico (9) in the preparation of a beverage made from maize gruel, which is sometimes lightly fermented with the addition of sugar or honey. The fruit pulp evidently contains a proteolytic enzyme that has potential value in the food industry (37).

The mature fruit flesh tends to be somewhat dry, fibrous, low in soluble solids, and white, indicating a low beta-carotene content. While the seeds have a high protein quality, as with the other squashes, the fruits have only a moderate quantity of carbohydrates and are low in vitamins and minerals (27). However, the daily intake of these cucurbits throughout the year may be very high in certain parts of the tropics. Although they are not nutrient dense, their contribution of daily vitamins, minerals, and calories should not be underestimated.

Outside of Latin America, *C. ficifolia* is primarily used for livestock feed. The French have sought to popularize the use of *C. ficifolia* fruits for human consumption and as an ornamental object. Their inventive proposals include using the placenta as a substitute for sauerkraut (10, 11, 38, 51). The vines with their rapid extensive growth provide excellent bowers and arbors (17, 64, 78).

The Dutch and Japanese use *C. ficifolia* as rootstocks for winter production of greenhouse cucumber (5, 49, 50, 77). This cultivation practice is for semi-forcing and improving the growth of cucumber scions earlier in the season under conditions of low soil temperature while saving greenhouse energy costs. The rootstock, furthermore, is somewhat resistant to fusarium wilt (34, 39, 77) and root-knot nematodes (34).

Cucurbita ficifolia contains genes potentially valuable for breeding purposes. The vine is resistant to several viruses, including papaya ringspot virus, watermelon mosaic virus 2, squash mosaic virus, and melon necrotic spot virus (13, R. Provvidenti, pers. comm., 1987), and appears to have some tolerance to powdery mildew (T. C. Andres and R. W. Robinson, unpublished observation). The plant, as already mentioned, has a unique ability to grow under cool, moist conditions, and the mature fruits may keep for two to three years without refrigeration (T. C. Andres, unpublished observation). Breeding efforts should be made both to introduce these traits into the other cultivated species and, in the reverse direction, to transfer genes such as those for higher carotene and soluble solid content into *C. ficifolia* to improve the quality and nutritional value of the fruit.

Literature Cited

1. Andres, T. C. 1981. A microscopy survey of *Cucurbita* pollen morphology. Texas Soc. Electron Microscop. 12: 23 (Abstract).
2. Andres, T. C. 1987. Hybridization of *Cucurbita foetidissima* with *C. pedatifolia*, *C. radicans*, and *C. ficifolia*. Cucurbit Genet. Coop. Rep. 10: 72–73.
3. Anonymous. 1891. Kew Bull., Appendix 2: 40.

116   Systematics and Evolution

4. Anonymous. 1896. El chilacayote. Progr. Mex. 3: 588–589.
5. Ashizawa, M. 1975. Local varieties of pumpkin and squash (*Cucurbita* spp.) in Japan. *In* T. Matsuo, ed., Gene Conservation: Exploration, Collection, Preservation, and Utilization of Genetic Resources. JIBP Synthesis. Vol. 5. Univ. Tokyo Press, Tokyo.
6. Bailey, L. H. 1929. The domesticated cucurbitas—first paper. Gentes Herb. 2: 61–115.
7. Bailey, L. H. 1937. The Garden of Gourds. Macmillan, New York.
8. Bailey, L. H. 1948. Jottings in the cucurbitas. Gentes Herb. 7: 447–477.
9. Berlin, B., D. E. Breedlove, and P. H. Raven. 1974. Principles of Tzeltal Plant Classification: An Introduction to the Botanical Ethnography of a Mayan-Speaking People of Highland Chiapas. Academic Press, New York.
10. Blin, H. 1942. La courge de Siam et ses utilisations. Rev. Hort. 114: 118–119.
11. Bois, D., and J. Gerome. 1920. La chilacayote du Mexique (courge de Siam): *Cucurbita ficifolia* Bouché (*C. melanosperma* Al. Braun). Bull. Mus. Hist. Nat. (Paris) 26: 675–678.
12. Bonavia, D. 1982. Los Gavilanes. Editorial Ausonia-Talleres Graficos, Lima.
13. Bos, L., H. J. M. van Dorst, H. Huttinga, and D. Z. Maat. 1984. Further characterization of melon necrotic spot virus causing severe disease in glasshouse cucumbers in the Netherlands and its control. Neth. J. Pl. Pathol. 90: 55–69.
14. Bouché, P. C. 1837. Verh. Vereins Beförd. Gartenbaues Konigl. Preuss. Staaten 12: 201–207.
15. Braun, A., J. F. Klotzsch, C. Koch, and P. C. Bouché. 1853. Appendix. Specierum Novarum et Minus Cognitarum quae in Horto Regio Botanico Berlinensi Coluntur, Auctoribus. P. 362. (Also published in 1854, Ann. Sci. Nat. Bot., Ser. 4, 1: 333–370).
16. Bukasov, S. M. 1930. The cultivated plants of Mexico, Guatemala and Colombia. Bull. Appl. Bot., Genet. Pl. Breed. (Suppl.) 47: 1–464.
17. Calvino, M. 1914. El chilacayote Mexicano. Bol. Fomento (San José) 4: 274–278.
18. Cardenas, M. 1969. Manual de las Plantas Economicas de Bolivia. IV. Cucurbitas. Imprenta Icthus, Cochabamba.
19. Cogniaux, A. 1881. Cucurbitacées. *In* A. De Candolle and C. De Candolle, eds., Monographiae Phanerogamarum. 3: 325–951. G. Masson, Paris.
20. Cohen, M. N. 1977. Population pressure and the origins of agriculture: an archaeological example from the coast of Peru. *In* C. A. Reed, ed., Origins of Agriculture. Mouton, The Hague.
21. Cutler, H. C., and T. W. Whitaker. 1961. History and distribution of the cultivated cucurbits in the Americas. Amer. Antiq. 26: 469–485.
22. Cutler, H. C., and T. W. Whitaker. 1969. A new species of *Cucurbita* from Ecuador. Ann. Missouri Bot. Gard. 55: 392–396.
23. Dammann & Co. (E. Dammann and Sprenger). 1890–1891. Dammann & Co. Seed Catalogue. 54: 42.
24. De Candolle, A. 1886. Origin of Cultivated Plants, 2nd ed. (reprinted 1959). Hafner, New York.
25. Dieterle, J. V. A. 1976. Cucurbitaceae. *In* D. L. Nash, ed., Flora of Guatemala. Fieldana, Bot. 24 (XI, 4): 306–395.

26. Drude, O. 1917. Erfahrungen bei Kreuzungsversuchen mit *Cucurbita pepo*. Ber. Deutsch. Bot. Ges. 35: 26–57.
27. Estrella, E. 1984. Ethno-historia de las cucurbitaceas nativas. Contributed paper. IV Intl. Congr. Andean Crops. Pasto, Narino, Colombia.
28. Filov, A. I. 1966. Ekologija i Klassifikatzija tykuy (in Russian). Bjull. Glaun. Bot. Sada 63: 33–41.
29. Flannery, K. V. 1973. The origins of agriculture. Ann. Rev. Anthropol. 2: 271–310.
30. Gasparrini, G. 1848. Observations morphologiques et physiologiques sur quelques espèces de courges cultivées. Ann. Sci. Nat. Bot., Ser. 3, 9: 207–218.
31. Granado, J. T. 1931. Plantas Bolivianas. Arnó Hermanos, La Paz.
32. Grebenscikov, I. 1965. Notulae Cucurbitologicae. VI. Über einige Artkreuzungen in der Gattung *Cucurbita*. Kulturpflanze 73: 145–161.
33. Herklots, G. A. C. 1972. Vegetables in South-east Asia. Hafner, New York.
34. Hönick, A. 1984. Neue Veredelungsunterlage für Gurken unter Glas. Gemuse 2: 45–46.
35. Hurd, P. D., Jr., E. G. Linsley, and T. W. Whitaker. 1971. Squash and gourd bees (*Peponapsis, Xenoglossa*) and the origin of the cultivated *Cucurbita*. Evolution 25: 218–234.
36. Hutchins, A. E., and L. Sando. 1941. Gourds—their culture, uses, identification, and relation to other cultivated Cucurbitaceae. Minnesota Agric. Exp. Sta. Bull. 356: 1–35.
37. Illanes, A., G. Schaffeld, C. Schiappacasse, M. Zuñiga, G. González, E. Curotto, G. Tapia, and S. O'Reilly. 1985. Some studies on the protease from a novel source: the plant *Cucurbita ficifolia*. Biotechnol. Lett. 7: 669–672.
38. Janson, M. 1923. Culture et utilisation de la courge de Siam. Rev. Int. Bot. Appl. Agric. Trop. 3: 551–552.
39. Kawaide, T. 1985. Utilization of rootstocks in cucurbits production in Japan. Jap. Agric. Res. Quart. 18: 284 289.
40. Martínez, M. 1979. Catálogo de Nombres Vulgares y Científicos de Plantas Mexicanas. Fondo de Cultura Económica, Mexico City.
41. Merrill, E. D. 1954. The botany of Cook's voyages. Chron. Bot. 14: 164–383.
42. Metcalf, R. L., A. M. Rhodes, R. A. Metcalf, J. Ferguson, E. R. Metcalf, and P.-Y. Lu. 1982. Cucurbitacin contents and diabroticite (Coleoptera: Chrysomelidae) feeding upon *Cucurbita* spp. Environ. Entomol. 11: 931–937.
43. Millan, R. 1945. Variaciones del zapallito amargo *Cucurbita andreana* y el origen de *Cucurbita maxima*. Revista Argent. Agron. 12: 86–93.
44. Morton, J. F. 1975. The sturdy Seminole pumpkin provides much food with little effort. Proc. Florida State Hort. Soc. 88: 137–142.
45. Naudin, C. 1856. Nouvelles recherches sur les caractéres spécifiques et les variétés des plantes du genre *Cucurbita*. Ann. Sci. Nat. Bot., Ser. 4, 6: 5–72.
46. Naudin, C. 1857. Les courges; leurs espèces et leurs variétés. Fl. Serres Jard. Eur. 12: 113–125.
47. Naudin, C. 1865. Nouvelles recherches sur l'hybridité dans les végétaux. Nouv. Arch. Mus. Hist. Nat. 1: 25–176.
48. Naudin, C. 1866. Cucurbitacées nouvelles cultivées au Muséum d'Histoire Naturelle en 1863, 1864 et 1865. Ann. Sci. Nat. Bot., Ser. 5, 5: 5–43.

49. Nijs, A. P. M. den. 1980. The effect of grafting on growth and early production of cucumbers at low temperature. Acta Hort. 118: 57–63.
50. Nijs, A. P. M. den. 1985. Rootstock-scion interactions in the cucumber: implications for cultivation and breeding. Acta Hort. 156: 53–60.
51. Paillieux, A., and D. Bois. 1892 and 1899. Le Potager d'un Curieux. Librairie Agricole de la Maison Rustique, Paris.
52. Robelo, C. A. 1904. Diccionario de Aztequismos. Published by the author, Cuernavaca.
53. Rose, J. N. 1899. Notes on useful plants of Mexico. Contr. U.S. Natl. Herb. 5: 209–259.
54. Sageret, A. 1826. Considérations sur la production des hybrides, des variantes et des variétés en général, et sur celles de la famille des cucurbitacées en particulier. Ann. Sci. Nat. Bot., Ser. 1, 8: 294–314.
55. Sageret, A. 1827. Deuxième mémoire sur les cucurbitacées, principalement sur le melon. Acad. Agric. France Mem. 1: 1–116.
56. Sahagun, F. B. 1956. Historia General de las Cosas de Nueva España. Tom. 1, Lib. 1, Cap. 21: 13. (New edition by A. M. Garibay K.) Editorial Porrua, Mexico City.
57. Sauer, C. O. 1969. Agricultural Origins and Dispersals. M.I.T. Press, Cambridge, MA.
58. Singh, D., and A. S. R. Dathan. 1972. Structure and development of seed coat in Cucurbitaceae. 6. Seeds of *Cucurbita*. Phytomorphology 22: 29–45.
59. Tachibana, S. 1987. Effect of root temperature on the rate of water and nutrient absorption in cucumber cultivars and figleaf gourd. J. Jap. Soc. Hort. Sci. 55: 461–467.
60. Towle, M. A. 1961. The Ethnobotany of Pre-Columbian Peru. Viking Fund Pub. Anthropol., No. 30, New York.
61. Urbina, M. 1902–1903. Notas acerca de los "ayotli" de Hernández, ó calabazas indígenas. Anales Mus. Nac. Mex., Ser. 1, 7: 353–390. (Also published in 1912; Naturaleza [Mexico City], Ser. 3, 1: 80–117.)
62. Van Eseltine, G. P. 1936. *Cucurbita* hybrids. Proc. Amer. Soc. Hort. Sci. 34: 577–581.
63. Vavilov, N. I. 1949–1950. The origin, variation, immunity and breeding of cultivated plants. (Trans. K. S. Chester.) Chron. Bot. 13.
64. Voigtländer, B. 1926. *Cucurbita ficifolia*. Gartenwelt 30: 605.
65. Walpers, W. G. 1857. *Cucurbita*. Ann. Bot. Syst. 4: 864–865.
66. Washek, R. L. 1983. Cucumber mosaic resistance in summer squash (*Cucurbita pepo* L.). Ph.D. dissertation, Cornell University, Ithaca, NY.
67. Watson, S. 1887. List of plants collected by Dr. Edward Palmer in the State of Jalisco, Mexico, in 1886. Proc. Amer. Acad. Arts 22: 414.
68. Weiling, F. 1955. Über die Interspezifische Kreuzbarkeit Verschiedener Kürbisarten. Zuchter 25: 33–57.
69. Weiling, F. 1959. Genomanalytische Untersuchungen bei Kürbis (*Cucurbita* L.). Zuchter 29: 161–179.
70. Whitaker, T. W. 1956. The origin of the cultivated *Cucurbita*. Amer. Naturalist 90: 171–176.

71. Whitaker, T. W. 1980. Cucurbitáceas americanas útiles al hombre. Comisión de Investigaciones Científicas, La Plata.
72. Whitaker, T. W. 1981. Archeological cucurbits. Econ. Bot. 35: 460–466.
73. Whitaker, T. W. 1981. Cucurbits in Andean pre-history. Abstract. XIII Intl. Bot. Congr., Sydney, Australia.
74. Whitaker, T. W., and W. P. Bemis. 1964. Evolution in the genus *Cucurbita*. Evolution 18: 553–559.
75. Whitaker, T. W., and G. W. Bohn. 1950. The taxonomy, genetics, production and uses of the cultivated species of *Cucurbita*. Econ. Bot. 4: 52–81.
76. Whitaker, T. W., and H. C. Cutler. 1971. Prehistoric cucurbits from the valley of Oaxaca. Econ. Bot. 25: 123–127.
77. Whitaker, T. W., and G. N. Davis. 1962. Cucurbits; Botany, Cultivation, and Utilization. Interscience, New York.
78. Wittmack, L. 1884. Der Schwarzsamige Kürbis, *Cucurbita melanosperma* Al. Br., zur Bekleidung von Veranden. Garten-Zeitung 3: 337–338.

10 | The Genus *Luffa*: A Problem in Phytogeography

Charles B. Heiser, Jr., and Edward E. Schilling

ABSTRACT. *Luffa* comprises seven species, four well-differentiated species of the Old World tropics and three rather similar species of the Neotropics. Artificial hybrids have been produced between most of the species. With one exception, the hybrids are sterile or nearly so. A cladistic analysis reveals that there are two main lines of evolution in the Old World species, one leading to *L. acutangula* and *L. aegyptiaca* and the other to *L. echinata* and *L. graveolens*. The New World species stem from the subline giving rise to *L. graveolens*. In addition to the disjunction of the Old and New World species, there are conspicuous disjunctions within two of the Old World species; *L. echinata* occurs in India and Africa, and *L. graveolens* is found in India and Australia. Whether the disjunctions are to be explained by dispersal or vicariance, or a combination of the two, is not clear, but a dispersalist interpretation may be the most plausible. Two of the species, *L. acutangula* and *L. aegyptiaca*, have domesticated varieties; the young fruits of both are used for food and the mature fibrous interior of the fruit of the latter species has a number of uses. Various parts of the plants of these species, as well as those of several of the wild species, are widely used in folk medicine.

The genus *Luffa* is well known because of the fibrous spongelike nature of its fruits and the wide cultivation of two of its species. In English-speaking countries the common names loofah, sponge gourd, rag gourd, and dishrag gourd are used for *L. aegyptiaca* Mill., the most extensively cultivated species. The names *Luffa* and loofah come from an Arabian word for the plant. In addition to its unusual fruits, there are several disjunctions in the distribution of the species that make the genus of considerable interest to botanists. The species are all vines, and in addition to the fibrous fruits, the other salient generic features are the operculate dehiscence of the fruit, three to five stamens, free petals, solitary pistillate flowers, racemose staminate flowers, and three- to five-fid tendrils.

120

Taxonomy

Luffa is a member of the subfamily Cucurbitoideae, tribe Benincaseae, subtribe Luffinae (17, 18). It is the only member of the subtribe, and unlike other genera of the tribe Benincaseae, which are confined to the Old World, it has species in both the Old and New Worlds. According to Jeffrey (18), in some respects it resembles members of the Cyclantherinae, an entirely New World subtribe of the Sicyeae. It thus may be the connecting link of these two tribes and hence is of considerable phylogenetic interest.

The last monographic treatment of the genus, in which eight species were recognized, was by Cogniaux and Harms (7). One of these species, *L. variegata* Cogn., has since been transferred to *Lemurosicyos* (20), and two others, *Luffa forskalii* Schwein. ex Harms and *L. umbellata* (Klein) M. J. Roem., have been reduced to synonymy or varietal status under *L. acutangula* (L.) Roxb. (13, 19). Five species, four in the Old World and one in the New World, have generally been accepted in recent years (13, 18). Heiser et al. (14), however, proposed that three species are present in the New World. The seven species now recognized can be distinguished by the following key:

Dioecious; fruits echinate; petals white or yellow . . . . *L. echinata*
Monoecious; fruits not echinate; petals yellow
    Fruits strongly ridged to smooth
        Fruits usually strongly ridged; seeds rugose, without wings; corolla primrose yellow, opening in the evening . . . . *L. acutangula*
        Fruits with sunken ribs or smooth; seeds smooth, winged; corolla deep yellow, opening during the day . . . . *L. aegyptiaca*
    Fruits tuberculate or spiny
        Fruits tuberculate, projections less than 1 mm long, scattered over surface; ecostate or nearly so; Old World . . . . *L. graveolens*
        Fruits spiny, spines 1–5 mm long, mostly confined to costae; New World
            Leaves deeply 5-lobed; fruits densely, finely pubescent; Mexico to Nicaragua . . . . *L. quinquefida*
            Leaves moderately to shallowly 3- to 5-lobed; fruits sparingly pubescent to glabrous; Panama and northern South America
                Fruits usually elongate-ovoid; leaves shallowly lobed; seeds rugose, 7.5–10 mm long . . . . *L. operculata*
                Fruits usually ovoid; leaves moderately to deeply 5-lobed; seeds smooth or rugose, 7–8 mm long . . . . *L. astorii*

The first four species are native to the Old World tropics and the last three to the Neotropics. The Old World species are well differentiated from each other and the American species. On the other hand, the American species are rather similar to each other. If, as we suppose, the genus had its origin in the Old World, the American species could well have been derived from a single introduction.

Luffa echinata Roxb., found in both Asia and Africa (Figure 1), readily stands apart from all of the others on the basis of the characters given in the key. It does not appear to be closely related to any other species, although it has not yet undergone detailed study. The extent of morphological differentiation of the Asian and African representatives is unknown, as are their fertility relationships. In their description of the species, Cogniaux and Harms (7) stated that petals are white, but some African floras indicate the

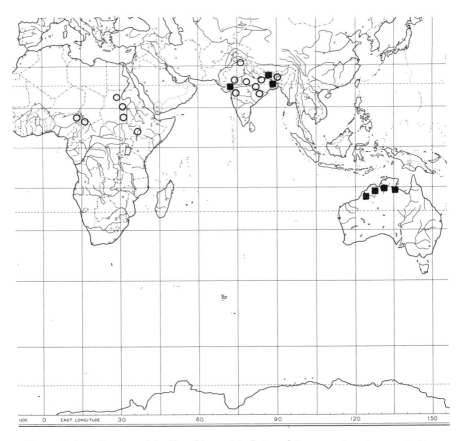

Figure 1. Distribution of *Luffa echinata* (circles) and *L. graveolens* (squares). Compiled from the following sources: Andrews (1); Briggs (2); Chakravarty (5); Cogniaux and Harms (7); Hutchinson and Dalziel (16); Keraudren (20). The localities for Ethiopia are based on specimens at Kew and were supplied by C. Jeffrey. Cogniaux and Harms also cite a specimen from Natal for *L. echinata*, but this species is not included in Ross (29). Goode base map no. 101M, University of Chicago; used with permission.

petals are yellow (15, 16) or yellow to white (1). Keraudren (20) reported only white. Whether both color morphs occur in Africa or if the yellow color results from aging in herbarium material is not known.

In *Luffa acutangula* three varieties are recognized: var. *acutangula*, the large-fruited, cultivated form whose probable origin was in India; var. *amara* (Roxb.) C. B. Clarke, a wild or feral form confined to India with smaller, extremely bitter fruits; and var. *forskalii* (Harms) Heiser & Schilling of Yemen. On the basis of the flavonoid chemistry (31) it seems possible that the last variety could have developed from escapes of the cultivated variety.

Luffa aegyptiaca comprises the domesticated variety, var. *aegyptiaca*, and a wild variety, var. *leiocarpa* (Naud.) Heiser & Schilling, which ranges from Burma and the Philippines to northeastern Australia and Tahiti. The domesticate differs from the wild form in its more deeply furrowed, less bitter, and larger fruits, at times reaching lengths of 50 cm. Its original place of domestication is unknown, but on the basis of the distribution of the wild variety, it is unlikely to have been India as has been suggested (26). The name to be used for this species is still controversial. *Luffa cylindrica* (L.) M. J. Roem. is in wide usage, and Jeffrey (17) maintains that it is the correct name. Schubert (32), however, has argued for *L. aegyptiaca*, and this name has been adopted by the United States Department of Agriculture (33).

Luffa graveolens Roxb. is found in India and northeastern Australia (Figure 1). Cogniaux and Harms (7) also reported the species in Samoa, based on a specimen collected by Whitmee (number *82*). We have been unable to locate this specimen. Parham (23), however, did not include this species in his flora of the islands but does report *L. aegyptiaca* var. *leiocarpa*. As in the case of *L. echinata*, *L. graveolens* has yet to receive detailed study.

In the Americas *Luffa quinquefida* (Hook. & Arn.) Seem., previously included in *L. operculata* (L.) Cogn., is readily separated morphologically from the other New World species and is geographically isolated from them, extending from western Mexico to Nicaragua (Figure 2).

Luffa operculata is the most widespread of the American species, ranging from Panama to Peru and Brazil, including Fernando Noronha (Figure 2). Although readily distinguished from *L. astorii* Svens. in living material, it is sometimes difficult to separate the two species in herbarium collections.

Luffa astorii, usually included in *L. operculata*, is reported from Venezuela (a single collection), western Ecuador including the Galapagos Islands, and northwestern Peru (Figure 2).

Cytogenetics

All of the species of *Luffa* are diploid ($n = 13$). A large number of interspecific hybrids has now been made, the earliest report being that of *L.*

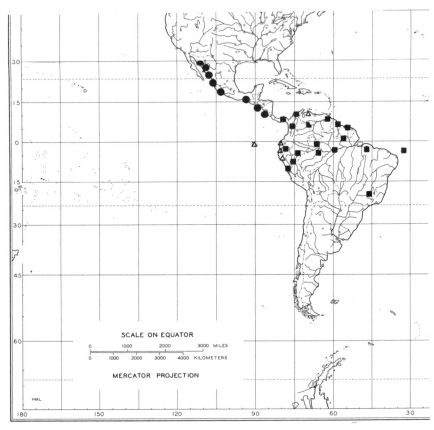

Figure 2. Distribution of *Luffa astorii* (open triangles), *L. operculata* (squares) and *L. quinquefida* (circles) from Heiser, Schilling and Dutt (14). Goode base map no. 101M, University of Chicago; used with permission.

acutangula × *L. aegyptiaca* by Naudin (22). This hybrid has since been formed many times, the most detailed study being that of Pathak and Singh (24). The hybrid shows great reduction in fertility. Dutt (8) and Dutt and Roy (9) made other hybrids of the Old World species. Hybrids of *L. graveolens* with *L. acutangula*, *L. aegyptiaca*, and *L. echinata* were all sterile and showed irregularities at meiosis.

Crosses between the New World species have been made in all possible combinations (14). After repeated efforts, a single hybrid was obtained between *L. astorii* and *L. operculata*. The plant was abnormal and sterile. Hybrids between *L. astorii* and *L. quinquefida* produced plants with 0 to 18% stainable pollen but set no seed. The one hybrid examined showed 13 pairs of chromosomes at metaphase. The most fertile interspecific hybrids are those between *L. operculata* and *L. quinquefida*. These hybrids showed 20 to 52% pollen stainability and produced some seed.

A number of crosses have also been attempted between the species of the Old and New Worlds (13, 14). Successful crosses were *L. aegyptiaca* with *L. astorii* and *L. operculata*; *L. echinata* with *L. operculata*; and *L. graveolens* with *L. operculata* and *L. quinquefida*. None of the hybrids produced seeds. The highest pollen stainability, 28%, was found in one plant of *L. graveolens* × *L. quinquefida*.

Although the species have not yet been crossed in all possible combinations, a general picture emerges that interspecific hybrids, with the exception of *L. operculata* × *L. quinquefida*, are sterile or show a very highly reduced fertility. Hybridization results, with the exception of *L. operculata* × *L. quinquefida*, might be interpreted as indicating that the species are not closely related, thus supporting the conclusions reached from morphological analyses.

Cladistics

A cladistic analysis of *Luffa* was reported earlier (13), but it included only one of the American species. Therefore, a new cladogram is presented here for all of the taxa, using a slightly refined methodology and a revised list of characters.

Table 1. List of characters for cladistic analysis of *Luffa*

1. Breeding system (0 = monoecious; 1 = dioecious)
2. Flowering time (0 = diurnal; 1 = nocturnal)
3. Tendril pubescence (0 = absent; 1 = present)
4. Leaf lobing (0 = strongly; 1 = intermediate; 2 = slightly)
5. Calyx nectaries (0 = absent; 1 = present)
6. Petal color (0 = deep yellow; 1 = light yellow or white)
7. Staminate flower arrangement (0 = solitary; 1 = racemose)
8. Filament color (0 = white; 1 = green)
9. Pollen mother cell inclusions (0 = absent; 1 = present)
10. Style color (0 = yellow; 1 = green)
11. Fruit pubescence (0 = glabrate;; 1 = dense)
12. Fruit shape (0 = ovoid; 1 = intermediate; 2 = elongate)
13. Fruit ridges (0 = absent; 1 = present)
14. Fruit surface (0 = smooth; 1 = not smooth)
15. Fruit echinate (0 = absent; 1 = present)
16. Fruit tubercule size (0 = less than 0.5 mm; 1 = more than 0.5 mm)
17. Distribution of tubercules (0 = throughout; 1 = on costae only)
18. Fruit operculum (0 = absent; 1 = present)
19. Fruit interior (0 = not fibrous; 1 = fibrous)
20. Seed size (0 = more than 9 mm; 1 = less than 9 mm)
21. Seed texture (0 = smooth or slightly rugose; 1 = quite rugose)
22. Seed wings (0 = absent or inconspicuous; 1 = present)
23. Flavones (0 = absent; 1 = present)
24. Flavonols (0 = absent; 1 = present)
25. 3'-OMe flavonoids (0 = absent; 1 = present)
26. 4'-glycosidic flavonoids (0 = absent; 1 = present)

Twenty-six morphological and chemical characters (Table 1) were used. Each character was coded into binary or multistate-ordered form. To avoid the necessity of polarizing characters initially, an outgroup of two other genera (*Lagenaria* and *Citrullus*) of the tribe Benincaseae was included in the analysis. The data matrix of character scores (Table 2) was analyzed using parsimony analysis, implemented with version 2.3 of the PAUP program written by D. W. Swofford of the Illinois Natural History Survey. Computations were performed on the VAX Cluster at the University of Tennessee Computing Center. The following options of the PAUP program were employed: the ROOT = OUTGROUP option results in the cladogram being rooted in the outgroup; the SWAP = GLOBAL and MULPARS options allow testing of alternative branch arrangements on a variety of possi-

Table 2. Data matrix of character scores

Character[1]	Taxon[2]											
	LA	LAA	LC	LCA	LE	LF	LG	LAS	LO	LQ	LAG	CIT
1	0	0	0	0	1	0	0	0	0	0	0	0
2	1	1	0	0	0	1	0	0	0	0	0	0
3	0	0	0	0	0	1	0	1	1	1	1	1
4	1	1	0	0	0	1	0	1	1	1	0	0
5	1	1	1	1	0	1	0	0	0	0	0	0
6	1	1	0	0	1	1	0	0	1	1	1	1
7	1	1	1	1	1	1	1	1	1	1	0	0
8	0	0	0	0	1	0	0	0	0	0	X	X
9	0	0	0	0	0	0	0	0	1	1	X	X
10	0	0	0	0	1	0	0	0	0	0	X	X
11	0	0	0	0	1	0	0	0	0	1	0	0
12	2	1	2	1	0	1	0	0	1	1	0	0
13	1	1	0	0	0	1	0	1	1	1	0	0
14	0	0	0	0	1	0	1	1	1	1	0	0
15	X	X	X	X	1	X	0	0	0	0	X	X
16	X	X	X	X	1	X	0	1	1	1	X	X
17	X	X	X	X	0	X	0	1	1	1	X	X
18	1	1	1	1	1	1	1	1	1	1	0	0
19	1	1	1	1	1	1	1	1	1	1	0	0
20	0	0	0	0	1	0	1	1	0	1	0	0
21	1	1	0	0	0	0	0	0	0	0	1	0
22	0	0	1	1	0	0	0	0	0	0	1	0
23	1	1	1	1	1	1	0	0	0	0	X	X
24	0	0	0	0	0	0	1	1	1	1	X	X
25	0	0	0	0	1	0	0	0	0	0	X	X
26	0	0	0	0	0	1	0	0	0	0	X	X

[1]See Table 1 for explanation of character states.

[2]LA, *L. acutangula*; LAA, *L. acutangula* var. *amara*; LC, *L. aegyptiaca*; LCA, *L. aegyptiaca* var. *leiocarpa*; LE, *L. echinata*; LF, *L. acutangula* var. *forskalii*; LG, *L. graveolens*; LAS, *L. astorii*; LO, *L. operculata*; LQ, *L. quinquefida*; LAG, *Lagenaria*; CIT, *Citrullus*. X, no comparison possible or no data available.

ble trees with the net effect of greatly increasing the chance that the shortest possible cladogram will be discovered.

Parsimony analysis of the *Luffa* data set with the SWAP = GLOBAL and MULPARS options produced a single minimum-length tree to connect the taxa of *Luffa* (Figure 3). This tree is very similar to the previous one (13). Two major evolutionary lines are indicated, one leading to *L. acutangula* and *L. aegyptiaca* and the other to the five remaining species. The three New World species form a monophyletic group as was expected. Among these species *L. astorii* appears to be the least derived and *L. quinquefida* the most derived compared with *L. graveolens*. This is in contrast to the crossing results, which suggest that *L. graveolens* may be most closely related to *L. quinquefida*.

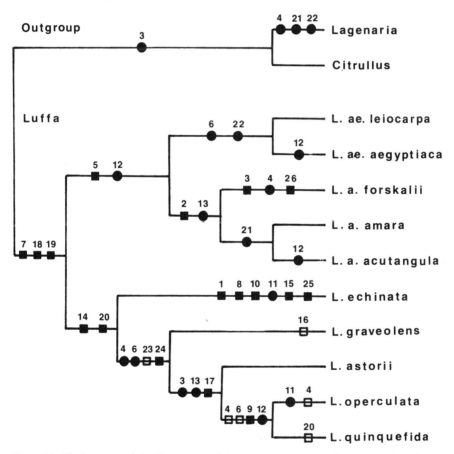

Figure 3. Cladogram of *Luffa* generated by PAUP using 26 characters (listed in Table 1). Apomorphies are indicated by closed squares, reversals by open squares, and parallelisms by closed circles.

Dispersal

The understanding of dispersal of *Luffa* is limited by few observations in the field. From a knowledge of morphology, however, certain inferences can be made. All species of *Luffa* are climbing vines, and the fruits are usually borne some distance from the ground. After the fruit matures, the operculum dehisces and the seeds are shaken from the fruit by the wind. The fibrous network of the fruit probably impedes the release of the seeds, thereby ensuring that the seeds are gradually dispersed. The fairly large size and weight of the seeds suggest that they are ordinarily carried only a short distance from the parent plant. What role, if any, animals play in the dispersal of seeds after they are on the ground is not known. The seeds are toxic to humans, and whether or not this is true for other animals is unknown. It seems unlikely that either animals, except for humans for the domesticated varieties, or winds play an important role in seed dispersal. Water, however, may be involved, at least for short distances.

At times the fruit may also serve as a dispersal unit. The fruits are not normally deciduous, but on the death of the plant or the plant on which they are climbing, some fruits would reach the ground and probably would still contain seeds. Our personal observations of *L. quinquefida* in Mexico reveal that while most of the fruits are borne aerially, some branches trail along the ground and produce fruits. Fruits on the ground could eventually reach streams or other bodies of water.

Guppy (11) observed fruits of *L. aegyptiaca* var. *leiocarpa* floating down streams in Fiji. His subsequent experiments showed that the fruits would not float more than a week in either fresh or salt water, but he stated that the seeds have an impervious coat and are well adapted to withstand prolonged immersion in salt water. He found that 60 out of 100 seeds were afloat after two months, but did not say whether they were still viable. He concluded that at times the seeds of this variety may have been dispersed by ocean currents.

A simple experiment was set up to test the ability of seeds and fruits to float (12). Twenty-five seeds and three fruits of all of the wild Old World taxa and *L. astorii* were placed in dishes of seawater. As the water evaporated distilled water was added to keep the level as constant as possible. Most of the seeds sank almost immediately. Those of *L. aegyptiaca* var. *leiocarpa* floated the longest. Nine floated for 29–32 days, and after they sank they were planted. Three of the seeds germinated. Fruits of *L. aegyptiaca* and *L. echinata* floated for less than two months, and none of the seeds of either germinated. One of the fruits of *L. graveolens* floated 89 days, and 2 of its 18 seeds germinated. One fruit of *L. astorii* floated 200 days, and 2 of 14 seeds germinated. One fruit of *L. acutangula* var. *forskalii* floated 307 days, but none of its 17 seeds germinated. Two fruits of *L. acutangula* var.

amara were still afloat after one year, at which time they were removed. The interior of one fruit was badly soaked, and only 2 of its 17 seeds germinated. Only a small amount of water had penetrated the second fruit and 10 of its 44 seeds germinated.

No far-reaching conclusions should be drawn from this experiment, since the ability of fruits to float in a dish of salt water is hardly an adequate test of the ability to stay afloat in the ocean. It should also be pointed out that only one accession was tested for four of the species and that there might be some variability within the species in regard to the ability of the fruits to float. This experiment, however, might indicate that short-distance dispersal in the ocean would be possible for some of the species, and that a considerable distance might be transversed by *L. acutangula* var. *amara*.

If the dispersal occurs as postulated above, it would appear that the external characters of the fruit (Figure 4) have no clear role in seed dispersal. Yet, some of the best taxonomic characters are found in the fruits. All of the species, with the possible exception of the two of South America, can readily be distinguished on characters of the fruit alone. What adaptive significance the fruit characters may have is not known. It is rather difficult to imagine

Figure 4. Fruits of *Luffa* species: a. *L. acutangula* var. *amara*, b. *L. acutangula* var. *forskalii*, c. *L. graveolens*, d. *L. echinata*, e. *L. quinquefida*, f. *L. operculata*, g. *L. astorii*, h. *L. aegyptiaca* var. *leiocarpa*.

that the long spines on the fruit of *L. echinata* do not have adaptive value. These fruits will readily stick to clothing; thus, animals may play some role in their dispersal. Detailed observations of fruit in nature are needed in order to learn more about their adaptive roles.

Geography

There are a number of disjunctions in the distribution of *Luffa*. The presence of wild varieties of *L. acutangula* in Yemen and India may result from human activities, as was pointed out above, and the presence of the two domesticated species around the world results from human dispersal. However, people are very unlikely involved in the spread of the other species. The presence of *L. echinata* in India and Africa, *L. graveolens* in India and Australia, and a group of species in the Americas with a disjunction between Central and South America defies any simple explanation. Various possibilities may be presented to account for these distributions. Perhaps the genus had its origin on Gondwanaland, and the species attained their distribution before the breakup of the southern continents and India, or perhaps long-distance dispersal over water occurred after the continents had assumed their present positions. A combination of the two kinds of events is also a possibility.

The occurrence of *Luffa* species in India, Australia, Africa, and South America is compatible with an origin on Gondwanaland. One of the questions that cannot be answered is whether or not the genus was in existence in the Jurassic before the breakup of Gondwanaland (28). Based on the existence of well-differentiated, strongly reproductively isolated species and their pattern of distribution, one is tempted to postulate that *Luffa* is, indeed, an old genus, but that would not necessarily mean that it had an origin in the Jurassic. It has, in fact, not even been established that the family Cucurbitaceae had an origin as early as the Cretaceous. If one accepts that the present disjunctions in *Luffa* are very ancient, there is another puzzling feature that requires explanation. Why is the same species present on two continents—*L. echinata* in India and Africa and *L. graveolens* in India and Australia—if the separation of the continents goes back to the Jurassic. Although morphological differentiation is not necessarily a function of time, it would seem surprising that there had been no morphological changes in either species on the different continents after 100 million years of isolation, even if the environments had remained the same.

The second possibility, i.e., long-distance dispersal, may offer a more likely explanation. Transport by ocean currents, however, involves the propagule reaching the sea, being capable of floating, resisting the action of salt water, and being deposited in a suitable climate and habitat for germination and establishment (10). Although it has been shown that some species, but not including *L. echinata*, have fruits that will float and retain seed viability, the time that they will float hardly seems adequate for a long voyage. It is possible, of course, that floating occurred on drifting wood. After all, the plants grow in trees. However, to have a species cross the ocean to give rise to the American species, and for species to go both to Africa and Australia from India is rather difficult to accept, particularly in view of the fact that the direction of the major ocean currents from India do not seem to be particularly favorable for any of these voyages. One might postulate that the dispersal occurred at an earlier time when the species had different areas or when the continents were closer together.

Ethnobotany

Two of the species, *L. acutangula* and *L. aegyptiaca*, include domesticated varieties. They are widely cultivated, especially *L. aegyptica*. The young fruits of both domesticates are cooked and used for food, particularly in India and southeast Asia. The older fruits become too fibrous and bitter to be so used. It is, however, the persistent fibrovascular bundles that have given the fruits a large number of other uses. This is particularly true of *L. aegyptiaca*, which has the stronger and more durable vascular network. Probably the greatest use is for washing and scrubbing utensils as well as the human body. Luffas, most of which are imported, have become quite popular as cosmetic sponges in the United States in recent years (12). Japan is the best source of superior sponges for this purpose. The seeds also furnish an edible oil (34). Porterfield (25) has reviewed the many uses of the domesticated species.

Luffa acutangula, *L. aegyptiaca*, and *L. echinata* are used as medicines in several parts of tropical Africa and Asia (3, 4, 6, 27, 30). Most frequently cited are uses as a laxative, purgative, and diuretic as well as for the treatment of skin ailments. In South America a number of medicinal uses have been reported for *L. operculata* (14, 21), particularly for the treatment of sinusitis.

Future Study

Future research on *Luffa* should concentrate on gathering additional information through field studies and on quantifying aspects of species relationships that bear on geographic distributions. As mentioned above, there are still large gaps in our knowledge concerning precise distributions for the Old World species. Knowledge of mechanisms of seed and fruit dispersal will be of particular interest, as will be studies of pollination biology and morphological variation of individual species. Laboratory analysis using biochemical markers, such as isozymes, that can quantify the relative amount of divergence between populations and species may help to establish a time frame for geographic disjunctions and thus provide the most reasonable explanation for them.

Literature Cited

1. Andrews, F. W. 1950. The Flowering Plants of the Anglo-Egyptian Sudan. Vol. 1. Buncle, Arbroath.
2. Briggs, B. (chair, edit. comm.) 1982. Flora of Australia. Vol. 8. Austral. Govt. Publ. Service, Canberra.
3. Burkill, H. M. 1985. The Useful Plants of West Tropical Africa, 2nd ed. Vol. I. Royal Botanic Gardens, Kew.
4. Burkill, I. H. 1966. A Dictionary of the Economic Products of the Malay Peninsula. Vol. II. Ministry of Agriculture and Cooperatives, Kuala Lumpur.
5. Chakravarty, H. L. 1959. Monograph on Indian Cucurbitaceae. Rec. Bot. Survey India 17: 1–234.
6. Chopra, R. N., R. Badhwar, and S. Ghosh. 1965. Poisonous Plants of India, 2nd ed. Vol. I. Indian Council of Agricultural Research, New Delhi.
7. Cogniaux, A., and H. Harms. 1924. *Luffa. In* A. Engler, ed., Pflanzenreich IV. 275. II. Wilhelm Engelmann, Leipzig.
8. Dutt, B. 1971. Cytogenetic investigations in Cucurbitaceae. I. Interspecific hybridization in *Luffa*. Genetica 42: 139–156.
9. Dutt, B., and R. P. Roy. 1969. Cytogenetical studies in the interspecific hybrid of *Luffa cylindrica* L. and *L. graveolens* Roxb. Genetica 40: 7–18.
10. Good, R. 1974. The Geography of the Flowering Plants, 4th ed. Longman, London.
11. Guppy, H. B. 1906. Observations of a Naturalist in the Pacific between 1896 and 1899. Vol. 2. Plant Dispersal. Macmillan, London.
12. Heiser, C. B. 1985. Of Plants and People. Univ. Oklahoma Press, Norman.
13. Heiser, C. B., and E. E. Schilling. 1988. Phylogeny and distribution of *Luffa* (Cucurbitaceae). Biotropica 20: 185–191.
14. Heiser, C. B., E. E. Schilling, and B. Dutt. 1988. The American species of *Luffa* (Cucurbitaceae). Syst. Bot. 13: 138–145.
15. Hooker, J. D. 1871. Cucurbitaceae. *In* D. Oliver, ed., Flora of Tropical Africa. Vol. 2. Reeve, London.
16. Hutchinson, J., and J. M. Dalziel. 1954. Flora of West Tropical Africa, 2nd ed. Vol. 1, pt. 1. Crown Agents, London.
17. Jeffrey, C. 1962. Notes on Cucurbitaceae, including a proposed new classification of the family. Kew Bull. 15: 337–371.
18. Jeffrey, C. 1980. A review of the Cucurbitaceae. J. Linn. Soc. Bot. 81: 233–247.
19. Jeffrey, C. 1980. Further notes on the Cucurbitaceae. V. The Cucurbitaceae of the Indian subcontinent. Kew Bull. 34: 789–809.

20. Keraudren, M. 1965. Cucurbitaceae. *In* A. Aubreville, ed., Flore du Cameroun. Vol. 6. Muséum National d'Histoire Naturelle, Paris.
21. Morton, J. F. 1981. Atlas of Medicinal Plants of Middle America. Charles C Thomas, Springfield, IL.
22. Naudin, C. 1862. Cucurbitacées cultivées au Muséum d'Histoire Naturelle en 1862. Ann. Sci. Nat., Ser. 4, 18: 159–208.
23. Parham, B. 1972. Plants of Samoa. New Zealand Dept. Sci. Indust. Res., Wellington.
24. Pathak, G. N., and S. N. Singh. 1949. Studies in the genus *Luffa*. I. Cytogenetic investigations in the interspecific hybrid *L. cylindrica* × *L. acutangula*. Indian J. Genet. 9: 18–26.
25. Porterfield, W. M. 1955. Loofah—the sponge gourd. Econ. Bot. 9: 211–223.
26. Purseglove, J. W. 1968. Tropical Crops. Dicotyledons. Vol. 1. Wiley, New York.
27. Quisumbing, E. 1951. Medicinal Plants of the Philippines. Bureau of Printing, Manila.
28. Raven, P. R., and D. I. Axelrod. 1974. Angiosperm biogeography and past continental movements. Ann. Missouri Bot. Gard. 61: 593–673.
29. Ross, J. H. 1972. The Flora of Natal. Mem. Bot. Survey South Africa. No. 39. Government Printer, Pretoria.
30. Sastri, B. N., ed. 1962. The Wealth of India. Vol. 6. Council of Scientific and Industrial Research, New Delhi.
31. Schilling, E. E., and C. B. Heiser. 1981. Flavonoids and the systematics of *Luffa*. Biochem. Syst. Ecol. 9: 263–265.
32. Schubert, B. G. 1975. Report of the standing committee on stabilization of specific names. Taxon 24: 171–177.
33. Terrell, E., S. Hill, J. Wiersema, and W. Rice. 1986. A Checklist of Names of 3,000 Vascular Plants of Economic Importance. USDA, Agricultural Handbook No. 505. Washington, DC.
34. Watt, J., and M. Breyer Brandwijk. 1862. The Medicinal and Poisonous Plants of Southern and Eastern Africa, 2nd ed. Livingstone, Edinburgh.

11 | Cytogenetics of the Old World Species of *Luffa*

Bithi Dutt and R. P. Roy

ABSTRACT. Cytogenetic studies of the Old World species of *Luffa*, including the wild *L. graveolens* and *L. echinata* and domesticated forms of *L. cylindrica* and *L. acutangula*, are presented. In each of the four species $2n = 26$, and the chromosomes regularly pair as 13 bivalents during meiosis. Among the species chromosome size is comparable. F_1 hybrids have been made between all species. These hybrids are largely intermediate morphologically between the parents. The hybrid between *L. cylindrica* and *L. acutangula* and its reciprocal are partially fertile; all others are sterile. Chromosome pairing in hybrids and derived amphidiploids shows a variable though relatively high frequency of bivalents. Univalents and multivalents are also present. Cytological data suggest that *L. cylindrica* and *L. acutangula* are closely related and probably were derived from *L. graveolens* or a common ancestor.

The genus *Luffa*, known by such common names as loofah, dishcloth gourd, and vegetable sponge, occurs in principally tropical regions of both the Old and New Worlds. It includes seven species, of which four are indigenous to the Indian subcontinent and beyond: *L. cylindrica* (L.) M. J. Roem., *L. acutangula* (L.) Roxb., *L. echinata* Roxb., and *L. graveolens* Roxb. The three remaining species, *L. operculata* (L.) Cogn., *L. quinquefida* (Hook. & Arn.) Seem., and *L. astorii* Svens, are native to the New World. Two of the seven species, *L. cylindrica* and *L. acutangula*, include domesticated varieties. Both domesticates are grown extensively for their fruits, which when immature are used for vegetables. When mature, those of *L. cylindrica* especially are harvested for their inner fibrous network, referred to as vegetable sponge or loofah, the uses of which have been reviewed by Porterfield (15), Chakravarty (this volume), and Heiser and Schilling (this

134

volume). The bitter fruits of the wild *L. echinata* provide the main source of the alkaloid luffein (18).

The systematic aspects of *Luffa* have been discussed by Chakravarty (4), and more recently by Heiser et al. (10), and Heiser and Schilling (this volume). Genetic studies have dealt with patterns of inheritance of monoecy and dioecy in *Luffa*, which appears to be controlled by the interactions of two independent, multi-allelic genes (5, 14, 16, 21), and inheritance of such characters as flower color, androecium type, fruit surface and taste, and time of anthesis, all of which exhibit monogenic segregation (22). Hybridization and cytological studies involving the New World species are summarized by Heiser et al. (10) and Heiser and Schilling (this volume). That of the Old World and their hybrids is the subject of this chapter. Cytological data are summarized and are used as a basis for discussing phylogenetic trends and relationships.

Cytology of the Species

The sporophytic chromosome number of *L. cylindrica*, *L. acutangula*, *L. graveolens*, and *L. echinata* in all cases is $2n = 26$ (6, 12, 18), as it is for the New World species (10). Polyploidy is unreported in the genus, (1, 17, 19, 24, A. K. Singh, this volume). Hence, *Luffa* is particularly well suited for the exploration and assessment of the role of genetic and structural alterations in speciation.

The total length of chromatin in *L. graveolens*, *L. acutangula*, *L. cylindrica*, and *L. echinata* was determined as 59.42, 59.40, 54.10, and 53.10 µm, respectively. The general similarity in chromatin lengths indicates that major additions or deletions of chromatin have not accompanied speciation in *Luffa*. The study of chromosome morphology/karyology was not especially fruitful. Because of their small size and poor stainability, the chromosomes of *Luffa* proved not to be suitable for making critical inferences, although chromosomes with primary and secondary constrictions could be recognized. Bhaduri and Bose (3) originally reported for *L. cylindrica* the presence of 12 chromosomes with distinct satellites and 4 others with indications of satellites. Our studies revealed only four pairs of chromosomes with secondary constrictions. *Luffa echinata* and *L. graveolens* had two such pairs. Otherwise the chromosomes of the four species exhibit a similar morphology.

Microsporogenesis in *Luffa* was first studied by Passamore (13), who reported only 11 bivalents for *L. cylindrica*, but McKay (11) reported 13 bivalents for the species (under the name *L. gigantea* auct.), and all subsequent studies (2, 3, 6, 9) have confirmed this degree of pairing and normal meiosis in all four species.

Hybridization in *Luffa*

Despite strong morphological differentiation within *Luffa* and wide-spread geographic dispersion, interspecific hybrids have been made in many combinations. Among the Old World species, crosses have been made between all four species (6, 7, 14, 20). Hybrids, for which cytological data are available, are summarized in Table 1. The New World species also have been crossed in all combinations, with varying levels of success (10, Heiser and Schilling, this volume). A single sterile hybrid was obtained between *L. astorii* and *L. operculata*. Hybrids of the cross between *L. astorii* and *L. quinquefida* had pollen fertility of 0 to 18%, as determined by pollen stainability, but set no seed, while those of the cross between *L. operculata* and *L. quinquefida* had pollen fertility ranging from 20 to 52% and set some seed. Last, in crosses involving Old and New World species, progeny were produced in crosses between *L. cylindrica* and both *L. astorii* and *L. operculata*, between *L. echinata* and *L. operculata*, and between *L. graveloens* and both *L. operculata* and *L. quinquefida* (11, Heiser and Schilling, this volume). Pollen fertility reached 28% in one plant of the *L. graveolens* and *L. quinquefida* cross. None of the hybrids produced seed.

In crosses involving the Old World species, only hybrids resulting from the cross between *L. cylindrica* and *L. acutangula* and its reciprocal exhibited pollen fertility. In these instances pollen fertility measured to 32%. The hybrids of other crosses were completely sterile. In crossing *L. graveolens* and *L. acutangula*, F$_1$ hybrids were obtained only when *L. acutangula* was the female parent. In amphidiploids derived from the hybrids of *L. cylindrica* with *L. echinata* and *L. acutangula* with *L. graveolens*, pollen fertility approached 70% in the former, but was 0 in the latter. Neither amphidiploid produced seeds or developed fruits.

Cytology of the Old World Hybrids

The results of a series of studies (6–8, 14, 20, 23) concerned with meiotic chromosomal pairing among hybrids of the Old World species of *Luffa* are summarized in Table 1. In crosses between the domesticated varieties of *L. cylindrica* and *L. acutangula*, and between these varieties and *L. graveolens*, bivalent formation was high. Univalents, trivalents, and quadrivalents occurred at low frequencies.

In contrast, in hybrids in which one parent was *L. echinata* the frequency of univalents increased markedly at the expense of bivalents. In crosses in which *L. acutangula* was the female parent, the progeny were weak and did not survive to maturity.

Meiosis in the amphidiploids formed from hybrids of the cross *L. cylin-*

Table 1. Chromosome associations at metaphase I in species hybrids and their amphidiploids

| Hybrid | | | Bivalents | | | | | Multivalents of more than 4 chromosomes |
Female	Male	Univalents	Rod	Ring	Total	Trivalents	Quadrivalents	
L. cylindrica × L. acutangula		Mean values not given by Pathak and Sing (14)						
L. acutangula × L. cylindrica		0.40	8.54	3.92	12.46	0.05	0.08	—
L. cylindrica × L. graveolens[1]		0.48	4.00	5.64	9.64	0.25	1.32	0.02
L. graveolens × L. cylindrica[1]		0.52	6.92	4.76	11.68	0.12	0.44	—
L. acutangula × L. graveolens[1]		0.68	8.16	4.28	12.44	0.04	0.08	—
L. graveolens × L. acutangula[1]		Not successful						
L. echinata × L. graveolens[1]		10.76	4.66	2.22	6.88	0.20	0.22	—
L. graveolens × L. echinata[1]		5.16	5.96	3.44	9.40	0.36	0.32	—
L. cylindrica × L. echinata[2]		12.20	3.50	2.30	5.80	0.08	0.08	—
L. echinata × L. cylindrica[2]		8.50	4.50	6.50	11.00	—	—	—
L. acutangula × L. echinata[2]		Plants died before reaching maturity						
L. echinata × L. acutangula[2]		8.56	5.04	3.56	8.60	0.08	—	—
L. cylindrica × L. echinata[3,4]		1.68	6.48	17.12	23.60	—	0.82	—
L. acutangula × L. graveolens[1,4]		5.83	8.13	13.53	21.66	0.40	0.37	0.03

Source: Data from [1]Dutt and Roy (6, 7, 8), [2]Roy et al. (20), [3]Trivedi and Roy (28).
[4]Amphidiploid.

137

drica with *L. echinata* showed a level of bivalent formation somewhat greater than that of amphidiploids derived from *L. acutangula* with *L. graveolens*. Multivalent formation remained low.

Discussion

Cytological data indicate that evolution in *Luffa* has not resulted either from changes in chromosome number or from major changes in chromosome structure, although it is recognized that the number of chromosomes bearing satellites differs between *L. cylindrica* and *L. acutangula*, on the one hand, and *L. graveloens* and *L. echinata*, on the other. In both of these regards *Luffa* seems typical of the intrageneric evolution seen in many of the Cucurbitaceae (A. K. Singh, this volume).

The similarity of chromatin content in the Old World species and the ability of these species to form F_1 hybrids when crossed in essentially all combinations suggests a relatively high level of chromosomal homology among the species. By extension, the generally high level of crossability among the American species of *Luffa* and that which occurs between the New and Old World species suggests that a similarly high level of chromosomal homology exists throughout *Luffa*.

Beyond a gross level of cytological similarity, however, it is evident that speciation has been accompanied either by cryptic structural differences, genic incompatibilities, or both. The relatively high level of bivalent formation among *L. cylindrica*, *L. acutangula*, and *L. graveolens* is indicative of chromosomal homology, suggesting that the complete or partial loss of fertility in the hybrids is genic in origin and action. Whether or not the lowered level of bivalent formation in hybrids derived from crosses involving *L. echinata* can be attributed to a higher level of structural differentiation in that species cannot be determined, nor is it possible to determine if the formation of trivalents, quadrivalents, and other multivalent associations resulted from translocations or are the result of alloautosyndetic pairing. Data from the amphidiploids may have bearing on questions concerning loss of fertility. The high percentage of fertile pollen in the *L. cylindrica*/*L. echinata* amphidiploid reinforces the hypothesis that structural incompatibilities between these two species are significant. Similarly, the sterility of the *L. acutangula*/*L. graveolens* amphidiploid suggests genic causes are important in species that show a high level of diploid pairing.

Heiser et al. (10) and Heiser and Schilling (this volume) present cladistic studies of the species of *Luffa*. Their analyses of morphological, biochemical, and breeding habit suggest two major lines of evolution in the genus: one represented by the wild forms and domesticated derivatives of *L. cylindrica* and *L. acutangula*, the other by the remaining species of the genus. In the latter line, *L. echinata* stands somewhat apart, and *L. graveolens* shows the closest affinities to the New World species.

Cytological and hybridization studies confirm the close relationship of *L. cylindrica* and *L. acutangula*, and, based on frequencies of bivalent formation, the relatively remote position of *L. echinata*. The higher frequency of bivalent formation in hybrids between *L. graveolens* and both *L. cylindrica* and *L. acutangula* led Dutt and Roy (7) to suggest derivation of those species from *L. graveolens*. While the particular cladistic analyses of Heiser and his coworkers does not support this view, it may well be that *L. graveolens* does retain, in large measure, the chromosomal structure of an ancestor to both lines of evolution.

Acknowledgments

We are grateful to Sunil Saran, Department of Botany, Patna University, Patna, for his helpful suggestions and discussions during preparation of the manuscript. Portions of this study were completed while R. P. Roy held the Jawaharlal Nehru Fellowship.

Literature Cited

1. Agarwal, P. K., and R. P. Roy. 1976. Natural polyploids in Cucurbitaceae. I. Cytogenetical studies in triploid *Momordica dioica* Roxb. Caryologia 29: 7–13.
2. Ahuja, M. R. 1955. Chromosome numbers of some plants. Ind. J. Genet. 15: 142–143.
3. Bhaduri, P. N., and P. C. Bose. 1947. Cytogenetical investigations in some common cucurbits. J. Genet. 48: 237–256.
4. Chakravarty, H. L. 1982. Cucurbitaceae. *In* K. Thothathri, ed., Fascicles of Flora of India. Fasc. 11. Botanical Survey of India, Howrah.
5. Choudhury, B., and M. R. Thakur. 1965. Inheritance of sex forms in *Luffa*. Ind. J. Genet. 25: 188–197.
6. Dutt, B., and R. P. Roy. 1969. Cytogenetic studies in the interspecific hybrid of *Luffa cylindrica* L. and *L. graveolens* Roxb. Genetica 40: 7–18.
7. Dutt, B., and R. P. Roy. 1971. Cytogenetic investigations in Cucurbitaceae. I. Interspecific hybridization in *Luffa*. Genetica 42: 139–156.
8. Dutt, B., and R. P. Roy. 1976. Cytogenetic studies in an experimental amphidiploid in *Luffa*. Caryologia 29: 15–25.
9. Hardas, N. W., and A. B. Joshi. 1954. A note on the chromosome numbers of some plants. Ind. J. Genet. 24: 47–49.
10. Heiser, C. B., E. E. Schilling, and B. Dutt. 1988. The American species of *Luffa* (Cucurbitaceae). Syst. Bot. 13: 138–145.
11. McKay, J. W. 1931. Chromosome studies in the Cucurbitaceae. Univ. Calif. Publ. Bot. 16: 339–350.

12. Naithani, S. P., and P. Das. 1947. Somatic chromosome numbers in some cultivated cucurbits. Curr. Sci. 16: 188–189.
13. Passmore, S. F. 1930. Microsporogenesis in Cucurbitaceae. Bot. Gaz. 90: 213–223.
14. Pathak, G. N., and S. N. Singh. 1949. Studies in the genus *Luffa*. Cytogenetic investigations in the interspecific hybrid *L. cylindrica* × *L. acutangula*. Ind. J. Genet. 9: 18–26.
15. Porterfield, W. M., Jr. 1955. Loofah—the sponge gourd. Econ. Bot. 9: 211–223.
16. Richharia, R. H. 1949. Genetical studies in *Luffa acutangula*. Ind. J. Genet. 9: 42–43.
17. Roy, R. P., and J. Ghosh. 1971. Experimental polyploids of *Luffa echinata* Roxb. Nucleus 14: 111–115.
18. Roy, R. P., A. R. Mishra, R. Thakur, and A. K. Singh. 1970. Interspecific hybridization in the genus *Luffa*. J. Cytol. Genet. 5: 16–20.
19. Roy, R. P., V. Thakur, and R. N. Trivedi. 1966. Cytogenetical studies in *Momordica* L. J. Cytol. Genet. 1: 30–40.
20. Roy, R. P., and R. N. Trivedi. 1961. Cytology of *Gomphogyne cissiformis*. Curr. Sci. 35: 420–421.
21. Singh, H. B., S. Ramanujam, and B. P. Pal. 1948. Inheritance of sex-forms in *Luffa acutangula*. Nature 161: 775–776.
22. Thakur, M. R., and B. Choudhury. 1966. Inheritance of some qualitative characters in *Luffa* species. Ind. J. Genet. 26: 79–86.
23. Trivedi, R. N., and R. P. Roy. 1976. Interspecific hybridization and amphidiploid studies in the genus *Luffa*. Genet. Iber. 28: 83–106.
24. Whitaker, T. W. 1950. Polyploidy in *Echinocystis*. Madrono 10: 209–211.

12 | Origin and Evolution of Chayote, *Sechium edule*

L. E. Newstrom

ABSTRACT. Chayote, *Sechium edule*, is a tropical and subtropical crop used for food, forage, and medicine. Consideration is given to its systematics relationships, origins, reproductive behavior, and variability in cultivation, and to the conservation of its germplasm. Evidence of various kinds suggests that chayote was domesticated in Mexico or Central America in pre-Columbian times, either from wild forms of the species or from its closest relative, *S. compositum*. Chayote is monoecious, self-compatible, and pollinated by a wide array of insect vectors. Ratios of male and female flowers vary seasonally, annually, and between cultivars and individuals. The fruits are viviparous. Genetic resources are being gathered and characterized and should play an important role in future breeding programs of the crop.

Chayote, *Sechium edule* (Jacq.) Swartz, a staple food for many Latin Americans, is now cultivated throughout tropical and subtropical regions of the world. All parts of the chayote plants are useful: the fruits and tubers are used as table vegetables, the shoots as pot herbs, the leaves as medicine and forage, and the stems as silver-white straw. The floral nectar is gathered by honey bees. The fruits are exported to temperate regions and perhaps there is greater potential in that regard.

Until recently, there have been few studies concerned with the biology of chayote. *Sechium* was considered monotypic and the single species a cultigen, for wild forms were unknown. The site of origin of the crop was generally thought to be Mexico or Central America, although the West Indies and South America have also been suggested. This chapter summarizes what is known about several aspects of chayote biology, focusing especially on the systematic relationships of *Sechium*, in the broad sense, and the origin and reproductive biology of chayote. In addition a summary and

141

discussion of the genetic resources available for maintaining and improving the crop are presented.

Systematic Relationships

Although earlier studies (10, 28) suggested that *Sechium* was not monotypic, it remained for Jeffrey (17) to redefine the genus, submerging in it the genera *Ahzolia*, *Polakowskia*, and *Frantzia*, and adding to it *Microsechium hintonii* P. G. Wils. As redefined, *Sechium* included seven species in two sections: section *Sechium* composed of five species of which each had a ten-pouched floral nectary, and section *Frantzia* in which each of the two species then known had a cushioned floral nectary. In a different approach Wunderlin (31, 32) recognized *Sechium* and *Frantzia* as distinct, but included *Polakowskia* in the latter.

Field studies were undertaken in Mexico and Central America to assess the affinities of *S. edule* to the other species included in *Sechium* by Jeffrey (17). Morphological analyses of collections made in the field and of herbarium materials suggest that *S. compositum* (J. D. Sm.) C. Jeffr. (*Ahzolia composita* J. D. Sm.), found in Mexico and Guatemala, is closest to *S. edule*. In fact, the two species are so similar that Standley and Steyermark (28) suggested that *A. composita* might be a wild form of chayote, and Dieterle (10) thought at the least it should be included in *Sechium*. In floral characters the species are essentially alike, both having a saucer-shaped hypanthium with a ten-pouched nectary, but the anther branches of *S. compositum* are narrower and lack the curl characteristics of *S. edule*. The fruits of the two species differ in the shape of the ridges and the arrangement of the spines. Those of *S. compositum* are bitter, and its tubers are used as a soap. There is evidence that the two species hybridize (22).

The other species included in section *Sechium* are *S. hintonii* (P. G. Wils.) C. Jeffr. from Mexico State, and *S. tacaco* (Pitt.) C. Jeffr. and *S. talamancensis* (Wunderlin). C. Jeffr., which would constitute the Costa Rican genus *Polakowskia*, if it were recognized. The last two named species have ten-pouched floral nectaries similar to those of *S. edule*; however, the nectaries are sunken in the base of a suburceolate hypanthium rather than one that is saucer-shaped. The anthers form a connate head instead of being free from one another on divaricate branches as in the other species of the section. The fruits of *S. tacaco* and *S. talamancensis* differ from those of *S. edule* in being ellipsoidal to fusiform and uncleft at the distal end.

Sechium hintonii was not found in the field, but examination of an isotype (UC) showed the male flowers to be similar to those of chayote, being five-merous and with anthers on branches. The fruits are covered with long spines bearing retrose barbs, a character not observed in any chayote cultivars. The other species of *Microsechium* are not included in *Sechium* because they lack floral nectaries and the male flowers are four-merous, although the female flowers are five-merous.

In Jeffrey's revision (17) section *Frantzia* consisted of two Costa Rican-Panamanian species. Not considered was a contemporaneously described species of *Frantzia* from Panama (32). An undescribed species has been noted recently from Mexico (22). Species of this section have a distinctive cushion nectary. In addition, the petals are thin and twisted, opening abruptly at midday, rather than early in the morning, as is characteristic of the species of section *Sechium*. The fruits are ellipsoid and the anthers are connate, both character expressions of the species that composed *Polakowskia*.

The systematics of *Sechium* can be considered at two levels. First, does *Sechium* in the sense of Jeffrey and in the context of the tribe Sicyeae constitute a monophyletic group? Second, if the genus is monophyletic, is the variation expressed in it best represented by recognizing a single genus or two or three segregate genera? The first question can be addressed only by understanding the Sicyeae as a whole, a problem that goes beyond the scope of these studies, although for the present, it is assumed that *Sechium* is monophyletic.

Patterns of floral and fruit variation within *Sechium* can be resolved in three groupings. One includes *S. edule*, *S. compositum*, and *S. hintoni*. The second *S. tacaco* and *S. talamancesis*; and the third the species of the section *Frantzia*. These groupings could be represented as the genera *Sechium*, *Polakowskia*, and *Frantzia*, or as sections of *Sechium*. In either case, the closest relatives to chayote are clustered in one taxon and are segregated from those more distant. Generic recognition tends to emphasize divergence between the taxa; while sectional recognition emphasizes their closeness. Recognition of three genera is favored, but additional comparative data are needed before a final decision is made.

Origins of Chayote

Most authors who have written about chayote agree that the crop was domesticated in Mexico and Central America, but some have suggested the West Indies, Venezuela, Colombia, or Peru as well (22, for references). The lines of evidence bearing on the origins of domesticated chayote are several, and include historical records, linguistics, centers of crop diversity, occurrence of wild forms and distribution of related wild species.

It is apparent that chayote was introduced outside of Latin American countries after the Spanish conquest. Hernandez (14) and the Florentine Codex recorded that the Aztecs used chayote in Mexico. The pre-Columbian presence of chayote in the West Indies was not reported in 1523 by the earliest historiographer, Oveido (23). Chayote is not an Andean plant according to Pittier (24, 25) and Horkeimer (16). Although documentary evidence can show that a listed plant was used, it does not prove that an unlisted plant was not used. An unmentioned plant may have been of minor importance and therefore undocumented. No records have so far been found that show chayote was indigenous to South America or the West Indies. Further research is needed to document whether or not the Aztecs, other indigenous groups, or the Spanish introduced chayote to South America and the West Indies.

Linguistic evidence can provide clues to plant introduction in the following way: a simple "unanalyzable" name such as "maize," in an indigenous language indicates an ancient existence and importance of a plant; a more descriptive, compound name allows the inference that the plant was introduced (B. Berlin, pers. comm.). In the case of chayote, DeCandolle (9) first used the linguistic evidence that chayotle is the Aztec name. This was later cited by many authors (5, 29, 27). But the use of an Aztec name or its derivations does not mean that the plant was introduced from Mexico because the name may have been adopted and spread when Spanish became the dominant colonial language. The linguistic evidence necessary to determine the origin of chayote is an analysis of the terms used in the indigenous languages of Latin America before colonization.

The center of diversity of chayote has not been considered in great detail, but Leon (18) and Bukasov (4) stated that Mexico and Central America have the highest diversity. Since diversification can be secondarily derived, this type of evidence is not strong, although it is in keeping with the other evidence.

The strongest evidence for the origin of a crop is usually found in the distribution of its wild relatives. In the case of chayote its wild relatives occur in Mexico and Central America, since all of the taxa constituting the Sechium line as described above are from that region. More specifically, however, two possibilities exist: first, that chayote was derived from wild populations of the same species, or second, that it was derived from a related wild species, that is, *S. compositum*. The evidence is equivocal, for the distinctions between truly wild and feral populations cannot be made with certainty, as is often the case with cucurbits (28).

According to Cook (5), writing at the turn of this century, chayote was unknown as a wild plant, although it is evident that escapes with the characteristics of cultivated forms are widely naturalized in the Americas and even elsewhere in the tropics, for example, on the islands of Java (2) and Reunion (7). In 1977 Brucher (3) reported a wild chayote from Venezuela, but this plant was most likely feral, since chayote was apparently introduced in that country (26). In 1982, Cruz (8) discovered two wild or feral populations in Veracruz, Mexico, although, the existence of such populations in Mexico had been known by Mexican botanists for some time (E. Hernández Xolocotzi, pers. comm.). Eventually three additional wild or feral populations were found by L. E. Newstrom (22) in Oaxaca, one by S. Montes and L. C. Merrick in Hidalgo, and one by M. Nee in Veracruz.

The plants of these seven wild or escaped populations all shared bitter fruits, but otherwise the populations could be classified into one of three types based on morphological and chemical analyses of the fruits and flowers (22). Four of the populations are either true wild forms or derivatives of the cultivated crop. The flowers of plants of these populations are indistinguishable from those of chayote and have thick, curled anther branches. The fruits are small, round, green, and spiny, as might be expected for a wild species. Experimental crosses between a plant of a population of this type from Tetla, Veracruz, and *S. edule* produced fertile fruits. Plants of one population from Veracruz combine characters of *S. edule* and *S. compositum* and may be of hybrid origin. The anther branches are thin like those of *S. compositum* and the fruits, which vary in size, shape, and degree of spininess, are similar to other cultivated forms of *S. edule*. Two populations from Oaxaca may represent a new species or they may be another hybrid between the cultivated crop and *S. compositum*. Plants of these populations have the straight, thin anther branches of *S. compositum* and the large, smooth, obovoid fruits quite similar to the cultivated chayote.

Reproductive Biology

The chayote fruits are viviparous. The developing cotyledons extract nutrients from the pericarp through haustoria (13). The selective advantage of vivipary in a cultivated crop such as chayote is not apparent, and the extent or advantage of vivipary in wild or feral populations or even in related species have not been examined. In fact, the relative rarity of vivipary in the plant kingdom, the mangroves, *Rhizophora* species, perhaps being the best known examples, raises interesting questions about its role in reproduction. In the feral or wild populations of *S. edule* and in *S. compositum* small wild mammals eat portions of the fruit, and since partly eaten fruit can resprout several times, if the embryo is not destroyed, these animals may act as dispersal agents.

Chayote depends on insect vectors to transfer pollen from male to female flowers (12, 22). Pollen transfer is essential since apomixis does not occur. The nearly rotate corollas permit many types of insects to pollinate successfully while foraging for pollen and nectar. In cultivated chayote pollination is achieved mainly by *Apis mellifera* L., bees of the genus *Trigona*, and various taxa of large bees and wasps (22, 30). In each male and female flower the ten pouchlike nectaries produce an average of one to nine μl of nectar per flower per day. The sugar concentration in the nectar rises from 15 to 20% w/w in the early morning to 55 to 80% w/w in the afternoon.

Since chayote is self-compatible and visited by a wide array of pollinators, relatively high fruit set might be expected. It is not known if fruit set differs between self- and cross-pollinated plants or what other factors might be involved in fruit set. Fruit abortion rates are variable but factors influencing this have not been studied.

An important factor in reproductive success is the ratio of female to male flowers. On each shoot, a solitary female flower opens at the most distal reproductive node. Occasionally, two female flowers open at the same time on adjacent nodes. Several nodes further toward the base of the shoot, from 3 to 20 nodes each bear one male inflorescence, which in turn, bears from 1 to 15 open male flowers. At any given time during the flowering cycle, each shoot may bear a total of from 3 to 45 male flowers and 1 or 2 female flowers. However, the ratio of male to female flowers in cultivated chayote varies throughout the season, between years, between plants and cultivars, and in different environments. At the beginning of the season, female buds are suppressed or sterile when male flowers are opening (21). In some cultivars, many additional male or female flowers are found in their usual position, or a female flower is produced on the normally male inflorescence (22). The flowering system in chayote, then, has three different methods for modifying the sex ratio: 1) suppression or abortion of floral buds, 2) addition of male flowers or female flowers, and 3) substitution of the opposite gender flower in a given position. Further work is needed to see how these patterns of modification vary genetically and environmentally.

Landraces, Cultivars, and Genetic Resources

As an ancient crop of indigenous peoples, much of the variation in cultivated chayote is expressed in landraces. Only recently has this variation and that of commercial cultivars begun to be documented and collections accumulated in gene banks. In 1901 Cook (5) described five cultivated types from Puerto Rico, but documentation has been most active only in the past decade. In cataloguing the genebank established by T. Leon and his collaborators at CATIE, Turrialba, Costa Rica, Maffoli (20) illustrated 69 types of Central American chayote, and later Engels (11) presented preliminary information on the geographic distribution of variation in chayote using the same genebank collection. More recently, Cruz and Querol (8) characterized 94 cultivated and two wild or feral populations from Mexico that are deposited in the Chapingo genebank collection in Huatusco, Veracruz. In addition to the collections of CATIE and Chapingo, chayote genebank are maintained by INIA in Celaya, Guanjuato, Mexico and EMBRAPA in Brazil. Newstrom (22) added 225 accessions to the genebanks in Huatusco and Celaya.

Cruz and Querol (8) were the first to systematically investigate the important character of flavor. Taste trials for 85 different Mexican introductions showed that flavor cannot be predicted from the external appearance of the fruits (22). Bland cultivars are used as fillers in baby foods and ketchup and as substitutes for apples (15) and artichokes (6). In Mexico, tasty cultivars are sold as cooked snacks in the market. The cultivar exported from Costa Rica is bland because it is intended for food industry use, but it is also sold as a table vegetable. The sale of chayote as a table vegetable would probably increase if the superior tasting cultivars were selected for commercial development.

Mexico, Costa Rica, and Puerto Rico export two commercial fruit types: the medium-sized obovoid, light green, smooth fruit, and the small, round, white, smooth fruit called *cocoros* in Costa Rica, *perulero* in Guatemala, and *chayotillo* in parts of Mexico. Selection of commercial cultivars is based on external appearance of the fruit because of consumer preferences. In the last ten years, vegetative propagation has been used for selections in the export industry in Costa Rica. In other countries, Mexico for example, selection has occurred mainly by discarding plants with undesirable fruits. Marketing demands have resulted in genetic erosion. In areas of Mexico serviced by commercial farms, the only cultivars found in the markets and kitchen gardens are the same two commercial types. In Costa Rica, the inexpensive rejects from the export industry flood the market so that distinctive local cultivars cannot compete.

Genetic erosion in chayote is serious because the commercial farms use highly inbred or cloned plants that suffer from viral and fungal diseases. Breeding programs for disease resistance are necessary if commercial development and export are to continue. The genetic resources for such programs have been partially collected and are now maintained in the four genebanks referred to above. Chayote genebanks must be maintained as living plants because the fruits are viviparous. In situ genebanks (1) could solve the problems of space and expense of maintaining large centralized living collections. Tissue culture technology now being developed could provide supplemental storage of chayote germplasm.

As a commercially valuable multipurpose plant, chayote merits further development. More detailed information on pollination biology and fruit set factors, as well as sex ratio determinants has both theoretical and practical value. The current genetic erosion and need for disease resistant plants emphasizes the importance of genetic resources and the timeliness of a breeding program. Further evaluation of the cultivars collected so far and exploration for new cultivars would be beneficial to such a program.

Literature Cited

1. Altieri, M. A., and L. C. Merrick. 1987. In situ conservation of crop genetic resources through maintenance of traditional farming systems. Econ. Bot. 41: 86–96.
2. Backer, C. A., and R. C. Bakhuizen van der Brink. 1963. Flora of Java. Vol. 1. Noordhoff-Groningen, The Netherlands.
3. Brucher, H. 1977. Tropische Nutzpflanzen: Ursprung, Evolution und Domestikation. Springer-Verlag, New York.
4. Bukasov, S. M. 1981. Las Plantas Cultivadas de Mexico, Guatemala, y Columbia. Proyecto CATIE-GTZ de Recursos Geneticos, Turrialba, Costa Rica.
5. Cook, O. F., 1901. The Chayote: A Tropical Vegetable. USDA, Div. Bot. Bull. No. 18. Washington, DC.
6. Cook, O. F. and G. N. Collins. 1903. Economic plants of Porto Rico. Contr. U.S. Natl. Herb. 8: 57–269.
7. Cordemoy, de, E. J. 1895. Flore de L'ile de la Reunion. Klincksieck, Paris.
8. Cruz Leon, A., and D. Querol Lipcovich. 1985. Catalogo de Recursos Geneticos de Chayote (*Sechium edule* Sw.) en el Centro Regional Universitario Oriente de la Universidad Autonomo Chapingo. Univ. Autonomo Chapingo, Chapingo.
9. De Candolle, A. L. 1886. Origin of Cultivated Plants. 2nd. ed. (reprinted 1959). Hafner, New York.
10. Dieterle, J. V. A. 1976. Cucurbitaceae. *In* D. L. Nash, ed., Flora of Guatemala. Fieldiana, Bot. 24 (XI, 4): 306–395.
11. Engels, J. M. 1983. Variation in *Sechium edule* Sw. in Central America. J. Amer. Soc. Hort. Sci. 108: 706–710.
12. Fidel, J., and E. Tristan. 1931. Bau und Bestaubung de Blüte von *Sechium edule* Sw. Biol. Gen. 7: 334–343.
13. Giusti, L., M. Resnik, T. del V. Ruiz, and A. Grau. 1978. Notas acerca de la biologia de *Sechium edule* (Jacq.) Swartz (Cucurbitaceae). Lilloa 35: 5–13.
14. Hernández, F. 1550. Historia de las Plantas de Nueva Espana. Vol. 3. Books 5, 6, 7. (Translation of 1790 ed.) Imprenta Universitaria, Mexico.
15. Hoover, L. G. 1923. The Chayote: Its Culture and Uses. USDA, Dept. Circular 286, Washington, DC.
16. Horkheimer, H. 1973. Alimentacion y Obtencion de Alimentos en el Peru Prehispanico. Univ. Nacional Mayor de San Marcos, Lima.
17. Jeffrey, C. 1978. Further notes on Cucurbitaceae. IV. Some new world taxa. Kew Bull. 33: 347–380.
18. Leon, J. 1968. Fundamentos Botanicos de los Cultivos Tropicales. Inst. Interam. de Cienc. Agric. de la OEA, San Jose, Costa Rica.
19. Lopes, J. F. 1979. Banco Ativo de Germoplasma de Chuchu. Congr. Bras. de Olericult., 19. Fpolis, SC, EM PASC.
20. Maffoli, A. 1981. Recurcos Geneticos de Chayote, *Sechium edule* (Jacq.) Swartz. (Cucurbitaceae). CATIE, Turrialba, Costa Rica.
21. Merola, A. 1955. Il gradiente sessuale in *Sechium edule* Sw. Delpinoa 8: 55–100.
22. Newstrom, L. E. 1986. Studies in Origin and Evolution of Chayote. Ph.D. dissertation, Univ. California, Berkeley.

23. Oviedo y Valdes, G. F. de. 1535. Historia General y Natural de las Indias. Sevilla. (First published entirely in 1851–1855.)

24. Pittier, H. 1910. New or noteworthy plants from Colombia and Central America. Contr. U.S. Natl. Herb. 13: 93–132.

25. Pittier, H. 1926. Manual de las Plantas Usuales de Venezuela. Litografia del Comercio, Caracas.

26. Pittier, H., T. Lasser, L. Shnee, and Z. de Febres y V. Badillo. 1947. Catalogo de la Flora Venezolana. Vol. 2. Vargas, Caracas.

27. Purseglove, J. W. 1968. Tropical Crops. Dicotyledons. Vol. 2. Longmans, London.

28. Standley, P. C., and J. C. Steyermark. 1944. Studies of Central American plants. IV. Field Mus. Nat. Hist., Bot. Ser. 23: 31–109.

29. Whitaker, T. W., and G. N. Davis. 1962. Cucurbits: Botany, Cultivation and Utilization. Interscience, New York.

30. Wille, A., E. Orozco. and C. Raabe. 1983. Polinizacion del chayote *Sechium edule* (Jacq.) Swartz en Costa Rica. Rev. Biol. Trop. 31: 145–154.

31. Wunderlin, R. P. 1976. Two new species and a new combination in *Frantzia* (Cucurbitaceae). Brittonia 28: 239–244.

32. Wunderlin, R. P. 1977. A new species of *Frantzia* (Cucurbitaceae) from Panama. Bull. Torrey Bot. Club. 104: 102–104.

13 | Reproductive Biology and Natural History of the Neotropical Vines *Gurania* and *Psiguria*

M. A. Condon and L. E. Gilbert

ABSTRACT. *Psiguria* and *Gurania* are large genera with complex life histories. Long thought to be dioecious, the genera are actually monoecious. We discuss the evolutionary and ecological significance of sexual patterns and review interspecific interactions that affect fitness. Some of those interactions include economically important pests. Understanding such networks helps provide insight into the structure of tropical communities and should aid the development of biological control and crop breeding programs.

We present an overview of the taxonomy, reproductive biology, and natural history of *Gurania* and *Psiguria* (syn. *Anguria*). The account is based on more than 15 years of work in Costa Rica and in greenhouses in Austin, Texas, and on four continuous years of field work in Guatopo National Park, Venezuela (5, 20). Also included is information from less intensive work in Mexico, Costa Rica, Ecuador, Peru, Trinidad, Tobago, Brazil, and other sites in Venezuela.

Taxonomy and Distribution

Gurania and *Psiguria* need thorough taxonomic revision (5, 6, 26, 31). Howard (26) accepted *Psiguria* over *Anguria*; some species of *Anguria*, e.g., *A. tabascensis* Donn. Sm. remain to be transferred to *Psiguria* (6). Cogniaux (4) recognized 73 *Gurania* species and 29 *Anguria* species, but Jeffrey (see Appendix) suggests there are only 40 species of *Gurania* and 12 of *Psiguria*. Since the nomenclature of the genera is volatile, and since we disagree with Jeffrey's concept of certain species (see 5, 6), we use names from published floras that treat areas near our study sites (7, 8, 40, 43).

150

Although the status of many species is tenuous, the cohesiveness of both genera is uncontested. Brilliant red, orange, and pink flowers distinguish *Gurania* and *Psiguria* from other New World Cucurbitaceae, and the two genera are easily distinguished from each other (Figures 1 and 2). *Gurania* has a succulent, brightly colored calyx tube and narrow, succulent, yellow petals; *Psiguria* has a green calyx and broad, spreading, colorful petals. Both produce pollen in tetrads (with the exception of one *Psiguria* species). *Gurania* and *Psiguria* are placed together with *Helmontia* in the subtribe Guraniinae of the tribe Melothrieae (30). *Helmontia* (2 species) resembles *Psiguria* but has white petals and single, separate pollen grains.

Gurania and *Psiguria* are found throughout the mainland Neotropics. *Psiguria* ranges as far north as Veracruz, Mexico, and *Gurania* to Chiapas, Mexico. *Psiguria* also occurs in the Antilles, where *Gurania* is absent. *Gurania* and *Psiguria* species show varying degrees of habitat specificity. Some species have tubers and can withstand drought, while other species are restricted to wetter forest. In Guatopo National Park, distributions of *Gurania* and *Psiguria* species are clearly defined by particular life zones (12) and stages of succession. In fact, three of the five species are restricted to single life zones: *P. umbrosa* (Kunth) C. Jeffr. to tropical dry forest, *P. racemosa* C. Jeffr. to premontane moist forest, and *G. acuminata* Cogn. to premontane wet forest. *Gurania spinulosa* (Poepp. & Endl.) Cogn. and *P. triphylla* (Miq.) C. Jeffr. are two of the most widespread species in the genera. They occur in both wet and moist forests and rarely along streams in dry forest. Both *P. triphylla* and *G. acuminata* are common in primary forest; *G. acuminata* often flowers in the lower levels of the canopy, but *P. triphylla* usually flowers only in clearings or in the upper canopy. *Gurania spinulosa* also occurs in primary forest, but mature individuals are common only along ridges above steep slopes where light intensity is predictably high.

Costa Rican species also segregate within and between life zones (25). *Psiguria warscewiczii* (Hook f.) Wunderlin is the only species that extends (via streambanks) into the dry habitats of Guanacaste. In the Atlantic rain forests, sexually active individuals of the four sympatric species occur in different microhabitats: *Psiguria bignoniacea* (Poepp. & Endl.) Wunderlin flowers in the high subcanopy to upper canopy of mature forests; *G. makoyana* (Lem.) Cogn., which we consider close to if not the same species as *G. spinulosa* (6), and *G. costaricensis* Cogn. flower both in the canopy and in successional patches within and along the edges of forests; and *P. warscewiczii* occurs primarily along edges of forest streams. Sterile individuals of these species, however, co-occur over a range of microhabitats. Species also appear restricted by particular edaphic conditions. *Psiguria bignoniacea*, for example, does not occur on sandy substrate in the Sirena area of Corcovado National Park.

Figure 1. Male flowers and inflorescences of *Gurania* and *Psiguria,* some with insect visitors. A. Male inflorescence of the Costa Rican *G. makoyana* visited by *Heliconius ethilla* in the greenhouse. The spearlike calyx lobes cause injury to the wings of *Heliconius.* B. Male inflorescence of *G. megistantha* from Rio Palenque, Ecuador, showing numerous scars from shed flowers. The inflorescences are cauliferous and often produced near the ground. At anthesis the pedicels bend away from the inflorescence. The flower shown has just closed. C. Longitudinal section of a male flower of a Peruvian *Gurania.* The constricted calyx tube prevents the removal of large pollen loads by *Heliconius.* The flower shown was visited by *H. cydno* in the greenhouse. D. Female *Blepharoneura* on male flowers of *G. costaricensis.* Previous ovipositions left circular white scars on flowers. E. Male flower of *P. racemosa* in Guatopo National Park, Venezuela. Ants on the corolla were feeding on nectar. F. *Heliconius charitonia,* a common visitor on male flowers of *Anguria tabascensis* in Veracruz, Mexico. Note numerous scars from previous flowers. G. Male inflorescence of *P. warscewiczii* with nymph of *Hyphinoe* sp. (Membracidae) in Corcovado National Park, Costa Rica.

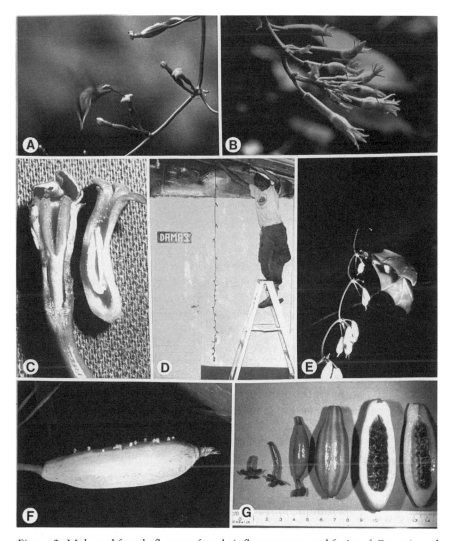

Figure 2. Male and female flowers, female inflorescences, and fruits of *Gurania* and *Psiguria,* some with animal visitors. A. *Phaethornis longuemareus* visiting a female flower of a *Psiguria triphylla* in Guatopo National Park, Venezuela. B. Female flowers of *G. costaricensis* at the terminal nodes of a branch in Monteverde, Costa Rica. C. Longitudinal section of female (left) and male flowers of *P. bignoniacea.* Male flower is approx. 1.5 cm long. D. Portion of a female branch *G. spinulosa* cut at the start of its third fruiting cycle. Branch was cut below the series of leafless, sterile nodes produced before the female flowers are formed. The total leafless hanging branch was at least 2 m longer than shown. E. *Phyllostomus hastatus* taking fruit from a *G. spinulosa* branch in Guatopo. Ripe fruit that are green, are borne on pendulous branches, and have sweet, juicy pulp are typical of "chiropterchorous" plants (37, 38). F. Oviposition sites, marked by hardened exudate on surface of *G. spinulosa* fruit, caused by seed predator *Blepharoneura* species "A" in Guatopo National Park, Venezuela. G. *Anguria tabascensis* from Veracruz, Mexico. Male and female flowers and fruit in various stages of development. Thick yellow rind of ripe fruit is atypical for Guraniinae and may be an adaptation to dispersal by birds rather than bats.

Sex Expression

Gurania and *Psiguria* are erroneously described as predominantly or entirely dioecious (4, 7, 8, 18, 43). Cogniaux (4) recognized a few monoecious species of *Gurania* and *Psiguria*, but considered most dioecious. Without long-term, detailed observations the monoecious condition of *Gurania* and *Psiguria* is difficult to detect (5, 6). We examined more than 300 herbarium specimens and found no more than six clearly monoecious specimens. Only 55 had female flowers or fruit.

Several factors probably account for such inadequate collections of monoecious and female specimens. First, if a branch is bisexual, the male inflorescences are usually separated by a meter or more from the first female flowers and generally have completed flowering before the female flowers reach anthesis. Second, flowering branches are often produced high in the forest canopy and are inaccessible to most collectors. Female flowers are often particularly inaccessible since femaleness is related to large plant size (5, 6). Third, female flowers are relatively rare compared with male flowers in many populations during much of the year (5, 18).

Male flowers are borne in corymbs on long peduncles produced in the axils of normal leaves and tendrils; each flowering branch usually bears several inflorescences. One flower per male inflorescence generally opens every other day for several months. Individual flowers are often shed the day after anthesis (Figure 1); however, in some species the floral display is enhanced by retention of closed flowers for several days following anthesis.

Large vines that will produce female flowers generally begin a flowering cycle by producing male inflorescences, but male flowering usually stops before anthesis of female flowers. Initiation of the female phase of a flowering cycle is marked by pronounced changes in branch morphology. First, the terminal meristem of an otherwise normal, climbing, leafy branch turns downward. The growing branch no longer climbs; instead, it produces a series of nodes with increasingly reduced leaves and tendrils and hangs from supporting vegetation. In some species the leafless branch drops over 2 m before the first female flowers are produced. One to six female flowers are produced at each of several terminal nodes of such hanging branches (Figure 2). Sexual branches develop synchronously within individual ramets. A female flowering period lasts about a week, after which two to three weeks elapse before fruits mature. Some individuals flower repeatedly during a single "season," and a single branch can bear as many as four fruit crops. Successive flowering periods begin on the same branches immediately after fruits of the preceding cycle ripen, and are not preceded by male flowers.

Are these patterns adaptive? If so, what is the adaptive significance of intraseasonal protandry, dichogamy, size-related sex expression, and spatial segregation of sexes?

Intraseasonal protandry is common among cucurbits, including many members of the Melothrieae (14). All species in the Guraniinae that we have observed are protandrous. Although protandry may not have evolved in response to particular conditions affecting *Gurania* and *Psiguria*, it may represent an important preadaptation that encourages faithful visitation by pollenivorous *Heliconius* (17, 18). For example, production of flowers with abundant pollen may effectively entrain butterflies to visit a site; butterflies so habituated may then "accidentally" visit female flowers. Visitation to female flowers may be reinforced by the tactile similarity of male and female flowers, since the stigmas and anthers in both *Gurania* and *Psiguria* have similar shape and texture (Figure 2C). This similarity may represent a form of mimicry (18). Did floral "mimicry" actually evolve as an adaptation to pollination by *Heliconius*? If stigma/anther similarity evolved in response to selective pressures exerted by *Heliconius*, then *Helmontia* (if primitive) should not have "mimetic" structures. Unfortunately, female flowers of *Helmontia* have not yet been described.

If protandry does in fact encourage visitation by *Heliconius*, and if the mimicry hypothesis is correct, then why are sexes temporally isolated? Given the tenets of a "mimicry" hypothesis, female flowers should be visited more frequently if they are produced simultaneously with male flowers as long as male flowers are more abundant than female flowers. Yet, in many *Gurania* and *Psiguria* species, female flowers open only after male flowering on a given branch ceases. Since degree of dichogamy varies within and among species and varies temporally within the individuals of certain species, there is ample opportunity to test hypotheses concerning the adaptive significance of dichogamy in the genera. At least two sets of hypotheses, which are not mutually exclusive, should be explored.

First, selection for outcrossing might favor dichogamy. Species we have self-pollinated (using asynchronous ramets of cloned individuals in greenhouses) produce viable offspring. The effect of cross- versus self-pollination on the fitness of offspring should be determined.

Second, dichogamy may affect the behaviors of pollinators, which in turn have various effects on plant fitness. For example, *Heliconius* butterflies might be more likely to visit female flowers of individuals that produce male and female flowers simultaneously. If such individuals had higher seed set as a result of increased visitation by *Heliconius*, then selection might favor temporal overlap, not dichogamy. Since patterns of visitation by *Heliconius* vary from site to site (5), further comparative study might reveal some correspondence between degree of dichogamy and relative importance of *Heliconius* as a pollinator.

Different factors may have influenced the evolution of size-related sex expression (3, 5, 15, 16, 33, 39). For example, in some plants, a high metabolic cost of seed production may set a lower limit to the size (biomass) that will sustain female reproduction (2, 21, 34). Availability of nutrients and light also affects the size at which some plants change from male to female phase (21, 22, 34).

The threshold for sex change in *Gurania* and *Psiguria* is related to basal stem diameter, but not total biomass (5). Although basal stem diameters and biomass are positively correlated in the field, garden experiments helped dissociate the two. In a garden in Guatopo, Condon cloned short sections of stems of varying diameters and monitored flowering. Only one clone flowered during the experiment. It produced a shoot that followed the typical pattern of female branch development, culminating in the production of fruit and fertile seeds. The estimated total biomass of that vine was less than the estimated total biomass of the smallest flowering male measured in the field. Clearly, vines with basal diameters well below the female threshold have a total biomass that could sustain fruit production (5).

If biomass is not a limiting factor, do different levels of light intensity or nutrients influence sex expression? We do not know whether these factors affect sex expression of *Gurania* and *Psiguria* at or above "threshold size"; however, small vines that are below the threshold size and grow in open, sunny areas do not produce female flowers (5). Analyses of soil taken from the bases of both male and female plants show no conspicuous differences in nutrient content (5).

Condon (5) suggested that sex expression is influenced by a developmental change associated with basal stem diameter and that the significance of the size-sex relationship lies with some size-related aspect of the vines' biology other than biomass. In the Guraniinae, sexually dimorphic branch morphology may have been a preadaptation that influenced the evolution of size-related sex expression. Thus, the adaptive significance of size-related sex change may lie in the relative advantage of climbing to vines of different sizes.

Gurania and *Psiguria* typically grow to the top of the canopy or subcanopy, where branching usually occurs. High light intensity may be the resource for which these vines climb; photosynthetic rate increases with light availability (G. Montes U. and M. A. Condon, unpublished observation). Selection may favor vines that do not expend energy in nonclimbing, terminal female branches until such vines attain a stable canopy position in a favorable light climate. Since *Gurania* and *Psiguria* are long-lived, those individuals that delay female branch production until they attain a stable canopy position, at which they have relatively constant, abundant light, may have higher reproductive success than small individuals that produce female branches in response to high light intensity in an ephemeral microhabitat, such as a clearing. Large vines in the canopy can alternate fruit production with production of climbing, photosynthetic branches without being trapped in a quickly shaded environment that might severely limit both vegetative growth and consequent reproductive success. Since mortality is much higher among small vines than large vines, the rate of vegetative growth has an obvious effect on fitness (5). In the context of this hypothesis, it would be interesting to examine *G. megistantha* Donn. Sm. (Figure 1B), a cauliflorous species that does not sacrifice climbing ability for female branch production. Although there are no other documented cases of size-related sex expression among dicotyledons (15), we suggest that many supposedly dioecious species should be more closely examined.

Interspecific Interactions

Other organisms play important, if not crucial roles in the lives of plants. Interspecific interactions can have major effects on plant fitness and affect the evolution of traits ranging from branch architecture, floral morphology, and sex expression to phytochemistry. Plant characters, in turn, can affect the evolution of other organisms. The net result is a dynamic state of coevolution that both directs and is directed by community structure (18, 19).

Gurania and *Psiguria* are critical participants in such coevolving interactions. Essentially the same set of organisms is associated with *Gurania* and *Psiguria* throughout the Neotropics. Although names and numbers of species may change from site to site, certain taxa are consistently represented, e.g., tephritid fruit flies of the genus *Blepharoneura* may have radiated within the Guraniinae. We characterize components of the *Gurania* and *Psiguria* fauna according to their effects on plant fitness as mutualists, predators, or herbivores. Since some predators and herbivores are economically important pests, our observations may be useful in the development of biological control and crop breeding programs.

Mutualists

Mycorrhizal fungi.   Janos (29) found a highly significant difference in growth between *Gurania* seedlings grown in soil innoculated with mycorrhizal fungi and sibling seedlings grown in sterile soil. Plant height, leaf length, and number of leaves of plants with mycorrhizae were four to five times greater than those of plants grown in sterile soil. These data strongly suggest that mycorrhizal fungi may be very important although not obligatory mutualists of *Gurania* and *Psiguria*.

Pollinators.   In most areas we have studied, the principal visitors to most *Gurania* and *Psiguria* flowers are hummingbirds and/or *Heliconius* butterflies (5, 10, 18). Both hummingbirds and *Heliconius* butterflies are capable of learning the location and schedule of nectar production of different plant species and will return to plants as long as a reward is available. Individuals of some species of hummingbirds and *Heliconius* often follow a foraging circuit or "trapline" among successive flowers or clumps of flowers (13, 18).

The degree to which different species of hummingbirds or *Heliconius* visit a given plant species varies with a number of factors: flower morphology, amount of nectar produced, and characteristics of the habitats and communities in which the interacting species occur. Distinct types of hummingbirds or butterflies are generally associated with different types of *Gurania* and *Psiguria* species, depending on calyx tube length and nectar production.

Gurania species with long calyces (Figure 1B) or long calyx lobes (Figure 1A) and/or constricted calyx openings that make extraction of large amounts of pollen by *Heliconius* difficult (Figure 1C), are visited by "high reward trapliners" (13), such as *Threnetes ruckeri* Bourcier in Ecuador. These hummingbirds have long bills that allow them to reach nectar that is probably inconveniently held, if not inaccessible, to hummingbirds with shorter bills and to *Heliconius*. Several species of *Psiguria* have relatively small flowers of low nectar reward and in some sites are visited primarily by *Heliconius* (Figures 1F, G; 2C, G). A great number of *Gurania* and *Psiguria* species have "intermediate" flowers that often produce abundant nectar but have shorter calyces or calyx lobes and allow both nectar and pollen extraction by *Heliconius* and hummingbirds of all sizes.

The degree to which different *Gurania* and *Psiguria* species attract hummingbirds or *Heliconius* is generally related to the amount of nectar produced. Interestingly, female flowers tend to produce more nectar than male flowers (5). Since female flowers on a branch are available for a shorter continuous time period than male flowers, greater reward may more quickly entrain and maintain visitors. Similarly, the first male flowers produced on inflorescences of many *Psiguria* species are larger and produce more nectar than later flowers of the same inflorescences.

In addition to hummingbirds and *Heliconius*, other insects occasionally visit certain *Gurania* and *Psiguria* species. Regular visits by non-*Heliconius* butterflies are infrequent local events. In Guatopo and Corcovado, *Trigona* bees sometimes visit *P. racemosa* and *P. warscewiczii* and prevent other visitors from gaining access to the flowers. Croat (7) reports *Trigona* bees as visitors to *P. warscewiczii* on Barro Colorado Island, Panama. In Guatopo, a brilliant green anthophorid (*Ceratina laeta* Spin) occasionally visits *P. triphylla*.

How does pollination of *Gurania* and *Psiguria* compare with other cucurbits? Although few detailed field studies of wild cucurbits exist, floral morphology suggests that most cucurbits are pollinated by bees, moths, or other insects (28). No other Neotropical genus has brightly colored, ornithophilous flowers, but three other genera are associated with vertebrate pollinators. *Elateriopsis* (32), *Calycophysum* (24), and some species of *Cayaponia* (7) have large whitish flowers, said to be visited by bats.

Helmontia, apparently the closest relative of *Gurania* and *Psiguria*, may provide a clue to the initial evolution of brightly colored flowers from the pale flowers characteristic of so many other cucurbits. White, night-flowering *Helmontia* species have a fairly long, tubular calyx and produce abundant nectar. Moths are probably its most important pollinators; however, plants grown in a garden in Guatopo, outside their normal distribution and habitat, attracted hummingbirds at dawn, when nectar was still available and before the flowers were fully closed. Nectar might have been abundant at dawn because little nocturnal visitation occurred. In this situation, selection would favor the evolution of bright colors if they more readily catch the attention of animals such as hummingbirds, which quickly incorporate conspecifics within a regular foraging route. Such plants would be readily visited in a community in which hummingbirds are resource-limited.

What conditions favor the evolution of floral characters that attract one pollinator but exclude another? More specifically, why have characters that exclude *Heliconius* evolved and other characters such as minimal nectar production also evolved? We speculate that *Heliconius* species were initially parasites (and some species still are) within a system that involved hummingbirds as primary pollinators. *Heliconius* butterflies actively collect pollen (17, 18). Pollen feeding allows both increased longevity and continuous egg production (9). Although all *Heliconius* species feed on pollen, their mode of foraging and potential effectiveness as pollinators vary. For example, *H. doris* L. often stays on individual flowers for a long time and therefore visits few flowers. This behavior reduces the amount of pollen available to other potential pollinators of those flowers. Other *Heliconius* species, e.g., *H. ethilla* Godt., collect pollen and nectar from numerous flowers on different plants (10).

If *Heliconius* are faithful visitors and effect pollination and if hummingbirds are only occasional visitors and rarely visit conspecifics, then selection might favor characteristics (e.g., "extra" pollen) that promote visitation by *Heliconius* or that make flowers less attractive to hummingbirds (e. g., reduced nectar). Conversely, if *Heliconius* steal a great deal of pollen from male flowers without carrying it to female flowers, and if hummingbirds effectively pollinate, then selection would favor mechanisms that inhibit pollen loss via *Heliconius* (Figures 1A–C).

Experience in different localities gives us different impressions of the relative importance of *Heliconius* as pollinators. Although *Heliconius* butterflies are the most frequent visitors to all species of *Psiguria* in Costa Rica and Mexico, hummingbirds are the most frequent visitors to both *Gurania* and *Psiguria* in northern Venezuela, Ecuador, and Tobago. Interestingly, we have different impressions of the relative importance of hummingbirds and *Heliconius* in Trinidad. Condon's observations suggest that hummingbirds visit *P. triphylla* as frequently as *Heliconius*, yet Gilbert never observed a hummingbird visit this species. We both made our observations at the same site on Andrew's Trace in Trinidad (10). While our impressions concerning *P. triphylla* differ, we agree that hummingbirds are frequent visitors to *G. spinulosa*. More attention should be given not only to such differences in visitation frequency but also to differences in the quality of pollination achieved.

Dispersal agents.   Extensive observations of the fates of ripe fruit in Guatopo National Park show that bats are the only animals that consistently remove only ripe fruit and carry it away from the fruiting branch (5). At least two species of bats, *Phyllostomus hastatus* and *P. discolor*, feed on *G. spinulosa* fruit in Guatopo. Some *P. hastatus* individuals forage over large areas, often several kilometers away from diurnal cave roosts (36). *Phyllostomus discolor* also forages over large areas (23), probably using diverse flyways, which would increase the probability that bats drop seeds in different microsites (23). We have never seen evidence of active destruction of seeds by bats. Condon and W. S. Perkins caught a *P. discolor* at a fruiting branch, caged the bat, and fed it ripe *G. spinulosa* fruit. The bat "gutted" the fruit, devoured the juicy pulp, but did not ingest the seeds. This observation is consistent with van der Pijl's observation (37) that bats spit out solid fruit material.

Predators

Seeds.   Both vertebrates and insects prey upon seeds. Mice eat both mature and immature seeds and can destroy entire seed crops. Squirrels and monkeys are also destructive. Saltators, which are large tropical finches, occasionally destroy ripe fruit by tearing holes in them and sucking out the juice, leaving seeds inside the fruit that will either fall (without being dispersed) or dry and die on the branch.

Several species of insects are even more devastating predators. The common pickleworm, *Diaphania nitidalis* Stoll (Pyralidae), is a serious pest of cultivated cucurbits (11) and attacks both *Gurania* and *Psiguria* at all study sites. In Guatopo the rate of attack increases early in peak fruit production season but later decreases markedly. When pickleworm larvae attack fruit, they typically chew a section out of the peduncle and fasten the fruit to the peduncle with web. This behavior prevents aborted fruit from falling and so facilitates movement of larvae from fruit to fruit. It may also affect translocation of defensive compounds. Larvae feed on fruit pulp and young developing seeds. Attacked fruit eventually rot on the vine.

Phthia lunata Fabr. (Coreidae) feeds on seeds in Guatopo and Trinidad. A fruit that has been attacked remains attached to the branch until the time it should mature. Then, unlike a normal mature fruit, which stays attached to the branch until dispersed, it falls to the ground and rots. Such fruits contain no viable seeds.

Blepharoneura species "A" (Tephritidae), feeds on seeds in Venezuela and Trinidad and causes more predispersal seed mortality in Guatopo than any other predator (5). This species oviposits directly into seeds a week or two before the fruit ripens, leaving a conspicuous bubble of sap on the surface of the fruit (Figure 2F). Larvae develop inside seeds and eventually crawl into the soft rind surrounding the juicy, ripe pulp. Tephritid damage does not cause abnormal fruit development or rot.

Postdispersal seed predation also occurs. Experiments in Guatopo showed that both insects and vertebrates prey upon seeds on the ground (M. A. Condon, unpublished observation).

Flowers. Pickleworms, perhaps other *Diaphania* species, and *Blepharoneura* eat flowers of *Gurania* and *Psiguria* and can have considerable effect on plant fitness. One *Diaphania* larva can devour all of the flowers on an inflorescence. *Blepharoneura* species have various effects on flowers; some cause little or no functional damage, others make nectar unpalatable, and still others cause abortion of female flowers. In Guatopo male and female flowers are attacked by different *Blepharoneura* species.

Blepharoneura species "C" feeds on female flowers of *G. spinulosa* in Guatopo, but larvae generally do not affect fertilization or fruit set. Maggots usually feed only on calyx tissue and drop to the ground to pupate, but some maggots destroy young buds, and others burrow into ovary tissue and cause abortion. Predators of the larvae are frequently attracted to damaged ovaries. Destructive larvae are patchily distributed; most plants are not affected, but numerous flowers of some plants are destroyed.

In Guatopo and Trinidad, *Blepharoneura* species "B" feeds on male flowers of *G. spinulosa*. Flies oviposit on the young, often green buds. Larvae feed primarily on the succulent calyx tissue, but sometimes feed on anther septae. Nectar production is not obviously affected by maggots. In contrast to *G. spinulosa*, male flowers of *G. acuminata* are never touched, even though they may occur next to *G. spinulosa* flowers that are attacked. Larval transfer experiments show that maggots can feed and reach maturity on *G. acuminata* male flowers (W. S. Perkins and M. A. Condon, unpublished observation).

A *Blepharoneura* species not found in Guatopo attacks the male flowers of *G. costaricensis* in Costa Rica (Figure 1D) and those of a *Psiguria* species (syn. *Anguria tabascensis* Donn. Sm.) in Mexico. Observations of pollinator activity on *P. warscewiczii* (D. Murawski, pers. comm.) suggest that flowers that contain tephritid maggots repel *Heliconius* butterflies.

Herbivores

In contrast to predators that kill seeds, herbivorous insects rarely kill plants, but they can affect growth by feeding on vegetative tissue.

Leaves. Both *Acalymma* and *Diabrotica* (Chrysomelidae) attack *Gurania* and *Psiguria*. In Costa Rica *Diabrotica* species are common on *Psiguria*. *Acalymma bivittula* Kirsch and *Diabrotica* species were abundant in a small garden of *Gurania* and *Psiguria* in a clearing in Guatopo but were relatively rare on plants in the forest understory. Both microhabitat and chemical cues probably affect pest distribution (1, 27).

Other organisms that eat *Gurania* and *Psiguria* leaves in Guatopo include *Diaphania hyalinata* L., *D. superalis* Quene, *D. niveocilia* Lederer, *Pseudoplusia includens* Walker (Noctuidae), an unidentified tettigoniid, and an unidentified snail, which is common in populations of *Psiguria umbrosa*. Early instar *Diaphania* larvae often feed gregariously and consume entire leaves of *G. acuminata* and *P. umbrosa* in the field.

In Costa Rica, two species of coreid bugs feed on *Psiguria* leaves: gregarious nymphs and adults of *Paryphes blandus* Horvath (Coreidae) in western Costa Rica (44), and an unidentified congener in the Atlantic lowlands.

Perhaps the most unusual leaf feeders are adults of various *Blepharoneura* species, which rasp the surfaces of young leaves. When mature, these leaves are laced with holes. This characteristic damage was observed at all study sites.

Stems. In Guatopo, three insects are stem feeders: *Piezosternum subulatum* (Thunberg) (Pentatomidae), *Blepharoneura* species "D," and unidentified cerambycid beetles. As many as 40 *P. subulatum* adults may remain on a stem for a week or more with no observed ill effects. In contrast to the apparently trivial effects of a mass of pentatomids sucking on a stem, one *Blepharoneura* larva can kill a shoot. Larvae of this species occur in young shoots of *G. acuminata* near the floor of primary forest in Guatopo. They feed only on young stems and are a primary cause of stem death.

Cerambycid beetles are probably another major source of meristem mortality. Stems with "clipped" tips are common, and often several crescent shaped scars precede the severed branch end. Such damage is characteristic of cerambycids. In addition to these stem specialists, pickleworms (*Diaphania nitidalis*) occasionally feed on stems, as they do on cultivated cucurbits (11).

Petioles, pedicels, and peduncles. In Guatopo, at least four different species of insect feed on *Gurania* peduncles and pedicels: an unidentified microlepidopteran (probably Pteriphoridae); *Hyphinoe asphaltina* (Fairmaire) (Membracidae); and *Blepharoneura* species.

Microlepidopteran larvae spin webs on the petioles of leaves, around the pedicels of male inflorescences, and around the bases of peduncles. Larvae feed on the tissue and make a troughlike hole beneath the web. The damage does not kill leaves or affect female flowers or fruit; however, male inflorescences may die as a result of microlepidopteran damage. Since adults have not been reared from all three microsites, we do not know if the similar larvae are the same species. All three types of damage occur on both *Gurania* species in Guatopo. We do not know if *Psiguria* is similarly attacked.

In Guatopo, the green nymphs of the membracid *Hyphinoe asphaltina* are common on peduncles and occasionally on stems of *Gurania*. Damage they cause is minimal; we see no effect on plants. Exoskeletons of the same or very similar membracids were found on *Gurania* in Rio Palenque, Ecuador. A similar, if not the same species also occurs on *P. warscewiczii* in Corcovado, Costa Rica (Figure 1G).

Adult *Blepharoneura* rasp the tips of pedicels. Sufficient tissue is sometimes ingested to make the contents of the abdomen appear noticeably green; however, the behavior has essentially no effect on the plant since the flies feed on pedicels after flowers have fallen.

Epilogue

We hope our effort reflects, at least in part, the predictable complexity that characterizes animal/plant interactions, especially in the tropics. To learn more about the role of such interactions in plant evolution, we need more long-term natural history studies and more thorough systematic studies that will generate phylogenies essential for understanding the history of characters and their adaptive significance.

Acknowledgments

We thank Luis Escalona, Superintendent of Guatopo National Park, José Rafael Garcia, Director, and José Ramon Orta, Resident Chief of the National Parks Institute of Venezuela for making work in Guatopo possible. The Costa Rican National Parks Service and staff at Corcovado National Park made work in Costa Rica possible. We give special thanks to Bill Perkins for providing more than two years of invaluable field assistance in Guatopo. Darlyne Murawski and Gary McCracken kindly provided unpublished observations. Charles Jeffrey provided both unpublished information and valuable encouragement. Richard Foote generously provided tentative identification of *Blepharoneura* specimens and holds vouchers.

Determinations of insects were provided by R. J. Gagné, D. C. Ferguson, R. W. Poole, R. White, G. Steyskal, J. L. Herring, J. P. Kramer, P. M. Marsh, T. J. Henry, and A. S. Menke of the Systematic Entomology Laboratory, USDA, Beltsville, Maryland.

NSF dissertation improvement grant DEB 76–80492 supported Condon's work in Guatopo from 1977 to 1979. An OAS Fellowship supported her work from 1979 to 1981. CONICIT Subvention s4–140 (Venezuela) to Charles Jeffrey and Balthazar Trujillo supported travel for three months (October-December, 1977). NSF DEB 79–06033 supported Gilbert's project in Corcovado National Park.

Literature Cited

1. Bach, C. E. 1980. The effects of plant density on diversity on the population dynamics of a specialist herbivore, the striped cucumber beetle, *Acalymma vittata* (Feb.) Ecology 61: 1515–1530.
2. Bierzychudek, P. 1981. Assessing "optimal" life histories in a fluctuating environment: the evolution of sex changing by jack-in-the-pulpit. Amer. Nat. 123: 829–840.
3. Charnov, E. L., and J. Bull. 1977. When is sex environmentally determined? Nature 266: 828–830.
4. Cogniaux, A. 1916. Cucurbitaceae: Fevilleae et Melothrieae. *In* A. Engler, ed., Pflanzenreich 66 (IV. 275. I.). Wilhelm Engelmann, Leipzig.
5. Condon, M. 1984. Reproductive biology, demography, and natural history of neotropical vines *Gurania* and *Psiguria* (Guraniinae, Cucurbitaceae): a study of the adaptive significance of size related sex change. Ph.D. dissertation, Univ. Texas, Austin.
6. Condon, M., and L. E. Gilbert. 1988. Sex expression of *Gurania* and *Psiguria* (Cucurbitaceae): Neotropical vines that change sex. Amer. J. Bot. 75: 875–884.
7. Croat, T. B. 1978. Flora of Barro Colorado Island. Stanford Univ. Press, Palo Alto, CA.
8. Dodson, C. H., and A. Gentry. 1978. Flora of Rio Palenque Science Center. Selbyana 4: 1–628.
9. Dunlap-Pianka, H. L., C. L. Boggs, and L. E. Gilbert. 1977. Ovarian dynamics in Heliconiine butterflies: programmed senescence versus eternal youth. Science 197: 487–490.
10. Ehrlich, P. R., and L. E. Gilbert. 1973. Population structure and dynamics of the tropical butterfly *Heliconius ethilla*. Biotropica 5: 69–82.
11. Elsey, K. D. 1981. Pickleworm: survival, development, and oviposition on selected hosts. Ann. Entomol. Soc. 74: 96–99.
12. Ewel, J. J., A. Madriz, and J. A. Tosi, Jr. 1976. Zonas de vida de Venezuela. Ministerio de Agricultura y Cria. Fondo Nacional de Investigaciones Agropecuarias, Caracas.
13. Feinsinger, P., and R. K. Colwell. 1978. Community organization among neotropical nectar feeding birds. Amer. Zool. 18: 779–795.

14. Frankel, R., and E. Galun. 1977. Pollination Mechanisms, Reproduction and Plant Breeding. Springer-Verlag, New York.
15. Freeman, D. C., K. T. Harper, and E. L. Charnov. 1980. Sex change in plants: old and new observations and new hypotheses. Oecologia 47: 222–232.
16. Ghiselin, M. T. 1969. The evolution of hermaphroditism among animals. Quart. Rev. Biol. 4: 189–208.
17. Gilbert, L. E. 1972. Pollen feeding and reproductive biology of *Heliconius* butterflies. Proc. Natl. Acad. USA 69: 1403–1407.
18. Gilbert, L. E. 1975. Ecological consequences of a coevolved mutualism between butterflies and plants. *In* L. E. Gilbert and P. H. Raven, eds., Coevolution of Animals and Plants. Univ. Texas Press, Austin.
19. Gilbert, L. E. 1980. Food web organization and the conservation of neotropical diversity. *In* M. Soulé and B. Wilcox, eds., Conservation Biology: An Evolutionary-Ecological Perspective. Sinauer, Sunderland, MA.
20. Gondelles, A. R., J. R. Garcia A., and J. Steyermark. 1977. Los Parques Nacionales de Venezuela. INCAFO, Madrid.
21. Gregg, K. B. 1973. Studies on the control of sex expression in the genera *Cycnoches* and *Catasetum*, subtribe Catasetinae, Orchidaceae. Ph.D. dissertation, Univ. Miami, Coral Gables, FL.
22. Gregg, K. B. 1975. The effect of light intensity on sex expression of species of *Cycnoches* and *Catasetum* (Orchidaceae). Selbyana 1: 101–113.
23. Heithaus, E., T. Fleming, and P. Opler. 1975. Foraging patterns and resource utilization in seven species of bats in a seasonal tropical forest. Ecology 56: 841–854.
24. Heywood, V. H. 1978. Flowering Plants of the World. Mayflower Books, New York.
25. Holdridge, L. R., W. C. Grenke, W. H. Hatheway, T. Liang, and J. A. Tosi, Jr. 1971. Forest Environments in Tropical Life Zones: A Pilot Study. Pergamon Press, Oxford.
26. Howard, R. A. 1973. The enumeratio and selectarum of Nicolaus von Jacquin. J. Arnold Arb. 54: 437–438, 440–442.
27. Howe, W. L., and A. M. Rhodes. 1976. Phytophagous insect associations with *Cucurbita* in Illinois. Environ. Entomol. 5: 747–751.
28. Hurd, P. D., Jr., E. G. Linsley, and T. W. Whitaker. 1975. Squash and gourd bees (*Peponapsis, Xenoglossa*) and the origin of the cultivated *Cucurbita*. Evolution 25: 218–234.
29. Janos, D. P. 1980. Vesicular-arbuscular mycorrhyizae affect lowland rainforest plant growth. Ecology 61: 151–162.
30. Jeffrey, C. 1962. Notes on Cucurbitaceae, including a proposed new classification of the family. Kew Bull. 15: 337–371.
31. Jeffrey, C. 1978. Further notes on Cucurbitaceae. IV. Some New World taxa. Kew Bull. 33: 347–380.
32. Jeffrey, C. 1980. A review of the Cucurbitaceae. J. Linn. Soc. Bot. 81: 233–247.
33. Lloyd, D. G. 1979. Parental strategies of angiosperms. New Zealand J. Bot. 17: 595–606.
34. Maekawa, T. 1924. On the phenomenon of sex transition in *Arisaema japonica* Bl. J. Coll. Agric. Hokkaido Imp. Univ. 13: 217–305.

35. Mather, K. 1940. Outbreeding and separation of the sexes. Nature 145: 484–486.
36. McCracken, G., and J. Bradbury. 1981. Social organization and kinship in the polygynous bat *Phyllostomus hastatus*. Behav. Ecol. Sociobiol. 8: 11–34.
37. Pijl, L. van der. 1957. The dispersal of plants by bats. Acta Bot. Neerl. 6: 618–641.
38. Pijl, L. van der. 1972. Principles of Dispersal in Higher Plants. Springer-Verlag, New York.
39. Policansky, D. 1981. Sex choice and the size advantage model in jack-in-the-pulpit (*Arisaema triphyllum*). Proc. Natl. Acad. USA 78: 1306–1308.
40. Steyermark, J., and O. Huber. 1978. Flora del Avila. Vollmer Foundation and MARNR, Caracas.
41. Stout, A. B. 1928. Dichogamy in flowering plants. Bull. Torrey Bot. Club 55: 141–153.
42. Thomson, J. D., and S. C. H. Barrett. 1981. Temporal variation of gender in *Aralia hispida* Vent. (Araliaceae). Evolution 35: 1094–1107.
43. Wunderlin, R. P. 1978. Cucurbitaceae. *In* R. E. Woodson and R. W. Shery, eds., Flora of Panama. Part IX. Ann. Missouri Bot. Gard. 65: 285–366.
44. Young, A. M. 1980. Notes on the interaction of the neotropical bug *Paryphes blandus* Horvath (Hemiptera: Coreidae) with the vine *Anguria warscewiczii* Hook. f. (Cucurbitaceae). Brenesia 17: 27–42.

14 | Coevolution of the Cucurbitaceae and Luperini (Coleoptera: Chrysomelidae): Basic and Applied Aspects

Robert L. Metcalf and A. M. Rhodes

ABSTRACT. The bitter and toxic cucurbitacins of the Cucurbitaceae are allomones that arose evolutionarily to protect these plants from herbivore attack. A large group of rootworm beetles use the cucurbitacins as kairomones promoting host selection through arrest and compulsive feeding. These beetles, including the common cucumber beetles and corn rootworms, sequester cucurbitacins as allomones to deter predators. Dependable *Cucurbita* hybrids with high cucurbitacin content have been developed as sources of these compounds for basic and applied research.

The tetracyclic triterpenoid cucurbitacins are intensely bitter and toxic secondary plant compounds that are characteristically present in the majority of the feral species of Cucurbiteae (18, 22, 35, 36, 51). From the viewpoint of chemical ecology the cucurbitacins play a comprehensive role, acting as both allomones and kairomones in regulating the attacks of invertebrate and vertebrate herbivores (6, 7, 9, 21, 24, 25, 43, 44, 50).

The evolutionary scenario for the behavioral and ecological interactions among Cucurbitaceae, cucurbitacins, and herbivores has been portrayed by Metcalf (24, 25) and Price (33) and can be summarized in the following way. Pre- or ancestral Cucurbitaceae carrying nonexpressive genes (*bi*) for cucurbitacins lacked bitterness and were heavily preyed upon by herbivores. Mutations to dominant alleles (*Bi*) formed bitter and toxic cucurbitacins that deterred herbivores and led to the evolution and diversification of the family. These events were countered by mutant ancestral Luperini rootworm beetles that developed detoxification and excretion pathways to neutralize the harmful effects of cucurbitacins, then expanded into new ecological niches by developing specific receptors for the detection of cucurbitacins and finally developed high blood and tissue levels of cucurbitacin conjugates for defense against predators.

167

The cucurbitacins are useful probes for exploring both the basic and applied aspects of the extensive coevolutionary association between plants and insects in the Cucurbitaceae. The development of dependable *Cucurbita* hybrids with relatively high yields of cucurbitacins make it possible to explore the applied ecology of the use of these kairomones to manipulate the behavior of the Diabroticite beetles that are important insect pests.

Cucurbitacins in the Cucurbitaceae

Cucurbitacins (Cucs), as the name suggests, are peculiarly associated with the Cucurbitaceae; they have been characterized in more than 100 species in at least 30 genera (18, 22, 29, 35). Cucurbitacins are also found in a few genera of Begoniaceae, Brassicaceae, and Datiscaceae (8, 32) and the Violiflorae (49), and occur in a few species of Euphorbiaceae and Scrophulariaceae (12). At least 20 chemically different cucurbitacins have been characterized from plants (18, 22).

Cucurbitacins typically are found in the roots, stems, leaves, and fruits of species of Cucurbitaceae. The concentrations in the roots increase with age, and in perennial plants of *Citrullus* and *Acanthosicyos* can reach levels of 1.0% (35). In 18 species of *Cucurbita*, Cucs B-D were detected in the roots of 7 species, and Cucs E-I in the roots of 6 species, at levels up to 0.4% (29). In the leaves of the same sample, Cucs B-D were found in 7 species, at concentrations up to 0.059% in *C. lundelliana* L. H. Bailey, and Cucs E-I in 6 species, at levels up to 0.1% in *C. okeechobeensis* L. H. Bailey (29). Rapidly growing young leaves of *Citrullus lanatus* (Thunb.) Matsum. & Nakai and *C. ecirrhosus* Cogn. contained only about 0.01% cucurbitacins, but the concentration ranged from 0.1% to 0.3% by the end of the vegetative season (35).

The fruits of a variety of taxa of Cucurbitaceae contain high concentrations of cucurbitacins, more than 1.0% in the fruits of *Citrullus colocynthis* (L.) Schrad., *C. ecirrhosus*, *Cucumis angolensis* Hook. f. ex Cogn., *C. longipes* Hook. f. (= *C. anguria* var. *longipes* (Hook. f.) Meeuse), *C. myriocarpus* Naud., and *C. sativus* L. (36). Of the 18 species of *Cucurbita* examined, the fruits of 7 contained Cucs B-D, at concentrations up to 0.31% in *C. andreana* Naud., and the fruits of 5 contained Cucs E-I, at levels reaching 0.23% in *C. foetidissima* HBK (29).

Of the 20 or so cucurbitacins thus far identified, Cuc B is the predominant form. It is found in about 91% of all species characterized and is followed by Cuc D (69%), Cucs G and H (47%), Cuc E (42%), Cuc I (22%), Cucs J and H (9%), and Cuc A (7%). Cucs C, F, and L were each found only in a single species (35, 36). Cuc B and Cuc E appear to be the primary Cucs, and the other Cucs are formed by enzymatic processes occurring during plant development and maturation. Cuc B can be metabolized to Cucs A, C, D, F,

G, and H and is characteristic of *Coccinia*, *Cucumis*, *Lagenaria*, and *Trochomeria* (36). Similarly, Cuc E can be metabolized to Cucs I, J, K, and L and is characteristic of *Citrullus* (35). In *Cucurbita* there are two groups of species characterized by either Cuc B or Cuc E (29).

The cucurbitacins of *Cucumis*, *Lagenaria*, and *Acanthosicyos* are present as free aglycones. In most species of *Citrullus*, *Echinocystis*, *Coccinia*, and *Peponium*, however, the cucurbitacins are present as glycosides (36). In *Cucurbita* cucurbitacins are present as aglycones in most species (29), but glycosides are found in *C. cylindrata* L. H. Bailey, *C. foetidissima*, *C. palmata* S. Wats, and *C. texana* A. Gray (3, 29). The presence of glycosides is related to the absence of β-glucosidase (elaterase), which may also be sequestered in intact plant tissues and released by crushing (13, 42).

The following generalizations can be made about the variable distribution of the complex of cucurbitacins in the Cucurbitaceae: 1) Cuc B and E are the parent substances found most widely and the other cucurbitacins, generally present in much lower quantities, are degradative products. 2) Cuc D is always associated with Cuc B, and Cuc I is always associated with Cuc E. 3) Cucs G and H are always associated with Cucs B and D. 4) Cucs J and K are always associated with Cucs E and I. 5) Cuc A is always associated with Cuc B. 6) Cuc C is singular and occurs alone in *Cucumis sativus* (15, 37). 7) Cucs B and E sometimes occur alone.

At least five independent genes are said to regulate the biosynthesis of cucurbitacins (40). Gene *Bi* regulates synthesis in *Cucurbita* seedlings. In *Citrullus* gene *su* apparently suppresses synthesis in fruits, and gene mo^{Bi} determines whether the cucurbitacin exists as a free aglycone or as a glycoside. In addition there is a gene that controls quantity and a gene that determines the chemical nature of the cucurbitacin formed.

Bitterness in the Cucurbitaceae was originally thought to be regulated by a single dominant gene *Bi* (10, 37) and that plants with nonbitter seedlings (*bibi*) did not synthesize cucurbitacins. However, later evidence indicated that nonbitter fruits could develop from bitter seedlings (37), thus suggesting that a single recessive suppression gene (*su*) prevented bitterness in watermelon fruits (5). Organospecific genes appear to control the qualitative and quantitative formation of cucurbitacins in roots, leaves, blossoms, and fruits (35). A single modifier gene Mo^{bi} that acts only in the presence of *Bi* and su^+ alleles apparently controls the quantity of Cuc E-glycoside (elaterinide) formed in bitter fruit (5).

Cucurbitacins as Allomones for the Cucurbitaceae

There is no ambiguity about the role of the cucurbitacins. They are semiochemicals that protect cucurbits against herbivore attacks. This is immediately apparent to anyone tasting a bitter squash, cucumber, or melon.

Cucurbitacins are the most bitter substances known and can be detected by humans at dilutions of 1 ppb (27). Trace amounts produce an almost paralytic response in lips and mouth and a persistent aftertaste. Moreover, the cucurbitacins are extremely toxic to mammals. Mice are affected intraperitoneally by Cucs A, B, and C at 1.2, 1.1, and 6.8 mg/kg, respectively (11); and orally by Cuc I and Cuc E glycoside at 5.0 and 40.0 mg/kg, respectively (48). Cattle and sheep that fed on bitter *Cucumis* and *Cucurbita* fruits during drought conditions have been severely poisoned (50), and outbreaks of severe human poisoning have resulted from ingestion of the fruits of *Cucurbita* cultivars that had reverted to the heterozygous *Bibi* state for bitterness (14, 18, 20, 41).

The cucurbitacins have been shown to be feeding deterrents for a number of arthropods, including the leaf beetles *Phyllotreta nemorum* L., *P. undulata* Kuts., *P. tetrastigma* Com., *Phaedon cochleariae* F., *P. cruciferae* Goeze, and *Cerotoma trifurcata* Forster, the stem borer, *Margonia hyalinate*, and red spider mites (10, 27, 30).

Cucurbitacins as Kairomones for Rootworm Beetles

The coevolutionary relationship between plants of the Cucurbitaceae and the rootworm beetles of the tribe Luperini (Coleoptera:Chrysomelidae) is a remarkable example of chemical ecology. This relationship has been effected by the cucurbitacins acting secondarily as kairomone cues for host plant selection by the beetles (6, 21, 24, 25, 27, 44). The scope of this association is presently worldwide. The Luperini comprise two very similar but geographically isolated subtribes: the Diabroticina of the New World, which contains about 933 species radiating from tropical America, and the Aulacophorina of the Old World, which contains about 535 species radiating from tropical Australasia (23, 24, 52). Maulik (23) described the similarity between these two large taxa of phytophagous beetles: "In the old world *Aulacophora* represents *Diabrotica*. . . . In larval, pupal, and adult structures, in breeding habits and in food plants, there is a remarkable resemblance between the two genera." The present distributions of the two subtribes of the Luperini was mapped by Metcalf (24).

Luperini beetles have a specific affinity for the Cucurbitaceae. These plants are recorded as the hosts for more than 50 species (4, 23, 24, 47, 52) and compose more than 80% of the available host records. The common names of notable pest species attest to the significance of this relationship. Among the Diabroticina are *Diabrotica balteata* LeConte, banded cucumber beetle; *D. (Paranapiacaba) connexa* LeConte, saddled cucumber beetle; *D. tibialis* Jacoby, painted cucumber beetle; *D. speciosa* Germar, cucurbit beetle; *D. undecimpunctata undecimpunctata* Mannerhein, western spotted cucumber beetle; *D. u. howardi* Barber, spotted cucumber beetle; *Acalymma trivittatum* Mannerhein, western striped cucumber beetle; and *A. vittatum* Fab., striped cucumber beetle. The Aulacophorina include *Aulacophora abdominalis* Fab., plain pumpkin beetle; *A. femoralis* Mots., cucurbit leaf beetle; *A. foveicollis* Lucas, red pumpkin beetle; and *A. hilaris* Boisd., pumpkin beetle.

The singular host plant association of Luperini beetles is cemented by the presence of cucurbitacins. This is illustrated by two studies relating cucurbitacin content of *Cucurbita* cotyledons to damage by rootworm beetles. The damage was rated on a five-point scale. For *Aulacophora foveicollis* attacking *Cucurbita moschata* (Duch. ex Lam.) Duch. ex Poir. cultivars there was a highly significant correlation (P = 0.001) between cucurbitacin content and feeding damage ($r = 0.62$, $n = 32$, $s = 0.14$). It was concluded that "low cucurbitacin content appeared to impart resistance" (31). In a similar study of *D. u. howardi* and *Acalymma vittatum* attacking *Cucurbita pepo* L. cultivars, there was a highly significant correlation (P = 0.001) between beetle damage and cucurbitacin content ($r = 0.78$, $n = 12$, $s = 0.20$). It was concluded that there was a 'strong positive correlation between seedling cucurbitacin content and Diabroticina beetle attacks" (15). Similar correlations were found between the average numbers of Diabroticina beetles feeding in crumpled leaves or sliced fruits of a variety of *Cucurbita* species and their total cucurbitacin content (29). For *D. u. howardi* the correlations for leaves were $r = 0.74$, $n = 16$, $s = 0.2$, P = 0.001; those for fruits were $r = 0.70$, $n = 11$, $s = 0.24$, P = 0.01. For *D. virgifera virgifera* Le Conte the correlations for leaves were $r = 0.64$, $n = 16$, $s = 0.24$, P < 0.01. For fruits they were $r = 0.58$, $n = 11$, $s = 0.27$, P = ca 0.05.

Diabroticina Beetle Sensitivity to Cucurbitacins

Cucurbitacins extracted from *Cucurbita* species by chloroform can be separated by thin-layer chromatography (TLC) on silica gel, using solvents such as chloroform:methanol, 95:5 (27, 29). These TLC plates represent the spectrum of cucurbitacins present in the plant. The cucurbitacins attractive to the various species of rootworm beetles can be identified by exposing the developed chromatograms for several days in cages containing about 100 rootworm beetles. The beetles eat the silica gel areas where cucurbitacins are present and produce "beetle prints" (15, 27, 29). This bioassay for cucurbitacins is both qualitative, by the R_f of the eaten spot compared with standard purified cucurbitacins, and semiqualitative, by the size of the area eaten. As shown in Table 1, the beetle bioassay using *D. u. howardi* is sensitive to nanogram quantities of Cuc B.

Table 1. Limit of response (LR) of Diabroticina beetles to pure cucurbitacins

Species	LR in μg of cucurbitacin							
	B	D	E	F	G	I	L	E gly
Diabrotica balteata	0.01	—	—	10.0	3.0	5.0	—	0.1
D. cristata	0.1	—	0.3	—	—	—	—	—
D. barberi	0.1	—	0.3	—	—	—	—	—
D. undecimpunctata howardi	0.001	0.03	0.01	1.0	3.0	0.1	0.01	0.05
D. undecimpunctata undecimpunctata	0.003	—	0.03	—	—	—	—	—
D. virgifera virgifera	0.01	0.1	0.3	0.1	3.0	0.3	1.0	0.03
Acalymma vittatum	0.3	—	10.0	—	—	—	—	50.0

Source: From Metcalf (24), reprinted with permission of the *Bulletin of the Illinois Natural History Survey.*

The beetle print bioassay technique has been used to characterize the spectrum of cucurbitacins present in the roots, leaves, and fruits of 18 species of *Cucurbita* (27, 29). The nature of the cucurbitacins present and the observed feeding responses of the beetles agreed reasonably well with evolutionary groupings in *Cucurbita* based on numerical taxonomy (38), cross-compatibilities (2), and isozyme analysis (34). On the basis of beetle prints, *Cucurbita* species fell into two principal groups corresponding to the presence of Cucs B-D or E-I. Among the species with Cucs B-D, *C. andreana* Naud. and *C. ecuadorensis* Cutler & Whitaker form one subgroup and *C. gracilior* L. H. Bailey, *C. palmeri* L. H. Bailey, and *C. sororia* L. H. Bailey another. *Cucurbita pedatifolia* L. H. Bailey stands apart, as does *C. lundelliana* L. H. Bailey, which has been allied with *C. martinezii* L. H. Bailey and *C. okeechobeensis* (Small) L. H. Bailey. The latter two species, however, are characterized by having Cucs E-I. Among the species with Cucs E-I, they form one subgroup. Another subgroup is composed of *C. cylindrata* and *C. palmata*. *Cucurbita foetidissima* stands apart. *Cucurbita texana*, which contains Cuc E-glycoside, is distinctive in its beetle print. Chromotograms of the five cultigens, *C. ficifolia* Bouché, *C. maxima* Duch. ex Lam., *C. mixta* Pang., *C. moschata*, and *C. pepo*, showed no discernable beetle feeding.

The beetle bioassay was used with five species of Diabroticina beetles to demonstrate an absence of qualitative differences among their feeding patterns on a spectrum of cucurbitacins, including B, C, D, E, I, and E-glycoside. Almost identical beetle prints were obtained with *Acalymma vittatum*, which is polyphagous on Cucurbitaceae and Fabaceae, *D. u. howardi*, which is polyphagous on Cucurbitaceae, Fabaceae, Convolvulaceae, and Poaceae, and the corn rootworms, *D. barberi* Smith & Lawrence and *D. virgifera virgifera*, which are essentially monophagous Poaceae specialists. *Dibrotica cristata* Harris, the larvae of which feed only on the roots of native prairie grasses, especially big bluestem, *Andropogon gerardii* Vitm. (53), also fed avidly on the TLC plates containing cucurbitacins and produced very similar beetle prints (24, 25, 27, 29).

These results demonstrate that since the divergence of *Acalymma* and *Diabrotica* (19, and unpublished research), there has been little change in the sensitivity of the various beetle species to the cucurbitacins over about 2.7 genetic distances or approximately 45 million years B.P. The demonstration that a functional cucurbitacin receptor is present in these and many other species of Luperini suggests that the entire group of rootworm beetles originally coevolved with the Cucurbitaceae and that preferences for other hosts, such as corn, must have been relatively recent (24, 47).

Limits of Diabroticina Species Response to Pure Cucurbitacins

The ultimate sensitivity of the Diabroticina species to pure crystalline cucurbitacins was determined by progressively decreasing the amounts of the various cucurbitacins present on silica gel TLC plates, thereby determining the limit of response (LR in µg) or the minimal amount producing a detectable feeding response. The limit-of-response values (Table 1) for seven species of rootworm beetles to eight cucurbitacins (24, 25) demonstrate the relative degree of complementarity of the cucurbitacins to the specific receptors on the maxillary palpi of the beetles (25, 27). The information in Table 1 shows that there are substantial differences in the LR values for the various Diabroticina beetles to the individual cucurbitacins. These values have both evolutionary and behavioral significance (24, 25). *Dibrotica u. undecimpunctata* and *D. u. howardi* consistently detected Cuc B at 0.001– 0.003 µg and Cuc E at 0.01–0.03 µg on the silica gel plates, while the LR values for the other species were somewhat higher (24, 25). This suggests that of the species investigated, these are the most primitive in evolutionary patterns.

A number of conclusions can be drawn from these data. 1) Cuc B was consistently detected in the lowest amount and is probably the parent Cuc to which Diabroticina receptors are attuned. 2) Cuc B was consistently detected at levels of about 0.1 that of Cuc E. 3) The acetoxy Cucs B and E were detected at levels about 0.1 those of the corresponding desacetoxy Cucs D and I, respectively. 4) Saturation of the desacetoxy Cucs at $C_{23}=C_{24}$ double bond (Cuc L) had little effect on the level of detection. 5) Sensitivity to the 2-OH, 3-C=O Cuc D was greater than to the 2-OH, 3-OH Cuc F. 6) The *D. undecimpunctata* subgroup (*D. balteata*, *D. u. unde-cimpunctata*, and *D. u. howardi*) is more sensitive to cucurbitacins than the *D. virgifera* subgroup (*D. barberi*, *D. cristata*, and *D. v. virgifera*). 7) *Acalymma vittatum* is substantially less sensitive to cucurbitacins than the *Diabrotica* species. It is of comparable evolutionary interest to note that *Aulacophora foveicollis* has also been shown to feed on crystalline Cuc E on filter paper (46).

Detoxication Mechanisms in Rootworm Beetles

The Diabroticite beetles are extraordinarily insensitive to the highly toxic cucurbitacins; they can grow, develop, and reproduce on bitter host *Cucurbita* plants, which contain as much as 0.32% fresh weight of cucurbitacins (29). Groups of 25 *D. u. howardi* or *D. v. virgifera* adults completely consumed 1 mg of pure Cuc B in 72 hours without any perceptible ill effects, indicating oral LD_{50} values of > 2000 mg/kg (27).

Paired experiments were conducted in which 50 newly emerged rootworm beetles fed exclusively on uniformly sized pieces of *C. andreana* $\times$ *C. maxima* F_1 hybrid fruit that contained 1–2 mg/g of Cucs B-D, or on *C. moschata* or *C. pepo* fruit devoid of cucurbitacins (17). Feeding on the bitter cucurbitacin fruit decreased the mean longevity of *D. v. virgifera* from 126 days to 59 days (P < 0.001) and of *D. balteata* from 129 days to 70 days (P < 0.001). With *A. vittatum* there was no significant difference in the longevity between beetles fed on sweet fruits (mean 136 days) and those fed on bitter fruits (mean 122 days) (17). The total cucurbitacin consumption over the lifetime of the beetles was estimated as 1–2 mg per beetle for *D. balteata*, 1.7–3.4 mg for *D. v. virgifera*, and 0.8–1.6 mg for *A. vittatum*. These effects on longevity can be interpreted in terms of stress related to energy requirements involving metabolism, excretion, and tissue storage of the cucurbitacins. The costs are greatest in *D. v. virgifera*, whose normal lifestyle as a Poaceae feeder does not expose it to cucurbitacins. They are less for *D. balteata*, which commonly feeds on cucurbits, and are barely perceptible for *A. vittatum*, which is monophagous on cucurbits (17).

The insensitivity of the Diabroticite beetles to the toxic cucurbitacins implies efficient mechanisms for detoxication, excretion, and tissue storage. Labeled Cuc B [^{14}C] was synthesized from DL [2-^{14}C] mevalonate in seedlings of *C. maxima* and purified by preparative thin-layer chromatography (17). The labeled Cuc B was fed to groups of 20 one- to two-week-old Diabroticite beetles over a 48 hour period, and the disposition of the ^{14}C was determined in excreta, gut, hemolymph, and body remainder. The total amounts of ^{14}C label excreted ranged from 67.2% in *A. vittatum* to 94.6% in *D. balteata*. In the five species of Diabroticites examined most of the ^{14}C was excreted as three polar metabolites. These constituted from 46.8% of the total ^{14}C excreted in *D. v. virgifera* to 91% in *D. balteata*. Cuc B was identified in the excreta, ranging from 1.7% in *D. u. howardi* to 30% in *D. v. virgifera*. A small proportion of cucurbitacin-conjugate, from 0.98–2.76% of the total ^{14}C, was permanently sequestered in the hemolymph and the remainder was retained, largely as conjugates, in the body and gut (17).

Hybridization of *Cucurbita* and Transfer of Cucurbitacin Controlling Genes

The possible utility of cucurbitacins in traps for monitoring corn root-worm and cucumber beetles and the use of dried, ground cucurbitacin-containing fruits for Diabroticina control (26, 28) have stimulated interest in the development of dependable, high-yielding *Cucurbita* cultivars that produce substantial amounts of cucurbitacins. A number of wild *Cucurbita* species hybrids were produced, and the cucurbitacin contents of their fruits were measured by extraction with chloroform, separation of the cucurbitacins by TLC, and quantitation by ultraviolet spectrometry at 210 nm, as shown in Table 2 (29). Under climatic conditions of Illinois, wild species of *Cucurbita* are not dependable producers of fruit since they are difficult to grow and yields are lower than those of domesticated species. In some cases fruiting is dependent on photoperiod. Therefore the transfer of cucurbitacin-controlling genes from wild bitter species to domesticated species was investigated. The initial crosses were made between the wild species *C. andreana*, *C. lundelliana*, and *C. ecuadorensis* with *C. maxima*; *C. texana* with *C. pepo*; and *C. gracilior*, *C. palmeri*, *C. texana*, and *C. okeecho-beensis* with *C. moschata*. The cucurbitacin contents of the fruits are shown in Table 3.

Two hybrid cultivars were selected as the most promising. The *C. andreana* × *C. maxima* hybrid produced long-vined plants with leaves, fruits, and blossoms resembling *C. maxima*. A six-plant plot produced 84 fruits with an average weight of 3.9 kg (39). The B-D cucurbitacin content (Table

Table 2. Cucurbitacins in fruits of wild *Cucurbita* spp. hybrids

Cross[1]	Cucurbitacins (mg/g fresh weight)				
	B	D	E	I	Gly
AND × ECU	0.40	0.62	—	—	—
(AND × ECU) × AND	0.46	0.90	—	—	—
AND × LUN	1.92	0.04	—	—	—
(AND × LUN) × AND	0.64	0.18	—	—	—
LUN × OKE	—	—	0.55	0.47	—
OKE × LUN	—	—	0.69	0.25	—
PAR × LUN	0.84	0.33	—	—	—
PAR × LUN F₂	0.18	0.58	—	—	—
PAR × GRA	0.46	0.22	—	—	—
GRA × TEX	—	—	0.36	0.15	0.23
PAR × TEX F₂	—	—	0.65	0.39	0.38
PAR × PED	0.12	0.21	—	—	—

[1]All species abbreviated by first three letters of name, except for *C. palmeri* (PAR). AND, *andreana*; ECU, *ecuadorensis*; GRA, *gracilior*; LUN, *lundelliana*; OKE, *okeechobeensis*; PED, *pedatifolia*; TEX, *texana*.

Table 3. Cucurbitacins in fruits of wild-domestic *Cucurbita* spp. hybrids

Cross[1]	Cucurbitacins (mg/g fresh weight)				
	B	D	E	I	Gly
AND × MAX	1.17	0.09	—	—	—
(MAX × ECU) × LUN	1.08	0.60	—	—	—
LUN × (LUN × MAX)F$_2$	0.30	0.12	—	—	—
(AND × ECU) × (AND × MAX)	1.06	0.74	—	—	—
(AND × LUN) × (AND × MAX)	0.18	0.54	—	—	—
OKE × MOS	—	—	trace	trace	—
LUN × MOS F$_2$	0.37	—	—	—	—
PAR × MOS	0.27	—	—	—	—
GRA × MOS F$_2$	0.13	0.18	—	—	—
(PAR × MOS) × GRA	<0.02	—	—	—	—
(GRA × MOS) × (OKE × MOS)	0.17	0.20	—	—	—
(GRA × MOS) × (OKE × MOS) F$_2$	0.07	0.40	—	—	—
TEX × PEP	—	—	0.23	0.09	0.16

[1]All species abbreviated by first three letters of name, except for *C. palmeri* (PAR). AND, *andreana*; ECU, *ecuadorensis*; GRA, *gracilior*; LUN, *lundelliana*; MAX, *maxima*; MOS, *moschata*; OKE, *okeechobeensis*; PEP, *pepo*; TEX, *texana*.

3) was an average 1.26 mg/g fresh weight. The *C. texana* × *C. pepo* hybrid produced semibush plants with leaves, fruits, and blossoms resembling *C. pepo* 'Zucchini'. A six-plant plot produced 98 fruits with an average weight of 0.73 kg. The content of E, I, and glycoside of E cucurbitacins (Table 3) was an average of 0.48 mg/g fresh weight (39).

It was soon evident that hybrids of *C. maxima* were not suitable for production of fruit containing high levels of cucurbitacin in Illinois. The blossoms of the *C. maxima* hybrids contained relatively high levels of cucurbitacins and were devoured by *D. u. howardi*, *D. v. virgifera*, and *A. vittatum* (1). In 1979 no fruit was set on these crosses until most of the corn rootworm beetles had migrated to nearby corn plots. In 1980 most of the plants were killed by beetles feeding on vines and leaves. Progress with this series of crosses is feasible only if these insects are controlled with timely applications of insecticides.

The *C. pepo* hybrids with *C. texana* have been carried successfully through the F$_2$ generation and have been grown in Arizona for production of cucurbitacin baits. These hybrids produce early fruit but are susceptible to virus diseases. Both the *C. andreana* × *C. maxima* and *C. texana* × *C. pepo* hybrids are highly susceptible to damage by the squash vine borer, *Melittia satyriniformis* Hubner.

The initial crosses of *C. moschata* with *C. gracilior*, *C. palmeri*, and *C. okeechobeensis* produced fruit of low cucurbitacin content (Table 3). Selections in advanced generations have shown increased cucurbitacin levels. In the 1980 season, when *C. maxima* hybrids were killed by beetle feeding, the *C. moschata* hybrids grew well and set fruit. *Cucurbita mixta* has been crossed with *C. gracilior*, but both species are more susceptible to powdery mildew than most other *Cucurbita* species grown in Illinois.

Dry, Bitter Cucurbita Fruits as Baits for Diabroticite Beetles

In 1939 Contardi (7) recognized that the cucurbit beetle, *D. speciosa*, strongly preferred to feed on the bitter fruits of *C. andreana* instead of those of *C. maxima*. Sharma and Hall (43) found *D. u. howardi* had a similar preference for *C. foetidissima* fruits over *C. pepo* fruits, and Howe et al. (21) showed that adult *D. v. virgifera* consistently preferred the fruits of *C. andreana* and *C. texana* to those of *C. maxima* and *C. pepo*. These observations and our own investigation of the effectiveness of sliced, bitter *Cucurbita* fruit sprinkled with a few milligrams of the insecticides trichlorfon and methomyl in killing thousands of corn rootworm beetles over periods of several weeks, led to consideration of the use of the bitter cucurbitacin kairomones in toxic baits for Diabroticina beetle control (24, 26, 28). Bitter *C. andreana* and *C. texana* fruits, completely dried and ground to pass a 2-mm screen, retained their arrestant quality and were exceedingly palatable to the several species of cucumber and corn rootworm beetles. Such dried baits containing 0.1% methomyl killed 80–100% of the rootworm beetles when broadcast in cucurbit and corn fields at rates of 11–33 kg/ha (26, 28).

The development of hybrid cultivars that produced dependable yields of fruit with relatively high cucurbitacin contents, such as those from *C. andreana* × *C. maxima* and *C. texana* × *C. pepo* (39), made it possible to experiment with toxic baits for control of rootworm beetles on a practical scale. It was determined that nearly all rootworm beetle adults could be killed by broadcast applications of dried, ground baits that contained methomyl or isofenphos, when applied at doses of 60–240 g total cucurbitacins and about 33 g insecticide per hectare (26). This dose represented a much more efficient use of insecticide compared with conventional spraying.

The dried, ground, bitter, cucurbitacin-containing baits proved to be very stable in storage. Bait from *Cucurbita* fruits grown and prepared in 1981 and stored at ambient temperatures in cardboard drums appeared to be fully effective for rootworm beetle control when broadcast in 1986 (26).

Hybridization produced both qualitative and quantitative changes in the cucurbitacin contents of the *C. andreana* × *C. maxima* and *C. texana* × *C. pepo* fruits (Table 4). The grinding and drying process liberated extracellular enzymes that converted Cuc B to D and Cuc E to I and also produced higher quantities of cucurbitacin glycosides (26). Nevertheless, as demonstrated in a variety of field experiments in cucurbits and corn, the F_1 and F_2 fruits of these crosses proved to be highly effective as arrestants in toxic baits for rootworm beetle control (26).

Table 4.   Cucurbitacins in fresh and dried *Cucurbita* fruits (26,39)

Fruit[1]	Cucurbitacins (mg/g)				
	B	D	E	I	Gly
C. andreana (fresh)	2.78	0.42	—	—	—
C. maxima (fresh)	none detected				
AND × MAX					
F$_1$ fresh	1.17	0.09	—	—	—
F$_1$ dried	0.90	0.64	—	—	2.92
F$_2$ dried	0.24	—	—	—	1.84
C. texana (fresh)	—	—	0.07	0.36	0.75
C. pepo (fresh)	—	—	none detected		—
TEX × PEP					
F$_1$ fresh	—	—	0.23	0.09	0.16
F$_1$ dried	—	—	1.44	1.36	3.15
F$_2$ dried	—	—	1.36	—	1.63

[1]Species abbreviations: AND, *andreana*; MAX, *maxima*; PEP, *pepo*; TEX, *texana*.

Cucurbitacins as Allomones for Diabroticite Beetles

The incredibly bitter taste of the cucurbitacins and their toxic effects on vertebrates after ingestion suggests a further role in the behavioral ecology of the rootworm beetles, i.e., as protective allomones against predators, as first suggested by Howe et al. (21).

Diabroticite beetles concentrate and sequester relatively large quantities of cucurbitacins in free and derivative form. A single elytron of *D. u. howardi* or *D. balteata* fed on bitter plants of *C. andreana* × *C. maxima* had an extremely bitter taste and was found to contain Cucs B and D at levels of 1–3 mg/g fresh body weight (16). The concentration of cucurbitacin-conjugate in the hemolymph of *D. balteata* fed on fruits of this hybrid increased very rapidly and was readily detected spectrophotometrically in 1 μl of hemolymph after one to two days of feeding. The concentration of cucurbitacin-conjugate reached a plateau of about 30 μg/μl after 30–40 days of feeding. Similar measurements of *D. u. howardi* collected from bitter squash plants in field plots indicated hemolymph concentrations of 20–26 μg/μl, or about 15 mg of cucurbitacins per gram of fresh body weight (16, 17).

The predatory Chinese mantis, *Tenodera aridifolia sinesis* Saussure, was first offered Diabroticite beetles reared on pollen and then beetles reared on bitter squash. Paired experiments showed that 72% of *D. balteata*, 46% of *D. u. howardi*, and 24% of *D. v. virgifera* fed on bitter squash were rejected. The results were highly significant: $P < 0.001$ to $P < 0.05$ (16). Mantids seizing beetles fed on bitter squash showed a typical rejection reaction by violently flinging away the beetle after a single bite on an elytron. Approximately 70% of the bitter beetles survived such encounters, but no beetles fed on pollen were ever rejected. The predator was obviously disturbed by the cucurbitacins and underwent a period of excessive grooming, unsteadiness, and regurgitation (16). The protective effects of cucurbitacins against predation appears to extend to larvae and to eggs of Diabroticites feeding on bitter *Cucurbita*. Such larvae had hemolymph that was distinctly bitter. Females of *D. u. howardi*, *D. balteata*, and *A. vittatum* reared on bitter squash laid eggs that contained substantial quantities of cucurbitacins as detected by beetle bioassay. These bitter eggs could be effective in discouraging ant predators (16).

Acknowledgments

Much of the research reviewed in this chapter was supported in part by Competitive Grants 5901–0410–8–0067–0 and 59–2171–1–1–659 from the USDA, SEA Grants Office. Any opinions, findings, and conclusions or recommendations are those of the authors and do not necessarily reflect the view of the USDA.

It is a pleasure to acknowledge the invaluable contributions of our co-workers R. A. Metcalf, E. R. Metcalf, J. E. Ferguson, D. C. Fischer, J. F. Andersen, R. L. Lampman, Po-yung Lu, Shen-yang Chang, Wayne Howe and Phil Lewis. Samples of pure cucurbitacins were kindly supplied by D. Lavie and P. R. Enslin.

Literature Cited

1. Andersen, J. F., and R. L. Metcalf. 1987. Factors influencing the distribution of *Diabrotica* spp. in the blossoms of cultivated *Cucurbita* spp. J. Chem. Ecol. 13: 681–699.
2. Bemis, W. P., A. M. Rhodes, T. W. Whitaker, and S. G. Carmer. 1970. Numerical taxonomy applied to *Cucurbita* relationships. Amer. J. Botany 57: 404–412.
3. Berry, J. W., J. C. Scheerens, and W. P. Bemis. 1978. Buffalo gourd roots, chemical composition and seasonal changes in steroid content. J. Agric. Food Chem. 26: 345–356.
4. Bogawat, J. K., and S. N. Pandey. 1967. Food preference in *Aulacophora* spp. Ind. J. Entomol. 29: 349–352.
5. Chambliss, O. L., H. T. Erickson, and C. M. Jones. 1968. Genetic control of bitterness in watermelon fruits. Proc. Amer. Hort. Sci. 93: 539–546.
6. Chambliss, O. L., and C. M. Jones. 1966. Cucurbitacins: specific insect attractants in Cucurbitaceae. Science 153: 1392–1393.
7. Contardi, H. 1939. Estudios geneticos en *Cucurbita* y consideraciones agronomicas. Physis 18: 331–347.

8. Curtis, P. S., and P. M. Meade. 1971. Cucurbitacins from Cruciferae. Phytochemistry 10: 3081–3083.
9. DaCosta, C. P., and C. M. Jones. 1971. Resistance in cucumber, *Cucumis sativus* L., to three species of cucumber beetles. HortScience 6: 340–342.
10. DaCosta, C. P., and C. M. Jones. 1971. Cucumber beetle resistance and mite susceptibility controlled by the bitter gene in *Cucumis sativus* L. Science 172: 1145–1146.
11. David, A., and D. K. Vallance. 1955. Bitter principles of Cucurbitaceae. J. Pharmacy Pharmacol. 7: 295–296.
12. Dryer, D. L., and E. K. Trousdale. 1978. Cucurbitacins in *Purshia tridentata*. Phytochemistry 17: 325–326.
13. Enslin, P. R., F. J. Joubert, and S. Rehm. 1956. Bitter principles of the Cucurbitaceae. III. Elaterase, an active enzyme for the hydrolysis of bitter principle glycosides. J. Sci. Food Agric. 7: 646–654.
14. Ferguson, J. E., D. C. Fischer, and R. L. Metcalf. 1983. A report of cucurbitacin poisoning in humans. Cucurbit Genetics Coop. Rep. 6: 73–74.
15. Ferguson, J. E., E. R. Metcalf, R. L. Metcalf, and A. M. Rhodes. 1983. Influence of cucurbitacin content in cotyledons of Cucurbitaceae cultivars upon feeding behavior of Diabroticina beetles (Coleoptera:Chrysomelidae). J. Econ. Entomol. 76: 47–51.
16. Ferguson, J. E., and R. L. Metcalf. 1985. Cucurbitacins: plant derived defense compounds for Diabroticina (Coleoptera:Chrysomelidae). J. Chem. Ecol. 11: 311–318.
17. Ferguson, J. E., R. L. Metcalf, and D. C. Fischer. 1985. Disposition and fate of cucurbitacin B in five species of Diabroticina. J. Chem. Ecol. 11: 1307–1321.
18. Guha, J., and S. P. Sen. 1975. The cucurbitacins—a review. Ind. Biochem. J. 2: 12–28.
19. Harvey, J. 1978. Biosystematics and hybridization in five species of Diabroticites. M.S. thesis, Univ. Illinois, Urbana-Champaign.
20. Herrington, M. E. 1983. Intense bitterness in commercial zucchini. Cucurbit Genet. Coop. Rep. 6: 75–76.
21. Howe, W. L., J. R. Sanborn, and A. M. Rhodes. 1976. Western corn rootworm adult and spotted cucumber beetle associations with *Cucurbita* and cucurbitacins. Envir. Entomol. 5: 1043–1045.
22. Lavie, D., and E. Glotter. 1971. The cucurbitacins, a group of tetracyclic triterpenes. Forts. Chemie Organishe Naturstoffe 29: 306–362.
23. Maulik, S. 1936. Coleoptera, Chrysomelidae (Galerucine). London, Taylor & Francis, London.
24. Metcalf, R. L. 1985. Plant kairomones and insect pest control. Bull. Ill. Nat. Hist. Survey. 33: 175–198.
25. Metcalf, R. L. 1986. Coevolutionary adaptations of rootworm beetles (Coleoptera:Chrysomelidae) to cucurbitacins. J. Chem. Ecol. 12: 1109–1124.
26. Metcalf, R. L., J. E. Ferguson, R. L. Lampman, and J. F. Andersen. 1987. Dry cucurbitacin containing baits for controlling Diabroticite beetles (Coleoptera:Chrysomelidae). J. Econ. Entomol. 80: 870–875.
27. Metcalf, R. L., R. A. Metcalf, and A. M. Rhodes. 1980. Cucurbitacins as kairomones for diabroticite beetles. Proc. Natl. Acad. Sci. USA 77: 3769–3772.

28. Metcalf, R. L., A. M. Rhodes, and E. R. Metcalf. 1981. Monitoring and controlling corn rootworm beetles with baits of dried bitter *Cucurbita* hybrids. Cucurbit Genet. Coop. Rep. 4: 37–38.
29. Metcalf, R. L., A. M. Rhodes, R. A. Metcalf, J. E. Ferguson, E. R. Metcalf, and P.-Y Lu. 1982. Cucurbitacin contents and Diabroticite (Coleoptera: Chrysomelidae) feeding upon *Cucurbita* spp. Environ. Entomol. 11: 931–937.
30. Nielson, J. K., M. Larsen, and H. J. Sorenson. 1977. Cucurbitacins E and I in *Iberis amara*: Feeding inhibitors for *Phyllotreta nemorum*. Phytochemistry 16: 1519–1522.
31. Pal, A. B., K. Srinivasan, G. Bharatan, and M. V. Chandravadana. 1978. Location of sources of resistance to the red pumpkin beetle *Rapidapalpa foveicollis* Lucas among pumpkin germ plasma. J. Entomol. Res. 2: 148–153.
32. Pohlmann, J. 1975. Die Cucurbitacine in *Bryonia alba* und *Bryonia dioica*. Phytochemistry 14: 1587–1589.
33. Price, P. 1975. Insect Ecology. Wiley, New York.
34. Pulchalski, J. T., and R. W. Robinson. 1978. Comparative electrophoretic analysis of isozymes in *Cucurbita* species. Cucurbit Genet. Coop. Rep. 1: 28.
35. Rehm, S. 1960. Die Bitterstoffe der Cucubitaceen. Ergeb. Biol. 22: 106–136.
36. Rehm, S., P. A. Enslin, A. D. J. Meeuse, and J. H. Wessels. 1957. Bitter principles of the Cucurbitaceae VII. The distribution of bitter principles in the plant family. J. Sci. Food Agric. 8: 679–686.
37. Rehm, S., and J. H. Wessels. 1958. Bitter principles of the Cucurbitaceae. VII. Cucurbitacins in seedlings—occurrence, biochemistry, and genetical aspects. J. Sci. Food Agric. 8: 687–691.
38. Rhodes, A. M., W. P. Bemis, T. W. Whitaker, and S. E. Carmer. 1968. A numerical taxonomic study of *Cucurbita*. Brittonia 20: 251–266.
39. Rhodes, A. M., R. L. Metcalf, and E. R. Metcalf. 1980. Diabroticite beetle response to cucurbitacin kairomones. J. Amer. Soc. Hort. Sci. 105: 838–842.
40. Robinson, R. W., H. M. Munger, T. W. Whitaker, and G. W. Bohn. 1976. Genes of the Cucurbitaceae. HortScience 11: 554–568.
41. Rymal, K. S., O. L. Chambliss, M. D. Bond, and D. A. Smith. 1984. Squash containing toxic cucurbitacin compounds occurring in California and Alabama. J. Food Protection 47: 270–271.
42. Schwartz, H. M., S. I. Biedron, M. M. von Holdt, and S. Rehm. 1964. A study of some plant esterases. Phytochemistry 3: 189–200.
43. Sharma, G. C., and C. V. Hall. 1971. Influence of cucurbitacins, sugars, and fatty acids on cucurbit susceptibility to spotted cucumber beetle. J. Amer. Soc. Hort. Sci. 96: 675–680.
44. Sharma, G. C., and C. V. Hall. 1973. Relative attractance of spotted cucumber beetles to fruits of fifteen species of Cucurbitaceae. Environ. Entomol. 2: 154–156.
45. Sharma, G. C., and C. V. Hall. 1973. Identifying cucurbitacins in cotyledons of *Cucurbita pepo* L. cv. Black Zucchini. HortScience 8: 136–137.
46. Sinha, A. K., and S. S. Krishna. 1970. Further studies on the feeding behavior of *Aulacophora foveicollis* on cucurbitacin. J. Econ. Entomol. 63: 333–334.
47. Smith, R. F. 1966. Distributional patterns of selected western North American insects. Bull. Entomol. Soc. Amer. 17: 106–110.

48. Stoewsand, G. S., A. Jaworski, S. Shannon, and R. W. Robinson. 1985. Toxicologic response in mice fed *Cucurbita* fruit. J. Food Protection 48: 50–51.
49. Thorne, R. F. 1981. Phytochemistry and angiosperm phylogeny, a summary statement. *In* D. A. Young and D. F. Seigler, eds., Phytochemistry and Angiosperm Phylogeny. Praeger, New York.
50. Watt, J. M., and M. G. Breyer-Brandwiyk. 1962. The Medicinal and Poisonous Plants of Southern and Eastern Africa, 2nd ed. Livingston, Edinburgh.
51. Whitaker, T. W., and W. P. Bemis. 1975. Origin and evolution of the cultivated *Cucurbita*. Bull. Torrey Bot. Club 102: 362–368.
52. Wilcox, J. A. 1972. Chrysomelidae Galerucinae, Luperini. Coleopterorum Catalogues Suppl. Pars. 78, Fasc. 2, 2nd ed. W. Junk's Gravenhagen, The Netherlands.
53. Yaro, N., and H. L. Krysan. 1986. Host relationships of *Diabrotica cristata* (Coleoptera:Chrysomelidae). Entmol. News 97: 11–16.

PART II

Comparative Morphology

15 | Embryology of the Cucurbitaceae

A. S. R. Dathan and Dalbir Singh

ABSTRACT. Basic features of cucurbitaceous embryology are described and summarized. Attention is given to the microsporangium and microsporgenesis and development of the male gameophyte and pollen tube. The megasporangium, megasporogenesis, fertilization, and the development of the embryo and endosperm are considered. Cucurbitaceous ovules are anatropous, bitegmic, and crassinucellar. The nucellus is flask-shaped and beaked, with the micropylar canal formed by the inner integument. The archesporium is unicellular or multicellular. The megaspore mother cell is deeply seated and undergoes normal meiosis. Generally, the chalazal megaspore develops into a *Polygonum* type embryo sac, although the bisporic *Allium* type has been observed. Somatic and generative apospory are noted. Pollen tubes are porogamous and persistent. Embryogeny in the subfamily Cucurbitoideae is of *Myosurus*, Onagrad type, although one instance of *Nicotiana*, Solanad type has been recorded. The mature embryo is straight, spatulate, and dicotyledonous. Endosperm development is nuclear and wall formation is centripetal. The endosperm forms a chalazal haustorium, which is unique among parietalean families, and enucleate cytoplasmic nodules, and exhibits unusual nuclear behavior.

Because of its economic importance, disputable taxonomic position, and morphological peculiarities, the family Cucurbitaceae has attracted the attention of embryologists during the past 150 years. The earlier investigations have been summarized by Kirkwood (38), Davis (26), and Singh (60). In this chapter attention is given to more recent embryological studies and an overview of the character and development of male and female reproductive structures.

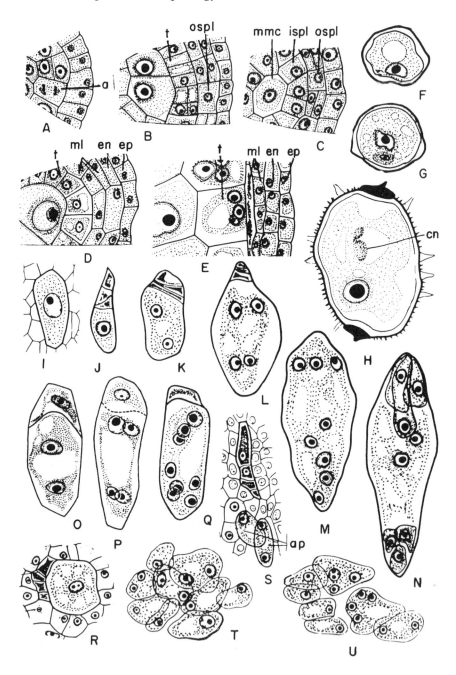

Figure 1. Anther wall, microsporogenesis, megasporogenesis, female gametophyte, and somatic apospory in cucurbits. A–E. *Momordica charantia.* F, G. *Citrullus*

Microsporangium and Development of the Male Gametophyte

Shridhar (52) recently studied anther development in 18 species of Cucurbitaceae. The anther lobes vary from five to eight and usually are distributed over three stamens in a ratio of 2:2:1. In *Coccinia grandis* (L.) Voigt and *Sechium edule* (Jacq.) Sw., however, the lobes appear in ratios of 1:1:1 and 2:1:1, respectively. *Cyclanthera pedata* (L.) Schrad. is exceptional in that the androecium forms a circular disk divided into single upper and lower chambers by a central partition.

Young anthers are composed of a homogeneous mass of cells with a procambial strand in the center. The male archesporium is hypodermal (Figure 1A) and is composed of one to three cells in each corner of the anther, although Heimlich (34, 35) reported three to five archesporial cells in *Cucumis sativus* L.

The development of anther walls is of the "dicotyledonous type" (Figure 1B–E). The anther wall is composed of an epidermis, a fibrous endothecium, two or three ephemeral middle layers, and a tapetum. During development, the epidermis persists as a depleted layer (52), but degeneration of the epidermal cells was reported in the dehisced anthers of *Cucumis sativus* (35). Degeneration of the inner middle layer precedes that of its outer layers. Rarely, the cells of the outer middle layer persist, in part, and develop fibrous thickenings.

The tapetum is usually composed of a single layer of cells (Figure 1E), but rarely it forms two layers in places, as in *Coccinia grandis*, *Cucurbita pepo* L., and *Momordica charantia* L. The tapetum is of dual origin (52). Both Kirkwood (40) and Heimlich (35) reported that the tapetum is only partly formed by the innermost parietal (ephemeral middle) layer. Although Heimlich (35) made no direct mention of the dual origin of the tapetum, his

---

lanatus. H. *Cucurbita pepo*. I–N. *Trichosanthes lobata*. O–Q. *Benincasa hispida*. R–U. *Cucumis metuliferus*. A. Portion of the developing anther wall showing archesporial cells. B–D. Stages in the development of the anther wall. E. Transverse section of part of the anther wall showing degeneration of the inner middle layer and two- or three-nucleate tapetal cells. F, G. One- and two-celled pollen grains, respectively. H. Uninucleate pollen grains showing enucleate cytoplasmic nodules. I–N. Stages in the development of monosporic female gametophyte. O–Q. Stages in the development of the bisporic embryo sac. A binucleate embryo sac together with an undivided micropylar dyad (O), four-nucleate (P), and eight-nucleate (Q) embryo sacs developed from the chalazal dyad. R, S. Longitudinal section of part of the nucellus showing enlargement of somatic cells and degeneration of the tetrad of megaspores. T, U. Developing aposporic embryo sacs. a, archesporium; ap, aposporic nucellar embryo sacs; cn, cytoplasmic nodule; en, endothecium; ep, epidermis; ispl, inner secondary parietal layer; ml, middle layer; mmc, microspore mother cell; ospl, outer secondary parietal layer; t, tapetum. A–H, after Shridhar (52), O–Q, after Chopra and Basu (14).

illustrations support that conclusion. The tapetum is secretory, except in *Luffa cylindrica* (L.) M. J. Roem., where the tapetal cell walls degenerate and a periplasmodium is formed in situ. Initially, the tapetum is uninucleate, but subsequently its cells become bi- or multinucleate (3, 6, 7, 35, 40, 45, 52; Figure 1E). Turala (65, 66) reported that the tapetum remains uninucleate in several species, but binucleate as well as uninucleate cells occurred in *Bryonia dioica* Jacq. He observed variation in the size of the nucleus of the tapetal cells and attributed it to endomitosis. Shridhar (52) observed fusion of nuclei in tapetal cells during advanced stages.

Sporogeneous tissue increases by mitotic divisions before the microspore mother cell enters meiosis. Normal meiotic steps are followed during karyokinesis (6, 8, 35, 45, 52), although Dzevaltovsky (32) reported numerous irregularities during both heterotypic and homotypic divisions in the anthers of triploid watermelon (*Citrullus lanatus* (Thunb.) Matsum. & Nakai). Cytokinesis is simultaneous and takes place through furrowing. Castelter (6) and Passmore (45) observed ephemeral cell plates in both hetero- and homotypic spindles during meiosis, but these plates did not participate in the quadripartitioning of pollen mother cells. The microspores are arranged tetrahedrally or isobilaterally. Dzevaltovsky (32) reported, in addition to normal tetrads of microspores, multinuclear microspores and microspores with restitutional and subrestitutional nuclei in triploid watermelon.

As the young microspores increase in size, their walls soon differentiate into intine and exine (Figure 1F). The nucleus is usually uninucleolate, although rarely it is multinucleolate in *Luffa cylindrica* and *Mukia maderaspatana* (L.) M. J. Roem. The nucleus divides to form the vegetative and generative cells (52; Figure 1G). The vegetative nucleus is usually uninucleolate, but multiple nucleoli have been observed in *Dactyliandra welwitschii* Hook. f., *Edgaria darjeelingensis* C. B. Clarke, *Luffa cylindrica*, *Mukia maderaspatana*, and *Trichosanthes dioica* Roxb. (52). The pollen grains are usually shed at the two-celled stage. They are monosiphonic or polysiphonic at germination. Shridhar (52) reported the formation of cytoplasmic nodules in pollen grains of *Cucurbita pepo*, *Trichosanthes tricuspidata* Lour. (as *T. bracteata* (Lam.) Voigt.), and *T. dioica*. The nodules appear as small balloonlike outgrowths entering the large central vacuole of the pollen grains from the tonoplast (Figure 1H).

Megasporogenesis and Development of the Female Gametophyte

The ovules of the Cucurbitaceae are anatropous, bitegmic, and crassinucellate. Details concerning the development and nature of ovules is reported elsewhere (4, 7, 13, 20–24, 37, 38, 53, 60, 69, Singh and Dathan, this volume).

Kirkwood (38) probably was the first to give a detailed account of megasporogenesis and development of the monosporic embryo sac in the family. The female archesporium is hypodermal and one-celled, but the occurrence of multiple archesporia has been reported in *Trichosanthes* (4) and *Benincasa* (13). Dzevaltovsky (27) reported that the female archesporium was potentially multicellular in most of the species that he investigated. In most instances the actual number of female archesporial cells remains obscure because of the meristematic nature of the cells of the young nucellus (25). A single archesporium was reported for *Coccinia grandis* (as *C. indica* Wight & Arn.) (25). Later in development, however, one or rarely two megaspore mother cells have been observed.

The archesporial cell divides periclinally, forming the primary parietal cell and the megaspore mother cell. Massive parietal tissue is produced by division of the primary parietal cell, and the megaspore mother cell becomes deep-seated (Figure 1I). The megaspore mother cell undergoes meiosis, forming a linear or T-shaped tetrad of megaspores (4, 7, 8, 14, 21–25, 27, 37, 38, 47, 53). Divisions in the dyad cells are usually synchronous, but rarely are nonsynchronous. In *Benincasa hispida* (Thunb.) Cogn., *Coccinia grandis*, *Cucumis melo* var. *agrestis* Naud. (as *C. melo* var. *pubescens* (Kurz) Willd.), and *Cucurbita pepo* the micropylar dyad cell fails to divide, and thus only triads are formed (14, 25, 53). Furthermore, in *B. hispida* the megaspore mother cell divides unequally, and the small micropylar cell degenerates while the chalazal cell directly forms a bisporic, *Allium* type embryo sac (13, 28; Figure 1O–Q). Earlier, Kirkwood (38) reported on the degeneration of the micropylar dyad cell and the formation of triads in several species. Twin megaspore mother cells and twin embryo sacs have also been observed (13, 25, 61).

Dzevaltovsky (30) reported uni- or multicelled archesporia and several irregularities during meiosis of triploid watermelon. Heterotypic divisions lead to the formation of dyads and triads, and their cells may be accompanied by micronuclei as well as a normal nucleus. Failure of cytokinesis after heterotypic division has also been reported. Homotypic divisions followed by cytokinesis results in the formation of tetrads, triads, and polyads. In *Cucurbita pepo* numerous bodies resembling micronuclei were observed in megaspore mother cells and in cells of dyads and tetrads, but these disappeared during later stages (14).

The chalazal megaspore is usually functional (Figures 1J, 3B), but Dzevaltovsky (27) observed embryo sac formation from an epichalazal megaspore in *Lagenaria siceraria* (Mol.) Standl. (as *L. vulgaris* Ser.), and Dathan and Singh (25) reported on divisions of the nucleus of the micropylar megaspore in *Citrullus colocynthis* (L.) Schrad. The functional megaspore nucleus undergoes three divisions to form eight nuclei, which organize to form a *Polygonum* type of embryo sac development. Similar development

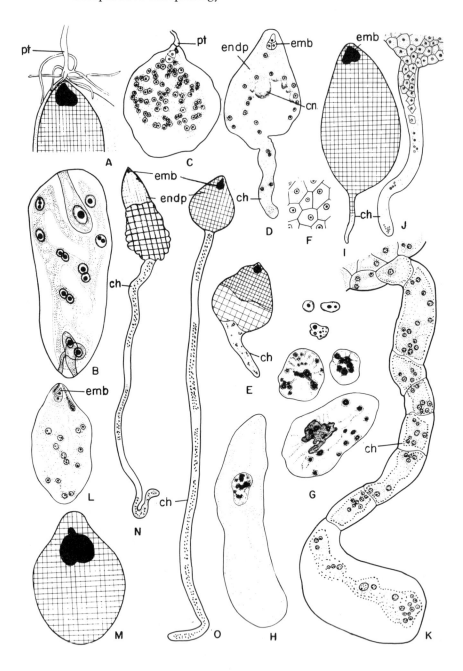

Figure 2. Persistent pollen tube, antipodals, and endosperm in Cucurbitaceae. A. *Cucurbita pepo.* Part of the cellular endosperm showing branched persistent pollen

has also been reported in *Benincasa hispida* (13, 28; Figure 1O–Q). Dimorphic embryo sacs have been observed in *Trichosanthes lobata* Roxb. (21). In this instance about 20% of the embryo sacs remained small (60 × 20 μm) as compared with normal embryo sacs (100 × 30 μm). In the former, the cells completely filled the cavity. Both types of embryo sac were observed in the same ovary, and there is evidence to suggest that both are functional. Dzevaltovsky (31) recorded a number of abnormalities during embryo sac development in triploid watermelon. He observed the presence of micronuclei, embryo sacs with broken polarity or absence of polarity, formation of a micropylar, haustorialike structure at the 8-nucleate embryo sac stage, formation of 16-nucleate embryo sacs because of the further division of the 8 nuclei, and a multinucleate condition in component cells of the organized embryo sac.

The mature organized embryo sac is eight-nucleate. The size and shape of component cells of the embryo sac vary in different species (25). Fusion of polar nuclei generally takes place at the time of fertilization, except in *Zehneria indica* (Lour.) Keraudren (syn. *Melothria leucocarpa* (Blume) Cogn.), where it may take place before fertilization. The synergids are smooth, pear-shaped or hooked, and of the same size as the egg, but egg cells larger than the synergids are seen in some species (25). In *Edgaria darjeelingensis, Herpetospermum pedunculosum* (Ser.) C. B. Clarke, and *Mukia maderaspatana*, one of the synergids persisted in fertilized embryo sacs having a several-celled embryo. The synergid enlarged and the nucleus hypertrophied in *E. darjeelingenesis* and *H. pedunculosum* (57). The antipodal cells vary in size in different species and may be small, medium, or large. They are usually ephemeral; only in *Citrullus lanatus* did one or two of them persist and undergo enlargement, forming tubular haustorial structures (Figure 2B).

In *Cucumis metuliferus* Naud., besides the *Polygonum* type of embryo sac, a *Hieracium* type of somatic apospory has also been reported (61),

---

tube. B. *Citrullus lanatus*. Embryo sac showing the zygote, endosperm nuclei, and persistent antipodal cells. C–H. *Marah marcrocarpus*. Development and structure of the endosperm and structure of endosperm and its chalazal haustorium (C–E). Note the persistent pollen tube in C and enucleate cytoplasmic nodule in D. Changes in endosperm cell size and its nuclei (F–H). I, J. *Cucumis ficifolius*. Endosperm with chalazal haustorium and the haustorium magnified respectively. K. *Trichosanthes tricuspidata*. Chalazal endosperm haustorium having multinucleate compartments. L, M. *Zehneria indica*. Nuclear endosperm (L). Cellular endosperm (M). Note the absence of a chalazal haustorium in M. N. *Cucurbita moschata*. Cellular endosperm with a coenocytic haustorium. O. *Trichosanthes lobata*. Cellular endosperm with a coenocytic haustorium. ch, chalazal endosperm haustorium; cn, cytoplasmic nodule; emb, embryo; endp, endosperm proper; pt, pollen tube.

being observed in distal ovules of the placental flanks. The aposporic embryo sacs were of nucellar origin (Figure 1S–U) and were slightly smaller than normal. Dzevaltovsky reported generative apospory in some ovules of *Momordica charantia* (29) and triploid watermelon (31). The embryo sac was formed directly from the archesporial cells and in triploid watermelon the aposporic embryo sacs were much larger than normal.

Pollen Tube and Fertilization

The persistence of the pollen tube attracted the attention of early workers, including Schleiden, whose erroneous observation that the embryo develops from its tip in *Cucurbita* led to the classical Amici-Schleiden controversy (2, 48). The pollen tube is porogamous (7, 22, 24, 37, 39, 49, 57; Figure 2A, C). Mesogamy was reported in *Cucurbita* (41), but has not been confirmed by other workers. In *Luffa acutangula* (L.) Roxb. (as *Luffa hermaphrodita* Singh & Bhandari) the pollen tube has not been observed entering the ovule through the micropyle. It does not always penetrate the nucellus at the tip, but instead may travel for some distance between the nucellus and the inner integument, finally penetrating the nucellus laterally (57). In the family, one pollen tube is usually associated with each ovule, but Guliajev (33) reported three or four pollen tubes in the micropyle of an ovule in *Citrullus lanatus*. The pollen tube is usually unbranched, but rarely may be branched in *Cucurbita pepo* (39, 43, 46, 62; Figure 2A).

Initially the pollen tube has uniform width, but later shows localized dilation before or after fertilization. Longo (42, 43) reported swelling of the pollen tube at the point where it reached the embryo sac in *Cucurbita pepo*. A number of branches emerged from the "bulla"; one of them led to fertilization, and the others ramified in the nucellus and integuments. Kirkwood (39) observed a similar swelling in *Citrullus*, *Cucumis* and *Cyclanthera*, but the pollen tube remained unbranched. Recently, Singh (57) recorded terminal dilation of pollen tubes in *Cyclanthera pedata*, *Dicoelospermum ritchiei* C. B. Clarke, *Edgaria darjeelingensis*, *Herpetospermum pedunculosum*, and *Mukia maderaspatana*. The dilation of pollen tubes was attributed to the presence of starch grains in the integuments (39, 43), but Singh (57) could not find any positive correlation between dilation and the occurrence of starch grains in the nucellus or in the integuments. Usual syngamy and triple fusion have been reported.

Embryo

Kirkwood (38) described the early zygotic divisions in several species. He reported linear, proembryonic tetrads, except in *Sicyos angulatus* in which the tetrads were T-shaped. In *Cucumis sativus*, Tillman (64) observed the vertical division of the apical cell and, subsequently, the formation of an epiphyseal cell from one of the daughter cells. Soueges (63) reported two pathways of embryo development in *Bryonia dioica*, depending on whether the basal cell segmented transversely or vertically. The embryogeny of several species has been described as *Myosurus* variation of the Onagrad type (16, 18–20, 22–24, 56, 63; Figure 3A–K, L–Q, R–W, X–BB, CC–II). Singh (53), however, described the formation of a linear, proembryonic tetrad as a result of transverse divisions of both the apical and basal cells in *Cucumis melo* var. *agrestis* (as var. *pubescens*) (Figure 3KK–PP). Embryologically, this is a *Nicotiana* variation of the Solanad type. The first few divisions in the zygote are transverse in *Fevillea cordifolia* L. (38) and *Actinostemma tenerum* Griff. (20), leading to the formation of a filamentous proembryo (Figure 3SS–VV). Embryogeny in the subfamily Zanonioideae, to which these two species belong, remains inadequately studied.

The mature cucurbit embryo is straight, spatulate, and dicotyledonous. It has a small hypocotyledonary root axis and lacks a suspensor. In *Marah macrocarpus* (Greene) Greene (24), frequently one or both cotyledonary primordia developed a notch and form hemitri- or hemitetracotyledonary embryos at maturity.

Polyembryony occurs in developing seeds of several species. It is nucellar in *Momordica charantia* (1) and *Cucumis melo* var. *agrestis* (as var. *pubescens*) (53; Figure 3QQ, RR), and of synergid origin in *Cucumis sativus* (44) and *Sicyos angulatus* (17). Twin embryos have been observed in *Dactyliandra welwitschii* (23; Figure 3JJ). Parthenogenesis has been reported in *Sicyos angulatus* (17).

Endosperm

Several early investigators (2, 5, 36, 38) reported a tubular chalazal extension of the endosperm, and Schnarf (49) described it as a chalazal endosperm haustorium. In recent years the development and structure of the endosperm, with special reference to the organization of the chalazal haustorium, has been studied in 57 species of 23 genera (9–12, 14, 15, 55, 58, 62, 68). The endosperm development is "nuclear" (Figure 2C–E). The formation of a chalazal cecum was reported for all investigated species (Figure 2E, I–K, N, O) except *Ctenolepis garcinii* (Burm. f.) C. B. Clarke, *Corallocarpus conocarpus* (Dalz. & Gibs.) C. B. Clarke, and *Zehneria indica* (as *Melothria leucocarpa*) (11, 15, 62; Figure 2L, M). The chalazal cecum acquires its shape after wall formation has taken place in the endosperm proper. The haustorium may be short or long, the longest being 19,000 μm in *Sechium edule*. It remains coenocytic (Figure 2N, O) or becomes partly

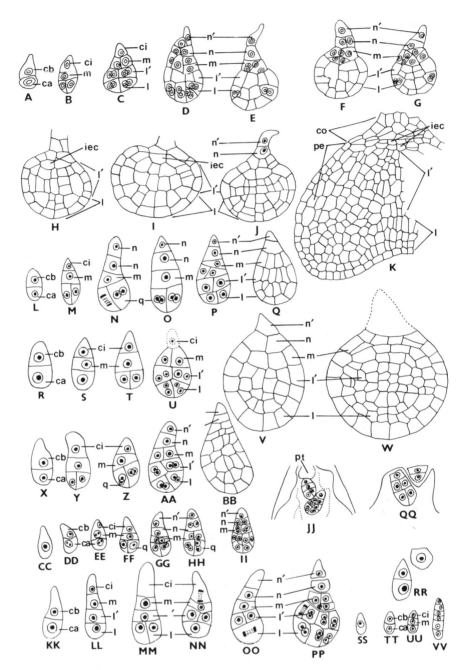

Figure 3. Stages in embryogeny of Cucurbitaceae. A–K. *Bryonia dioica.* L–Q. *Dicoelospermum ritchiei.* R–W. *Citrullus colocynthis.* X–BB. *Mukia maderaspatana.* CC–II. *Zehneria indica.* JJ. Twin embryos in *Dactyliandra welwitschii.* KK–RR. *Cucumis melo* var. *agrestis.* Note the initial polyembryony in QQ and RR. SS–VV.

(Figure 2I, J) or completely cellular (Figure 2K). Its tip may be acute or dilated, and its base is simple or bulbous. The haustorium is coenocytic as well as cellular in *Cucumis ficifolius* A. Rich, *Trichosanthes bracteata*, *Citrullus lanatus*, *C. colocynthis*, and *Corallocarpus epigaeus* (Rottl.) C. B. Clarke (55, 58, 62).

Singh (55) proposed that the length of the haustorium determines its coenocytic or cellular nature. Chopra and Agrawal (12) grouped the chalazal endosperm haustoria into two arbitrary categories: haustoria up to 1500 μm, which remain coenocytic or become cellular, and haustoria longer than 1500 μm, which are always coenocytic. This division has been only partly supported by subsequent studies (58, 62). It is now evident that size alone does not determine the nature of the haustorium. The chalazal endosperm haustorium of *Luffa echinata* Roxb. is only 240–400 μm long, but it always remains coenocytic.

The coenocytic haustorium usually remains healthy until the cordate embryo stage (9–12, 14, 15, 55, 58, 62). From that point its contents are lost gradually, and the haustorium shrinks but remains attached to the endosperm proper. In contrast, the cellular haustorium remains healthy even during later stages, and aborts only with the endosperm proper. As the activity of the haustorium declines, the peripheral cells of the cellular endosperm in the chalazal region bulge or form short mounds. Cells of these regions show large vacuoles and granular cytoplasm, suggesting their haustorial nature. They have been termed *secondary haustoria* (55, 58, 62).

Another interesting feature of cucurbitaceous endosperm is the formation of enucleated cytoplasmic nodules in the coenocytic endosperm (54, 59, 62). The nodule formation has been observed in the endosperm proper (Figure 2D) and in the coenocytic haustorium. The nodules arise from the inner lining of the endosperm cytoplasm and bulge into its central vacuole. They enlarge, fuse with one another, and finally merge with the general cytoplasm of the endosperm. The significance and function of the cytoplasmic nodules still remains unsettled, but Singh (59) and Singh and Dathan (62) consider it to be an active method of endosperm growth, resulting in obliteration of the central vacuole of the embryo sac.

The endosperm cells are initially small, polygonal, and richly protoplasmic, each containing a small (5–7 μm) round or oval nucleus (Figure 2F). During advanced stages of development, the cells of the central part of endosperm show enormous enlargement (Figure 2H). The nuclei of these cells undergo interesting changes before they are absorbed (11, 50, 51, 55,

---

Actinostemma tenerum. ca, apical cell of bicelled proembryo; cb, basal cell of the bicelled proembryo; ci, lower daughter cell of cb; ca, cap; iec, initials of the root cortex; 1, upper daughter cell of q; 1′, lower daughter cell of q; m, upper daughter cell of cb; n, upper daughter cell of ci; n′, lower daughter cell of ci; o, upper daughter cell of n′; pe, periblem; q, quadrants. A–K, after Souégés (63).

58, 62). They enlarge, vacuolate, and become irregular in shape. The nucleolus develops intranucleolar vacuoles and become amoebiform, from which segments break off, making the nucleus multinucleolate (Figure 2G). Turala (67) reported the endosperm nucleus of *Echinocystis lobata* (Michx.) Torr. & A. Gray showed considerable increase in size because of endomitotic polyploidization. She recorded a high degree of endopolyploidy, up to $3072n$ in parts of the endosperm proper and $24n$ in its chalazal haustorium.

Systematic Implications

The relationships of the Cucurbitaceae to other parietalean families, based on characters of the ovule and seed, are considered by Singh and Dathan (this volume) and are not repeated in this chapter. Members of the Cucurbitaceae are distinct in having a nucellar beak and strongly differentiated outer and inner integuments, of which the outer are vascularized. The endosperm of cucurbits is unique among the parietalean families in having a chalazal haustorium.

Literature Cited

1. Agrawal, J. S., and S. P. Singh. 1957. Nucellar polyembryony in *Momordica charantia*. Sci. Cult. 22: 630–631.
2. Amici, G. B. 1824. Observations microscopiques sur diverses espèces des plantes. Ann. Sci. Nat. (Paris) 2: 41–70, 211–248.
3. Asana, J. J. and R. N. Sutaria. 1932. Microsporogenesis in *Luffa aegyptiaca* Mill. J. Indian Bot. Soc. 11: 181–187.
4. Banaeji, I., and M. C. Das. 1937. A note on the development of the embryo sac in *Trichosanthes dioica* Roxb. Curr. Sci. 5: 427–428.
5. Brongniart, A. 1827. Mémoire sur la generation et développment de l'embryon dans les végétaux phanérogames. Ann. Sci. Nat. (Paris) 12: 14–53, 145–172, 225–296.
6. Castelter, E. F. 1926. Cytological studies in the Cucurbitaceae: Microsporogenesis in *Cucurbita maxima*. Amer. J. Bot. 13: 1–10.
7. Chakravorti, A. K. 1947. The development of female gametophyte and seed of *Coccinia indica* W. & A. J. Indian Bot. Soc. 26: 95–104.
8. Chauhan, S. V. S. 1970. Micro and megasporogenesis in *Luffa echinata* Roxb. Agra Univ. J. Res. Sci. 19: 37–42.
9. Chopra, R. N. 1953. The endosperm in some Cucurbitaceae. Curr. Sci. 22: 283–384.
10. Chopra, R. N. 1954. Occurrence of endosperm haustoria in some Cucurbitaceae. Nature 173: 352–353.
11. Chopra, R. N. 1955. Some observations on endosperm development in Cucurbitaceae. Phytomorphology 5: 219–230.

12. Chopra, R. N., and S. Agrawal. 1958. Some further observations on the endosperm haustoria in the Cucurbitaceae. Phytomorphology 8: 194–201.

13. Chopra, R. N., and S. Agrawal. 1960. The female gametophyte of *Benincasa cerifera* Savi. Bot. Not. 113: 192–202.

14. Chopra, R. N., and B. Basu. 1965. Female gametophyte and endosperm of some members of Cucurbitaceae. Phytomorphology 15: 217–223.

15. Chopra, R. N., and P. N. Seth. 1977. Some aspects of endosperm development in Cucurbitaceae. Phytomorphology 27: 112–115.

16. Crété, P. 1958. Embryogénie des Cucurbitacées. Développment de l'albumen et de l'embryon chez le *Sicyos angulata*. Compt. Rend. Hebd. Seances Acad. Sci. 246: 456–459.

17. Crété, P. 1958. La parthénogénèse chez le *Sicyos angulata*. Bull. Soc. Bot. France 105: 18–19.

18. Crété, P. 1960. Embryogénie des Cucurbitacées. Développement de l'embryon chez l'*Ecballium elaterium* (L.) A. Rich. Compt. Rend. Hebd. Seances Acad. Sci. 251: 968–970.

19. Crété, P. 1962. Embryogénie des Cucurbitacées. Développement de l'embryon chez le *Cyclanthera explodens* Naud. Compt. Rend. Hebd. Seances Acad. Sci. 254: 3411–3412.

20. Dathan, A. S. R. 1971. Structure and development of seeds of Cucurbitaceae and some of its related families. Ph.D. dissertation, Rajasthan Univ., Jaipur.

21. Dathan, A. S. R. 1974. Occurrence of dimorphic embryo sacs in *Trichosanthes lobata* Roxb. Curr. Sci. 43: 91–92.

22. Dathan, A. S. R. 1974. Embroyology of *Melothria leucocarpa* (Blume). Cogn. Sci. Cult. 40: 119–122.

23. Dathan, A. S. R. 1974. A contribution of the embryology and seed development of *Dactyliandra welwitschii* Hook. f. Proc. Indian Acad. Sci. 80B: 207–216.

24. Dathan, A. S. R., and D. Singh. 1971. Morphology and embryology of *Marah macrocarpa* Greene. Proc. Indian Acad. Sci. 73: 241–249.

25. Dathan, A. S. R., and D. Singh. 1979. Structure and development of female gametophyte in Cucurbitaceae. *In* S. S. Bir, ed., Recent Researches in Plant Sciences. Kalyani Publications, New Delhi.

26. Davis, G. L. 1966. Systematic Embryology of the Angiosperms. Wiley, New York.

27. Dzevaltovsky, A. K. 1963. Cytoembryological investigations in a number of Cucurbitaceae species (in Russian). Ukrajins'k Bot. Zhurn. 20: 16–19.

28. Dzevaltovsky, A. K. 1963. Development of female gametophyte in *Benincasa hispida* Thunb. (in Russian). *In* D. K. Zarov, ed., Nutrition, Physiology, Cytoembryology and Ukranian Flora. Ukrajins'k Bot. Inst. RSR, Kiev.

29. Dzevaltovsky, A. K. 1963. Peculiarities of development of the female gametophyte in *Momordica charantia* L. (in Russian). Ukrajins'k Bot. Zhurn. 20: 53–60.

30. Dzevaltovsky, A. K. 1967–68. Megasporogenesis in triploid watermelon. J. Mysore Univ. Golden Jubilee Vol., Sec. B: 11–17.

31. Dzevaltovsky, A. K. 1967–68. Abnormalities in the embryo sac formation of triploid watermelon. J. Mysore Univ. Golden Jubilee Vol., Sec. B: 18–23.

32. Dzevaltovsky, A. K. 1970. Some peculiarities of meiosis in microsporogenesis in an experimental polyploid form of watermelon (in Russian). Ukrajins'k Bot. Zhurn. 27: 425–430.

33. Guliajev, V. A. 1961. Notes on fertilization in *Citrullus vulgaris* Schrad. (in Russian). Buyll. Vsesojuzn. Inst. Rasteniev. 9: 27–30.
34. Heimlich, L. F. 1927. Microsporogenesis in cucumber. Proc. Natl. Acad. USA. 13: 113–115.
35. Heimlich, L. F. 1927. The development and anatomy of the staminate flower of the cucumber. Amer. J. Bot. 14: 13–115.
36. Hofmeister, W. 1849. Die Entstehung des Embryo der Phanerogamen, eine Reihe mikroskopischer Untersuchungen. Leipzig.
37. Johri, B. M., and C. R. Chowdhary. 1957. A contribution to the embryology of *Citrullus colocynthis* Schrad. and *Melothria maderaspatana* Cogn. New Phytol. 56: 51–60.
38. Kirkwood, J. E. 1904. The comparative embryology of the Cucurbitaceae. Bull. New York Bot. Gard. 3: 313–402.
39. Kirkwood, J. E. 1906. The pollen tube in some of the Cucurbitaceae. Bull. Torrey Bot. Club. 33: 327–341.
40. Kirkwood, J. E. 1907. Some features of pollen formation in the Cucurbitaceae. Bull. Torrey Bot. Club. 34: 221– 242.
41. Longo, B. 1901. La mesogamia nella commune zucca *Cucurbita pepo*. Compt. Rend. Acad. Lincei, Roma 10: 168–172.
42. Longo, B. 1902. La nutrizione delle, embrione delle *Cucurbita* operate per mezzo del tubetto pollinico. Ann. Bot. (Rome). 1: 71–74.
43. Longo, B. 1904. Richerche sulle Cucurbitaceae e il significato del percorso intercellulare (endotropico) del tubetto pollinico. Mem. Reale Accad. Lincei 4: 523–549.
44. Mihov, A., and L. Zagorcheva. 1966. A case of apogamy in *Cucumis sativus*. L. Compt. Rend. Bulg. Sci. 19: 527–529.
45. Passmore, S. 1930. Microsporogenesis in the Cucurbitaceae. Bot. Gaz. 90: 213– 222.
46. Poddubnaja-Arnoldi, V. A. 1936. Beoachtungen über die Keimung des Pollens einiger Pflanzen auf. künstlichem Nährboden. Planta 25: 502–529.
47. Schagen, R. 1956. Embryologische Untersuchungen am Feigenblattkürbis (*Cucurbita ficifolia* Bouché) nach Bestäubung mit Pollen den Gartenkürbis (*C. pepo* L.). Flora 143: 91–126.
48. Schleiden, M. J. 1845. Ueber Amci's letzten Beitrag zur Lehre von der Befruchtung der Pflanzen. Flora 28: 593–600.
49. Schnarf, K. 1931. Vergleichendle Embryologie der Angiospermen. Gebruder Borntraeger, Berlin.
50. Scott, F. M. 1944. Cytology and microchemistry of nuclei in developing seed of *Echinocystis macrocarpa*. Bot. Gaz. 105: 330–338.
51. Scott, F. M. 1953. The physical consistency of the endosperm nucleus of *Echinocystis macrocarpa*. Phytomorphology 3: 66–76.
52. Shridhar, 1979. Studies on the development of anther and palynology of Indian Cucurbitaceae. Ph.D. dissertation, Rajasthan Univ., Jaipur.
53. Singh, D. 1955. Embryological studies in *Cucumis melo* var. *pubescens* Willd. J. Indian Bot. Soc. 34: 72–78.
54. Singh, D. 1955. Cytoplasmic nodules in the endosperm of some Cucurbitaceae. Nature 175: 607–608.

55. Singh, D. 1957. Endosperm and its chalazal haustorium in Cucurbitaceae. Agra Univ. J. Res. Sci. 6: 75–89.
56. Singh, D. 1961. Development of embryo in the Cucurbitaceae. J. Indian Bot. Soc. 40: 620–623.
57. Singh, D. 1963. Studies on the persistent pollen tubes of Cucurbitaceae. J. Indian Bot. Soc. 42: 208–213.
58. Singh, D. 1964. A further contribution to the endosperm of the Cucurbitaceae. Proc. Indian Acad. Sci. 60B: 399–413.
59. Singh, D. 1964. Cytoplasmic nodules in the endosperm of angiosperms. Bull. Torrey Bot. Club. 91: 86–94.
60. Singh, D. 1970. Comparative embryology of angiosperms. Bull. Indian Nat. Sci. Acad. 41: 212–219.
61. Singh, D., and A. S. R. Dathan. 1972. Development of generative and aposporic embryo sacs in C. *metuliferus* E. May. ex. Schrad. Curr. Sci. 41: 32–35.
62. Singh, D., and A. S. R. Dathan. 1973. Further studies in the endosperm of Cucurbitaceae. Pl. Sci. 5: 42–51.
63. Souégés, R. 1939. Embryogénie des Cucurbitacées. Développement de l'embryon chez le *Bryonia dioica*. Compt. Rend. Hebd. Seances Acad. Sci. 208: 227–229.
64. Tillman, O. J. 1906. The embryo sac and embryo of *Cucumis sativus*. Ohio Nat. 6: 423–430.
65. Turala, K. 1958. Endomitosis in the tapetal cells of *Cucurbita pepo*. Acta Biol. Cracov., Ser. Bot. 1: 25–35.
66. Turala, K. 1963. Studies in the endomitotical processes during the differentiation of the tapetal layer of the Cucurbitaceae. Acta Biol. Cracov., Ser. Bot. 61: 87–102.
67. Turala, K. 1966. Endopolyploidie im endosperm von *Echinocystis lobata*. Oesterr. Bot. Z. 113: 235–244.
68. Weiling, F., and R. Schagen. 1955. Über die Präparation und Gestalt des Endospermhaustoriums bei den gross-Samigen Kürbitsarten. Ber. Deutsch. Bot. Ges. 68: 2–10.
69. Zahur, M. R. 1962. Early ontogeny of the ovule in *Coccinia indica*. Biologia (Lahore) 9: 179–197.

16 | Palynology of the Indian Cucurbitaceae

Shridhar and Dalbir Singh

ABSTRACT. The palynology of 69 of the 108 species belonging to 34 genera of Cucurbitaceae occurring in India was studied. Palynologically the subfamily Zanonioideae is remarkably homogeneous. The pollen grains are small, three-zonocolporate, and prolate to subprolate with a striate exine. In the subfamily Cucurbitoideae pollen grains are variable in size, shape, and exine characteristics, and are operculate or not. They are predominantly three-zonocolporate or zonoporate, although four, five, seven, and eight to ten colpi characterize certain genera. Pollen characters are constant in most genera and some tribes and subtribes, but occasionally are variable in these taxa. Nine morphological groups of pollen are recognized on the basis of the number, position, and character of the apertures. The pollen groups are placed in an evolutionary sequence, and the taxonomic implications of pollen characters are discussed.

Pollen characters in the Cucurbitaceae are useful taxonomic indicators. In this chapter, characteristics of the pollen grains of the Cucurbitaceae occurring in India are summarized. The representation of taxa from this broad sample gives the work particular value in considering the range of variation in pollen grains in the family, serving as a basis for defining types of pollen grains, and giving insights into systematic and evolutionary relationships.

Previous Palynological Studies

Investigations on pollen morphology of the Cucurbitaceae were intiated in the 1920's with the pioneering studies of Zimmermann (37). Subsequently, pollen of the family was studied by a number of workers (1–9, 11,

13, 15, 17, 18, 22, 23, 27, 34, 35, 36). Of special note are the studies of Erdtman (13), who described the pollen morphology of 30 species in 26 genera and five tribes of the family, and Marticorena (22), who dealt with 23 species of Zanonioideae and 176 of Cucurbitoideae, covering all of the tribes and subtribes recognized by Jeffrey (19). Straka and Simon (34) placed 24 genera of the family in nine groups on the basis of type and number of pollen grain apertures and exine ornamentation, while Aleshina (5) in describing the pollen of some 100 species in 90 genera, recognized 11 basic pollen groupings.

In India studies of the palynology of Cucurbitaceae were initiated by Awasthi (6, 7). She described the great variation in the size, apertures, and exine excrescences of the pollen grains of *Cucurbita moschata* (Duch. ex Lam.) Duch. ex Poir. and *C. pepo* L. and reported normal pollen grains as large as 304 μm in diameter in *C. moschata*, perhaps the largest in the angiosperms (6). In an illustrated account of the pollen grains of 18 species belonging to 12 genera, she presented a key for their identification (7).

Nair and Kapoor (25) described pollen grains of 14 species in 9 genera, and Parveen and Bhandari (26) did so for 21 species in 13 genera, while also briefly discussing taxonomic relationships. Shridhar (29) reported the development of protoplasmic protuberances in pollen of four species, in which pollen tube initiation takes place before anthesis. Recently, Shridhar (30) investigated the pollen of 62 species of 28 genera, placed in 7 tribes and 11 subtribes of both Cucurbitoideae and Zanonioideae. In addition to usual palynological information, data were also recorded about minipollen grains, pollen sterility, and in situ pollen grain germination.

Results

The data collected in this chapter summarize and extend our knowledge of the pollen of indigenous and exotic Indian cucurbits. Observations were made of fresh and preserved pollen grains, with or without acetolysis, using both light and scanning electron microscopes.

Apertures of several types were recorded, including three-zonocolpate (*Indofevillea khasiana* Chatterj.), eight- to ten-zonocolpate (*Sechium*), three-zonocolporate (*Momordica, Benincasa, Citrullus, Coccinia, Lagenaria, Luffa, Ctenolepis, Dactyliandra, Corallocarpus, Melothria,* and all members of the Zanonioideae), five-zonocolporate (*Cyclanthera*), three-zonoporate, nonoperculated (*Cucumis* and *Trichosanthes*) or operculated (*Diplocyclos palmatus* (L.) C. Jeffr., *Biswarea, Edgaria,* and *Herpetospermum*), and pantoporate, operculated (*Cucurbita*). Plants with three-zonoporate pollen grains usually included some that were four-zonoporate. The pores were highly variable in size and shape, ranging from elliptical to circular to transversely elliptical. In a few instances (*Cucumis*), the pores were annulate. In *Gymnopetalum, Melothria*, and *Thladiantha* both porate and colporate pollen grains were present. The colpus or pore area was usually simple but rarely was crustate.

Pollen grain size, in terms of both polar and equatorial diameters, varied from about 30 to 60 μm in most Cucurbitoideae; however, larger grains are present in the Luffinae, Herpetosperminae, Hodgsoninae, and Cucurbiteae, with grains up to 195 μm observed in *Cucurbita pepo*. In the Zanonioideae pollen grains were consistently small, with those of *Gynostemma pedata* Blume less than 20 μm in polar and equatorial dimensions. In addition to normal-sized pollen grains, minipollen grains were observed in cultivated *Benincasa hispida* (Thunb.) Cogn., *Praecitrullus fistulosus* (Stocks) Pang., *Coccinia grandis* (L.) Voigt, *Lagenaria siceraria* (Mol.) Standl., *Cucurbita moschata, Luffa cylindrica* (L.) M. J. Roem., and a few wild species, including *Diplocyclos palmatus* and *Zehneria thwaitesii* (Schweinf.) C. Jeffr. (syn. *Melothria zeylanica* C. B. Clark).

In shape, pollen grains usually varied from oblate- to prolate-spheroidal. Those of the Cucurbiteae and Herpetosperminae were spheroidal. In a few cases, suboblate to oblate grains were observed (*Coccinia, Cyclanthera,* and *Cucumis*), while subprolate or prolate grains characterized the Zanonioideae.

Reticulate exine ornamentation dominated species of the sample, yet pollen grains of *Hodgsonia* were readily distinguished by their heterobrochate (macro- and microreticulations) exine. Spinescent exine surfaces occurred in *Cucurbita, Diplocyclos, Praecitrullus*, and *Sechium*. In *Cucurbita* the size and shape of the spines was variable, with both minute spinules and spines present. In other species of Cucurbitoideae the exine was smooth, punctate, honeycombed, budded, or warty. The exine surface of Zanonioideae was striate.

A lidlike structure covering the pore, termed the *operculum*, characterized the pollen grains of *Diplocyclos palmatus* and those of Cucurbiteae and Herpetosperminae. In the Cucurbiteae the operculum was spined (Figure 1A, B).

Pollen sterility was evident in a large number of cultivated and wild species. In situ germination of a few pollen grains was recorded in a number of species.

Taxonomic Distribution of Pollen Types

The Cucurbitaceae are eurypalynous, but the range of variability in pollen morphology within most subtribes and even tribes is of limited extent. Furthermore, data of Awasthi (6, 7), Nair and Kapoor (25), Parveen and

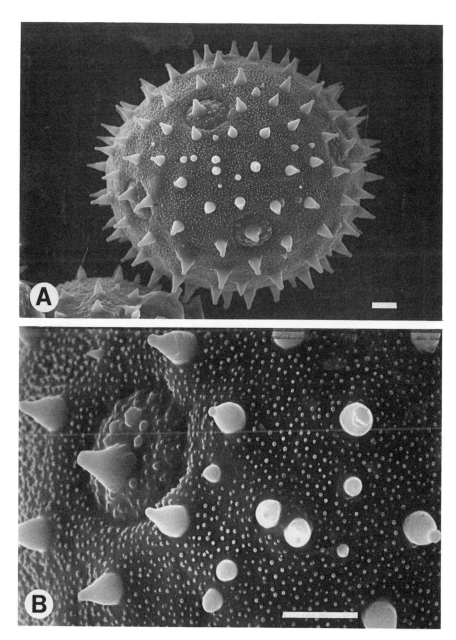

Figure 1. Scanning electron micrographs of *Cucurbita pepo*. A. A pollen grain. Bar represents 10 μm. B. A pollen grain surface showing the spinate operculum. Bar represents 10 μm.

Bhandari (26), and Shridhar (30) clearly reveal that most genera are palynologically uniform. Thus, pollen morphology has considerable taxonomic value in defining genera and higher level taxa (13, 14, 20, 36).

In the Joliffieae, of which only the subtribe Thladianthinae is represented in India, pollen grains are usually three-zonocolporate and rarely, e.g., *Thladiantha cordifolia* (Blume) Cogn. (syn. *T. calcarata* Cogn.), three-zonoporate or three-zonocolpate.

In *Benincasa, Bryonia, Citrullus, Praecitrullus, Coccinia,* and *Lagenaria* of the Benincaseae, pollen grains are three-zonocolporate (7, 26, 30). Nair and Kapoor (25) reported mostly three-colpate pollen grains in *Benincasa hispida,* but others found them to be three-zonocolporate (5, 22, 30). Jeffrey (20) noted that three-zonocolporate, more or less prolate, and predominantly reticulate pollen grains are typical of the subtribe Benincasinae. One principal exception is found in *Diplocyclos,* where *D. palmatus* and two other species have spherical, operculated, spinescent, three-zonoporate pollen grains (22, 30). The subtribe Luffinae also has three-zonocolporate pollen grains with a reticulate exine, but they are larger than those of the Benincasinae (7, 25, 26, 30).

In the Cucurbiteae the pollen grains are porate, operculated, spherical, and spinous (6, 25, 30). Similar pollen grains are recorded in the Trichosantheae subtribe Herpetosperminae (*Biswarea, Edgaria,* and *Herpetospermum*) and in *Diplocyclos,* as mentioned above.

The only member of the Sicyeae subtribe Cyclantherinae represented in India, *Cyclanthera pedata* (L.) Schrad., possesses five-zonocolporate pollen grains (26, 30). Marticorena (22) showed that, except for *Echinopepon wrightii* (A. Gray) S. Wats. (as *Echinocystis wrightii* A. Gray) with colpate pollen grains, pollen grains of the subtribe are colporate. In *Sechium* (Sicyeae subtribe Sicyinae) the pollen grains are eight- to ten-zonocolpate, while in *Schizopepon* (Schizopeponeae) they are three-zonocolporate.

In the Trichosantheae subtribe Trichosanthinae, three-zonoporate pollen grains are associated with *Trichosanthes* (7, 25, 26, 30), although Marticorena (22) reported three-zonocolporate in *T. beccariana, T. kirilowii* var. *japonica* (Mig.) Kitamura (as *T. japonica* (Miq.) Regel), *T. pedata,* and an unidentified species. Of the two species of *Gymnopetalum, G. chinensis* (Lour.) Merr. has three-zonocolporate pollen grains, but those of *G. wightii* Arn. are three-zonoporate. These species also differ in their pollen size, 54–60 μm and 36–42 μm in polar view, and exine characteristics, smooth and reticulate, respectively. Differences of this magnitude suggest that the genus may be polyphyletic.

Of the other three subtribes of Trichosantheae, the Hodgsoniinae have three-zonocolporate, reticulate pollen grains (7), and the Herpetosperminae have three-zonoporate, operculated pollen grains (7, 22). Ampelosicyinae are not represented in the Indian flora.

Two major pollen types are found in the subtribes Dendrosicyinae and Cucumerinae of the Melothrieae, being three-zonocolporate in *Corallocarpus* and *Kedrostis* of the former and three-zonoporate in *Cucumis* of the latter. Jeffrey (20), using the subtribal name Melothriinae, reported that the pollen grains are usually three-colporate in the Cucumerinae, but that *Cucumis*, *Cucumella*, *Mukia*, and *Oreosyce* form a distinct subgroup with three-zonoporate, oblate to suboblate pollen grains. Similar pollen grains occur in *Dicoelospermum*, a genus now placed in the subtribe (21).

In redefining *Melothria*, Jeffrey (19) transferred species from it to *Mukia*, *Solena*, and *Zehneria*. Pollen characters, like those of seeds (32, Singh and Dathan, this volume) tend to support the segregation of *Mukia* and *Solena*, but not that of *Zehneria*. Pollen grains in *Mukia* are three-zonoporate, while those of *Melothia* and *Zehneria* are three-zonocolporate. *Solena amplexicaulis* (Lam.) Gandhi, which also has colporate pollen grains, stands apart in its verrucate exine.

The uniformity in pollen morphology in the subfamily Zanonioideae correlates with the uniformity of its vegetative and floral morphology (21) and seed coat structure (31, Singh and Dathan, this volume).

Pollen Morphoforms and Their Relationships

Aleshina (5) identified 4 basic types and 11 groups or subtypes of pollen in the Cucurbitaceae, based on exine ornamentation and aperture characteristics. She proposed four lines of evolution from the basic colpate pollen grain. The first four groups, represented by *Sicyos*, *Praecitrullus*, *Cucurbita*, and *Diplocyclos*, respectively, have spinate exine and colpate-colporate-porate (zoned and dispersed porate) aperture characters. The remaining seven groups, including two intermediary types, possess nonspinous grains. Three of these groups, represented by *Herpetospermum*, *Trichosanthes*, and *Cucumis*, respectively, have porate pollen grains; three groups, as shown by *Zanonia*, *Bryonia*, and *Lagenaria*, have colporate pollen grains; and one group, of which *Cyclanthera* is an example, has both colporate and porate grains.

Based on information concerning the number, position, and character of pollen grain apertures described in the literature (5, 22, 30) and observed in this study, a somewhat different classification of pollen types emerges. Pollen grains seem to fall into nine principal groups or morphoforms:

1. Three-zonocolpate, subprolate—*Indofevillea*, *Thladiantha cordifolia*
2. Multicolpate, spherical—Sicyeae
3. Three-zonoporate, nonoperculated—Trichosanthinae, Cucumerinae

4. Multizonoporate, nonoperculated—*Zehneria peneyana* (Naud.) Aschers. & Schweinf.
5. Three-zonoporate, operculated—*Diplocyclos*, Herpetosperminae
6. Pantoporate, operculated—Cucurbiteae
7. Three-zonocolporate—most common throughout the family
8. Multizonocolporate—Cyclantherinae
9. Multicolporate—*Rytidostylis* (syn. *Elaterium*)

In keeping with generally accepted views of pollen evolution (12, 14, 16, 24, 33), pollen grain morphoforms in the Cucurbitaceae could be interpreted phylogenetically in the following way. From the basal three-zonocolpate pollen grains of *Indofevillea* and *Thladiantha* of the Joliffieae, evolution might have progressed along three lines. The first line may have led to the multicolpate pollen characteristic of the Sicyeae, while the second line resulted in the three-zonoporate, nonoperculated pollen of the Trichosanthinae, Cucumerinae, and Guraniinae. This second lineage, in turn, may have given rise to the multizonoporate, nonoperculated pollen of *Zehneria peneyana*, on the one hand, and to three-zonoporate, and eventually pantoporate, operculated pollen grains. The third line of evolution involved the development of composite apertures, i.e., three-zonocolporate, which further evolved along two pathways—one with small grains of the Zanonioideae and the other the larger grains in the Cucurbitoideae. The Cucurbitoideae of evolution apparently terminates in the multizonocolporate grains of the Cyclantherinae, perhaps with an offshoot to multicolporate grains, as recorded in *Rytidostylis*. Phylogenetically, pollen characters do not support subfamilial segregation of the Zanonioideae.

In the Cucurbitaceae, evolution of pollen morphoforms seems to have proceeded from simple colpate apertures in zoned positions, then to pores, and finally to the colporate condition. Further evolution took place toward a greater number of simple aperatures in zoned and eventually dispersed positions and in the organization of the operculum in the porate lines.

Literature Cited

1. Aleshina, L. A. 1964. On the pollen grains of the Cucurbitaceae (in Russian). Bot. Zhurn. SSSR 49: 1773–1776.
2. Aleshina, L. A. 1966. Palynological data on the systematics of the tribe Fevilleae Pax. (Cucurbitaceae Juss.) (in Russian). Bot. Zhurn. SSSR 51: 244–250.
3. Aleshina, L. A. 1966. Morphology of the pollen grains of family Cucurbitaceae Juss. (in Russian). *In* The Importance of Palynological Analysis for Stratigraphic and Paleofloristic Investigations. Acad. Sci. USSR, Moscow. Pp. 15–22. For the 2nd Intl. Conf. Palynology, Utrecht.
4. Aleshina, L. A. 1967. The morphology of pollen grains of Himalayan Cucurbitaceae (subtribe Herpetosperminae C. Jeffr.) (in Russian). Bot. Zhurn. SSSR 52: 865–867.

5. Aleshina, L. A. 1971. The palynological data on systematics and phylogeny of the family Cucurbitaceae Juss. (in Russian). *In* L. A. Kuprianova and M. S. Yakulev, eds., Morphologie du Pollen. Izdat Nauka, Leningrad.
6. Awasthi, P. 1961. On the morphology of pollen grains of the two species of *Cucurbita* L. Pollen Spores 4: 263–268.
7. Awasthi, P. 1962. Palynological investigations of Cucurbitaceae. Proc. Natl. Inst. Sci. India B, Biol. Sci. 28: 285–496.
8. Batalla, M. A. 1940. Estudia morphologico de los granos de polen de las plantas vulgares des valle de Mexico. Anales Inst. Biol. Univ. Nac. Mexico 11: 129–161.
9. Campso, M. 1962. Pollen grains of plants of the Cerrado. IV. Revista Brasil. Biol. 22: 307–315.
10. Chakravarty, H. L. 1959. Monograph on Indian Cucurbitaceae. Records Bot. Survey India. 17: 1–234.
11. Cranwell, L. 1942. New Zealand pollen studies 1. Key to the pollen grains of families and genera in the native flora. Rec. Auckland Inst. Mus. 2: 260–308.
12. Doyle, J. A. 1969. Cretaceous angiosperm pollen of the Atlantic coastal plain and its evolutionary significance. J. Arnold Arbor. 50: 1–35.
13. Erdtman, G. 1952. Pollen Morphology and Plant Taxonomy: Angiosperms. Chronica Botanica, Waltham, MA.
14. Erdtman, G. 1963. Palynology. *In* R. D. Preston, ed., Advances in Botanical Research. Vol. 1. Academic Press, London.
15. Erdtman, G., B. Berglund, and J. Praglowski, 1961. An introduction to a Scandinavian pollen flora. Grana Palynol. 2: 1–92.
16. Faegri, K., and J. Iverson, 1964. Textbook of Pollen Analysis. Munksgaard, Copenhagen.
17. Horigome, K. 1961. Forms of pollen grains in Cucurbitaceae. Collect. Breed. 23: 271–275.
18. Ikuse, M. 1956. Pollen Grains of Japan. Hirokawa, Tokyo.
19. Jeffrey, C. 1961. Notes on Cucurbitaceae, including a proposed new classification of the family. Kew Bull. 15: 337–371.
20. Jeffrey, C. 1964. A note on pollen morphology on Cucurbitaceae. Kew Bull. 17: 473–477.
21. Jeffrey, C. 1966. On the classification of Cucurbitaceae. Kew Bull. 20: 417–426.
22. Marticorena, C. 1963. Material para una monografia de la morfología del polen de Cucurbitaceae. Grana Palynol. 4: 78–91.
23. Melhem, T. S. 1966. Pollen grains of plants of Cerrado. XII. Cucurbitaceae, Menispermaceae and Moraceae. Anais Acad. Brasil. Ci. 38: 195–203.
24. Nair, P. K. K. 1970. Pollen Morphology of Angiosperms: An Historical and Phylogenetic Study. Vikas, New Dehli.
25. Nair, P. K. K., and S. Kapoor. 1974. Pollen morphology of Indian vegetable crops: Cucurbitaceae. *In* P. K. K. Nair, R. K. Grover, and T.M. Varghese, eds., Glimpses in Plant Research. Vol. 2. Vikas, New Delhi.
26. Parveen, F., and M. M. Bhandari. 1977. Pollen morphology of Indian desert Cucurbitaceae. Trans. Isdt. Ucds. 2: 212–222.
27. Saad, S. I. 1964. Pollen morphology of some Egyptian Cucurbitaceae. Pollen Spores 6: 113–124.
28. Selling, O. 1947. Studies in Hawaiian pollen statistics. Part II. The pollen of the Hawaiian Phanerogams. Special Publ. Bernice Pauhi Bishop Mus. 38. Honolulu.

29. Shridhar. 1973. Protoplasmic germinal processes in some cucurbits. Curr. Sci. 42: 397–398.
30. Shridhar. 1979. Studies on the Development of Anther and Palynology of Indian Cucurbitaceae. PhD. dissertation, Rajasthan Univ., Jaipur.
31. Singh, D., and A. S. R. Dathan. 1973. Structure and development of seed coat in Cucurbitaceae. XI. Seeds of Zanonioideae. Phytomorphology 23: 138–148.
32. Singh, D., and A. S. R. Dathan. 1974. Structure and development of seed coat in Cucurbitaceae. V. Seeds of *Melothria* Linn. Bull. Bengal Bot. Soc. 28: 47–56.
33. Sporne, K. R. 1972. Some observations on the evolution of pollen types in dicotyledons. New Phytol. 71: 181–185.
34. Straka, H., and A. Simon. 1969. Palynologia Madagassica et Mascarenica. Cucurbitaceae. Pollen Spores 11: 311–325.
35. Tarnavschii, I. T., and D. Redulescu. 1961. Contributii la cunoasterea morphologiei des microsporilor de Cucurbitaceae. Stud. Cercet. Biol. (Cluj). 13: 29–47.
36. Wodehouse, R. P. 1935. Pollen Grains. Their Structure Identification and Significance in Science and Medicine. McGraw Hill, New York.
37. Zimmermann, A. 1922. Die Cucurbitaceen. 2 Vols. Fischer, Jena.

17 | Structure, Ontogeny, Organographic Distribution, and Taxonomic Significance of Trichomes and Stomata in the Cucurbitaceae

J. A. Inamdar, M. Gangadhara, and K. N. Shenoy

ABSTRACT. The structure, ontogeny, and organographic distribution of trichomes and stomata in 10 genera and 17 species of Cucurbitaceae were studied. Depending on their form, structure, ontogeny, and contents, trichomes are classified into three categories: eglandular, glandular, and glandular-cum-eglandular. Eglandular trichomes are subdivided into three main types, i.e., unicellular, bicellular, and multicellular, and 17 subtypes. Glandular trichomes are classified into 10 subtypes. All trichomes originate from a single trichome initial. Stomata are anomocytic, paracytic, anisocytic, tetracytic, cyclocytic, or staurocytic, or with a single subsidiary cell. The ontogeny of all stomatal types, except paracytic and stomata with a single subsidiary cell, is perigenous. The ontogeny of stomata with a single subsidiary cell is either perigenous or mesogenous, while that of the paracytic type is mesoperigenous or perigenous. Variations, including single guard cells and the presence of contiguous stomata are described. The taxonomic significance of trichomes and stomata is discussed.

This chapter summarizes information about trichomes and stomata in the Cucurbitaceae, integrating into the summary data from 17 recently studied taxa. Comments are also presented concerning the significance of trichomes and stomata in understanding the systematics of the family.

Previous Studies

Research concerned with trichomes and stomata of the Cucurbitaceae has been carried on by several workers. Seven types of foliar trichomes and extrafloral nectaries in the Cucurbitaceae were recorded by Metcalfe and Chalk (15). Zimmermann (27) and Schroedter (22) studied the physiology

209

of trichomes, and extrafloral glands were studied by Chakravarty (3–5). Pant and Banerji (16) described the development of capitate glandular hairs and uniseriate multicellular trichomes in *Coccinia grandis* (L.) Voigt (as *C. cordifolia* auct.). The structure, ontogeny, classification, and organographic distribution of trichomes in the Cucurbitaceae were investigated by Inamdar and Gangadhara (12).

Metcalfe and Chalk (15) found ranunculaceous stomata either confined to the lower surface or present on both surfaces of leaves in the Cucurbitaceae. Pant and Banerji (16) studied the ontogeny of stomata in five species and reported that the stomata were either perigenous or meso-perigenous in their development. Stomatal variations in the anther epidermis of *Momordica charantia* L. were reported by Rao and Ramayya (20). According to Ramayya and Rao (19), stomata are anomocytic in *Cucumis melo* L. (as *C. pubescens* Willd.), and some of the stomata tend to be anisocytic and are mesoperigenous in their ontogeny. Recently, Inamdar and Gangadhara (13) investigated the structure and ontogeny of stomata in 12 species of cucurbits.

Methods and Materials

Preparations for the study of both trichomes and stomata were made from fresh and herbarium materials. Epidermal peels were stained with haematoxylon, washed in distilled water, and mounted in glycerin.

Taxa included in this study were *Citrullus colocynthis* (L.) Schrad., *Coccinia grandis*, *Cucumis melo* var. *agrestis* Naud. (syn. *C. callosus* (Rottl.) Cogn.), *C. sativus* L., *Cucurbita maxima* Duch. ex Lam., *Lagenaria siceraria* (Mol.) Standl., *Luffa acutangula* var. *amara* (Roxb.) C. B. Clarke, *L. aegyptiaca* Mill. (syn. *L. cylindrica* (L.) M. J. Roem.), *Momordica charantia*, *M. cochinchinensis* (Lour.) Spreng., *M. dioica* Roxb. ex Willd., *Mukia maderaspatana* (L.) M. J. Roem. (syn. *Melothria maderaspatana* (L.) Cogn.), *Sechium edule* (Jacq.) Sw., *Trichosanthes cucumerina* L. var. *cucumerina* and var. *anguina* (L.) Haines, *T. dioica* Roxb., and *Zehneria scabra* (L. f.) Sond. (syn. *Melothria perpusilla* sensu Chakravarty).

Trichomes

Types of Trichomes

Trichomes of the Cucurbitaceae fall into three major types: eglandular, glandular, and glandular-cum-eglandular. Within these types, they vary from unicellular to multicellular. A summary of the morphology of trichome types, illustrated in Figures 1 and 2, follows.

I. Eglandular. Eglandular trichomes range from those of simple structure or form to those that are complex. The nearly universal appearance of multicellular, unbranched, conical trichomes (Figure 1J), suggests that this may be the basic type in the family from which both reduced and more elaborate trichomes were derived. In the following characterizations, unless otherwise noted, the outer walls of the trichomes are thin and straight, convex, or concave. The surface is smooth, and the cuticle and internal cross walls are thin.

    A. Unicellular (Figure 1A–D). Trichomes unbranched, conical (A), spinelike (B), hooked (C), or wartlike (D); outer walls and cuticle thin or thick; surface smooth or rough; lumen narrow to wide; base surrounded by two to many epidermal cells.

    B. Bicellular (Figure 1E–G). Trichomes unbranched, conical (E), spinelike (F), or hooked (G); outer walls, cuticle, and cross walls thin or thick; surface smooth or rough; base broad, simple or compound.

    C. Multicellular

        1. Simple filiform (Figure 1H). Trichomes with simple foot, uniseriate stalk, and head either of uniform thickness throughout or tapered slightly to the apex; surface smooth or rough.

        2. Branched (Figure $1I_1$, I_2). Trichomes with simple foot and multicellular, branched body; outer walls mostly thin and often constricted at the joints.

        3. Conical (Figure 1J–L). Trichomes with a broad, bulbous, simple (J_1) or compound foot (J_2) and a uniseriate, conical body consisting of two to many cells; apex unbranched, acute, obtuse, or hooked (K), or bifurcated (L); outer walls often nodulose at the joints; cross walls thin or thick.

        4. Cylindrical (Figure 1M). Trichomes uniseriate, cylindrical; foot simple or compound; apex obtuse; outer walls convex or concave, constricted at the joints.

        5. Shaggy (Figure 1N, O). Trichomes with a simple or compound foot and a multicellular, unbranched (N) or branched (O) body.

        6. Scalelike (Figure 1P). Trichomes scalelike, one cell thick, differentiated into a broad base and a flat, multicellular body; tip of the scales thickened, hard, and acute, or unthickened, obtuse. These ramentalike scales occur on the anthers of a number of species and rarely on the filaments of *Mukia maderspatana* and *Luffa acutangula* var. *amara*. Unicellular, scalelike structures were noted on the stigma of *Momordica cochinchinensis*.

        7. Fusiform (Figure 1Q). Trichomes with a simple foot and multicellular, fusiform body.

II. Glandular. Classes of glandular trichomes, as those of nonglandular types, tend to intergrade and could be related in various ways. Here distinctions are made on the degree to which the secretory head is defined, its

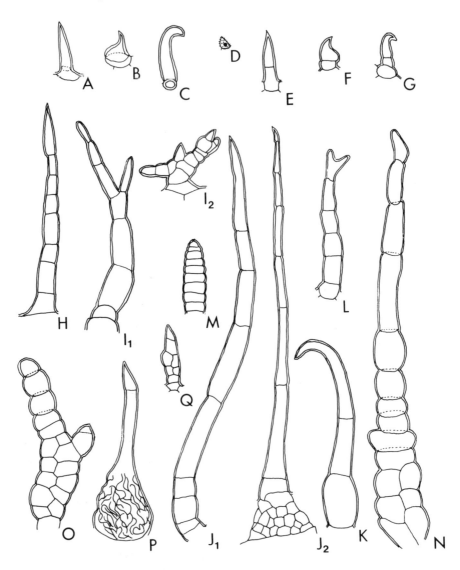

Figure 1. Eglandular trichome types in the Cucurbitaceae. A–D. Unicellular, conical (A), spinelike (B), hooked (C), and wartlike (D). E–G. Bicellular, conical (E), spinelike (F), and hooked (G). H. Multicellular, simple filiform. I_1, I_2. Multicellular, branched. J–L. Multicellular, conical with simple foot (J_1), conical with compound foot (J_2), apex hooked (K), and apex bifurcated (L). M. Multicellular, cylindrical. N, O. Multicellular, shaggy, unbranched (N), branched (O). P. Multicellular, scalelike. Q. Multicellular, fusiform. (A, E, I_2, *Trichosanthes cucumerina* var. *anguina*; B, O, *Momordica dioica*; C, K, *Cucurbita maxima*; D, *Momordica charantia*; F. *Zehneria scabra*; G, P, *Trichosanthes cucumerina* var. *cucumerina*; H, *Trichosanthes dioica*; I_1, N, *Luffa acutangula* var. *amara*; J_1, *Momordica cochinchinensis*; J_2, *Cucumis sativus*; L, Q, *Luffa aegyptiaca*; M, *Cucumis melo* var. *agrestis*.)

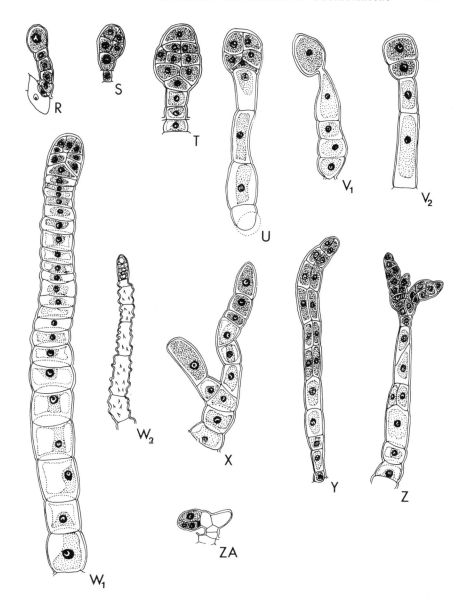

Figure 2. Glandular trichome types in the Cucurbitaceae. R. Capitate with unicellular head. S–U. Capitate with unicellular (S), bicellular (T) and multicellular (U) stalks, single foot cell. V. Explosive, with unicellular (V_1) and bicellular (V_2) heads. W. Uniseriate, filiform with convex walls (W_1) and walls with projections (W_2). X. Uniseriate, filiform, branched. Y. Fusiform, with uniseriate stalk and multicellular head. Z. Branched with multicellular stalk and heads. ZA. Glandular-cum-eglandular. (R, T, *Momordica dioica;* S, *Citrullus colocynthis;* U, Y, Z, *Luffa aegyptiaca;* V_1, *Momordica charantia;* V_2, W_2, *Cucumis melo* var. *agrestis;* W_1, *Cucurbita maxima;* X, *Trichosanthes dioica;* ZA, *Zehneria scabra.)*

persistence or lack of it, and the number and arrangement of cells in the stalk and head. All glandular trichomes have a single foot cell and thin cross walls and cuticle.

A. Capitate (Figure 2R–U). Trichomes with a uniseriate stalks composed of one (S), two (T), or more cells (U) and a unicellular (R), bicellular, or multicellular head (S–U); outer cell walls thin, convex, concave, or rarely straight; surface smooth.

B. Explosive (Figures $2V_1$, V_2). A modification of the capitate type, in which the unicellular or bicellular head dehisces at maturity to leave only a stumplike stalk.

C. Uniseriate filiform (Figure $2W_1$, W_2, X). Trichomes with a uniseriate, filiform, unbranched (W_1, W_2) or branched (X) stalk; head terminal, not expanded, multicellular or unicellular; outer walls straight, convex, concave, or sinuous; surface smooth or rough with projections (W_2). Trichomes of this type occur commonly on floral organs, especially the calyx and corolla.

D. Fusiform (Figure 2Y). Trichomes with a uniseriate stalk and multicellular, unbranched, fusiform head; outer walls straight, convex, or concave.

E. Branched with multicellular stalk and head (Figure 2Z). Trichomes with a multicellular stalk and branched, multicellular head; outer walls straight, convex, or concave. Trichomes of this type probably have arisen from the fusiform type. Both types occur in *Luffa aegyptica*, although only the branched type has been recorded in *Trichosanthes dioica*.

III. Glandular-cum-eglandular (Figure 2ZA). Trichomes branched with one arm glandular, the other eglandular. Apparently a modification of the preceding branched trichome, it has been observed only in *Zehneria scabra*.

Organographic Distribution of Trichomes

The distribution of trichomes is amazingly complex and difficult to characterize in a meaningful way. Complexity results from 1) the diversity in types of trichomes, 2) marked differences in the density of trichomes types, taken individually and collectively, and 3) differences in the distribution of trichome types on different parts of a given plant, e.g., upper and lower surfaces of the leaves, lower and upper parts of the calyx and corolla, or anthers.

Within the space limitation of this presentation the data cannot be presented fully, yet Tables 1 and 2 provide illustrative examples. The types of trichomes observed in each of the 17 species examined are summarized in Table 1. A sample of five taxa selected to demonstrate variability in the type, density, and distribution of trichomes in the Cucurbitaceae is given in Table 2.

Table 1. Occurrence of trichome types in 17 taxa of Cucurbitaceae

Taxon	Trichome types[1]	
	Eglandular	Glandular
Melothrieae		
Cucumis melo var. *agrestis*	A, B, M, J, P	R, T, U, V, W
Cucumis sativus	B, E, J, M, P	S, T, U, V, W
Mukia maderaspatana	J, K, P	T, W
Zehneria scabra	A, B, E, F, I, J, N	S, T, U, ZA
Joliffieae		
Momordica charantia	D, H, J, K, Q	U, V, W
Momordica cochinchinesis	H, J, P, Q	S, T, U, W
Momordica dioica	A, B, C, E, J, K, N, O, Q	R, S, T
Trichosantheae		
Trichosanthes cucumerina		
var. *anguina*	A, E, I, J, K, M, P	S, T, U, V
var. *cucumerina*	B, G, I, J, M, P	S, T, U, V, W
Trichosanthes dioica	H, J	S, T, Z
Benincaseae		
Citrullus colocynthis	H, J	S, T, U
Coccinia grandis	J	S
Lagenaria siceraria	E, F, H, J, P	S, T, U, W
Luffa acutangula var. *amara*	F, H, I, J, N, P	S, T, V, W
Luffa aegyptiaca	B, E, H, I, J, K, M	S, T, U, W, Y, Z
Cucurbiteae		
Cucurbita maxima	A, E, F, I, J, K	S, T, U, V, X
Sicyeae		
Sechium edule	E, F, I, J, K, M, P	S, U, V, W

[1]The key to the letters and trichome types is presented in the text and illustrated in Figure 1 (eglandular) and Figure 2 (glandular). The illustrations are representative only.

General Development of Trichomes

Both glandular and eglandular trichomes originate from a single initial, which can be easily distinguished from the adjacent epidermal cells by its larger size, prominent nucleus, dense staining properties, and papillate nature. The hair initial divides further to give rise to different types of vegetative and floral, glandular and eglandular, trichomes.

In the eglandular trichomes the hair initial may not divide or more usually it divides periclinally to give rise to a basal and a terminal cell. The terminal cell undergoes one or more periclinal, sometimes anticlinal or irregular, divisions to form different types of eglandular trichomes.

In the glandular trichomes the hair initial undergoes two periclinal divisions to produce basal, stalk, and head cells. The basal cell forms a simple foot. The stalk cell may or may not undergo periclinal divisions. The head cell may remain undivided or may undergo divisions to produce a capitate, or fusiform glandular head. The hair initial sometimes divides periclinally to

Table 2. Organographic distribution of trichomes

Taxon	Organ[1]	Type of trichome[2] and frequency		
		Common	Occasional	Rare
Coccinia grandis	LL		J, S	
	LU		J, S	
	P		J, S	
	S		J, S	
	T		J, S	
	CL		J, S	
	CU		J, S	
Luffa aegyptica	LL	J	E, K, S, T	
	LU	J	E, K, S, T	
	P	J	K	E
	S	J		E, K
	T		J	
	CL	J	E	
	CU	J	E	
	H	J	E	
	Pd	H, S	T	
	Ca L	J, M, S, T, U	B, Q	
	Ca U	J, U, Z, ZA	B	M
	Co L	H, I, U, W		L
	Co U	U		
	F			H
	Pe	I	S	
Momordica dioica	LL	J	E, T	R
	LU	J	E, T	
	P		T	
	S			T
	T			T
	BL	A, N, S	K, T	U
	BU	B, N	K	O
	Ca L	R	Q	B
	Ca U			
	Co L		C, R	
	Co U			
	St	J		B
	Pe	J	T	B
Sechium edule	LL	E, J, S, T		
	LU	E, J	U	S, T
	Pd	S, U, V		T
	Ca L	U, V		S
	Ca U	E, U, V		S
	Co L	E, I, J, S, T, U, V		F
	Co U	E, J, S		K, M
	A	P		

Table 2. (Continued)

Taxon	Organ[1]	Type of trichome[2] and frequency		
		Common	Occasional	Rare
Trichosanthes cucumerina var.	LL	J, V		S, T
anguina	LU	J, V		
	P	J, V		
	S	J, V		S, T, U
	Pd	J, V	A	E, S
	Ped	J, V		K, S, U
	Ca L	V	A	E, K, S, T, U
	Ca U	V		K, S
	Co L	I, U	M, V	
	Co U	U		S
	A	P		

[1]LL, lower leaf surface; LU, upper leaf surface; P, petiole; S, stem; T, tendril; CL, lower cotyledon surface; CU, upper cotyledon surface; H, hypocotyl; Pd, peduncle; Ped, pedicel; BL, lower bract surface; BU, upper bract surface; Ca L, lower calyx; Ca U, upper calyx; Co L, lower corolla; Co U, upper corolla; F, filament; A, anther; St, style; O, ovary; Pe, pericarp.
[2]See Figures 1 and 2 for representative illustrations of trichome types.

give rise to a basal cell, which forms a simple foot, and a terminal cell. The terminal cell undergoes few to several periclinal divisions to produce an uniseriate, filiform, glandular trichome. Cystoliths are often observed in the base of the glandular trichomes.

The glandular-cum-eglandular trichome is branched and originates from a single hair initial. One arm of the trichome is glandular and the other is eglandular.

Stomata

Types of Stomata

The stomatal apparatus in Cucurbitaceae consists of two guard cells bounding a lenticular pore, the orientation of which is either parallel to or at right angles to the guard cells. This apparatus is surrounded either by typical epidermal cells or by one or more subsidiary cells. Cuticular thickenings around the pore sometimes extend beyond the guard cells. The majority of stomata are anomocytic or occasionally have only a single subsidiary cell. Other types are paracytic, tetracytic, staurocytic, anisocytic, or cyclocytic. These and other aspects of stomata observed in the taxa of this study are illustrated in Figure 3.

Variations in stomatal types included the presence of single guard cells with or without a pore (Figure 3G, I), or two or three contiguous stomata, which were juxtaposed, superimposed, or obliquely oriented (Fig-

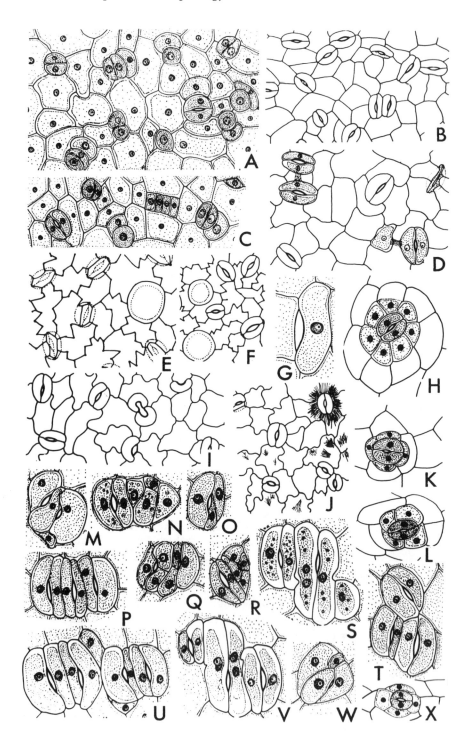

ure 3B, J). Also noted were the degeneration of one or both of the guard cells of a stoma (Figure 3B, D), the presence of cytoplasmic connections between guard cells of neighboring stomata or between guard cells and adjoining epidermal cells (Figure 3D), and the occurrence of divisions among the guard cells.

Stomata on the anthers of *Momordica charantia* exhibited a wide range of variations. These included two to four variously oriented, contiguous stomata (Figure 3P, T, U); longitudinal, transverse, and oblique divisions of one or both guard cells (Figure 3M, O, Q, R, W), although without formation of miniature stomata, as reported by Rao and Ramayya (20); single guard cells with a pore (Figure 3G); one normal stoma contiguous with a single guard cell with a pore, an apparatus that could be termed one and one half stomata (Figure 3M); elongation and notching of guard cells, binucleate guard cells, and miniature stoma formed by divisions of adjacent epidermal or subsidiary cells (all shown in Figure 3S, V); and normal stoma contiguous with a group of initials resulting from divisions of adjacent cells (Figure 3M).

Epidermal cells were polygonal, isodiametric, or elongated in various directions with straight (Figure 3A, C), sinuous (Figure 3I, J), arched (Figure 3B, D), or articulated (Figure 3E, F) anticlinal walls. Abundant needlelike crystals of calcium oxalate occured mostly in groups, or rarely were solitary in the epidermis of all organs of *Sechium edule* (Figure 3J).

Organographic Distribution of Stomata

Stomatal distribution varies among species and from organ to organ on a given plant. In general, however, stomata are common on leaves, although always more so on the lower surface than the upper, which in some instances may lack stomata altogether. They are expressed more sporadically on other organs and often are absent. Stomata are oriented in irregular patterns on foliar organs, but on axes they are oriented parallel to, at right angles to, or obliquely to the axis.

Figure 3. Stomatal characteristics in the Cucurbitaceae. A. *Coccinia grandis,* leaf, lower surface. B. *Momordica dioica,* leaf, lower surface. C. *Luffa aegyptica,* leaf, lower surface. D. *Coccinia grandis,* cotyledon, upper surface. E. *Trichosanthes cucumerina* var. *anguina,* mature leaf, lower surface. F. *Trichosanthes cucumerina* var. *cucumerina,* mature leaf, lower surface. G. *Momordica charantia,* anther. H. *Trichosanthes cucumerina* var. *anguina,* petiole. I. *Coccinia grandis,* mature leaf, lower surface. J. *Sechium edule,* leaf, lower surface. K–L. *Trichosanthes cucumerina* var. *cucumerina,* petiole. M–W. *Momordica charantia,* anther. X. *Trichosanthes cucumerina* var. *anguina,* stem.

Among the taxa of the study sample, notable distributional features are the following. The leaf blades and cotyledons are amphistomatic, except in *Zehneria scabra* and *Trichosanthes cucumerina* var. *anguina* in which they are hypostomatic, as are the bracts in the former. The sepals are amphistomatic, except in *Luffa aegyptica*, *Zehneria scabra*, *Momordica cochinchinensis*, and *M. dioica*, where they are hypostomatic. The petals are hypostomatic in *Citrullus colocynthis*, *Luffa aegyptica*, *L. acutangula* var. *amara*, *Lagenaria siceraria*, and *Trichosanthes cucumerina* var. *anguina*, and astomatic in the remaining species. In *Mukia maderaspatana* only the lower surface of sepals and petals is stomatic. The staminal filaments of *Citrullus colocynthis*, *Cucurbita maxima*, *Luffa aegyptica*, *Trichosanthes cucumerina* vars. *anguina* and *cucumerina*, *Sechium edule*, and *Momordica cochinchinensis* are stomatic, but are astomatic in *Lagenaria siceraria* and *Zehneria scabra*. The anthers of *Citrullus colocynthis*, *Cucumis melo* var. *agrestis*, *Luffa acutangula* var. *amara*, *L. aegyptica*, *Mukia maderaspatana*, *Trichosanthes cucumerina* vars. *anguina* and *cucumerina*, *Sechium edule*, and *Momordica cochinchinensis* are astomatic. They bear stomata in *Cucurbita maxima*, *Lagenaria siceraria*, *Momordica charantia*, *M. dioica*, and *Zehneria scabra*. The stomata are mostly confined to the connective region. The style is stomatic in all three species of *Momordica*, and the stigma is as well in *M. cochinchinensis*.

Other measures of stomatal variation are found in their frequency per unit area and their abundance relative to the number of epidermal cells. Both of these measures are presented in Table 3, the first as the number of stomata per square millimeter, for both the upper and lower leaf surfaces, and the second as stomatal indices, again from upper and lower leaf surfaces. The stomatal index is derived from the equation

$$SI = \frac{S}{S + E} \times 100$$

where S is the number of stomata and E the number of epidermal cells per unit area.

Development of Stomata

Stomata are initiated from protodermal cells, which characteristically are uninucleate, polygonal, and isodiametric with mostly straight walls and uniform staining properties. Meristemoids are cut off toward the corner or side of protodermal cells and can be distinguished by their smaller size, shape, prominent nucleus, and dense stain properties. They are solitary or occur in groups of two or three (Figure 3A, C). Further development is perigenous in all cases, but in stoma with a single subsidiary cell and paracytic stomata, it also may be mesogenous or mesoperigenous.

Table 3. Occurrence and frequency of stomata in 17 taxa of Cucurbitaceae

		Stomatal index[2]		Stomata/mm2	
	Stomatal types[1]	LL[3]	LU	LL	LU
Melothrieae					
Cucumis melo var. *agrestis*	A, C, SS	30	16	464	149
Cucumis sativus	A, C, P, SS	34	14	544	157
Mukia maderaspatana	A, C, P, SS	29	10	352	69
Zehneria scabra	A, P, SS, T	22	—	240	—
Joliffieae					
Momordica charantia	A, SS	29	10	688	24
Momordica cochinchinensis	A, P, SS	17	10	442	144
Momordica dioica	A, P, SS	18	10	312	122
Trichosantheae					
Trichosanthes cucumerina					
var. *anguina*	A, AN, C, P, T	17	—	208	—
var. *cucumerina*	A, AN, C, ST, T	17	—	320	—
Trichosanthes dioica	A, P, SS	18	—	285	—
Benincaseae					
Citrullus colocynthis	A, P, SS	31	25	536	339
Coccinia grandis	A, P, SS	22	7	256	59
Lagenaria siceraria	A, P, SS	21	9	316	72
Luffa acutangula var. *amara*	A, P, SS	28	7	331	48
Luffa aegyptiaca	A, C, P, SS	21	12	294	137
Cucurbiteae					
Cucurbita maxima	A, P, SS	27	8	621	198
Sicyeae					
Sechium edule	A, P, SS	17	6	339	86

[1] A, anomocytic; AN, anisocytic; C, cyclocytic; P, paracytic; SS, stoma with a single subsidiary cell; ST, staurocytic; T, tetracytic.
[2] See text for explanation.
[3] LL, lower leaf surface; LU, upper leaf surface.

Stoma with a single subsidiary cell.   In the mesogenous pathway, the meristemoid enlarges and is divided by a slightly curved wall to form two subequal cells. The larger cell differentiates as a single subsidiary cell, and the smaller one functions as a guard mother cell. The guard mother cell is divided by a straight wall, which is either parallel to the subsidiary cell or oblique to it, to produce the guard cells. The mature stoma is flanked by a single mesogene subsidiary cell (Figure 3A). In perigenous development, the meristemoid functions directly as the guard mother cell and divides to form the pair of guard cells. A subsidiary cell is derived from an adjacent epidermal cell or from the protoderm cell originally cut from the meristemoid. The mature stoma is flanked by a single perigene subsidiary cell.

Paracytic.   Stomata of this type may develop either mesoperigenously or perigenously. In the former the meristemoid divides unequally to produce a larger and a smaller cell. The larger cell differentiates as a subsidiary cell and the smaller gives rise to the pair of guard cells. A second, parallel subsidiary cell is cut off from an adjacent epidermal cell and the mature stoma is flanked by one mesogene and one perigene subsidiary cell. In perigenous development the subsidiary cells develop in one of two ways: 1) the two parallel subsidiary cells are cut off from adjacent epidermal cells, or 2) one of the subsidiary cells is derived from an adjacent epidermal cell, the other from the protodermal cell that gave rise to the meristemoid. The mature stoma is flanked by two parallel perigene subsidiary cells.

Anomocytic.   The meristemoid increases in size and divides by a straight wall to give rise to two guard cells. A lenticular pore develops between the two guard cells. The mature stoma is surrounded by three to seven ordinary epidermal cells (Figure 3A–C, F, J). Development of the succeeding types of stomata is similar to that of the anomocytic type except in the formation of the subsidiary cells and their number and orientation.

Anisocytic.   The three subsidiary cells are cut off from adjacent epidermal cells, of which one is distinctly smaller than the other two.

Tetracytic.   Two polar and two lateral subsidiary cells are cut off from the adjacent epidermal cells that surround the stoma (Figure 3K, X).

Cyclocytic.   A narrow ring of perigene subsidiary cells is cut off from adjacent epidermal cells to surround the stoma (Figure 3H).

Staurocytic.   Four perigene subsidiary cells are cut off from adjacent epidermal cells and are arranged in cruciate fashion around the stoma (Figure 3I). This particular type of stoma was not included in the ontogenetic summaries of Fryns-Claessens and Van Cotthem (7, 8).

Discussion

Previous, although as yet relatively limited studies, have documented the presence of considerable variation in trichomes and stomata of Cucurbitaceae. This study has extended these studies by characterizing more fully the range of trichome and stomatal types.

The significance of trichomes and stomata in assessing taxonomic relations is varied among plant taxa and has been widely discussed (1, 9–11, 16–18, 23–25). Of interest in the Cucurbitaceae is the degree to which they may be used 1) to shed light on the relationship of the family to other dicotyledonous families, and 2) to provide insights concerning intrafamilial relationships.

The Cucurbitaceae are generally allied with the Passifloraceae. Cronquist (6) placed both families in the Violales. Takhtajan (26) recognized the Cucurbitaceae as the single family of the Violales suborder Cucurbitineae and pointed out its relationship to the Passifloraceae, a relationship also recognized by Bentham and Hooker (2) and Rendle (21). Jeffrey (14) suggested that the Cucurbitaceae and Passifloraceace are remote. Stomata in the Passifloraceae are anomocytic, cyclocytic, paracytic, anisocytic with a single subsidiary cell. In cucurbits the stomata are dominately anomocytic, paracytic, or with single subsidiary cells, but include cyclocytic, anisocytic, and other types as well. The fact that the Cucurbitaceae and Passifloraceae share stomatal types may be indicative of a phylogenetic relationship, but the lack of specificity and general distribution of these types of stomata indicate that stomata are not definitive in this case.

Following Jeffrey's classification of the Cucurbitaceae (14, Appendix, this volume), all the genera represented in this study are members of the subfamily Cucurbitoideae. They are distributed through six of the seven tribes of the subfamily. Although the sample is small, it suggests in the array of trichome and stomatal variants that neither will be of major significance in defining tribes, subtribes, or genera. All the species of this sample bear both eglandular and glandular trichomes. Yet, despite the complexity in trichome expression, a relatively small number are basic. All species share multicellular conical, simple, eglandular trichomes and most have glandular hairs with a uniseriate stalk and multicellular head. Similarly, among stomata, anomocytic, paracytic, or those with a single subsidiary cell predominate and may be considered the basic types of the family.

Species exhibit some rather striking differences in trichome characteristics, even within genera, and these may prove to be useful as species markers; however, constancy in trichome expressions, in terms of form and distribution, will continue to remain uncertain until a broader sample has been examined. Similarly, species differ in a variety of stomatal characters, including those that are developmental, as well as those that relate to frequency and distribution. Further studies will reveal the extent to which the observed patterns are constant or labile.

Literature Cited

1. Bachmann, O. 1866. Untersuchungen über die Systematische Bedeutung der Schildhaare. Flora 69: 387–400.
2. Bentham, G., and J. D. Hooker. 1867. Genera Plantarum. Vol. 1. Reeve, London.
3. Chakravarty, H. L. 1937. Physiological anatomy of leaves of Cucurbitaceae. Philipp. J. Sci. 63: 409–431.
4. Chakravarty, H. L. 1948. Extrafloral glands of Cucurbitaceae. Nature 162: 576.

5. Chakravarty, H. L. 1951. Glandular product in *Coccinia cordifolia* Cogn. Sci. Cult. 17: 225–226.
6. Cronquist, A. 1981. An Integrated System of Classification of Flowering Plants. Columbia Univ. Press, New York.
7. Fryns-Claessens, E., and W. Van Cotthem. 1965. Stomata structuur bij de Marcgraviaceae. Biol. Jaarb. Dodonaea 33: 357–364.
8. Fryns-Claessens, E., and W. Van Cotthem. 1973. A new classification of ontogenetic types of stomata. Bot. Rev. 39: 71–138.
9. Hutchinson, J. 1969. Tribalism in the family Euphorbiaceae. Amer. J. Bot. 56: 738–758.
10. Inamdar, J. A. 1967. Studies on the trichomes of some Oleaceae: structure and ontogeny. Proc. Indian Acad. Sci. 66: 164–177.
11. Inamdar, J. A., D. C. Bhatt, R. C. Patel, and V. H. Dave. 1973. Structure and development of stomata in vegetative and floral organs of some Passifloraceae. Proc. Indian Natl. Sci. Acad. 39: 553–560.
12. Inamdar, J. A., and M. Gangadhara. 1975. Structure, ontogeny, classification and organographic distribution of trichomes in some Cucurbitaceae. Feddes Repert. 86: 307–320.
13. Inamdar, J. A., and M. Gangadhara. 1976. Structure, ontogeny and taxonomic significance of stomata in some Cucurbitaceae. Feddes Repert. 87: 295–312.
14. Jeffrey, C. 1967. On the classification of Cucurbitaceae. Kew Bull. 20: 417–426.
15. Metcalfe, C. R., and L. Chalk. 1950. Anatomy of Dicotyledons. Vols. 1 and 2. Clarendon, London.
16. Pant, D. D., and R. Banerji. 1965. Ontogeny of stomata and hairs in some cucurbits and allied plants. J. Indian Bot. Soc. 44: 191–197.
17. Patel, R. C., and J. A. Inamdar. 1972. Studies in the trichomes and nectaries of some Gentianales. *In* V. Puri, ed., Biology of Land Plants. Sarita Prakasan, India.
18. Rajagopal, T., N. Ramayya, and B. Bahadur. 1972. The development of organographically variable stomata in two species of *Turnera* L. J. Indian Bot. Soc. 51: 201–208.
19. Ramayya, N., and R. B. Rao. 1968. On the classification of certain angiospermous stomata. Curr. Sci. 37: 662–664.
20. Rao, R. B., and N. Ramayya. 1967. Stomatal abnormalities in two dicotyledons. Curr. Sci. 13: 357–358.
21. Rendle, A. B. 1952. The Classification of Flowering Plants. II. Dicotyledons. Cambridge Univ. Press, London.
22. Schroedter, K. 1926. Zur physiologischen Anatomie der Mittelzelle drusiger Gebilde. Flora 120: 19–86.
23. Singh, V., and D. K. Jain. 1975. Trichomes in Acanthaceae. I. General structure. J. Indian Bot. Soc. 54: 116–127.
24. Singh, V., and D. K. Jain, and M. Sharma. 1974. Epidermal studies in Berberidaceae and their taxonomic significance. J. Indian Bot. Soc. 53: 271–276.
25. Solereder, H. 1908. Systematic Anatomy of Dicotyledons. Vol. 2. Clarendon, London.
26. Takhtajan, A. L. 1980. Outline of the classification of the flowering plants (Magnoliophyta). Bot. Rev. 46:225–359.
27. Zimmermann, A. 1922. Die Cucurbitaceae. Fischer, Jena.

18 | Seed Coat Anatomy of the Cucurbitaceae

Dalbir Singh and A. S. R. Dathan

ABSTRACT. The Cucurbitaceae broadly follow a common pattern of seed coat development. The mature seed coat is elaborate and largely derived from the outer epidermis of the outer integuments. The inner integument is nonplicate and degenerates in fertilized ovules. The outer epidermis undergoes two periclinal divisions to form three cell layers. In the subfamily Cucurbitoideae the innermost layer usually divides only anticlinally, but in the Zanonioideae it divides periclinally as well. The mature seed coat in the Cucurbitoideae usually consists of five zones: epidermis, hypodermis, main sclerenchymatous region, aerenchyma, and inner chlorenchymatous/parenchymatous region. In the Zanonioideae the main sclerenchymatous region is poorly or not demarcated from the hypodermal sclerenchyma. The aerenchyma is well developed, and the cells have characteristic, lignified thickenings. The inner region is more strongly plicate than in the Cucurbitoideae. Variation in cell characteristics, especially in the hypodermis, are useful in delimiting taxa. The general similarities in seed coat development in the two subfamilies suggest they have a common origin, but one that remains obscure within the Polypetaleae.

In this chapter, data concerning the development and structure of the ovules and seed coats of the Cucurbitaceae are summarized. The implications of these data for considering systematic problems of the family, especially its phylogenetic position among the flowering plants, are discussed.

Literature Review

In 1829 Mirbel (23) provided the first description of ovule structure and development in the Cucurbitaceae. This and similar research through the early 1920's was reviewed by Netolitzky (24). During this period, Fickel

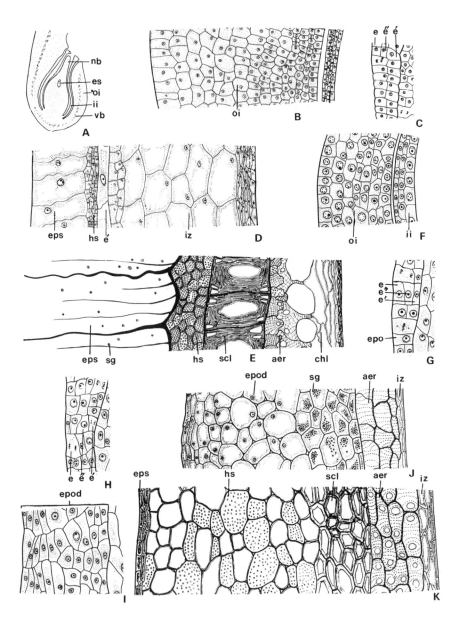

Figure 1. Seed coat development in the Cucurbitaceae. A–E. *Cucurbita pepo* (Cucurbitoideae). A. Longitudinal section of the ovule. B. Outer and inner integuments. C. Outer region of the outer integument showing initial layers. D, E. Developing and mature seed coats. F–K. *Actinostemma tenerum* (Zanonioideae). F. Outer and inner integuments. G–I. Outer regions of the outer integuments, showing successive development from periclinal divisions. J, K. Developing and mature seed

(10), Godfrin (11), Barber (1), Kratzer (19), and Reiche (26) presented observations on the seed coat anatomy of species in the genera *Abobra, Benincasa, Bryonia, Citrullus, Cucumis, Cucurbita, Cyclanthera, Diplocyclos* (as *Bryonopsis*), *Ecballium, Echinocystis, Lagenaria, Luffa, Momordica, Sechium, Sicyos, Thladiantha, Trichosanthes*, and *Zehneria* (as *Melothria*). Von Hohnel (54) noted that cucurbit seeds fall into two categories, those in which the carpel endothelium is firmly attached to the seed coat, e.g., *Cucurbita pepo* L. and *Lagenaria siceraria* (Mol.) Standl. (as *L. vulgaris* Ser.), and those in which it is loose and does not form part of the seed coat, e.g., *Cucumis sativus* L.

In the late 1940's and 1950's, studies involving the seed coats of *Coccinia grandis* (L.) Voigt, *Citrullus* species, and cultivated *Cucurbita* species, including the naked seed mutant of *C. pepo*, were reported by Chakravorti (5), Rutkovskaya (28) and Heinisch and Ruthenberg (13). Two papers by B. Singh (30, 31) marked the beginning of an intensive period of investigation of cucurbitaceous seed coats, subsequently carried on largely by D. Singh and his coworkers (32–41). In these studies comprehensive accounts of the development and structure of ovules and seed coats were presented and used to evaluate taxonomic and phylogenetic relationships of the family. These features and inferences drawn from them are summarized in the remainder of this chapter.

Ovules

Throughout the Cucurbitaceae, each placenta, either directly or after branching, bears anatropous, bitegmic, crassinucellate ovules (1, 10, 29–31, 33, 35–51, 55; Figure 1A). In the subfamily Cucurbitoideae ovules are commonly borne in a horizontal position, but there are exceptions. For example, in *Raphanocarpus* (= *Momordica*) both pendulous and erect ovules occur in the same ovary, or all ovules are erect. In the Trichosantheae subtribe Herpetosperminae ovules are horizontal in *Biswarea* but pendulous in *Edgaria* and *Hepetospermum*. A similar situation exists in the Melothrieae subtribe Cucumerinae, where the pendulous ovules of *Dicoelosper-*

coats. aer, aerenchyma; chl, chlorenchyma; e, e″, e′, layers formed by periclinal divisions of the outer epidermis of the outer integument; emb, embryo; end, endosperm; epl, large cells of epidermis; epo, outer epidermis of the outer integument; epod, derivatives of the outer integument; eps, seed epidermis; epu, seed epidermis of uniform size; es, embryo sac; hs, seed hypodermis; hsp, parenchymatous cells of hypodermis; ii, inner integument; iz, inner zone; k, calcium oxalate crystals; nb, nucellar beak; nu, nucellus; oi, outer integument; scl, sclerenchymatous layer; sg, starch grains; vb, vascular bundle; w, wing of the seed. A–E, after Singh and Dathan (44), F–K, after Singh and Dathan (46).

mum stand in contrast to the horizontal ovules of the other genera (47–49). In the Sicyeae the ovules are erect and ascending in the subtribe Cyclantherinae, whereas in *Sechium* and *Sicyos* of the subtribe Sicyinae they are solitary and pendulous (31, 36). Members of the subfamily Zanonioideae are characterized by pendulous ovules with the micropyle facing the stylar end of the ovary.

Throughout the family the outer integument is massive and the inner integument is usually two- or three-layered. In *Bryonia* and *Sechium* (19, 36), however, the inner integument eventually is four- or five-layered along the sides and six- to ten-layered near the base. The micropyle is formed by the inner integument alone.

The vascular supply to the ovule is usually a single, unbranched bundle that extends to the apex in the post-chalazal region of the outer integument. Branching of this bundle has been observed in *Momordica* (including *Raphanocarpus*) (34, 39), *Trichosanthes* (31, 50), *Cyclanthera* (31, 32), *Marah* (45), *Edgaria* (37), and *Actinostemma* (47). A branched vascular supply has also been observed in seeds of *Fevillea peruviana* (Huber) C. Jeffr., in the chalazal wing of *Gerrardanthus grandiflorus* Gilg. ex Cogn., and in abnormal seeds of *Cucurbita moschata* (Duch. ex Lam.) Duch. ex Poir. (44). Finally, the solitary ovule of *Sechium edule* (Jacq.) Sw. (36) and *Sicyos angulatus* L. (25) receives three or four vascular traces, which branch profusely in their downward course, then fuse before reaching the chalaza.

Development of the Seed Coat

In broad terms, members of the Cucurbitaceae follow a common pattern in their seed coat development. The outer integument alone contributes to the formation of the elaborate seed coat, while the inner integument, which is nonplicate, degenerates in fertilized ovules.

In the subfamily Cucurbitoideae the outer epidermis of the outer integument undergoes two periclinal divisions to form, in succession, three layers, e, e″, and e′ (Figure 1C). Layer e′ usually divides anticlinally to form the main mechanical or sclerenchymatous layer (Figure 1D). Layers e and e″ divide anticlinally and periclinally during further development to form the seed coat epidermis and hypodermis (Figure 1D). Interior to the sclerenchymatous layer and derived from its initials are aerenchymatous then chlorenchymatous or parenchymatous layers.

In the Zanonioideae cell divisions in the outer epidermis initiate as in the Cucurbitoideae, again forming three layers (Figure 1G), but further divisions in each layer are both anticlinal and periclinal. The innermost derivative of the e layer forms the principal mechanical layer, although it is inconspicuous in most of the genera investigated (Figure 1J, K). The hypodermal and other layers of the outer integument compose a broad zone of aerenchyma and the innermost parenchymatous zone (Figure 1K).

Mitotic activity in the epidermis and its derivative layers shows some variation in different parts of the ovules, but it is fairly constant for each species in any given region (37, 41, 42). Since mitotic activity varies among species, the number of hypodermal layers characteristic of species also varies. As the seed coat continues to mature, the cells enlarge and the walls thicken. Cells of the epidermis and main sclerenchymatous layer show characteristic tangential or radial elongation in the Cucurbitoideae but not in the Zanonioideae. The hypodermal or subepidermal cells become wholly or partly thick-walled. Cells of the aerenchyma become stellate and remain thin-walled, or become thick-walled with characteristic bands of thickenings. This aerenchymatous region is especially elaborate in *Fevillea peruviana*. The innermost layers are composed of narrow, compressed, thin-walled cells, which become chlorenchymatous in some taxa (Figure 1E).

Mature Seed Coat

The mature seed coat in the Cucurbitoideae generally consists of five identifiable zones: an epidermis; a hypodermis, which may be sclerotic or both sclerotic and parenchymatous; a main sclerenchymatous zone; an aerenchymatous zone; and a parenchymatous or chlorenchymatous zone. The inner epidermis, which has sometimes been recognized as a sixth zone, is indistinguishable from the adjacent thin-walled cells. In *Momordica* (39), *Gymnopetalum* (32), *Ctenolepis* (51), and *Dactyliandra* (51), the main sclerenchymatous layer cannot be identified clearly in sections, but macerations reveal the presence of cells of this layer.

Usually only four primary zones can be recognized in the seed coats of the Zanonioideae. Because of periclinal divisions in the initial layer e′, the main sclerenchymatous layer is mostly indistinct (Figures 1J; 2A, B, D–F). An exception is found in *Gerrardanthus grandiflorus*, in which the cells of the innermost layer of the sclerenchymatous zone are radially enlarged. Further complexity in the seed coats of the subfamily results from subdifferentiation of the subepidermal and aerenchymatous layers, especially *Fevillea peruviana*. In general, seed coats of the Zanonioideae are characterized by small, horizontal epidermal cells, polyhedral brachysclereids, and aerenchyma with lignified, fibrous, thickened bands. Brachysclereids are unknown in the Cucurbitoideae, although banding in the aerenchyma of some xerophytic and cultivated taxa has been recorded.

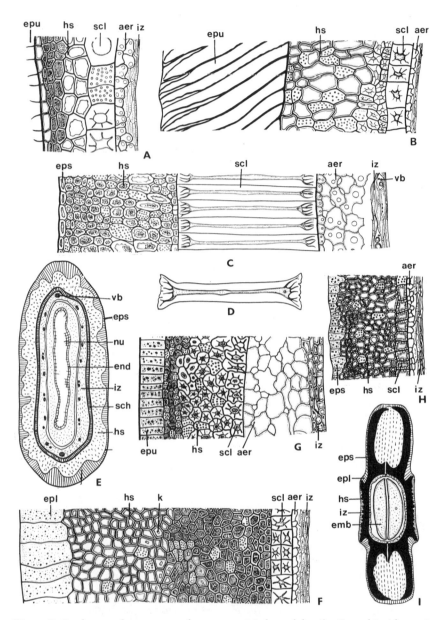

Figure 2. Seed coat characters and structure in the subfamily Cucurbitoideae. A. *Schizopepon bryoniifolius* (Schizopeponeae), mature seed coat. B. *Benincasa hispida* (Thunb.) Cogn. (Benincaseae), mature seed coat. C, D. *Marah macrocarpus* (Greene) Greene (Cucurbiteae). C. Mature seed coat. D. A single osteosclereid. E, F. *Trichosanthes lobata* (Trichosantheae), developing seed and seed coat detail. G. *Trichosanthes dioica*, seed coat detail. H, I. *Trichosanthes ovigera* Blume, seed coat detail

Variations in the Seed Coats of the Cucurbitoideae

The Cucurbitoideae exhibit considerable heterogeneity. In the Joliffieae subtribe Thladianthinae, Benincaseae subtribes Benincasinae and Luffinae, and Cucurbiteae, the epidermal cells are erect and either homocellular or heterocellular. The main sclerenchymatous layer consists of astroscloreids (Figures 1E; 2A, B). Yet, the seed coats of *Ecballium elaterium* (L.) A. Rich. and *Bryonia dioica* Jacq. differ from other members of the Benincasinae in having osteoscloreids in the main sclerenchymatous layer (31), and in *Luffa* the epidermis consists of erect cells with rodlike thickenings. The main sclerenchymatous layer is characterized by narrow, palisadelike osteoscloreids. *Raphanocarpus* (= *Momordica*) of the Thladianthinae also has osteoscloreids rather than astroscloreids in the main mechanical layer (39). From SEM studies, Lott (21) described the seed coat characters of *Cucurbita maxima* Duch. ex Lam. His description of the aerenchyma is excellent but that of the outer zones is at variance with observations summarized in this chapter.

Small, horizontal, homocellular or heterocellular epidermal cells and an osteoscloreid-sclerenchymatous layer are found in the Sicyeae subtribe Cyclantherinae (Figure 2C, D). In the subtribe Sicyinae, *Sechium* (36) and *Sicyos* (31) exhibit variable features. The seed coat of *Sechium edule* is unlignified and composed of horizontal, thin-walled cells, which are filled with starch grains. In contrast, the epidermis in *Sicyos* is composed of upright, thick-walled cells. The main sclerenchymatous layer is made up of osteoscloreids. In *Schizopepon* of the tribe Schizopeponeae, the epidermis is also of upright, thick-walled cells, but the main sclerenchymatous layer is composed of astroscloreids (Figure 2A).

In the Trichosantheae seed coat characters are consistent within, but different between, the subtribes Trichosanthinae and Herpetosperminae. In the Trichosanthinae, the epidermis consists of upright, heterogeneous cells, and the astroscloreids form the main mechanical layer (Figure 2E–I), while in the Herpetosperminae the sclerenchymatous layer is composed of osteoscloreids.

Seed coat anatomy in the tribe Melothrieae also encompasses variations in the epidermal and mechanical layers. For example, astroscloreids form the main mechanical layer throughout the Trochomeriinae, in *Melothria*, *sensu lato* (48), and *Cucumis* (4) of the Cucumerinae, but osteoscloreids are characteristic of *Kedrostis*, *Ibervillea*, most species of *Corallocarpus* (49), and *Apodanthera undulata* A. Gray (32) of the Dendrosicyinae.

and cross section of seed showing embryo confined to the central region. See Figure 1 for abbreviations. B, after Singh and Dathan (51), C–D, after Singh and Dathan (45), E–I, after Singh and Dathan (50).

Systematic Considerations

Species and Generic Relationships

Comparative studies of the features of the epidermis, hypodermis, and sclerenchymatous layer, together with exomorphic characters of the seeds have been used to characterize species and genera and to shed light on relationships. Examples are the investigations of species in *Momordica* (39), *Luffa* (42), *Cucurbita* (44), *Cucumis* (47), *Trichosanthes* (50), and *Melothria* (48). *Corallocarpus boehmii* (Cogn.) C. Jeff. may be distinguished from *C. epigaeus* (Rottl.) C. B. Clarke and *C. conocarpus* (Dalz. & Gibs.) C. B. Clarke by its thin-walled epidermis and sclerenchymatous layer of astrosclereids, which contrast with the thick-walled epidermal cells and osteosclereids of the latter two species. In India floras *Cucumis trigonus* Roxb. and *C. callosus* (Rottl.) Cogn. are usually considered conspecific, but differences in seed coat anatomy support the view that the latter is properly *C. melo* var. *agrestis* Naud. The seed coat of *C. trigonus* has moderately thick-walled, relatively short (80–100 µm), upright epidermal cells and a five-layered, heavily lignified hypodermis; whereas *C. melo* var. *agrestis* is known by its thick-walled, elongate (165–180 µm) epidermal cells and one-layered hypodermis.

The generic position of *Momordica cymbalaria* Fenzl. ex. Hook. f. is confirmed by seed coat studies. This species, once placed in *Luffa* as *L. tuberosa* Roxb., differs strongly from other *Luffa* species. Its ovules are either erect or pendulous, the vascular supply is branched, and the sclerenchymatous layer is composed of astrosclereids. In *Luffa* the ovules have a horizontal orientation, the vascular supply is unbranched, and the sclerenchymatous layer is composed of osteosclereids.

The seed coat anatomy of 12 species of *Melothria, sensu lato* (48), exhibits three distinct patterns and supports the reestablishment of the segregate genera *Solena* and *Mukia*, but not that of *Zehneria*. Seed coat characters of *Melothria pendula* L., the New World type species, show no differences in comparison with the compressed seeds of investigated oriental species, i.e., *M. leucocarpa* (Blume) Cogn. (= *Zehneria indica* (Lour.) Keraudren), *M. wallichii* C. B. Clarke (= *Z. wallichii* (C. B. Clarke) C. Jeffr.), *M. mucronata* auct. (= *Z. maysorensis* (Wight & Arn.) Arn.), *M. indica* Lour. (= *Z. indica*), *M. zeylanica* C. B. Clarke (= *Z. thwaitesii* (Aschers. & Schweinf.) C. Jeff.), *M. purpusilla* auct. (= *Z. scabra* (L. f.) Sond.).

The seed coats of *Melothria, sensu stricto*, are compressed and smooth. They have upright, narrow, and elongate epidermal cells with thickened radial walls. The hypodermis is one to three layers thick. In species segregated as *Mukia*, the seeds are tumid and sculptured. The epidermal cells are thin-walled and upright and heteromorphic. The hypodermis is multilayered and is composed of both parenchyma and sclerenchyma.

In *Melothria heterophylla* (Lour.) Cogn. (= *Solena amplexicaulis* (Lam.) Gandhi) the epidermis consists of small isodiametric or horizontal, thick-walled cells, and the hypodermis is multilayered and sclerotic. These characters provide one basis for the recognition of *Solena* (48).

Status of the Subfamily Zanonioideae

Jeffrey (17) and Singh and Dathan (46) summarized data concerning the relationship of the Zanonioideae and Cucurbitoideae. Salient morphological, palynological, and embryological features of the two subfamilies are compared in Table 1.

Table 1. Comparison of morphological and embryological characters of the subfamilies of the Cucurbitaceae

Characters	Cucurbitoideae	Zanonioideae
Tendril	Simple or branched, branching proximal, mostly spiralling above the point of the point of branching	Bifid, bifurcation distal, branches short, spiralling above and below the point of branching
Stamens	Inserted on the receptacle tube, free from disk when present	Inserted on or above disk
Style	One	Three, free
Pollen grains	Heterogeneous, colporate, colpate, porate, and aperturate	Small, 3-colporate, prolate
Ovule	Horizontal, ascending, erect or pendulous, anatropous, bitegmic, crassinucellate, nucellus flask-shaped, vascular supply up to the tip of the outer integument, unbranched or branched	Pendulous, anatropous, bitegmic, crassinucellate, nucellus flask-shaped, vascular supply up to the tip of the outer integument, unbranched or branched
Embryo sac	Polygonum type	Polygonum type
Endosperm	Nuclear, chalazal haustorium usually present	Nuclear, chalazal haustorium present
Seed coat	Outer integument, inner degenerates	Outer integument, inner degenerates
Behavior of outer epidermis and its derivatives	Divides periclinally forming e, e″, e′ layers, e′ usually does not divide periclinally and forms sclerenchymatous layer	Early mitosis forms e, e″, e′ layers, cells in all the layers divide anticlinally
Sclerenchyma layer	Usually distinct, rarely inconspicuous, various forms of large osteo- and astrosclereids	Usually inconspicuous, rarely distinct, small brachysclereids
Aerenchyma	One to many layered, mostly thin-walled, rarely thick-walled	One to many layered, usually strongly thick-walled and lignified with bands of thickenings
Seed	Nonwinged, rarely winged	Winged or nonwinged

Source: From Singh and Dathan (46).

Although seed structure has been studied in only five species of the Zanonioideae, *Actinostemma tenerum* Griff., *Gerrardanthus grandiflorus*, *Fevillea peruviana*, *Sicydium tamnifolium* (HBK) Cogn., and *Zanonia indica* L., the characteristics are uniform and reveal a common pattern of organization (46). The plicate nature of the outer integumental epidermal layer (e) is more pronounced than that observed in any of the Cucurbitoideae. Similarly, the generally inconspicuous nature of the sclerenchymatous layer and banded aerenchyma seem to be indicative of the Zanonioideae. Otherwise, however, the degeneration of the inner integument, the plicate nature of the outer epidermis of the outer integument, and the elaborate organization of the seed coat are features of more or less common pattern, indicating a probable common origin for the two subfamilies. They certainly stand closer together than does either to other families. Whether recognized as distinct families or as subfamilies, they logically form a unit.

Systematic Position of the Cucurbitaceae

The phyletic relationships of the Cucurbitaceae are uncertain. Classifications have placed the family in the Polypetalae (3, 4, 6–8, 12, 14, 22, 27, 53), Sympetalae (2, 9, 15, 20, 57), and, in one instance, the Apetalae (55). The more modern treatments (6, 7, 22, 53) have placed the family in polypetalous groups but with significant differences. Melchior (22) gave the family ordinal status and placed it near his Violales. Takhatajan (53) included the family in the Violales, as the only family of the suborder Cucurbitineae. He allied it to the Passifloraceae. Cronquist (6, 7) placed it questionably in the Violales, which includes some 24 other parietalean families.

The Parietales of Engler includes 29 families, of which 17 have been surveyed regarding the development and structure of the seed coat (43). The ovules are bitegmic and crassinucellar, except in the Guttiferae, Actinidiaceae, Marcgraviaceae, and Loasaceae. In these families the ovules are tenuinucellar and either unitegmic in the Actinidiaceae and Loasaceae or bitegmic in the Guttiferae and Marcgraviaceae. The seed coat is formed by both integuments. The outer epidermis of the inner integument forms the main mechanical layer in the Bixaceae, Caricaceae, Cochlospermaceae, Cistaceae, Elatinaceae, Flacourtiaceae, Guttiferae, Passifloraceae, Turneraceae, and Violaceae. The seed coat is thin and formed of the outer epidermis of the outer integument in the Begoniaceae and Datiscaceae. The Loasaceae, with unitegminal and tenuinucellar ovules, cellular endosperm with micropylar and chalazal haustoria, and seed coat formed by the peripheral layers of the integument, stands apart among other parietalean families, as does the Cucurbitaceae. Comparison of characters of the ovary, ovule, and seed of the Cucurbitaceae with families of the Parietales reveals a lack of significant resemblances. The family is isolated and deserves ordinal status.

Singh (32, 34) has shown that many features of the ovule and seed of the Cucurbitaceae compare with those of the Leguminosae. In both families the ovules are crassinucellar and bitegmic. They have an integumentary vascular supply and nuclear endosperm with a chalazal cecum, which is either coenocytic or cellular. The seeds either lack endosperm or have scant endosperm. The seed coat is formed by the outer integument, and the main sclerenchymatous layer develops from the outer epidermis. Cucurbitaceous seed development differs from that of the legumes in the formation of the outer epidermal layers and in derivation of the main sclerenchymatous layer. The resemblances between the Cucurbitaceae and Leguminosae perhaps are best considered indicative of parallel development, yet the data are suggestive and provide additional support for the origin of the Cucurbitaceae from a calycifloral plexus, to which the Leguminosae also belong, as suggested by Wernham (56). Singh (40) was drawn to a similar conclusion and proposed that Cucurbitaceae are an advanced, terminal taxon of Polypetalae.

Literature Cited

1. Barber, K. G. 1909. Comparative histology of fruits and seeds of certain species of Cucurbitaceae. Bot. Gaz. (Crawfordsville) 47: 263–310.
2. Benson, L. 1957. Plant Classification. Heath, Boston.
3. Bentham, G., and J. D. Hooker. 1867. Genera Plantarum. Vol. 1. Reeve, London.
4. Bessey, C. E. 1915. The phylogenetic taxonomy of flowering plants. Ann. Missouri. Bot. Gard. 2: 109–164.
5. Chakravorti, A. K. 1947. The development of female gametophyte and seed of *Coccinia indica* W. & A. J. Indian Bot. Soc. 26: 95–104.
6. Cronquist, A. 1968. The Evolution and Classification of Flowering Plants. Houghton Mifflin, Boston.
7. Cronquist, A. 1981. An Integrated System of Classification of Flowering Plants. Columbia Univ. Press, New York.
8. De Candolle, A. P. 1828. Prodromus Systematis Naturalis Regni Vegetabilis. Vol. 3. Treuttel et Wurtz, Paris.
9. Engler, A., and E. Gilg. 1924. Syllabus der Pflanzenfamilien. Neunten und Zehnten Auflage. Gebruder Borntraeger, Berlin.
10. Fickel, J. 1876. Über die Anatomie und Entwicklungs-geschichte der Samenschalen einiger Cucurbitaceae. Bot. Zeitung (Berlin) 34: 737–744, 753–760, 769–776, 785–792.

11. Godfrin, F. J. 1880. Etude histologique sur les teguments seminaux des angiosperme. Bull. Soc. Sci. Nancy 4: 160–163.
12. Gundersen, A. 1950. Families of Dicotyledons. Chronica Botanica, Waltham, MA.
13. Heinisch, O., and M. Ruthenberg. 1950. Die Bedeutung der Samenschale für die Züchtung des Öl-kürbis. Z. Pflanzenzucht. 29: 159–174.
14. Hutchinson, J. 1959. The Families of Flowering Plants. Vol. I. Dicotyledons. Macmillan, London.
15. Janchen, E. 1932. Entwurf eines Staumbaumes der Blutenpflanzen nach Richard Wettstein. Oesterr. Bot. Z. 81: 161.
16. Jeffrey, C. 1962. Notes on Cucurbitaceae, including a proposed new classification of the family. Kew Bull. 15: 337–371.
17. Jeffrey, C. 1966. On the classification of Cucurbitaceae. Kew Bull. 20: 417–426.
18. Johansen, D. A. 1950. Plant Embryology. Chronica Botanica, Waltham, MA.
19. Kratzer, J. 1918. Die Verwandtschaftlichen Beziehungen der Cucurbitaceen auf Grund ihrer Samenentwicklung. Flora 110: 275–343.
20. Lawrence, G. H. M. 1951. Taxonomy of Vascular Plants. Macmillan, New York.
21. Lott, J. N. A. 1973. A scanning electron microscope study of *Cucurbita maxima* seed coat structure. Canad. J. Bot. 51: 1711–1714.
22. Melchior, H., ed. 1964. Angiospermae. *In* A. Engler's Syllabus der Pflanzenfamilien, 12th ed. Vol 2. Gebruder Borntraegen, Berlin.
23. Mirbel, C. F. B. 1829. Nouvelles recherches sur la structure et les développement de l'ovule vegetale. Ann. Sci. Nat. 17: 302–318.
24. Netolitzky, F. 1926. Anatomie der Angiospermen-samen. Gebruder Borntraeger, Berlin.
25. Puri, V. 1954. Studies in floral anatomy. VII. On placentation in the Cucurbitaceae. Phytomorphology 4: 278–299.
26. Reiche, V. K. 1921. Zur Kenntnis von *Sechium edule* Sw. Flora 14: 232–248.
27. Rendle, A. B. 1952. The Classification of Flowering Plants. Vol. I. Cambridge Univ. Press, London.
28. Rutkovskaya, C. B. 1949. On the anatomical constructions of the seed coat of the genus *Citrullus* (in Russian). Coll. Sci. Res. Works of Azovochirnamorskova 12 (Novacherkank).
29. Schnarf, K. 1931. Vergleichendle Embryologie der Angiospermen. Gebruder Borntraeger, Berlin.
30. Singh, B. 1952. Studies on the structure and development of seeds of Cucurbitaceae. I. Seeds of *Echinocystis wrightii* Cogn. Phytomorphology 2: 201–209.
31. Singh, B. 1953. Studies on the structure and development of seeds of Cucurbitaceae. Phytomorphology 3: 224–239.
32. Singh, D. 1959. Studies on endosperm and development of seed in Cucurbitaceae and some of its related families. Ph.D. dissertation, Agra Univ., Agra.
33. Singh, D. 1961. Studies on endosperm and development of seeds in Cucurbitaceae and some of its related families. Agra Univ. J. Res. Sci. 10: 117–124.
34. Singh, D. 1964. Seed structure in the classification of the Cucurbitaceae. Proc. 10th Intl. Bot. Congr., Edinburgh. Pp. 516–517 (Abstract).

35. Singh, D. 1965. Ovule and seed of *Dicoelospermum* C. B. Clarke together with a note on its systematic position. J. Indian Bot. Soc. 44: 183–190.
36. Singh, D. 1965. Ovule and seed of *Sechium edule*. A reinvestigation. Curr. Sci. 34: 696–697.
37. Singh, D. 1967. Structure and development of seeds of the Cucurbitaceae. I. Seeds of *Biswarea* Cogn., *Edgaria* Clarke and *Herpetospermum* Hook. f. Proc. Indian Acad. Sci. 6: 267–276.
38. Singh, D. 1968. Structure and development of seed coat in Cucurbitaceae. III. Seeds of *Acanthosicyos* Hook. f. and *Citrullus* Schrad. Proc. 55th Indian Sci. Congr., Varanasi, Part 3: 347 (Abstract).
39. Singh, D. 1968. Structure and development of seed coat in Cucurbitaceae. IV. Seeds of *Momordica* and *Raphanocarpus* Hook. f. Proc. 55th Indian Sci. Congr. Varanasi, Part 3: 347–348 (Abstract).
40. Singh, D. 1969. Seed structure and the systematic position of Cucurbitaceae. Proc. 11th Intl. Bot. Congr., Seattle. P. 200 (Abstract).
41. Singh, D. 1970. Cucurbitaceae. Comparative embryology of angiosperms. Bull. Indian Nat. Sci. Acad. 41: 212–219.
42. Singh, D. 1970. Structure and development of seed coat in Cucurbitaceae. II. Seeds of *Luffa* Mill. J. Indian Bot. Soc. 50A: 208–215.
43. Singh, D. 1975. Development and structure of seed coat in Parietales and its taxonomic significance. Proc. Symp. Form, Structure and Function. Sardar Patel Univ., Anand. Pp. 99–101 (Abstract).
44. Singh, D., and A. S. R. Dathan. 1972. Structure and development of seed coat of Cucurbitaceae. VI. Seeds of *Cucurbita* L. Phytomorphology 22: 29–45.
45. Singh, D., and A. S. R. Dathan. 1972. Structure of development of seed coat in Cucurbitaceae. VIII. Seeds of *Marah* Kell. Bull. Torrey Bot. Club 99: 239–242.
46. Singh, D., and A. S. R. Dathan. 1973. Structure and development of seed coat in Cucurbitaceae. XI. Seeds of Zanonioideae. Phytomorphology 23: 138–148.
47. Singh, D., and A. S. R. Dathan. 1974. Structure and development of seed coat in *Cucumis* L. New Botanist 1: 8–22.
48. Singh, D., and A. S. R. Dathan. 1974. Structure and development of seed coat in Cucurbitaceae. V. Seeds of *Melothria* Linn. Bull. Bengal Bot. Soc. 28: 47–56.
49. Singh, D., and A. S. R. Dathan. 1974. Structure and development of seed coat in Cucurbitaceae. IX. Seeds of *Corallocarpus* Welw. ex. Hook. f., *Kedrostis* Medik and *Ibervillea* Greene. Bull. Torrey Bot. Club 101: 78–82.
50. Singh, D., and A. S. R. Dathan. 1976. Structure and development seed coat in Cucurbitaceae. X. Seeds of *Trichosanthes* L. J. Indian Bot. Soc. 55: 160–168.
51. Singh, D., and A. S. R. Dathan. 1978. Structure and development of seed coat in Cucurbitaceae. XII. Seeds of the subtribes Benincasinae and Trochomeriineae. *In* C. P. Malik et al., eds., Physiology of Sexual Reproduction in Flowering Plants. Kalyani, New Dehli.
52. Soifer, V. N. 1962. Evolution of anatomic seed structures in the squash family (in Russian). Bjull. Moskovsk. Obsc. Isp. Prir., Otd. Biol. 67: 147.
53. Takhatajan, A. L. 1980. Outline of the classification of the flowering plants (Magnoliophyta). Bot. Rev. 46: 225–359.
54. Von Hohnel, F. 1876. Morphologische Untersuchungen über die Samenschalan

der Cucurbitaceen und einiger verwandten Familien. Sitzb. Akad. Wiss. Wien. Math.-Naturwiss. 73: 297–337.

55. Vuillemin, P. 1923. Recherches sur les Cucurbitacées type anomalies affinities. Ann. Sci. Nat. Bot. 5: 5–19.

56. Wernham, H. F. 1911. Floral evolution, with particular reference to the sympetalous dicotyledons. New Phytol. 10: 73–83, 107–120, 293–305.

57. Wettstein, R. 1911. Handbuch der Systematischen Botanik. Deuticke, Leipzig.

19 | Epidermal and Anatomical Features of the Succulent Xerophytic Cucurbitaceae of Madagascar: Adaptive and Phylogenetic Aspects

Monique Keraudren-Aymonin

ABSTRACT. A comparative account of the anatomy of the photosynthetic organs of the Madagascan endemic succulent genera *Xerosicyos* and *Seyrigia* is presented, with special attention given to epidermal features. Stomatal densities are shown to be comparable with those of succulent tropical xerophytes of other families and much lower than those reported for nonsucculent cucurbits. Adaptive and evolutionary implications are briefly discussed.

Of the 27 genera of Cucurbitaceae in Madagascar, 8 are endemic and include succulent species without an equivalent elsewhere. Succulence is strongly developed in *Xerosicyos* and *Seyrigia*, although in different ways. In *Xerosicyos* the adult stems are woody and the leaf blades are succulent; in *Seyrigia* the plants are completely leafless in the adult stages and the stems are succulent. These two genera occur in the southwestern region of Madagascar, an area distinguished phytogeographically by spiny thickets. These thickets indeed contain many spiny plants, but their dominant character lies in the simultaneous existence of both succulence and spinosity in a great number of species.

Xerosicyos and *Seyrigia* are common in the hot, dry zone southwest of a line between Morombe (north of Toliara, previously Tuléar) and Tolañaro (Fort-Dauphin), on both sides of the Tropic of Capricorn. *Xerosicyos danguyi* Humbert, *X. perrieri* Humbert, *Seyrigia gracilis* Keraudren, and *S. multiflora* Keraudren are widely distributed throughout the zone, but the other species, *X. pubescens* Keraudren, *X. decaryi* Guillaum. & Keraudren, *S. bosseri* Keraudren, *S. humbertii* Keraudren, and *S. marnieri* Keraudren, are much more restricted.

Two species have been deliberately omitted from this study. *Xerosicyos*

239

pubescens, a species with weakly succulent leaves covered on both surfaces with very dense woolly hairs, is represented in herbaria by very little material. It was described from a single male specimen, from a now vanished plant that was cultivated in the Tsimbazaza Botanic Garden, Antananarivo (Tananarive), and is unknown in the living condition. The other species, X. *decaryi*, differs from X. *perrieri* only in its more elongate-elliptic leaves, and its specific distinctness is doubtful. Seeds of X. *perrieri*, collected in 1970 from a plant showing all the characters of that taxon, gave rise to plants with much more elongated leaves than the mother plant, when cultivated in Paris, indicating that this character may be unstable.

Leaf, Stem, and Stomatal Characters of the Madagascan Species

Xerosicyos

Members of the genus *Xerosicyos* are suffrutescent lianalike plants with woody stems that bear fleshy, ovate to suborbicular, glabrous (except in X. *pubescens*) leaves. In X. *danguyi* the leaf blades are 35–55 mm long and 25–50 mm wide and up to 5 mm thick (slightly less at the margins). In the fresh state the veins, apart from the mid-rib, are imperceptible. In X. *perrieri* the leaf blades are 18–20 mm long and 16–18 mm wide, crescentic in cross-section, and not more than 2 mm thick at the center.

Xerosicyos danguyi (Figure 1A–C). A transverse section of the adult leaf blade shows an epidermal layer on both surfaces of very elongated, narrow cells, about 50 μm long, with very thick longitudinal walls. The strongly cutinized external wall is covered with a layer of wax. The wax is continuous on adult leaves, but on young leaves it forms discontinuous patches. The epidermis is underlain by a chlorophyllous hypodermis of rather regular cells, which contain mucilage; then by a watery mesophyll, which surrounds fibrovascular bundles. Apart from the bundles forming the three to five basal veins, which are apparent in dried leaves in the herbarium, the vascular system is weakly developed. The fibrovascular bundles are accompanied on the posterior face by a crescent of sclerenchyma.

The stomata of X. *danguyi* are of the anomocytic type, as defined by Metcalfe and Chalk (3), with the guard cells surrounded by ordinary, regularly distributed epidermal cells. A few actinocytic stomata, as defined by Van Cotthem (5), with six to eight cells in a stellate arrangement around the stoma, are sometimes found. There are about 60 stomata per square millimeter on the upper surface and 75 per square millimeter on the lower surface of the blades. Toward the margin of the blade the density diminishes to nearly none. The stomata are not sunken or scarcely sunken and communicate with the substomatal chamber by a narrow canal between the long epidermal cells. The wax layer breaks up at the level of the ostiole. The anticlinal walls of the epidermal cells are somewhat sinuous or almost wavy.

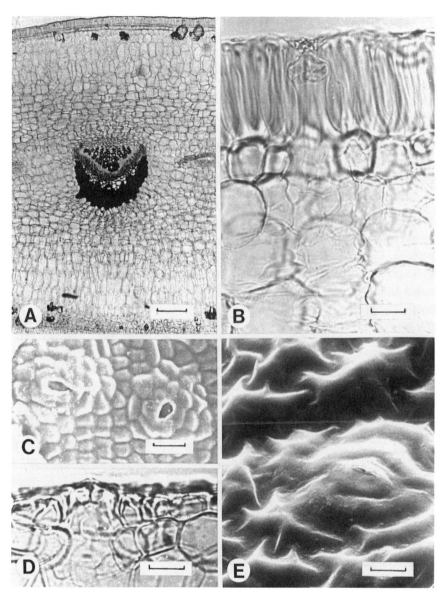

Figure 1. A. Transverse section of a leaf of *Xerosicyos danguyi* (upper surface above). Bar represents 0.2 mm. B. Transverse section of the upper surface of a leaf of *X. danguyi.* Detail of the epidermis in the region of a stoma. Bar represents 20 μm. C. Surface of the epidermis of the stem of *X. danguyi* (SEM). Bar represents 50 μm. D. Transverse section of the upper surface of a leaf of *X. perrieri.* Detail of the epidermis in the region of a stoma. Bar represents 20 μm. E. Surface of the epidermis of the leaf-blade of *X. perrieri:* stoma surrounded by "jig-saw puzzle" cells. Bar represents 10 μm.

The peculiarities of the leaf structure are an epidermis similar on the two faces, formed of long, thickened cells; stomata not sunken or slightly sunken and somewhat less abundant on the upper surface; an absence of islets of sclerenchyma on the lower side, although they occur in the hypodermis of the petioles; and mesophyll without intercellular spaces, composed of large cells with wrinkled walls, susceptible to changes in volume. In young leaves of seedlings, the epidermal layer is reduced to small quadrangular cells without thickened walls.

Transverse sections of the stems show strongly cutinized epidermal cells of the same type as those of the leaf blades, but a little less elongated. There are two continuous rings of sclerenchyma: one in the parenchyma and the other at the level of the pericycle.

Xerosicyos perrieri (Figure 1D, E). Sections of the adult blade show an epidermis formed of cells that are about 8 μm in diameter. The sinuosity of the anticlinal walls of the epidermis brings about a complex interlocking of the cells. This jigsaw puzzle structure is especially pronounced on the upper surfaces of the blades. In very young leaves cutinization of the outer surface is discontinuous, but in adult leaves it becomes continuous. There is no apparent hypodermis. The mesophyll is watery and without intercellular spaces. It is formed of smooth-walled cells, which reach 50 μm in diameter. The vascular bundles, which are small except for the median bundle, are borne in the mesophyll.

There is an average of 45 stomata per square millimeter on the upper surface and 70 on the lower surface. The globular guard cells rise a little above the neighboring epidermal cells, i.e., in section, the guard cells appear above the epidermis. As in *X. danguyi*, the wax layer breaks up at the level of the ostiole and there is no sclerenchyma in the hypodermis; however, the vascular bundles do not have a crescent of supportive sclerenchyma. The stems are covered with wax and have regular, not jigsawlike, epidermal cells.

Seyrigia

The genus *Seyrigia* includes herbaceous climbers, leafless in the adult condition, with little-branched stems reaching 7–8 m long.

Seyrigia bosseri (Figure 2B).   The stems are 6–7 mm in diameter and irregularly angular-ribbed. The surface is closely pubescent, mainly in the furrows. The anatomical structure is similar to the other *Seyrigia* species, and the stomata are in longitudinal bands, separated by narrow bands of elongated (almost 100 μm long) cells. There are about 45–50 stomata per square millimeter. These stomata are closer to one another than in the other species and are sometimes contiguous.

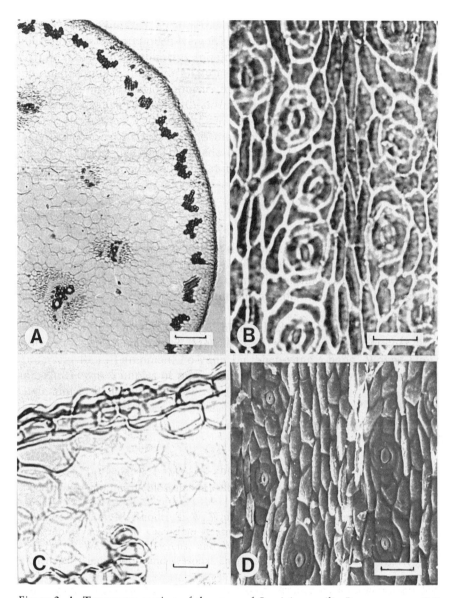

Figure 2. A. Transverse section of the stem of *Seyrigia gracilis*. Bar represents 0.2 mm. B. Surface of the epidermis of the stem of *S. bosseri* (photograph of a replica or imprint). Bar represents 50 μm. C. Detail of a transverse section of the stem of *S. multiflora;* epidermal cells covered by the cuticular layers and the position of a stoma, showing the flange of cuticle above the stomatal vestibule. Bar represents 20 μm. D. Surface of the epidermis of the stem of *S. marnieri*, showing the alignment of the cells and the stomata (SEM photograph). Bar represents 50 μm.

Seyrigia gracilis (Figure 2A). The subcylindrical stems are about 1.5–2 mm in diameter. A transverse section shows an epidermis of small regular cells about 16 μm long, with a well-developed cuticle on the outer face, covered with a thin layer of wax. Islets of sclerenchyma form a very discontinuous ring in the parenchyma, and the vascular bundles are disposed in a single ring. The central parenchyma, without intercellular spaces, is composed of large cells reaching 120 μm in diameter.

The stomata, always anomocytic, are almost at the level of the epidermis. They are aligned in longitudinal bands, 45 per square millimeter, the ordinary epidermal cells being distinctly more elongated than those around the stomata. The guard cells are about 18–20 μm long in surface view. In transverse section, the cuticle, covering the external walls of the guard cells and ending in a flange delimiting a stomatal vestibule, can be distinguished.

Seyrigia humbertii. The stems are ribbed stems with five to six furrows and reach 5 mm in diameter. They are covered with very dense white woolly hairs, which form a protective mantle. The anatomical structure is essentially the same as that of the other species, but the epidermis is covered with a cuticle 6–8 μm thick and with a discontinuous layer of wax having a laminated appearance. Examination of the surface by SEM shows the stomata, 45 to 50 per square millimeter, again arranged in longitudinal bands parallel with stomata-free bands of elongated cells. The contiguous guard cells end in a surface flange.

Seyrigia marnieri (Figure 2D). This species is similar to *S. gracilis*, but the stems are marbled on the surface, thicker (5 mm in diameter), and entirely glabrous. The stomata, rather few in number, 20 per square millimeter, are grouped in narrow longitudinal bands, and the cells of the stomata-free bands are distinctly more elongated.

Seyrigia multiflora (Figure 2C). The adult stem is 3–4 mm in diameter, angular-sinuous in outline, and comparable to *S. gracilis* in structure. Stomata, 50 per square millimeter, are arranged in parallel, longitudinal bands alternating with zones devoid of stomata. The peristomatic cells are again less elongated than the others. In section, the thickness of the cuticle (which can reach nearly 8 μm) and the guard cells forming a stomatal vestibule are clearly visible.

Comparative Discussion

Observations of the epidermis of the succulent Cucurbitaceae of Madagascar reveal the presence of isodiametric cells and elongate cells in the five species of *Seyrigia*, with anticlinal walls generally straight or a little curved, especially in the cells bordering the guard cells and their surrounding cells. Elongate cells with wavy anticlinal walls in *Xerosicyos danguyi* pass into the jigsaw type of *X. perrieri*. (Of interest is the fact that the eastern Asiatic *Neoalsomitra sarcophylla* (Wallich) Hutch., like *Xerosicyos* a member of the Zanonieae subtribe Zanoniinae and also with succulent leaves, does not have the same epidermal cell features.) Although they exhibit some distinct characteristics, the epidermal cells of *Seyrigia* and *Xerosicyos* are in accord with and complement the epidermal types described by Inamdar and Gangadhara (1).

Various authors have reported the presence of stomata of the paracytic type in *Coccinia*, *Lagenaria*, *Luffa*, *Momordica*, and *Trichosanthes*. Not previously noted are the anomocytic type and its derivative, the actinocytic, which are found in the two Madagascan genera.

Stomatal densities are often considered to be interesting indicators of adaptation. When one compares stomatal numbers in *Xerosicyos* and *Seyrigia* with those mentioned by Yasuda (9) for nonsucculent cucurbits, it is apparent that they are much lower in the Madagascan genera. On the basis of the few species studied, it appears that cucurbit adaptation to dry conditions is accompanied by a decrease in stomatal density. This is an agreement with the observations of Zemke (10), as quoted by Walter (6). Although not including species of Cucurbitaceae, Zemke described comparable features in the succulent species of the Namib Desert.

Perrot and Guerin (4) noted that stomata are relatively few and are confined to the lower leaf surface of *Didierea*, an endemic Madagascan genus of the Didiereaceae, which occupies habitats similar to those of *Xerosicyos* and *Seyrigia*. Killian and Lemée (2) demonstrated that there are about 100 stomata per square millimeter in nonsucculent xerophytes and 50 or fewer in succulents. A stomatal density of the same order is likewise associated with succulence in the species of Arizona (6). Wood (7, 8) stated that chenopodiaceous species of the Australian desert (*Bassia* and *Kochia*), with small cylindrical succulent leaves, have a stomatal density not exceeding 80 stomata per square millimeter. The epidermal characteristics of *Xerosicyos* and *Seyrigia* studied here clearly corroborate these observations, and it may reasonably be concluded that lowered stomatal numbers are adaptive in nature. Zemke (10) emphasized the importance of cuticular development for plants of the Namib Desert, where the cuticle may attain a thickness of 6–8 μm, as in the Madagascan cucurbits.

An adaptive character very often emphasized with respect to xerophytes is the existence of sclerenchymatous structures in the hypodermis, but this feature is not found in all xerophytes and seems not to exist in the leaves of many succulents, particularly *Xerosicyos*. On the other hand, the leaf of *X. danguyi* is distinctive in having the epidermal layer formed of very elongated, more or less prismatic cells. Somewhat comparable cells have been described by Yasuda (9) in the epidermis of the pericarp of the cucumber (*Cucumis sativus* L.), but they do not appear to have been reported again and are not found in *X. perrieri*.

Conclusions

In their low stomatal densities *Xerosicyos* and *Seyrigia* show a character state that corresponds to that found in other succulent tropical xerophilous taxa including, although in a less extreme manner, another endemic Madagascan cucurbit, *Trochomeriopsis diversifolia* Cogn., a species of the south and the driest zones of the west. Other endemic Madagascan cucurbit genera, e.g., *Ampelosicyos*, that occupy mesophilous or hygrophilous habitats show much greater stomatal densities.

The presumed adaptive nature of lowered stomatal number, epidermal characteristics, and thickened cuticle in *Xerosicyos* and *Seyrigia* must be tempered with caution for the following reasons. There are no nonxerophilous taxa among the species of *Xerosicyos* and *Seyrigia* with which comparison can be made, nor do these genera, with the exception of *Seyrigia* to *Dendrosicyos* of Socotra, exhibit a relationship with the other endemic xerophils, such as *Acanthosicyos*, of the African continent. From a biogeographical and phylogenetic viewpoint it may be noted that the cucurbitaceous genera best represented in the flora of Madagascar or the Afro-Madagascan group have not furnished the taxa most strongly adapted to the xerophilous conditions of southern Madagascar. Each of the two subfamilies, Cucurbitoideae and Zanonioideae, provide one genus. These two genera belong to phylogenetic lines remote from each other: *Xerosicyos* to the Zanonieae of Southeast Asia, tropical Africa, and tropical America; *Seyrigia* (like *Trochomeriopsis*) to the Dendrosicyinae with a somewhat similar distribution.

Literature Cited

1. Inamdar, J. A., and M. Gangadhara. 1970. Structure, ontogeny, and taxonomic significance of stomata in some Cucurbitaceae. Feddes Repert. 87: 293–310.
2. Killian, C., and G. Lemée. 1956. Les xérophytes: Leur économie d'eau. Encycl. Pl. Physiol. 3: 787–824.
3. Metcalfe, C. R., and R. Chalk. 1950. Anatomy of the Dicotyledons. Vol. 1. Oxford Univ. Press, Oxford.
4. Perrot, E., and P. Guerin. 1903. Les *Didierea* de Madagascar. J. Bot. (Morot) 17: 240–244.
5. Van Cotthem, W. R. J. 1970. A classification of stomatal types. J. Linn. Soc. Bot. 63: 235–246.
6. Walter, H. 1964. Die Vegetation der Erde in öko-physiologischer Betrachtung, 2nd ed. Vol. 1. Fischer, Stuttgart.
7. Wood, J. G. 1924. The relation between distribution structure and transpiration in arid south Australian plants. Trans. Proc. Roy. Soc. South Australia 48: 226–235.

8. Wood, J. G. 1937. The Vegetation of South Australia. Handbook of the Flora and Fauna of South Australia. Gov. Printer, Adelaide.
9. Yasuda, A. 1903. On the comparative anatomy of the Cucurbitaceae wild and cultivated in Japan. J. Coll. Sci. Imp. Univ. Tokyo 18: 1–56.
10. Zemke, E. 1939. Anatomische Untersuchungen an Pflanzen der Namibwüste. Flora 133: 365–416.
11. Zimmermann, A. 1922. Die Cucurbitaceen. Fischer, Jena.

PART III

Sex Expression

20 | Sex Expression in the Cucurbitaceae

R. P. Roy and Sunil Saran

ABSTRACT. Cucurbits display unique diversity in sex expression. Monoecious species are in the majority, but dioecious species and those with intermediate sex types also occur. The modes of sex expression in the Cucurbitaceae at the genetic, chromosomal, and biochemical levels are presented with a particular emphasis on the origin and evolution of dioecy, in which three distinct stages are noted. The first stage is represented by species whose sexual dimorphism is based entirely upon genic differentiation favoring development of one or the other type of sex expression. In the second stage, an incipient type of cytological differentiation is evident, and dioecy can be visualized in the abnormal behavior of a pair of chromosomes at microsporogenesis. The third stage is reached in *Coccinia* with an X/Y type of chromosomal sex mechanism, comparable with the system of dioecy displayed by *Melandrium*. The effects of the environment and chemicals on sex expression reveal the role of certain plant regulators, especially gibberellic acid and ethylene, in the development of staminate and pistillate flowers.

Most cucurbit species are monoecious, bearing staminate and pistillate flowers on the same plant, but many are dioecious with separate male and female plants, and a few species bear hermaphroditic flowers. Intermediate sex forms are also quite common in the Cucurbitaceae. Plants may be andromonoecious, with both hermaphroditic and staminate flowers, and gynomonoecious, with both hermaphroditic and female flowers. Occasional plants may be subgynoecious, with the female plant of a dioecious species bearing some flowers that are either male or hermaphroditic, or gynodioecious, in which female plants and those with hermaphroditic flowers occur. Such assemblages of diverse sex forms within closely related genera and species provide excellent material for investigations of the genetic, cytological, and biochemical basis of sex expression and furnish valuable

251

clues about the origin and evolution of dioecy in the Cucurbitaceae and other plant families. In this chapter the major contributions concerned with sex expression in the Cucurbitaceae are summarized and discussed.

Genetic Basis of Sex Expression

Bryonia

One of the earliest genera studied to understand the mechanism of sex determination was *Bryonia*. Correns (18), in a classical experiment, reported a 1:1 ratio of female to male plants in the intracrossed progeny of dioecious *B. dioica* Jacq. When *B. dioica* females and monoecious *B. alba* L. males were crossed, the offspring were all females, and on crossing *B. alba* females with *B. dioica* males, the offspring gave a ratio of 1:1 females and males. The hybrids of *B. dioica* and *B. alba* were sterile. This breeding behavior was explained on the assumption that the male was the heterogametic sex in *B. dioica*, and dioecism was dominant.

Later investigators included two more species in their studies, the dioecious *B. multiflora* Boiss. & Heldr. and the monoecious *B. macrostylis* Heilb. & Bilge (= *B. aspera* Stev. ex Ledeb.). Unlike Correns' F_1 hybrids, which were all sterile, Heilbronn (25–27) obtained *dioica* × *alba* hybrids that could be backcrossed to *B. alba*. Furthermore, crosses between two dioecious species could yield monoecious offspring, and crosses between two monoecious species could give rise to dioecious progeny. Westergaard (84) inferred from the studies in *Bryonia* that the male-promoting Y chromosome in the dioecious species carried sex determining genes that prevented female development, whereas the female-promoting X chromosome contained sex-determining genes that prevented formation of an androecium.

An explanation of the genetic basis of sex inheritance in *Bryonia* along with a tentative genetic map (Figure 1A, B) of the chromosomes that determine sex were presented by Williams (86). He explained that the monoecious *B. alba* was homogametic and carried both male- and female-promoting factors in all their gametes, but when these gametes were united with the female-promoting gametes from the dioecious *B. dioica*, their male-promoting factors were suppressed and the offspring were all females. In the same manner, when the male or the heterogametic sex of the dioecious species was involved in a cross with the female of the monoecious species, the female-promoting factors of the monoecious species were suppressed by the male-promoting factors of the dioecious species, and based upon segregation in male gametes, the offspring contained males and females in the ratio of 1:1. He suggested that the two factors, *F* and *M* or *m*, in the

Monoecious   Dioecious

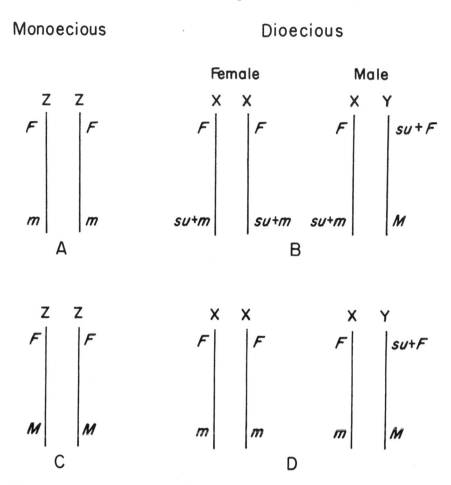

Figure 1. Tentative genetic maps of the sex chromosomes in *Bryonia* and *Ecballium*. A. *Bryonia alba*. B. *Bryonia dioica*. su^+F suppresses carpel development by *F*; su^+m suppresses anther development by m; su^+m is ineffective against *M* carried on the Y chromosome of dioecious species. C. *Ecballium elaterium* var. *elaterium*. D. *Ecballium elaterium* var. *dioicum*. See text for explanation. After Williams (86).

monoecious species promote female and male development, respectively. A male suppressor gene, su^+m, present in the female of the dioecious species, is specific against the *m* allele, but not against *M*. In the male plant, *F* is suppressed by su^+F, which is carried on the Y chromosome. These linked combinations of male- and female-promoting genes, along with their respective suppressors, control sex expressions in *Bryonia*. The Z chromosome represents the monoecious "wild type" state.

Ecballium

Genetic studies by Galan (21) with two varieties of *Ecballium elaterium* (L.) A. Rich. revealed a different type of interaction between genes determining monoecious vs. dioecious types. When intracrossed, the monoecious *E. elaterium* var. *elaterium* (as var. *monoicum* Batt. & Trab.) produced only monoecious progeny, and the dioecious *E. elaterium* var. *dioicum* Batt. produced males and females in a ratio of 1:1. When a female plant was fertilized by a monoecious plant, a monoecious plant resulted, in contrast to *Bryonia*, which produced a female plant from such a cross.

To interpret these results, Galan (21) assumed three alleles at one locus, a^D, a^+, and a^d in descending order of dominance, with the genotypes of different sex forms being monoecious (a^+a^+), female (a^da^d), and male (a^Da^D). No female suppressor was visualized. Westergard (84) and Williams (86) questioned such a simple model of sex expression because in all other known cases of established dioecy, X and Y chromosomes differed in large segments. Although Galan's multiple allelic hypothesis was not wholly discounted, an alternative scheme (Figure 1C, D) was suggested (86) based on the assumption that the "X" chromosome in the monoecious variety carries the dominant male-promoting factor M. In the dioecious variety, this factor on the "X" chromosome is the recessive m. The "Y" chromosome was assumed to be carrying the male-promoting factor M and also a female suppressor su^+F to reinforce dioecy. As in *Bryonia*, the Z chromosome represents the monoecious state.

Cucumis

Through detailed studies in *Cucumis melo* L., Poole and Grimball (54) attempted to explain the genetic basis of sex expression in muskmelons. They reported six different sex forms, i.e., monoecious, andromonoecious, gynomonoecious, hermaphrodite, trimonoecious, and gynoecious, but considered trimonoecious and gynoecious forms to be transitory expressions caused by environmental fluctuations. The interaction of two genes, A and G, was assumed to determine the inheritance of the various sex forms. Gene A was believed to suppress development of the androecium in female flowers, and gene G was believed to inhibit development of the gynoecium in male flowers. The recessive combination of the two genes allowed the development of perfect flowers with fully fertile male and female parts.

The genotypes suggested by Poole and Grimball (54) for the different sex forms of muskmelon were

Monoecious: +/+ +/+, +/a +/+, or +/+ +/g
Hermaphrodite: a/a g/g

Andromonoecious or androecious: a/a $+/+$ or a/a $g/+$
Gynomonoecious or gynoecious: $+/+$ g/g or $+/a$ g/g
Trimonoecious: $+/a$ $+/g$

Magdum et al. (41) reported segregation of the androecious sex form in the F_2 hybrid of hermaphrodite × monoecious muskmelon, and suggested that the androecious form may be genotypically similar to andromonoecious.

The cucumber, *Cucumis sativus* L., is predominantly monoecious but includes other sex variants. Furthermore, the monoecious type consists of strictly male, mixed male and female, and strictly female phases in a progressive order. Tkachenko (80) reported a rare gene that was often associated with complete femaleness, causing developmentally persistent gynoecism. Shifriss and Galun (72) provided evidence that this gene did not control gynoecism, but simply accelerated the rate of sex conversion, particularly femaleness. They designated this gene *Acr* (70), but the symbol *F* (80) takes precedence. Two gynodioecious races were synthesized by incorporating gene *F* into diverse genetic backgrounds (73). From these gynodioecious lines, true dioecious races were developed through the incorporation of known genes for determinate habit of growth and strong inhibition of lateral branches. The gene for determinate habit (*de*) eliminated the female phase from the main stem and transformed a monoecious plant into a strictly male one. Similarly, complete inhibition of laterals could lead to a wholly male plant because the branches possessed a strong feminizing tendency. Different combinations of three genes in *Cucumis sativus*, *a* (androecious), *F* (gynoecious), and *m* (andromonoecious), can produce an array of sex forms from the predominantly monoecious "wild" type.

Luffa

Luffa cylindrica (L.). M. J. Roem. and *L. graveolens* Roxb. are strictly monoecious, *L. echinata* Roxb. is dioecious, and *L. acutangula* (L.) Roxb. is monoecious except for the hermaphrodite cultivar 'Satputia' (syn. *L. hermaphrodita* Singh & Bhandari).

The first reports on the genetic basis of sex expression in *Luffa* appeared in 1948 (56, 74). Based on crosses between monoecious and hermaphroditic cultivars of *L. acutangula*, Richharia (56) reported F_2 segregation as 12 monoecious, 3 gynoecious, 1 hermaphrodite. He explained the breeding results on a digenic basis, assuming that two independent genes (*A* and *B*) were involved. Gene *A* determined the inheritance of both sexes, whereas gene *B* controlled the inheritance only of the female. Gene *A* was also assumed to be epistatic. Plants recessive for both genes bore perfect flowers.

To explain inheritance of various sex forms in *Luffa*, a multiple allelic series, *A*, *a'*, *a* and *G*, *g'*, *g*, in a two-gene scheme was assumed by Singh et al.

(74). Different combinations were designated AG (monoecious), $a'G$ (andromonoecious), aG (androecious), Ag' (gynomonoecious), Ag (gynoecious), and ag, $a'g'$, or ag' (hermaphroditic).

Choudhury and Thakur (15) extended studies in *Luffa* by making interspecific crosses, using *L. cylindrica* (monoecious) as the female parent and *L. acutangula* 'Satputia' (hermaphroditic) as the male parent. To explain the appearance of five sex forms in the F_2 hybrid, they assumed that two genes, A and G, control sex expression. Gene A suppresses the androecium in the solitary female flowers, whereas gene G suppresses the gynoecium in the male floral racemes. Plants recessive for both genes exhibit hermaphroditism. Genotypes of various sex forms were proposed (15) as $A/- G/-$ (monoecious or trimonoecious), $a/a G/-$ (andromonoecious or androecious), $A/- g/g$ (gynomonoecious or gynoecious), and $a/a g/g$ (hermaphroditic). These conclusions were derived by combining trimonoecious with monoecious, andromonoecious with androecious, and gynomonoecious with gynoecious sex forms, based on the assumption that androecious and gynomonoecious forms were not known in this genus. This explanation of sex expression in *Luffa* suffers because androecious and gynomonoecious forms were excluded and certain sex forms were considered to be transitory and triggered by environmental fluctuations without any experimental evidence to define the nature of environmental factors involved and the manner in which they affect sex expression.

Studies by Mishra (44) and Roy et al. (63) used the reciprocal cross, *L. acutangula* 'Satputia' as the female parent and *L. cylindrica* as the male parent. The F_1 hybrids were fertile and monoecious and were backcrossed to the hermaphrodite 'Satputia'. In the progeny from this backcross, only three plants reached maturity, although a large number of seeds were sown. Each of the surviving plants had a different sex form: monoecious, trimonoecious, and hermaphroditic. The plants were interbred, and from the resulting population all seven possible sex forms were recovered, including a wholly male plant.

On the basis of these results, the genetics of sex expression in *Luffa* were explained by assuming a multiple allelic series involving two independent genes, A and G, which function as follows: gene A suppresses development of the androecium in the flower borne in a leaf axil; a' permits occasional development of the androecium, and a fails to suppress its development. Similarly, gene G suppresses development of the gynoecium in flowers borne in racemes; g' permits occasional development of the gynoecium, and g fails to suppress its development. The order of dominance of A and G was assumed as A, a', a; G, g', g. The genotypes of varous sex forms were designated as follows:

Monoecious: $A/a' G/g$, $A/a G/g$, $A/a g/g'$, $A/a G/g'$
Andromonoecious: $a/a' G/g'$, $a/a' G/g$, $a/a g/g'$, $a/a G/g$

Gynomonoecious: A/a' g/g, A/a g/g
Hermaphrodite: a/a' g/g, a/a g/g', a/a g/g
Gynoecious: A/a' g/g'
Trimonoecious: a/a' g/g'
Androecious: a/a G/G, a/a' G/G, a'/a' G/g'

Dioecy in Cucurbits

Although the majority of plant species are hermaphroditic, dioecy occurs through a large number of widely separated families. About 5% of all plant families are entirely dioecious, and about 75% contain some dioecious species (90). A survey of the British flora showed that 92% of all species were hermaphroditic, 5% completely monoecious, and only 2% completely dioecious (39). A similar survey of the Indian flora revealed a slightly higher incidence of dioecism, with about 7% of the known species being completely dioecious (59). In contrast, the Cucurbitaceae shows a high incidence of dioecy. In one sample of some 48 genera and 135 species (8, 9, 10, 85), 67 species, scattered over 19 genera, were described as dioecious. Of the 19 genera with dioecious species, 12 are monotypic, suggesting dioecy has arisen independently in many evolutionary lines.

Cytogenetic studies in this laboratory on a number of dioecious cucurbits revealed three major determinants of dioecy. These are 1) entirely genic inheritance of sexual dimorphism without any cytological manifestation in chromosome structure or behavior; 2) chromosomal complements that are homomorphic, but with the behavior of one pair of chromosomes during microsporogenesis suggestive of an incipient type of sexual differentiation; and 3) presence of a distinctly heteromorphic chromosome pair in the male, with an X/Y type of sex determination.

The first category includes species such as *Luffa echinata, Momordica dioica* Roxb., *Edgaria darjeelingensis* C.B. Clarke, and *Solena amplexicaulis* (Lam.) Gandhi (syn. *Melothria heterophylla* (Lour.) Cogn.). Critical examination of somatic chromosomes in male and female plants and behavior of chromosomes during microsporogenesis clearly rule out chromosomally related dioecy, suggesting that the inheritance of sexual dimorphism is entirely genic in these species.

The second type of sex determination is displayed by *Trichosanthes dioica* Roxb. In an early study of this species (51), a pair of heteromorphic chromosomes was reported to be associated with its dioecism, a conclusion that was later questioned (84). Recently, a thorough study of the mitotic and meiotic chromosomes has shown that all chromosome pairs are homomorphic, and there is no cytological evidence to suggest heteromorphy as a basis for dioecy (62). Yet during microsporogenesis of *T. dioica*, one pair of chromosomes was found to behave aberrantly (62). Some of the peculiarities noted were 1) heteropycnosis of a pair of chromosomes at pachytene; 2) precocious separation of chromosomes in a bivalent, either at diakinesis or metaphase I (Figure 2A); 3) clear distinction of a single chromosome at each of the two poles of anaphase I (Figure 2B); and 4) delayed disjunction or nondisjunction of sister chromatids in one of the chromosomes at anaphase II, resulting in the formation of $n + 1$ gametes.

The aberrant behavior of only one pair of chromosomes suggests that these chromosomes primarily carry sex-deciding genes, one for maleness and the other for femaleness, without any external heteromorphy. It can be inferred that *T. dioica* represents an intermediate stage in chromosomal differentiation of dioecism, in which the pair of chromosomes associated with sexual dimorphism appears homomorphic, but carries very short homologous segments and large differential ones, leading to precocious separation.

Heteromorphic Sex Chromosomes and Policy

The presence of a heteromorphic pair of sex chromosomes in a dioecious cucurbit was recorded in *Trichosanthes kirilowii* var. *japonica* (Miq.) Kitam. (as *T. japonica* (Miq.) Regel) by Sinoto (76). This was confirmed by Nakajima (46), who also reported a similar heteromorphic pair in *T. ovigera* Blume (as *T. cucumeroides* Maxim.), but these studies were not extended further to reveal the nature of the X/Y system.

Among the investigated dioecious cucurbits with heteromorphic sex chromosomes, *Coccinia grandis* (L.) Voigt. (syn. *C. indica* Wight & Arn.) has proved to be the most interesting. Studies in this species (3, 11, 33, 34) recorded $2n = 24$ chromosomes with a cytologically demonstrable XY pair in the male. Investigations by Roy and his coworkers (59–61) provided extensive information about the chromosomal basis of sex expression in this species, and suggested that it represents the culminating point in the evolution of dioecy in the Cucurbitaceae.

Study of somatic chromosomes in *Coccinia grandis* confirmed $2n = 24$, with 2A + XX and 2A + XY constitutions for the female and male plants, respectively. The Y chromosome is conspicuously longer. During meiosis, X and Y chromosomes form a heteromorphic bivalent with very short pairing contact, indicating a short homologous and a long differential segment. A series of aneuploids and euploids provided important information on the nature of genes carried on the Y chromosome and their strength in determining maleness. A diploid trisomic male (2A + XYY) obtained from a cross between diploid and triploid plants showed an increase in the number of male flowers borne in clusters and certain deformities in the leaf. It was further discovered that the Y chromosome was fairly potent in determining

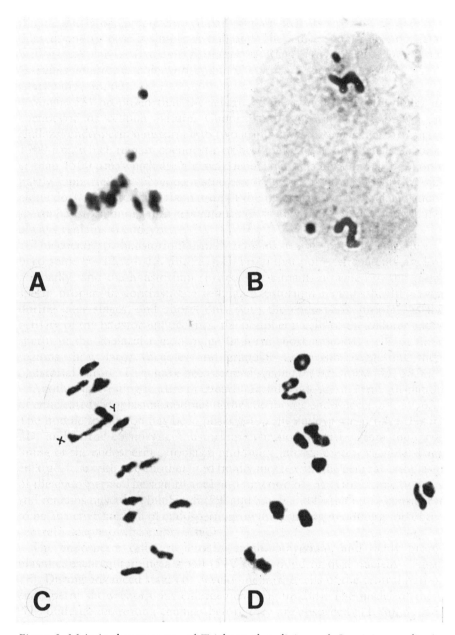

Figure 2. Meiotic chromosomes of *Trichosanthes dioica* and *Coccinia grandis*. A. Metaphase I of *T. dioica*, showing precocious separation of a chromosome pair. B. Late anaphase I of *T. dioica*. C. Metaphase I of *C. grandis*, showing the heteromorphic bivalent. D. Metaphase I of an irradiated hermaphrodite flower of *C. grandis*.

maleness, as was evident from the observation that plants with the chromosomal constitutions of 3A + XXY (triploid), 3A + XXXY (triploid tetrasomic), and 4A + XXXY (tetraploid) were all males.

Seeds of *Coccinia grandis* were irradiated in order to induce breakages in the Y chromosome so that its segments associated with specific male-determining functions could be ascertained. Among the progeny was a plant with normal pistillate flowers and some hermaphroditic flowers. The hermaphroditic flowers were fully fertile, and the pollen mother cells possessed the chromosomal complement of female plants, having two X chromosomes. The somatic chromosomes were also normal. Interestingly, the chiasma frequency in the pollen mother cells of the hermaphroditic flowers was much higher than the chiasma frequency observed in the pollen mother cells of staminate plants with a chromosome complement of 2A + XY (Figure 2C, D).

Kumar and Vishveshwaraiah (34) reported a gynodioecious plant of *Coccinia grandis* with the chromosome constitution of a normal female (2A + XX), but it was sterile. Although anthers developed in the hermaphroditic flowers, no pollen grains were formed. It was suggested that the androecium could develop in this species, even in the absence of the Y chromosome, but the chromosome carried a certain factor or factors responsible for the formation and fertility of male gametes.

The fully fertile androecium in a subgynoecious plant reported from this laboratory (60) indicated that the basic sex trigger in *Coccinia* could be similar to that of *Melandrium*, with the Y chromosome possessing a dominant female suppressor gene and at least two more genes. Development of occasional hermaphroditic flowers on gynoecious plants could best be explained on the basis of a genetic interaction between particular sex-promoter and sex-suppressor genes in conjunction with the accumulation or depletion of a specific growth regulator substance. The concentration level of this regulator appears to channelize sex expression, tilting it in favor of one sex form or the other. This function may reside with a locus controlling endogenous accumulation of a plant regulator, such as ethylene, which causes suppression of the androecium and development of pistillate flowers. If this gene is rendered partly ineffective as a result of irradiation, occasional development of male organs through endogenous accumulation of another plant regulator, such as gibberellin, could occur and lead to the formation of perfect flowers. Such action could account for the gynodioecy observed by Kumar and Vishveshwaraiah (34) in *Coccinia* and the stimulation of anther development in *Melandrium* females infected by *Ustilago* (83, 84).

The Y chromosome of *Coccinia grandis* is fairly potent in its male-determining characteristics. The male-promoting effect of the Y chromosome is dominant over the female-producing effect of the X chromosome, as in *Melandrium*. In both *Coccinia* and *Melandrium*, plants containing a single Y chromosome in opposition to three X chromosomes develop into males. When the number of X chromosomes exceeded three in *Melandrium*, the single Y chromosome failed to ensure strict maleness, leading to varying expressions of hermaphroditism (82, 84). The effect of a single Y chromosome in a polyploid combination of more than three X chromosomes is not known in *Coccinia*.

Effects of Environment and Chemicals in Sex Expression

Sex expression in cucurbits results from an interaction between genetic constitution, environment, and the chemical makeup of the plant. High light intensity and long days favor development of staminate flowers in cucumber and squash plants (48, 79). Differences occur between and within species in response to environmental factors (40, 65, 66). Depending upon the genetic background, cultivars of *Cucumis sativus* could be classified as day-length insensitive, short-day sensitive, or long-day sensitive.

Sex expression in cucumbers is affected by other factors as well. A correlation between sex form, on the one hand, and paramagnetic resonance and intensity of luminiscence on the other was reported by Nabokov (45). Mining and Matzekevitch (43) found that high soil moisture promoted the development of female flowers, while Takahashi and Suge (77) demonstrated that the number of female flowers increased in monoecious cultivars subjected to mechanical stress.

The effect of environmental factors on sex expression of *Cucumis* species is mediated through alterations in hormonal balance (64, Rudich, this volume). It was established that ethylene is the endogenous factor regulating femaleness in *C. sativus* and *C. melo* . High evolution of ethylene is associated with femaleness, low evolution with maleness. Day length and CO_2 concentration affect ethylene evolution. Short days enhanced ethylene evolution, favoring the expression of femaleness. Enrichment of the atmosphere with CO_2, a competitive inhibitor of ethylene, resulted in an increase in the number of staminate flowers.

Ethephon, an ethylene-releasing compound, was found to be highly effective in the early induction of femaleness in cucumbers by Robinson et al. (58). A number of workers have used ethephon for enhancing femaleness in cucumber and squash plants (2, 17, 20, 52, 67). Shannon and Robinson (69) reported that the application of 600 ppm of ethephon was most effective in reducing the number of staminate flowers in *Cucurbita pepo* L.

Although studies on the effect of chemicals (35–38), especially the growth regulators, on sex expression in cucurbits were initiated in the early 1950's, they have vastly expanded and diversified in the past twenty years. Laibach and Kribben (35–38) were mainly concerned with the application of auxins in controlling sex expression in cucurbits. Ito and Saito (29, 30) reported that appearance of pistillate flowers in cucumbers was markedly enhanced by treatments with napthaleneacetic acid (NAA), indoleacetic acid (IAA), and 2,4-dichlorophenoxyacetic acid (2,4-D). Early studies (87) revealed that the proportion of male flowers decreased when young cucumber plants were sprayed with NAA at 100 ppm or 2,3,5-triiodobenzoic acid (TIBA) at 25 ppm. A similar effect of TIBA was reported by Rehm (55) in watermelons. Galun (22) confirmed that spraying cucumbers with NAA at a young stage accelerated the appearance of pistillate flowers, but repeated applications of gibberellic acid (GA) resulted in the production of more male flowers. Similarly, Peterson and Anhder (53) observed a significant rise in maleness in cucumbers with application of GA at concentrations of 100–5000 ppm. Female flowers occurred earlier and in larger numbers when GA was applied at low concentrations (5–10 ppm) to cucumbers (13) and watermelons (24). Concentrations of GA at 25 ppm and above caused an increase in the number of male flowers. This showed that there might be a concentration relationship of GA with sex expression. Although the general trend was toward an increase in maleness with the application of GA in most cucurbits, Ghosh and Basu (23) recorded a strong feminizing effect of GA in *Momordica charantia*. Bisaria (6) reported earlier that NAA promoted femaleness in this species.

Maleic hydrazide (MH) has also been found to be effective in altering sex expression in cucurbits (68). Choudhury and Patil (14) and Bhalla (4) reported that MH at 200 ppm caused an increase in the number of female flowers in cucumbers. A complete suppression of maleness was recorded in Acorn squash (*Cucurbita pepo*) (87) when plants were treated with higher concentrations (250–1000 ppm) of MH.

Flurenol and chlorflurenol are also known to suppress maleness and promote femaleness (7). Foliar sprays of 25 mg of either compound, applied four times at weekly intervals beginning with the two-leaf stage, increased the number of female flowers in *Cucumis melo* (31). Soaking seeds in the same solution for four to five days at 5°C decreased the number of staminate flowers while increasing those of females.

Boron at 3.5 ppm caused an increase in the number of female flowers in *Citrullus lanatus* (Thunb.) Matsum. & Nakai (19) and *Lagenaria siceraria* (Mol.) Standl. (12). Exogenous application of abscisic acid (ABA) also induced a feminizing effect in cucumbers (64, 81).

Growth regulators introduced through the roots in the early stages of development caused sex shifts in cucumbers, GA favoring maleness and 6-benzylaminopurine favoring femaleness (32). An increase in the number of staminate flowers was also obtained when GA was applied in combination with silver nitrate to gynoecious cucumbers. In another study to the determine comparative effects inter se of gibberellin, silver nitrate, and aminoethoxyvinylglycine (AVG), it was found that all three reagents induced staminate flowers in gynoecious cucumbers (1). Several other reports suggested that male sex expression is favored, in general, by GA, silver nitrate, and AVG (16, 28, 47, 49, 50, 71).

It is evident that sex expression in cucurbits is influenced through alterations in the endogenous ethylene activity. It has been suggested that treatment with AVG inhibits ethylene action or retards its metabolism, resulting in the formation of staminate flowers (1). The phthalimides, 1-(1-cyclohexane-1,2-dicarboximide) cyclohexanecarboximide, 1-(3-chlorophthalimide)-cyclohexanecarboximide, and a related hydrophthalimide induced staminate flowers in gynoecious plants or delayed formation of pistillate flowers in cucumbers, through ethylene inhibition (88, 89). The mechanism by which GA induces staminate flowers is not fully understood. It is, however, well demonstrated that sex expression in cucumbers is regulated by a balance of ethylene, auxin, abscisic acid, and giberellin, in which the first three promote femaleness and gibberellin favors maleness. Androecism is generally associated with a relatively low rate of ethylene evolution, which may be due to either a deficiency or a partial block of ethylene production (64). Application of ethephon, an ethylene-releasing compound, or 1-amino cyclopropane-1-carboxylic acid (ACC) (78), an ethylene precursor, induces the production of pistillate flowers, even on androecious plants.

The majority of studies dealing with chemical alteration of sex expression was directed toward obtaining a higher yield of fruits through an increase in the number of pistillate flowers in cultivated cucurbits. Attempts to explain the physiological or biochemical behavior of plants on the basis of their genotypes, in terms of sex expression, were very sketchy

Conclusions

In cucurbits, dioecy appears to be as widespread as monoecy. In contrast, hermaphroditism is represented only by a few cultivars of *Luffa*, *Cucumis*, and *Benincasa*. It is probable that hermaphroditism developed in these taxa secondarily from monoecy. The consensus favors the origin of dioecious species from bisexual progenitors, as suggested by Correns (18) and others. According to Mather (42), however, an incompatibility system is far more efficient as a breeding method for obligate outcrossing than dioecy because it does not involve a waste of reproductive potential. As pointed out by Williams (86), dioecy is conditioned primarily as a system, developed in response to evolutionary demands for outcrossing. But in comparison to those hermaphrodites in which inbreeding is reduced as a result of incompatibility, sexual dimorphism is less efficient and is likely to lead to extinction or a reversion to hermaphroditism. The high frequency of dioecious species in the Cucurbitaceae indicates that this form of sex expression has been fully exploited in the family even if it is a result of a short-term evolutionary response to outbreeding. In this interpretation, a few cultivars with perfect flowers could be taken to represent an evolutionary step toward ultimate incorporation of an incompatibility system.

Acknowledgments

It is a pleasure to thank Bithi Dutt, D. P. Srivastava, Chitra Kumar, and Harsh Kumar for their valuable assistance in the preparation of this chapter. Studies discussed in the chapter were conducted by the senior author during the tenure of a Jawaharlal Nehru Fellowship.

Literature Cited

1. Atsmon, D., and C. Tabbak. 1979. Comparative effects of gibberellin, silver nitrate, and aminoethoxyvinylglycine on sexual tendency and ethylene evolution in the cucumber plants (*Cucumus sativus* L.). Pl. Cell Physiol. 20: 1547–1555.
2. Baker, E. C., and G. A. Bradley. 1976. Effects of ethephon on yield and quality of winter squash, *Cucurbita maxima* Duch. HortScience 11: 140–142.
3. Bhaduri, P. N., and P. C. Bose. 1947. Cytogenetical investigations in some common cucurbits with special reference to fragmentation of chromosomes as a physical basis of speciation. J. Genet. 48: 237–256.
4. Bhalla, S. C. 1962. Varietal response to plant regulator sprays on floral biology, sex expression and fruit development in cucumber (*Cucumis sativus* L.). M.S. thesis, Indian Agric. Res. Inst., New Delhi.
5. Bilge, E. 1955. Researches morphologiques, anatomique et genetique sur *Bryonia macrostylis* Heilb. et Bilge. Rev. Fac. Sci. Univ. Istanbul, Ser. B, 20: 121–146.
6. Bisaria, A. K. 1974. The effect of foliar spray of alpha naphthalene acetic acid on the sex expression in *Momordica charantia* L. Sci. Cult. 40: 78–80.
7. Bisaria, A. K. 1977. Influence of flurenol and chlorflurenol on growth, flowering and sex expression in muskmelon (*Cucumis melo* L.). Israel J. Bot. 26: 209–214.
8. Chakravarty, H. L. 1959. Monograph on Indian Cucurbitaceae. Records Bot. Survey India. 17: 1–234.
9. Chakravarty, H. L. 1966. Monograph on the Cucurbitaceae of Iraq. Tech. Bull. 133. Ministry of Agriculture, Baghdad.
10. Chakravarty, H. L. 1968. Cucurbitaceae of Ghana. Bull. I'I.F.A.N., Ifan, Dakar.
11. Chakravorti, A. K. 1948. Cytology of *Coccinia indica* with reference to the behavior of its sex chromosomes. Proc. Indian Acad. Sci. B27: 74–86.
12. Choudhury, B., and Y. S. Babel. 1969. Studies on sex expression, sex ratio and fruit set as affected by different plant regulator sprays in *Lagenaria siceraria* (Molina) Standl. HortScience 1: 61.
13. Choudhury, B., and S. C. Pathak. 1959. Sex expression and sex ratio in cucumber (*Cucumis sativus* L.) as affected by plant regulator sprays. Indian J. Hort. 16: 162–169.

14. Choudhury, B., and A. V. Patil. 1962. Effect of plant regulator sprays on sex, fruit set and fruit development in cucumber (*Cucumis sativus* L.). Proc. Bihar Acad. Agric. Sci. 9–10: 28–34.

15. Choudhury, B., and M. R. Thakur. 1965. Inheritance of sex forms in *Luffa*. Indian J. Genet. 25: 188–197.

16. Christopher, D. A., and J. B. Loy. 1980. Influence of exogenously applied hormones on sex expression in watermelon. HortScience 15: 381 (Abstract).

17. Churata-Masca, M. G. C., and M. Awad. 1974. The effect of 2–chloroethoxyphosphonic acid (ethephon) on flowering and fruiting of cucumber. Rev. Ceres 21: 284–293.

18. Correns, C. 1928. *In* K. Baur and M. Hartman, eds., Handbuch der Verbungwissenschaft. Vol. 2. Gebruder Borntraeger, Berlin.

19. Ekholy, E. 1968. Chemical sex modification in watermelon (*Citrullus vulgaris* Schrad.). M.S. thesis, Indian Agri. Res. Inst., New Delhi.

20. Friedlander, M., D. Atsmon, and E. Galun. 1977. Sexual differentiation in cucumber. Pl. Cell Physiol. 18: 261–269.

21. Galan, F. 1951. Analyse génétique de la monoecie et de la dioecie zygotiques et de leur différence dans *Ecballium elatarium*. Acta Salmanticensia. Ser. Cienc. Biol. 1: 7–15.

22. Galun, E. 1959. Effects of gibberellic acid and NAA on sex expression and some morphological characters in cucumber plant. Exp. Bot. 13: 1–8.

23. Ghosh, S., and P. S. Basu. 1983. Hormonal regulation of sex expression in *Momordica charantia*. Physiol. Pl. (Copenhagen) 57: 301–305.

24. Gopalkrishnan, P. K., and B. Choudhury. 1978. Effect of plant regulator sprays on modification of sex, fruit, set, and development in watermelon (*Citrullus lanatus* (Thunb.) Mansf.). Indian J. Hort. 35: 235–241.

25. Heilbronn, A. 1948. Über die Genetik von Monicie and Getrenntgeschlechttigkeit bei *Bryonia* Basterden. Proc. 8th Cong. Genet. Hereditas, Suppl. Vol. Pp. 590–591.

26. Heilbronn, A. 1953. Über die Rolle des Plasmas bei der Geschlechtbestmmung der Bryonien. Rev. Fac. Sci. Univ. Istanbul, Ser. B, 18: 205–207.

27. Heilbronn, A., and M. Basarman. 1942. Über die F$_2$ der *Bryonia* Basterde und ihre Bedeutung für das Problem der Geschlechtsrealisation. Rev. Fac. Sci. Univ. Istanbul, Ser. B, 7: 138–144.

28. Hunsperger, M. H., D. G. Helsel, and L. R. Baker. 1982. Patterns of staminate flower expression in hermaphroditic pickling cucumber induced by silver nitrate. HortScience 17: 33 (Abstract).

29. Ito, H., and T. Saito. 1956. Factors responsible for sex expression of Japanese cucumber. III. The role of auxin on plant growth and sex expression. J. Hort. Assoc. Jap. 25: 101–110.

30. Ito, H., and T. Saito. 1956. Factors responsible for sex expression in Japanese cucumber. IV. The role of auxin on plant growth and sex expression. J. Hort. Assoc. Jap. 25: 141–151.

31. Kaushik, M. P., and A. K. Bisaria. 1972. Effect of foliar spray and chemical vernalisation with morphactin on the sex ratio in muskmelon (*Cucumis melo*). Pl. Sci. 4: 57–59.

32. Khranin, V. N., and M. K. Chailakhyan. 1979. Effect of growth regulator on sex. Nauk. Imeni V.I. Lenia 1: 10–13.

33. Kumar, L. S. S., and G. B. Deodikar. 1940. Sex chromosomes of *Coccinia indica* Wight & Arn. Curr. Sci. 9: 128–130.

34. Kumar, L. S. S., and S. Vishveshwaraiah. 1952. Sex mechanism in *Coccinia indica* Wight and Arn. Nature 170: 330–331.

35. Laibach, F., and F. J. Kribben. 1950. Der Einfluss von Wuchsstoff auf die Bildung Männlicher und Weiblecher Blüten bei einer monözischen Pflanze (*Cucumis sativus* L.) Ber. Deutsch. Bot. Ges. 62: 53–55.

36. Laibach, F., and F. J. Kribben. 1950. Der Einfluss von Wuchsstoff auf das Geschlecht der Blüten bie einer monözischen Pflanze. Beitr. Biol. Pflanzen 28: 64–67.

37. Laibach, F., and F. J. Kribben. 1951. Die Bedeutung des Wuchsstoff für die Bildung and Geschlechtsbestimmung der Blüten. Beitr. Biol. Pflanzen 28: 131–144.

38. Laibach, F., and F. J. Kribben. 1951. Über die Bedeutung der β-Indolylessigsäure für die Blütenbildung. Ber. Deutsch. Bot. Ges. 63: 119–120.

39. Lewis, D. 1942. The evolution of sex in flowering plants. Biol. Rev. Cambridge Philos. Soc. 17: 46–67.

40. Lower, R. L., J. D. McCreight, and O. S. Smith. 1976. Photoperiod and temperature effects on growth and sex expression of cucumber. HortScience 10: 318 (Abstract).

41. Magdum, M. B., N. N. Shinde, and V. S. Sheshadiri. 1982. Androecious sex forms in muskmelon. Cucurbit Genet. Coop. Rep. 5: 24–25.

42. Mather, K. 1940. Outbreeding and separation of the sexes. Nature 145: 484–486.

43. Mining, E. G., and P. O. Matzekevitch. 1944. Sexualization of plants as affected by different moisture conditions of the medium. C. R. (Dokl.) Acad. Sci. URSS 42: 309–312.

44. Mishra, D. P. 1975. Cytogenetical investigations in the Cucurbitaceae. Ph.D. dissertation, Patna Univ., Patna.

45. Nabokov, I. G. 1977. Method of determining sex in cucumber. Ref. Zhurn. Biol. 4(55): 263.

46. Nakajima, G. 1937. Cytological studies in some dioecious plants. Cytologia, Fuji Jubilee Vol. Pp. 282–292.

47. Nijs, A. P. M. den. 1980. Effectiveness of AVG for inducing staminate flowering on gynoecious cucumber. Cucurbit Genet. Coop. Rep. 3: 22–23.

48. Nitsch, J. P., E. B. Kurtz, J. L. Liverman, and F. W. Went. 1952. The development of sex expression in cucurbit flowers. Amer. J. Bot. 39: 32–42.

49. Owens, K. W., C. E. Peterson, and G. E. Tolla. 1980. Induction of perfect flowers on gynoecious muskmelon by $AgNO_3$ and aminoethoxyvinylglycine. HortScience 15: 654–655.

50. Owens, K. W., G. E. Tolla, and C. E. Peterson. 1980. Induction of staminate flowers on gynoecious cucumber by aminoethyoxyvinylglycine. HortScience 15: 256–257.

51. Patel, G. I. 1952. Chromosome basis of dioecism in *Trichosanthes dioica* Roxb. Curr. Sci. 21: 343–344.

52. Pathak, S. C., and D. J. Cantliffe. 1976. Persistence of ethephon to induce female flowering in cucumber. HortScience 11: 27–28.

53. Peterson, C. E., and L. D. Anhder. 1960. Induction of staminate flowers on gynoecious cucumber with gibberellin A₃. Science 131: 1673–1674.
54. Poole, C. F., and P. C. Grimball. 1939. Inheritance of sex forms in *Cucumis melo* L. J. Heredity 30: 21–25.
55. Rehm, S. 1952. Male sterile plants by chemical treatment. Nature 170: 38–39.
56. Richharia, R. H. 1948. Sex inheritance in *Luffa acutangula*. Curr. Sci. 17: 359.
57. Risser, G., and J. C. Rode. 1979. Inducing staminate flowers in gynoecious plants of melon (*Cucumis melo* L.). Ann. Amelior. Pl. 29: 349–352.
58. Robinson, R. W., S. Shannon, and M. D. DeLaguardia. 1969. Regulation of sex expression in cucumber. BioScience 19: 141–142.
59. Roy, R. P. 1974. Sex mechanism in higher plants. J. Indian Bot. Soc. 53: 141–155.
60. Roy, R. P., and P. M. Roy. 1971. Mechanism of sex determination in *Coccinia indica*. J. Indian Bot. Soc. 50A: 391–400.
61. Roy, R. P., S. Saran, and B. Dutt. 1973. Speciation in relation to the breeding system in the Cucurbitaceae. *In* Y. S. Murty, ed., Advances in Plant Morphology. Sarita Prakashan, Meerut.
62. Roy, R. P., S. Saran, and B. Dutt. 1981. Sex mechanism in *Trichosanthes dioica* Roxb. *In* S. C. Verma, ed., Contemporary Trends in Plant Sciences. Kalyani, New Dehli.
63. Roy, R. P., S. Saran, and D. P. Mishra. 1975. Genetic basis of sex expression in *Luffa*. 2nd Congr. Cytol. Genet.
64. Rudich, J., and L. R. Baker. 1976. Phenotypic stability and ethylene evolution in androecious cucumber. J. Amer. Soc. Hort. Sci. 101: 48–51.
65. Saito, T. 1977. Sex differentiation in Cucurbitaceae. I. Agric. Hort. 52: 1337–1341.
66. Saito, T. 1977. Sex differentiation in Cucurbitaceae. II. Agric. Hort. 52: 1471–1474.
67. Sams, C. E., and W. A. Krueger. 1977. Ethephon alteration of flowering and fruit set pattern of summer squash. HortScience 12: 162–164.
68. Schoene, D. L., and L. O. Hoffmann. 1949. Maleic hydrazine, a unique growth regulator. Science 109: 588–590.
69. Shannon, S., and R. W. Robinson. 1978. Genetic differences in sex expression and response to ethephon in summer squash, *Cucurbita pepo* L. Cucurbit Genet. Coop. Rep. 1: 33.
70. Shifriss, O. 1961. Sex control in cucumber. J. Heredity 52: 5–12.
71. Shifriss, O. 1985. Origin of gynoecism in squash. HortScience 20: 889–891.
72. Shifriss, O., and E. Galun. 1956. Sex expression in the cucumber. Proc. Amer. Soc. Hort. Sci. 67: 479–486.
73. Shifriss, O., W. L. George, and J. A. Quinones. 1964. Gynodioecism in cucumbers. Genetics 49: 285–291.
74. Singh, H. B., S. Ramunajam, and B. P. Pal. 1948. Inheritance of sex forms in *Luffa acutangula* Roxb. Nature 161: 775–776.
75. Singh, N., and B. Choudhury. 1977. Further studies on nutrient uptake as affected by growth regulator sprays in relation to sex modification in cucumber (*Cucumis sativus* L.). Indian J. Hort. 34: 56–59.

76. Sinoto, Y. 1929. Chromosome studies in some dioecious plants, with special reference to allosomes. Cytologia 1: 109–191.
77. Takahashi, H., and H. Suge. 1980. Sex expression in cucumber plant as affected by mechanical stress. Pl. Cell Physiol. 21: 303–310.
78. Takahashi, H., and H. Suge. 1982. Sex expression and ethylene production in cucumber plants as affected by 1–aminocyclopropane-1–carboxylic acid. J. Jap. Soc. Hort. Sci. 51: 51–55.
79. Tiedjens, V. A. 1928. Sex ratio in cucumber flowers as affected by different conditions of soil and light. J. Agric. Res. 36: 731–736.
80. Tkachenko, N. K. 1935. Preliminary results of a genetic investigation of the cucumber (*Cucumis sativus* L.). Bull. Appl. Bot. Genet. Pl. Breed. 9: 311–356.
81. Verma, V. K., and B. Choudhury. 1980. Chemical sex modification in cucumber through growth regulators and chemicals and their effect on yield. Indian J. Agr. Sci. 50: 231–235.
82. Warmke, H. E. 1946. Sex determination and sex balance in *Melandrium*. Amer. J. Bot. 33: 648–660.
83. Westergaard, M. 1953. Über den Mechanismus der Geschlechtsbestimmung bei *Melandrium album*. Naturwissenschaften 40: 253–260.
84. Westergaard, M. 1958. The mechanism of sex determination in dioecious flowering plants. Adv. Genet. 9: 217–281.
85. Whitaker, T. W., and G. N. Davis. 1962. Cucurbits: Botany, Cultivation, and Utilization. Interscience, New York.
86. Williams, W. 1964. Genetical Principles and Plant Breeding. Blackwell, Oxford.
87. Wittwer, S. H., and I. G. Hillyer. 1954. Chemical induction of male sterility in cucurbits. Science 120: 893–894.
88. Xu, S.-Y., and M. J. Bukovac. 1981. Modification of sex expression in cucumber with phthalimides. HortScience 16: 457 (Abstract).
89. Xu, S.-Y., and M. J. Bukovac. 1983. Phthalimide inhibition of the ethylene effect on sex expression in monoecious cucumber plant. J. Amer. Soc. Hort. Sci. 108: 282–284.
90. Yampolsky, C., and H. Yampolsky. 1922. Distribution of sex forms in the phanerogamic flora. Biblio. Genet. 3: 1–62.

21 | Biochemical Aspects of Hormonal Regulation of Sex Expression in Cucurbits

Jehoshua Rudich

ABSTRACT.   Sex expression in the Cucurbitaceae is affected by day length, temperature, and genetic and hormonal controls. Ethylene may be the primary hormone affecting femaleness in cucumbers, muskmelons, and squash, an involvement suggested by the detection of ethylene evolution from flower buds and apices. Gynoecious cucumber cultivars were found to have a higher ethylene evolution than other cucumber genotypes. Environmental factors promoting femaleness enhanced ethylene evolution. Ethylene biosynthesis and action inhibitors, such as aminoethoxyvinylglycine, carbon dioxide, and silver nitrate served as tools for additional understanding of the involvement of ethylene in sex expression. Factors involved in the biosynthetic pathways of ethylene from methionine may explain the role of auxin in femaleness. Gibberellin$_4$ and gibberellin$_7$ are probably the natural hormones involved in male flowering. High gibberellin activity was found in genotypes with strong male tendencies. Gibberellin activity was found to be influenced by length of day. The possible involvement of day length in sex expression and its relationship to the biosynthetic pathway of gibberellin are discussed. Sex expression in watermelons seems to be controlled by a different mechanism than for cucumbers, muskmelons, and squash.

Economic hybrid seed production in the Cucurbitaceae is facilitated by hormonal regulation of sex expression. The use of gibberellic acid (gibberellin, GA) to maintain and propagate gynoecious cucumber (*Cucumis sativus* L.) parent lines was a major breakthrough of combined research in hormonal regulation and genetic control of sex expression. Present information on the genetic control of sex expression in cucurbits has been ascertained primarily from studies with cucumbers (15, 29, 56). Less knowledge exists about the genetics of sex expression in muskmelon (*Cucumis melo* L.)

269

(47), squash (*Cucurbita* spp.), and watermelon (*Citrullus lanatus* (Thunb.) Matsum. & Nakai).

Cucurbit species exhibit great diversity in sexual types. In cucumbers, the most common sexual types are gynoecious and monoecious, although androecious, andromonoecious, and hermaphrodite types exist. The latter three are quite useful in genetic studies. In muskmelons, most cultivars are andromonoecious, although androecious and hermaphrodite types also occur. Gynoecious and gynomonoecious muskmelon cutivars have been developed with limited success. Most squash cultivars are monoecious, and all watermelon cultivars are either monoecious or andromonoecious.

The expression of this sexual diversity in cucurbits is affected by the interaction of environment and hormonal responses to it. Environmental influences on sex expression were first studied by observations on the effect of planting season on flower sex ratio (15). More detailed studies found that length of day and temperature were the major factors governing sex expression. In general, short days and low temperatures promote femaleness, while long days and high temperatures promote maleness (15, 37), but there are exceptions. Matsuo (31), using Japanese cucumber cultivars, showed varying sexual responses to changes in day length and temperature. Several cultivars were found to be temperature sensitive, while others were temperature insensitive. Likewise, some cultivars were day-length sensitive, while others were day-length neutral. The most interesting cultivar discovered was one in which short days and low temperatures promoted maleness. Little information is known about the environmental effects on sex expression in muskmelons, squash, and watermelons (42, 54).

A thorough understanding of the interactions between environmental conditions and hormonal responses and their influences on different sexual genotypes is necessary for successful hybrid seed production. This chapter will attempt to summarize the available knowledge relevant to the biochemical regulation of hormonal response to environmental and genetic interactions.

Hormones Affecting Sex Expression

There are two possible approaches in studying the effects of hormones on sex expression. One approach is to observe the effect of external applications of growth regulators on plant development (4, 5, 8, 10, 12–20, 27, 30, 32, 40, 41, 43, 46, 49, 50, 57, 58); the other is to study the biosynthesis, transport, and degradation of endogenous hormones (3, 9, 17, 21, 24, 25, 26, 48, 52, 53, 55). For practical reasons, these approaches are often pursued separately. A proper explanation of hormonal regulation in sex expression, however, must incorporate the findings from both approaches.

Present information on plant hormonal control is based primarily on the

external application of growth regulators. Several groups of growth regulators have been demonstrated to affect sex expression in cucurbits. Ethylene
and ethylene-releasing compounds have been shown to promote femaleness
in cucumbers (5, 32, 33, 45, 50), muskmelons (9, 30, 50) and squash (11,
46, 50), while promoting maleness in watermelons (10, 42, 44). Inhibitors
to the biosynthesis and action of ethylene, such as silver nitrate, carbon
dioxide, and aminoethoxyvinylglycine (AVG), increase maleness in cucumbers and muskmelons (1, 30, 41). Indoleacetic acid (IAA) was shown to
enhance femaleness in cucumbers and muskmelons (15, 19, 20, 22), although this enhancement was found to be less effective than the response to
ethylene. Application of the gibberellins GA_3, GA_4, or GA_7 changes male
flower tendencies by the conversion of gynoecious lines into predominantly
male lines (12, 16, 18, 20, 27, 43, 49, 57, 58). GA_{4+7} was demonstrated to
be 10 to 20 times more effective than GA_3 in promoting male flowers.
Growth retardants such as chlorocholin chloride, Alar, Amo 1618, Phosphon D, and ancymidol increase female tendencies (1, 15, 23, 27, 42, 44,
51). Abscisic acid (ABA) was found to act as a gibberellin inhibitor in
gynoecious cucumber cultivars, promoting female flower production (16,
49), although in monoecious cucumber cultivars, external application of
ABA was shown to delay the appearance of the first female flower (16).

Endogenous hormonal levels have been studied in monoecious and
gynoecious cucumber isolines (53). Ethylene evolution varied with different
sexual types (48, 52). Low ethylene levels were found in androecious
cucumber cultivars, while high ethylene levels were found in the apices of
gynoecious cultivars. In monoecious lines, low ethylene evolution occurred
during female flower initiation. Auxin levels, as determined by bioassay or
by gas liquid chromatography, were present in higher concentrations in the
apices of gynoecious lines (15, 21, 53). During cucumber plant development, an enhanced female tendency was found to correspond to an increase
in auxin levels (21). The gibberellin level increased within the plant early in
the development of monoecious cucumber cultivars (17, 24), but decreased
after the third leaf stage. Abscisic acid and other inhibitors were found in
high concentrations in gynoecious cucumber cultivars (53).

Sex expression in cucurbits is not only a response to internal and external
hormonal levels but also an interaction with the surrounding plant environment. Short days have been correlated with increased levels of ethylene,
auxin, and ABA, while lowering the level of GA and GA-active compounds
within the plant. Conversely, long days were found to increase GA levels.
Both day length and temperature influence responses to growth regulator
applications. Cucumbers were less responsive to treatment with GA under
short days and low temperatures. Higher GA concentrations were necessary
to exhibit similar flower production under optimal conditions (16). This
interaction between day length, temperature, and growth regulator con-

centrations could explain, in part, the conflicting results on hormonal sex expression in cucurbits mentioned in the literature.

Ethylene Biosynthesis and Biosynthesis Inhibitors

The mechanism of ethylene biosynthesis in the Cucurbitaceae may be similar to the pathway proposed by Adams and Yang for apple tissue (2). The proposed pathway (2, 59) contains three steps, with methionine the precursor of S-adenosylmethionine (SAM), 1-aminocyclopropane-1-carboxylic acid (ACC), and ethylene.

The synthesis of ethylene from ACC is easily accomplished in most plant tissues (59, 60). This biosynthetic step requires oxygen and is inhibited by cobalt ions (Co^{++}) and temperatures above 40°C. Information is needed on the effect of these variables, particularly cobalt ions, on ethylene evolution in cucurbits.

Compounds exist that inhibit or enhance ethylene biosynthesis and that also influence sex expression. The synthesis of ACC from SAM is strongly inhibited by rhizobitoxin and its analog AVG (59). Not surprisingly, AVG has been found to be an important inhibitor of female flower production in cucumbers and muskmelons (4, 30, 41). Alpha-aminoxyacetic acid has a similar effect on flowering in cucumber (Table 1).

Ethylene-Auxin Interactions

As early as the 1950's, auxin was known to effect female flower production in cucurbits (19). Since then, extensive studies support the view that auxin and ethylene are principle factors in the control of sex expression.

Table 1. Effect of foliar application of silver nitrate and α-aminoxyacetic acid on sex expression in two cucumber cultivars 'GY3' and 'Biet Alpha'

	'GY3'			'Biet Alpha'		
	Node of first pistillate flower	No. of flowers in first 10 nodes		Node of first pistillate flower	No. of flowers in first 10 nodes	
Treatment		Pistillate	Staminate		Pistillate	Staminate
AgNO₃ (100 ppm)	8.4	1.8	6.0	1.5	5.0	5.0
α-Aminoxyacetic acid						
200 ppm	9.5	1.7	7.0	7.0	7.0	1.4
100 ppm	7.8	3.4	5.8	9.6	5.6	3.2
50 ppm	5.0	7.0	2.0	6.8	6.2	2.2
Control	2.3	8.0	1.0	2.0	9.0	0

Several possible theories on the specific effects and interactions of auxin and ethylene and their involvement with other compounds have been suggested (1), although none has been clearly shown to exist in cucurbits.

Recent investigations have shown that auxin interferes directly with ethylene biosynthesis (59, 60). Indoleacetic acid was found to enhance the production of ACC from SAM at the same step in ethylene biosynthesis that is inhibited by AVG. Yu and Yang (60) proposed that IAA affected the activity of the enzyme ACC synthetase in mung beans, but this effect needs to be demonstrated in cucurbits. Application of ethylene was found to decrease the auxin concentration in muskmelons (53). This phenomenon may be the result of a reduction in auxin transport or an increase in auxin conjugation or peroxidation.

Ethylene may also affect the ability of auxin to bind to specific receptor sites (39). The binding of auxin to these sites is believed by some to be necessary for activity and transport (28). Thus, if ethylene inhibits auxin binding, both auxin activity and transport would be reduced. Osborne and Mulling (39) suggested that the carrier of auxin transport can bind both ethylene and auxin. Accordingly, the binding of either ethylene or auxin precludes the binding of the other hormone. Thus, competition with ethylene may indirectly affect auxin levels.

A second hypothesis concerning the effect of ethylene and auxin was suggested by Osborne (38). Hormone sensitive "target" cells were classified according to their response to ethylene and auxin applications. Three types of cells were found: 1) those that enlarge and elongate only with auxin, 2) those that enlarge and elongate only with ethylene, and 3) those affected by both hormones. This target cell concept may help explain the effect of auxin and ethylene on the development of female flowers. One hormone could be necessary for differentiation, while the other might be necessary for cell growth and enlargement. In cucumber, however, it is not clear whether ovule development in female flowers requires only ethylene for enlargement or requires both hormones.

More research is obviously necessary to clarify whether the effect of ethylene is on the biosynthesis, binding, transport, or degradation of auxin. Considering the relative activity and the complexity of the interaction of auxin and ethylene, auxin could be proposed as a secondary hormone affecting femaleness in cucurbits.

Ethylene Action Inhibitors

Silver nitrate was demonstrated to inhibit actively female flower production in gynoecious cucumbers (41) and increase ethylene evolution (4). Silver ions were found to interfere with ethylene action (7), probably because silver nitrate binds to the site of ethylene action. Carbon dioxide caused a similar inhibition of female flower production (53). This inhibition is a result of competition with ethylene for the site of action (1, 59). Silver nitrate has a stronger effect than CO_2 and is currently used to induce male flowers in gynoecious cucumber and muskmelon cultivars (6, 30, 41).

Gibberellin-Ethylene Interactions

Gibberellic acid and ethylene individually exhibit opposite effects on sex expression in cucurbits. When GA_4 and GA_7 were applied with ethephon at the first leaf stage in cucumbers and squash, a synergistic effect on female flower production was observed. Thus, gibberellin and ethylene seem to have different sites of hormone action (18, 27). Ethylene production has not been found to be affected by the application of gibberellin (49).

The shift in the development from potentially staminate to pistillate buds has been demonstrated directly in floral bud cultures (22) and by observations of intersexed flowers in cucumber, muskmelon, and squash. Single GA_{4+7} treatments induced the appearance of staminate floral buds in several consecutive nodes on the main stem of a genetically female cucumber line (12). Staminate buds appeared next to pistillate buds that were in various stages of degeneration. Repeated GA treatments on hermaphoditic plants also induced the appearance of staminate flowers next to bisexual flowers. These results suggest that the GA-induced staminate buds did not develop from sexual reversion but were adventitious buds that normally do not develop in genetically female cucumber lines (18).

Gibberellin Metabolism

The influence of gibberellin on sex expression in cucurbits could be explained by differences in transport and biosynthesis, or by the conversion of biologically active forms to inactive forms via conjugation or degradation. Hypocotyl segments from a gynoecious cucumber line were found to transport more H^3-GA_1 than hypocotyl segments from an androecious cucumber line (55). This result suggests that it is unlikely that sex expression in cucumbers is controlled by differences in GA transport.

Very little is known about the origin and the physiological significance of conjugated or bound gibberellins. It was suggested that bound GA could be considered as either a reserve or storage form which releases GA, as inert GAs that have no function, or as a transport form (34). There is only meager evidence available concerning GA conjugates in cucumber seeds (26).

Gibberellin$_1$ is present in high concentrations in cucumber seeds and plant tissues, with lesser quantities of GA_3 and traces of GA_4 and GA_7

present (24, 25). Gibberellin$_4$ and GA$_7$ in relatively low concentrations were shown to strongly influence sex expression in cucumbers (12). Gibberellin$_4$ and GA$_7$ have been proposed as the biologically active forms of gibberellin, while GA$_1$ is a less active form (13, 14). The metabolism of GA$_4$ to GA$_1$, and GA$_1$ to GA$_8$ might function to inactivate GA activity. Evidence in cucumbers (54) strongly suggests that GA$_1$ is the precursor of GA$_8$, and therefore GA metabolism in cucumbers is similar to that in other plant species (14). The conversion of GA$_4$ to GA$_1$ by a single hydroxylation (14, 35, 36) may be relevant, because both GA$_4$ and GA$_1$ are present in cucumbers (24).

Gibberellins differ markedly in their potency (12, 13, 14). The particular GA present in a tissue or organ at any one time is not necessarily constant, so that metabolic interconversion of gibberellins must occur (34). The interconversion of gibberellins in the cucumber plant has received only limited attention (55).

Gibberellin Biosynthesis Inhibitors

Several plant growth retardants have been synthesized that inhibit the growth of cucumbers (15, 23, 49), muskmelons (51) and watermelons (42, 44). These growth retardants affect sex expression and inhibit the biosynthesis of gibberellin from mavalonic acid to GA$_{12}$. Particular attention has recently been given to growth retardants because of their practical utility in growth inhibition and their inhibition of GA activity.

AMO 1618 is the most specific inhibitor of GA biosynthesis. It inhibits the cyclization of geranylgeranlyl pyrophosphate to copalyl pyrophosphate and the synthesis of sterols and other triterpenes. CCC (2-chloroethyltrimethyl-ammonium chloride) inhibits GA biosynthesis after the formation of kaurene and other reactions (34). CCC is less effective than AMO 1618 in the control of sex expression. Ancymidol blocks the conversion of kaurene to kaureuol, and the subsequent two mixed function oxidase catalyzed reactions (34).

Growth retardants affect femaleness in cucumbers and muskmelons. Alar was found to increase female flower production and decrease gibberellin activity, as indicated by bioassay in muskmelons (51). AMO 1618, Phosphon D, and CCC were shown to promote femaleness in cucumbers (15, 23).

Abscisic Acid and Other Natural Inhibitors

Various bioassays revealed the existence of a high level of inhibitors in cucumber plants (53), which were influenced by environmental conditions, genotypic variation, and ethephon treatment (53). The only inhibitor so far identified in cucumbers is abscisic acid (17, 53). The lack of identification and observations of the action of other inhibitors make it difficult to interpret their involvement in sex expression.

Under short day conditions, ABA was shown to be in high concentrations in the leaves of gynoecious cucumber lines (16, 53). An even higher level of ABA was obtained after application of ethephon. In gynoecious cucumbers, the application of ABA to the apices or roots in water culture enhanced the development of female flower buds. Conversely, ABA was found to enhance male flower development in monoecious cucumber cultivars (16). These results suggest that ABA in gynoecious cucumber cultivars affects bud development by preventing the abortion of female flower buds. Abscisic acid is antagonistic to gibberellin by restricting the action of GA on ovary development in female flower buds (49). Thus, ABA may influence flower development, not flower differentiation.

The action of ABA can also be interpreted by optimal concentration curves (16). According to this hypothesis, the level of ABA in gynoecious cucumber cultivars is found in that part of the curve at which increasing ABA levels enhance female flower production. The high ABA concentration in monoecious cucumbers is superoptimal for female flower development but is optimal for male flower development (16).

Abscisic acid is biosynthesized from mevalonate in the isoprenoid pathway after parsnel pyrophosphate (34). The isoprenoid pathway is common to the initial sites in both ABA and gibberellin biosynthesis. Since ABA and GA are known to be antagonistic (34, 49), it is important to identify the enzymatic reaction determining the rate of ABA and BA biosynthesis.

The involvement of ABA in sex expression needs clarification through further studies, especially in other cucurbits. A general interpretation of the action of ABA is at present impossible. A knowledge of the metabolism of ABA in buds and the identification of other inhibitors and their interrelationships with ABA could aid in the understanding of ABA involvement in sex expression.

Conclusions

In the Cucurbitaceae the effects of day length and genotype on sex expression are influenced by hormonal activity. The interrelationships of these factors, determined for cucumbers, are probably similar in muskmelons and squash. Present knowledge indicates that watermelons may have a different control mechanism for sex expression (10, 42, 44). A simple explanation in watermelons might be that ethylene affects male flower differentiation, and ethylene biosynthesis and action inhibitors enhance femaleness (10). Further study of the control mechanism in watermelon will clarify the general view of the control of sex expression in plants.

Ethylene can be considered as the main hormone affecting femaleness in cucumbers (52), squash (11, 46) and muskmelons (9). The precise mechanism is unclear. Is this a direct effect on the differentiation of growth of specific cells or the inhibition of specific primordia organs? There is some indication that the primary effect of ethylene is on ovary development. Careful examinations of the interrelationship between ethylene and auxin and auxin changes during female bud differentiation will improve the understanding of the determination of femaleness. Progress made recently on understanding the biosynthetic pathway of ethylene, the discovery of ACC as an intermediate product, and the identification of enzymes affecting the rate of ACC biosynthesis (49) have improved the understanding of the role of ethylene in sex expression. Study of the involvement of auxin on ACC synthetase will help to clarify the interrelationships between ethylene and auxin.

The effect of day length in cucumbers is mediated by phytochrome. There are important interactions between phytochrome and the metabolism and action of plant hormones. Further studies are necessary for an understanding of these interactions and their influence on sex expression.

A unique opportunity for the study of the physiological genetics of sex expression exists with the development of isogenic lines differing only in the genes or alleles that control sex expression. Further studies on the biochemistry of plant hormone biosynthesis and metabolism, with specific attention to their interaction with the genetic control mechanism, will contribute information on the general effect of gene action and hormonal regulation.

Literature Cited

1. Abeles, F. B. 1973. Ethylene in Plant Biology. Academic Press, New York.
2. Adams, D. O., and S. F. Yang. 1979. Ethylene biosynthesis: identification of 1-aminocyclopropane-1-carboxylic acid as an intermediate in the conversion of methionine to ethylene. Proc. Natl. Acad. USA 76: 170–174.
3. Atsmon, D., A. Lang, and E. N. Light. 1968. Contents and recovery of gibberellins in monoecious and gynoecious cucumber plants. Pl. Physiol. 43: 806–810.
4. Atsmon, D., and C. Tabback. 1979. Comparative effects of gibberellin, silver nitrate and aminoethoxyvinylglycine on sexual tendency and ethylene evolution in cucumber plant (*Cucumis sativus* L.). Pl. Cell Physiol. 20: 1547–1555.
5. Augustine, J. J., L. R. Baker, and H. M. Sell. 1973. Female flower induction in an androecious cucumber, *Cucumis sativus* L. J. Amer. Soc. Hort. Sci. 98: 197–199.

6. Beyer, E. 1976. Silver ion: A potent antiethylene agent in cucumber and tomato. HortScience 11: 195–196.

7. Beyer, E. 1979. Effect of silver ion, carbon dioxide and oxygen on ethylene action and metabolism. Pl. Physiol. 63: 169–173.

8. Bukovac, M. J., and S. H. Wittwer. 1961. Gibberellin modification of flower sex expression in *Cucumis sativus* L. Adv. Chem. Series, Gibberellins 28: 80–88.

9. Byers, R. E., L. R. Baker, H. M. Sell, R. C. Herner, and D. R. Dilley. 1972. Ethylene: a natural regulator of sex expression in *Cucumis melo* L. Proc. Natl. Acad. USA 69: 717–720.

10. Christopher, D. A., and J. B. Loy. 1982. Influence of foliar applied growth regulators on sex expression in watermelon. J. Amer. Soc. Hort. Sci. 107: 401–404.

11. Chrominski, A., and J. Kopcewicz. 1972. Auxin and gibberellins in 2-chloroethyl phosphonic acid-induced femaleness of *Cucurbita pepo* L. Z. Pflanzenphysiol. 68: 184–189.

12. Clark, R. K., and D. S. Kenney. 1969. Comparison of staminate flower production on gynoecious strains of cucumbers, *Cucumis sativus* L., by pure gibberellins (A$_3$, A$_7$, A$_{13}$) and mixtures. J. Amer. Soc. Hort. Sci. 94: 131–132.

13. Crozier, A., C. C. Kuo, R. C. Durley, and R. P. Pharis. 1970. The biological activities of 26 gibberellins in nine plant bioassays. Canad. J. Bot. 48: 867–877.

14. Durley, R. C., and R. P. Pharis. 1973. Interconversion of gibberellin A$_4$ to gibberellins A$_1$ and A$_{34}$ by dwarf rice cultivar Tan-ginbozu. Planta 109: 357–361.

15. Frankel, R., and E. Galun. 1977. Pollination Mechanisms and Their Application in Plant Breeding. Springer-Verlag, Heidelberg.

16. Friedlander, M., D. Atsmon, and E. Galun. 1977. Sexual differentiation in cucumber: the effects of abscisic acid and other growth regulators on various sex genotypes. Pl. Cell Physiol. 18: 261–269.

17. Friedlander, M., D. Atsmon, and E. Galun. 1977. Sexual differentiation in cucumber: abscisic acid and gibberellic acid of various sex genotypes. Pl. Cell Physiol. 18: 681–691.

18. Fuchs, E., D. Atsmon, and A. H. Halevy. 1977. Adventitious staminate flower formation in gibberellin-treated gynoecious cucumber plants. Pl. Cell Physiol. 18: 1193–1201.

19. Galun, E. 1959. The role of auxins in the sex expression of the cucumber. Physiol. Pl. 12: 48–61.

20. Galun, E. 1959. Effects of gibberellic acid and napthaleneactic acid on sex expression and some morphological characters in cucumber. Phyton 13: 1–8.

21. Galun, E., S. Izhar, and D. Atsmon. 1965. Determination of relative auxin content in hermaphroditic and andromonoecious *Cucumis sativus* L. Pl. Physiol. 40: 321–326.

22. Galun, E., Y. Young, and A. Lang. 1963. Morphogenesis of floral buds of cucumber cultured in vitro. Develop. Biol. 6: 370–387.

23. Ghosh, M. S., and T. K. Bose. 1970. Sex modification in cucurbitaceous plants by using CCC. Phyton 27: 131–135.

24. Hayashi, F., D. R. Beorner, C. E. Peterson, and H. M. Sell. 1971. The relative content of gibberellin in seedlings of gynoecious and monoecious cucumber (*Cucumis sativus*). Phytochemistry 10: 57–62.

25. Hemphill, D. D., L. R. Baker, and H. Sell. 1972. Isolation and identification of gibberellins of *Cucumis sativus* and *Cucumis melo*. Planta 103: 241–248.
26. Hemphill, D. D., L. R. Baker, and H. M. Sell. 1974. Isolation of novel conjugated gibberellins from *Cucumis sativus* seed. Canad. J. Biochem. 51: 1647–1653.
27. Iwahori, S., J. M. Lyons, and O. E. Smith. 1970. Sex expression in cucumber plants as affected by chloroethylphosphonic acid, ethylene and growth regulators. Pl. Physiol. 46: 412–415.
28. Jacobs, M., and H. Hertel. 1978. Auxin binding to subcellular fraction from *Cucurbita* hypocotyls: in vitro evidence for an auxin transport carrier. Planta 142: 1–10.
29. Kubicki, B. 1969. Investigation of sex determination in cucumber (*Cucumis sativus* L.). Genet. Polon. 10: 3–143.
30. Loy, J. B., T. S. Natti, C. D. Zack, and S. K. Fritz. 1979. Chemical regulation of sex expression in gynomonoecious line of muskmelon. J. Amer. Soc. Hort. Sci. 104: 100–101.
31. Matsuo, E. 1968. Studies on the photoperiodic sex differentiation in cucumber, *Cucumis sativus* L. I. Effect of temperature and photoperiod upon the sex differentiation. J. Fac. Agric. Kyushu Univ. 14: 483–506.
32. McMurray, A. L., and C. H. Miller. 1968. Cucumber sex expression modified by 2-chloroethanephosphonic acid. Science 162: 1397–1398.
33. Minina, E. G., and L. G. Tylkina. 1947. Physiological study of the effect of gases upon sex differentiation in plants. Dokl. Acad. Nauk. SSSR 55: 165–168.
34. Moore, T. C. 1979. Biochemistry and Physiology of Plant Hormones. Springer-Verlag, Heidelberg.
35. Nadeau, R., and L. Rappaport. 1972. Metabolism of gibberellin A_1 in germinating bean seeds. Phytochemistry 11: 1611–1616.
36. Nadeau, R. L., L. Rappaport, and C. F. Stolp. 1972. Uptake and metabolism of ^{3}H gibberellin GA_1 by barley aleurone layers: response to abscisic acid. Planta 107: 315–324.
37. Nitsch, J. P., E. B. Kurtz, Jr., J. L. Liverman, and F. W. Went. 1952. The development of sex expression in cucurbit flowers. Amer. J. Bot. 39: 32–43.
38. Osborne, D. J. 1977. Auxin and ethylene and the control of cell growth. Identification of three classes of target cells. *In* P. E. Pilet, ed., Proc. 9th Int. Conf. on Plant Growth Regulation, Lausanne. Springer-Verlag, Heidelberg.
39. Osborne, D. J., and M. G. Mullins. 1969. Auxin, ethylene and kinetin in a carrier-protein model system for the polar transport of auxins in petiole segments of *Phaseolus vulgaris*. New Phytol. 58: 977–991.
40. Ota, T. 1963. Studies on BCB (bromochloline bromide). V. Effects of BCB on the growth and flowering of cucumber plants. J. Hort. Assoc. Jap. 31: 329–336.
41. Owens, K. W., C. E. Peterson, and G. E. Tolla. 1980. Induction of perfect flowers on gynoecious muskmelon by silver nitrate and aminoethoxyvinylglycine. HortScience 15: 654–655.
42. Peles, A. 1975. Factors effecting sex expression in watermelon (*Citrullus lanatus*), and possibilities of hybrid seed production. M.S. thesis, Hebrew Univ. of Jerusalem, Rehovot.

43. Peterson, C. E., and L. D. Anhder. 1960. Induction of staminate flowers on gynoecious cucumbers with gibberellin A$_3$. Science 131: 1673–1674.
44. Regev, Y. 1977. Hormone effects on sex expression in watermelon (C. *lanatus*) and their application in hybrid production. M.S. thesis, Hebrew Univ. of Jerusalem, Rehovot.
45. Robinson, R. W., S. Shannon, and M. D. de la Guardia. 1969. Regulation of sex expression in the cucumber. BioScience 19: 141–142.
46. Robinson, R. W., T. W. Whitaker, and G. W. Bohn. 1970. Promotion of pistillate flowering in *Cucurbita* by 2-chloroethylphosphonic acid. Euphytica 19: 180–183.
47. Rowe, P. R. 1969. The genetics of sex expression and fruit shape, staminate flower induction, and F$_1$ hybrid feasibility of gynoecious muskmelon. Ph.D. dissertation, Michigan State Univ., East Lansing.
48. Rudich, J., L. R. Baker, H. M. Sell, and J. W. Scott. 1976. Phenotypic stability and ethylene evolution in androecious cucumber. J. Amer. Soc. Hort. Sci. 101: 48–51.
49. Rudich, J., and A. H. Halevy. 1974. Involvement of abscisic acid in regulation of sex expression in cucumber. Pl. Cell Physiol. 15: 635–642.
50. Rudich, J., A. H. Halevy, and N. Kedar. 1969. Increase in femaleness of three cucurbits by treatment with ethrel, an ethylene releasing compound. Planta 86: 69–76.
51. Rudich, J., A. H. Halevy, and N. Kedar. 1972. Interaction of gibberellin and SADH on growth and sex expression of muskmelon. J. Amer. Soc. Hort. Sci. 97: 369–372.
52. Rudich, J., A. H. Halevy, and N. Kedar. 1972. Ethylene evolution from cucumber plants as related to sex expression. Pl. Physiol. 49: 998–999.
53. Rudich, J., A. H. Halevy, and N. Kedar. 1972. The level of phytohormones in monoecious and gynoecious cucumbers as affected by photoperiod and ethephon. Pl. Physiol. 50: 585–590.
54. Rudich, J., and A. Peles. 1976. Sex expression in watermelon as affected by photoperiod and temperature. Scientia Hort. 5: 339–344.
55. Rudich, J., H. M. Sell, and L. R. Baker. 1976. Transport and metabolism of ^{3}H-Gibberellin A$_1$ in dioecious cucumber seedlings. Pl. Physiol. 57: 734–737.
56. Scott, J. W., and L. R. Baker. 1975. Inheritance of sex expression from crosses of dioecious cucumber. J. Amer. Soc. Hort. Sci. 100: 452–461.
57. Wittwer, S. H., and M. J. Bukovac. 1962. Staminate flower formation on gynoecious cucumbers as influenced by the various gibberellins. Naturwissenschaften 49: 305–306.
58. Wittwer, S. H., and M. J. Bukovac. 1962. Quantitative and qualitative differences in plant response to the gibberellins. Amer. J. Bot. 49: 524–529.
59. Yang, S. F. 1980. Regulation of ethylene biosynthesis. HortScience 15: 238–243.
60. Yu, Y. B., and S. F. Yang. 1979. Auxin-induced ethylene production and its inhibition by aminoethoxyvinylglycine and cobalt ions. Pl. Physiol. 64: 1074–1077.

22 | Mechanism of Male Sterility in Some Cucurbitaceae

S. V. S. Chauhan

ABSTRACT. A comparative morphological study of genic and chemically induced male-sterile plants of *Cucumis melo, Cucumis sativus, Cucurbita maxima, Luffa cylindrica*, and *Momordica charantia* revealed that male sterility is associated with tapetal abnormalities both in premeiotic and postmeiotic stages. Tapetal cells either degenerate or become abnormally enlarged. This is followed by degeneration of sporogenous cells or pollen grains. Histochemical observations indicated deficiencies in total carbohydrates, insoluble polysaccharides, DNA, histones, and total proteins and reduced acid phosphatase activity in all anther parts, including the tapetum of male-sterile plants. Anthers of completely male-sterile plants are devoid of proline.

Male sterility is useful in commercial production of hybrid seed (9, 12), and thus cytohistological understanding of cytoplasmic and genic male-sterile plants is of interest (11, 16). Genic male sterility has been reported in *Cucumis melo* L. (4) and *Cucurbita maxima* Duch. ex Lam. (14). In the present investigation a comparative morphological, histochemical, and biochemical study of male-sterile and male-fertile plants of *Cucumis melo, C. sativus* L., *Cucurbita maxima, Luffa cylindrica* (L.) M. J. Roem., and *Momordica charantia* L. was made.

Materials and Methods

In order to study the effect of some gametocides on pollen sterility and anther development, male-fertile plants of the aforementioned species were subjected to foliar sprays with maleic hydrazide (1,2-dihydropyridiazine 3,5-dione) at 0.25 and 0.35% in aqueous solution; FW-450 (sodium 2,3-

281

dichloroisobutyrate) at 0.3, 0.45 and 0.6%; and Dalapon (2,2-dichloro-propionic acid) at 0.1, 0.2 and 0.3% (Table 1).

Pollen viability in male-fertile, genic male-sterile, and chemically treated plants was tested at regular intervals, using Alexander's method (2). Fixed floral buds were dehydrated, cleared, and embedded by customary methods. For morphological studies, sections were stained with a combination of Heidenhan's iron alum haematoxylin and safranin fast green. For localization of total carbohydrates of insoluble polysaccharides (PAS test), DNA (Feulgen reaction), total proteins (Ninhydrin-Schiff's test), histones (alkaline fast green test), and enzyme acid phosphatase (lead sulfide method), procedures described by Jensen (15) were applied to microtomed sections. For localization of the enzyme acid phosphatase, fresh sections were cut with a sledge microtome using dry ice. Free proline in the anthers was quantitatively estimated by the colorimeteric method of Bates et al. (3).

Observations

Early anther ontogeny in genic and chemically induced male-sterile plants was more or less similar to their fertile counterparts until differentiation of the wall layers and sporogenous cells.

Anther Ontogeny in Genic Male-Sterile Plants

Cells in the endothecial layer failed to elongate radially, and the development of fibrous bands on their radial walls was inhibited. The tapetal cells in such anthers enlarged radially and became hypertrophied at the sporogenous tissue stage in *Cucurbita maxima* and at the vacuolate pollen grain stage in *Cucumis melo*. In *Cucurbita maxima* the sporogenous cells were

Table 1. Classification of male sterility in chemically treated plants

Growth regulator	Growth regulator treatments in each sterility class[1]			
	Normal fertility (N)	Low degree of male sterility (LS)	High degree of male sterility (HS)	Complete male sterility (CS)
Dalapon	0.10 (T_1,T_2) 0.20 (T_1) 0.30 (T_1)	0.10 (T_3) 0.20 (T_2)	0.20 (T_3) 0.30 (T_2, T_3)	—
FW-450	0.30 (T_1,T_2) 0.45 (T_1)	0.30 (T_3) 0.45 (T_2) 0.60 (T_1)	0.35 (T_3) 0.60 (T_2)	0.60 (T_3)
Maleic hydrazide	0.25 (T_1,T_2) 0.35 (T_1)	0.25 (T_3) 0.35 (T_2)	0.35 (T_3) 0.45 (T_3)	—

[1]Plants were treated with the indicated percentages of growth regulator at T_1 (pre-floral), T_2 (post-floral bud initiation), or T_3 (anthesis) stages.

crushed before the onset of meiosis (Figure 1A). Degeneration of tapetal cells followed. In *Cucumis melo* pollen mother cells underwent normal meiosis, and microspore tetrads developed normally. The tapetal cells in such anthers elongated radially but became hypertrophied at the vacuolate pollen stage (Figure 1B). The pollen grains remained cemented together, despite dissolution of the common callose mother wall, and formed an irregular mass possessing scanty cytoplasm, degenerated nuclei, and rudimentary exine devoid of characteristic reticulate sculpturing. The growth of hypertrophied tapetal cells was uninterrupted, with the result that the anther sac was occluded and the degenerated microspores were lost (Figure 1C). The encroaching tapetal cells were highly vacuolated and contained scanty cytoplasm and degenerated nuclei. On complete degeneration of the microspores the tapetal cells collapsed. The vascular strand in such anthers failed to differentiate.

Anther Ontogeny in Chemically Treated Plants

Treated plants were classified into four groups: normal (N), low degree of sterility (LS), high degree of sterility (HS), and complete sterility (CS) (Table 1). The pollen sterility for each of these groups was 0–25, 25–50, 51–95, and 96–100%, respectively. Untreated plants (MF) were classified in group N and genic male-sterile plants (GMS) in group CS.

Anther development in N and LS plants was more or less normal. In the LS type, however, the degeneration of tapetal cells was delayed from the early vacuolate pollen stage to the late vacuolate pollen stage. Degeneration was followed by normal development of fibrous thickenings in the endothecial cells. In the anthers of HS type plants, tapetal cells elongated radially and remained intact up to anthesis. Characteristic fibrous thickenings in the endothecial cells of such anthers failed to develop. The extent of radial elongation was closely related to the degree of sterility.

The tapetal cells in the anthers of CS plants treated three times with 0.6% FW-450 either degenerated before meiosis, as seen in a limited number of anthers of *Cucurbita maxima*, *Luffa cylindrica* and *Momordica charantia*, or elongated radially and became hypertrophied at the sporogenous tissue stage, as in *Cucurbita maxima*, or at the vacuolate pollen stage in other species. In any case, the tapetal cells collapsed after the degeneration of sporogenous tissue or pollen grains. Vascular differentiation in these anthers was much inhibited.

Histochemical Reactions

Histochemical reactions in different parts of anthers of MF, GMS, and CS plants at the sporogenous tissue stage were low or moderate, but with age the intensity of the reactions in MF anthers increased gradually and reached

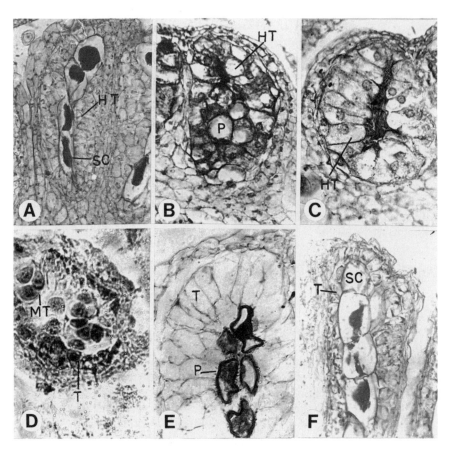

Figure 1. Anthers and sterility effects in cucurbits. A. Genic male-sterile *Cucurbita maxima,* with crushed sporogenous cells and hypertrophied tapetal cells. B-C. Genic male-sterile *Cucumis melo,* with hypertrophied tapetal cells and occluded, vacuolate pollen grains (B), and highly hypertrophied tapetal cells with central, degenerated microspore masses (C). Tapetal cells with scanty cytoplasm and degenerated nuclei. D. Genic male-sterile *Cucumis melo,* with localization of the enzyme acid phosphatase at the microspore tetrad stage. Tapetal cells and microspores marked by high enzyme activity. E. Chemically induced male-sterile *Cucurbita maxima.* PAS reaction with localized total carbohydrates of insoluble polysaccharides. Exine of nonviable pollen grains is highly reactive, but tapetal cells are deficient in insoluble polysaccharides. F. Chemically induced male-sterile *Momordica charantia.* Localization of total protein at the sporogenous tissue stage is low in sporogenous and tapetal cells. All ×230. HT, hypertrophied tapetal cells; MT, microspore tetrad; P, pollen; SC, sporogenous cells; T, tapetal cells.

a maximum at the microspore tetrad stage (Figure 1D). In subsequent stages, the intensity of the reactions declined in all parts except the microspores, where they showed further enhancement. Mature pollen grains exhibited highly intense staining reactions. In contrast, anthers of GMS and CS plants failed to show any significant change in various histochemical reactions in subsequent stages. Mature anthers of these plants were markedly deficient in total carbohydrates of insoluble polysaccharides (Figure 1E), DNA, histones, proteins (Figure 1F), and acid phosphatase.

Biochemical Activity

The proline content in the young anthers was quite low (1.1–2.3 μg/g fresh weight) in MF, GMS, and CS plants. However, in the anthers of MF plants, there was a gradual increase of proline in subsequent stages, until it reached a maximum (340–520 μg/g fresh weight) at the pollen grain stage. In GMS and CS plants proline failed to show any significant increase at any stage, and mature anthers were totally devoid of it.

Discussion

In plants of *Cucurbita maxima*, *Luffa cylindrica*, and *Momordica charantia* treated with FW-450, tapetal cells degenerate at the sporogenous tissue stage in a limited number of anthers followed by degeneration of sporogenous cells before meiosis. Electron microscopic studies have shown that the sporogenous cells have plasmodesmatal connections with the tapetal cells before the onset of meiosis (17). It is suggested here that degeneration of the immature tapetum disturbs the normal communication between these two tissues and leads to male sterility.

Tapetal cells enlarge radially and become hypertrophied in the premeiotic stage in genic and chemically induced male-sterile plants of *Cucurbita maxima*. This is followed by the degeneration of pollen mother cells, causing complete sterility. However, in genic and chemically induced male-sterile plants of *Cucumis melo*, the pollen mother cells enter normal meiosis but produce microspores that degenerate, followed by the degeneration of the occluding hypertrophied tapetal cells. There are two possible explanations for hypertrophy: 1) the tapetum receives more nutrients from outer tissues than it is capable of supplying to the developing microspores; 2) the tapetal cells receive little or no nutrition from elsewhere and therefore develop haustorially. Growth toward the anther cavity probably reflects the lack of cellular resistance in this direction. The final degeneration of tapetal cells seems certainly to reflect a lack of nutrition.

The morphological findings are corroborated by histochemical studies that indicate a deficiency in total carbohydrates, proteins, DNA, and histones in various parts of an anther, including the hypertrophied tapetal cells. Poor enzyme acid phosphatase activity in the tapetal cells also reflects upon their low metabolic state. A deficiency of various important metabolities in the anthers of sterile plants has been recorded (1, 6–8, 10, 18, 19), and according to these authors, serves as a limiting factor in the normal development of pollen grains.

The anthers of genic and chemically induced male-sterile plants also exhibited a marked deficiency of free proline. Similar results have also been recorded in wheat and maize (1) and sugar beets (13). The presence of proline, a major amino acid in the pollen of most plants (20), is connected with pollen fertility and is of physiological significance because of its presumed function in the sexual process in flowering plants (5).

Acknowledgments

I am grateful to Bahadur Singh for his criticisms of this work, to T. W. Whitaker and W. P. Bemis for seeds of male-sterile strains, and to S. N. Chaturvedi and Roshan Singh for their support and use of their facilities.

Literature Cited

1. Alam, S., and P. C. Sandal. 1969. Electrophoretic analysis of anther proteins from male-fertile sudan grass, *Sorghum vulgare* var. *sudanese* (Piper) A. S. Hitchc. Crop Sci. 9: 157–159.
2. Alexander, M. P. 1969. Differential staining of aborted and nonaborted pollen. Stain Tech. 44: 117–122.
3. Bates, L. S., R. P. Waldren, and I. D. Teare. 1973. Rapid determination of free proline for water-stress studies. Pl. Soil 39: 205–207.
4. Bohn, C. W., and T. W. Whitaker. 1949. A gene for male sterility in *Cucumis melo* L. Proc. Amer. Soc. Hort. Sci. 53: 309–314.
5. Britikov, E. A., N. A. Mustava, S. V. Valdimirtseva, and M. A. Protsenko. 1964. Proline in reproductive systems of plants. *In* H. F. Linskens, ed., Pollen Physiology and Fertilization. North Holland, Amsterdam.
6. Chauhan, S. V. S. 1979. Histochemical localization of enzyme acid phosphatase in the anthers of male-fertile, cytoplasmic, genic and induced male-sterile plants. Curr. Sci. 48: 35–38.
7. Chauhan, S. V. S., and T. Kinoshita. 1979. Histochemical localization of histones, DNA and proteins in the anthers of male-fertile and male-sterile plants. Jap. J. Breed. 29: 287–293.
8. Chauhan, S. V. S., and R. K. S. Rathore. 1980. Histochemical evaluation of PAS reaction in the anthers of male-fertile and male-sterile plants. Curr. Sci. 49: 433–436.

9. Duvick, D. N. 1968. Use of male sterility in the breeding of outcrossing crops. Proc. 12th Intl. Congr. Genet. II: 228–229.
10. Fukasawa, H. 1954. On free amino acids in anthers of male-sterile wheat and maize. Jap. J. Genet. 29: 135–137.
11. Gottschalk, W., and M. L. H. Kaul. 1974. The genetic control of microsporogenesis in higher plants. Nucleus 17: 133–166.
12. Heyen, E. G., and R. W. Livers. 1968. Use of male sterility in the breeding of self-pollinating crops. Proc. 12th Intl. Congr. Genet. II: 228–229.
13. Hosokawa, S., C. Tsuda, and T. Takeda. 1963. Histochemical studies on the cytoplasmic male sterility of sugar beets. Jap. J. Breed. 13: 425–454.
14. Hutchins, A. E. 1944. A male and female variant in squash (*Cucurbita maxima* DC.). Proc. Amer. Soc. Hort. Sci. 44: 494–496.
15. Jensen, W. A. 1962. Botanical Histochemistry. Freeman, San Francisco.
16. Laser, K. D., and N. R. Lersten. 1972. Anatomy and cytology of microsporogenesis in cytoplasmic male-sterile angiosperms. Bot. Rev. 38: 425–454.
17. Mascarenhas, J. P. 1975. The biochemistry of angiosperm pollen development. Bot. Rev. 41: 259–291.
18. Nakashima, H., and S. Hosokawa. 1971. Histochemical studies on the cytoplasmic male sterility of some crops. II. Changes of DNA, polysaccharides and proteins and translocation of photosynthetic products of sugar beets. Mem. Fac. Agric. Hokkaido Univ. 8: 1–4.
19. Saini, S. S., and G. N. Davis. 1969. Male sterility in *Allium cepa* and some species hybrids. Econ. Bot. 23: 37–49.
20. Tupy, J. 1964. Metabolism of proline in styles and pollen tubes of *Nicotiana alata*. *In* H. F. Linskens, ed., Pollen Physiology and Fertilization. North Holland, Amsterdam.

23 | Comparative Ontogeny of Male and Female Flowers of *Cucumis sativus*

Martin C. Goffinet

ABSTRACT. This study of *Cucumis sativus* describes and compares development, under field conditions, of homologous male and female flowers from the time of initiation to anthesis, and attempts to determine if each is bisexual very early in its development. Node-four primary male flowers of the monoecious cultivar 'SMR 58' were compared with their female homologs in the gynoecious line 'GY 3'. There were no obvious differences between sexes until one to two days after carpel initiation. Thereafter, stamens developed fully while carpels formed only a nectariferous pistillodium in males. Conversely, carpels developed to maturity while stamens remained vestigial in females. The initial size and elongation rates for the stamens and for the stigma-style complex were similar in the two sexes, but subsequently the growth rate of the functional structure exceeded that of the nonfunctional. When log stigma-style length was plotted against log stamen length, i.e., allometrically, the pattern of allometric growth in female flowers was a mirror image of the male pattern. There thus was a fundamental morphogenetic difference in the two sexes expressed as a shift in emphasis between the growth rates of functional and nonfunctional sex organs. Flowers were therefore sexually committed before there was any gross morphological difference.

Several sexual systems are known in cucurbitaceous plants. Particular species may produce male, female, or perfect flowers in varying proportions on a given plant (20–26, 36, 42). This variation has stimulated great interest in the morphogenetic control mechanisms of sex expression in the family, especially as they can be manipulated by hormonal treatments (Roy and Saran, Rudich, this volume).

Most cultivars of the cucumber, *Cucumis sativus* L., are monoecious (8),

This work constitutes a portion of Ph.D. dissertation research completed in the Department of Botany, University of Minnesota, St. Paul, MN 55108.

288

producing separate male and female flowers at fairly predictable locations on a plant of known genotype in a given environment. Thus, monoecious individuals generate two sexually distinct kinds of flowers under the control of a single genetic constitution. Male and female flowers are commonly thought to arise from potentially bisexual primordia that subsequently develop along separate organogenetic and histogenetic pathways to ultimately reach anthesis as dimorphic structures (1, 10, 27, 30).

Understanding the basis of sex expression and floral morphogenesis in the Cucurbitaceae requires a better understanding of floral development from the primordial stage through anthesis. Although a number of descriptive studies have been made of cucumber flower development (6, 11, 12, 18, 19, 27, 33, 41), the present account directly compares male and female organogenesis. The objectives of this study were 1) to document the major qualitative changes that accompany the differentiation of homologous male and female flowers in this species, 2) to determine when and where sexual differences first become evident by time-course and allometric analyses, and 3) to evaluate the hypothesis that the floral primordium in the cucumber is fully bisexual in its early ontogeny.

Materials and Methods

To secure homology of male and female flowers, observations were confined to the first-formed, or primary, flower of leaf axil four on the main axis of field-grown plants (Figure 1). The gynoecious line 'GY 3' was selected to ensure exclusively female flowers at this position. The monoecious cultivar 'SMR 58', which generally produces male flowers in leaf axil four, was selected for comparison.

For three years (1973–75) seeds of the two cultivars were planted on June 4 at St. Paul, Minnesota. Ten plants of each cultivar were sampled thereafter at one- to three-day intervals. The node-four axillary complex was dissected out and fixed in formalin-acetic acid-ethanol. Some specimens were later embedded in paraffin for microtomy and microscopy, some were prepared for scanning electron microscopy (SEM), and still others were prepared as dissections by the method of Sattler (35). In addition, stamen length, stigma-style length, and total carpel length (stigma-style plus ovary) were measured in longitudinally sectioned primary flowers, prepared in sequence from time of organ initiation to anthesis, as described in Goffinet (11).

Observations

The primary flower meristem arose at node four approximately ten days after planting seed of 'GY 3' and 11 days after planting for 'SMR 58'. The

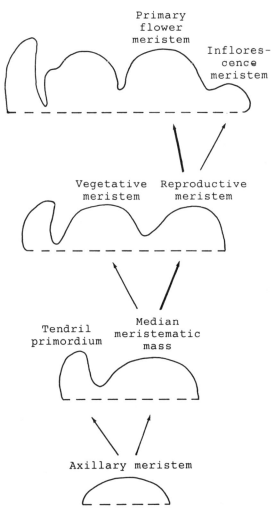

Figure 1. Origin of node-four axillary structures in *Cucumis sativus*. Vegetative and reproductive axes diverge quite early. The lineage of the primary flower is denoted by dark arrows. Dashed horizontal lines denote insertions of subtending leaf. Not to scale.

average time taken to reach anthesis in node-four primary flowers of 'GY 3' and 'SMR 58' was, respectively, 22 days and 23 days after flower initiation, i.e., 32 and 34 days after planting. The calyx in either sex is composed of five small green sepals. The showy yellow corolla has five petal lobes, which expand above a corolla tube. The pronounced sexual dimorphism between primary male ('SMR 58') and female ('GY 3') flowers at anthesis is illustrated in Figure 2.

The gynoecium of the female flower consists of an inferior ovary and a three-lobed stigma on a columnar style. The style is ringed at its base by a nectariferous gland. The ovary contains three carpels, each with many

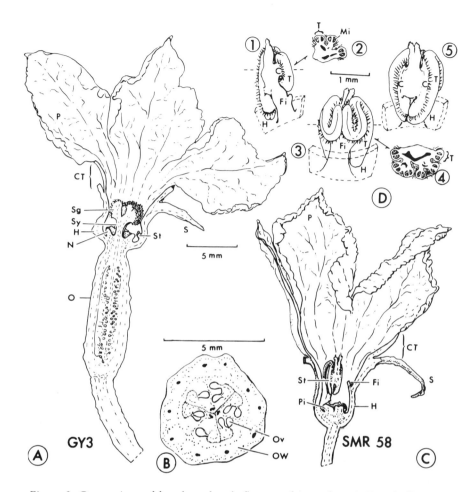

Figure 2. Comparison of female and male flowers of cucumber. A. Female flower at anthesis, longitudinal section. B. Tricarpellate ovary, transection at anthesis. C. Male flower at anthesis, longitudinal section. D. Simple stamen (1, 2) and two compound stamens (3-5) at microsporogenesis. C, anther connective; CT, corolla tube; Fi, filament; H, hypanthium ("floral cup"); Mi, microsporangium; N, nectary; O, ovary; Ov, ovule; OW, ovary wall; P, petal; Pi, pistillodium; S, sepal; Sg, stigma; St, stamen or staminodium; Sy, style; T, theca.

ovules, surrounded by ovary wall (Figure 2B). The entire stigma-style complex is surrounded by the "floral cup" or hypanthium. Three vestigial stamens, or staminodia, are inserted on the inner surface of the hypanthium.

The male flower has a much-reduced nectariferous pistil (pistillodium) centrally located in the basin of the "floral cup" (Figure 2C). The androecium is composed of one small and two large stamens (Figure 2D). The number of microsporangia in the small stamen is half that of a large stamen. This stamen arrangement is common to a number of cultivated species in several cucurbitaceous genera (3). The serpentine thecae (pollen sacs) of each stamen are separated by connective tissues, and each anther is supported by a rather short but broad filament inserted on the hypanthium wall. Microsporangia (two per theca) dehisce extrorsely along longitudinal sutures.

Floral Ontogeny

After the differentiation of the major components of the node-four axillary complex (Figure 1), the apex of the primary flower broadened, then became concave, as greater meristematic activity occurred about its periphery (Figure 3). Five sepal primordia arose in a 2/5 phyllotactic sequence at the periphery. Petal primordia thereafter arose centripetal to and alternate to the sepal lobes (Figure 4). A 2/5 phyllotactic pattern for petal origin could sometimes be seen, as was reported by Heimlich (12). As the perianth members elongated, they were also elevated by persistent meristematic activity beneath them. In this way the hypanthium originated (Figure 4), and the floral apex occupied an ever deepening concavity.

Stamen primordia, then carpel primordia, were initiated in turn at the inner periphery of this concavity. The five stamen primordia were initiated in the sepal radii. Four quickly fused into two pairs opposite petal radii, leaving the fifth in its original antesepalous position (Figure 5). The paired stamens developed as compound stamens, while the simple stamen developed alone. By five days after the primary flower was initiated, the compound stamens were composed of three meristematic regions (Figure 6): the two free tips derived from the original two stamen primordia, a central region forming thecae and connectives, and the short filament attaching the stamen to the hypanthium. The simple stamen had similar growth centers but did not achieve the bilateral symmetry of the compound stamens.

The three carpel primordia arose last and alternate to the three stamens as low protuberances on the flanks of the floral apex (Figure 7). As the carpels began growth, their margins extended tangentially and their bases fused with adjacent carpel margins in the stamen radii (Figure 8). Stamens and carpels were, in turn, subtended by newly intercalated hypanthium tissues, and the proximal portions of the petals soon coalesced in the radii of sepals to initiate the corolla tube (Figure 7).

Female Flower Development

The first obvious evidence that the primary flower would be female was an increase in incipient ovary wall tissues at six to seven days after flower

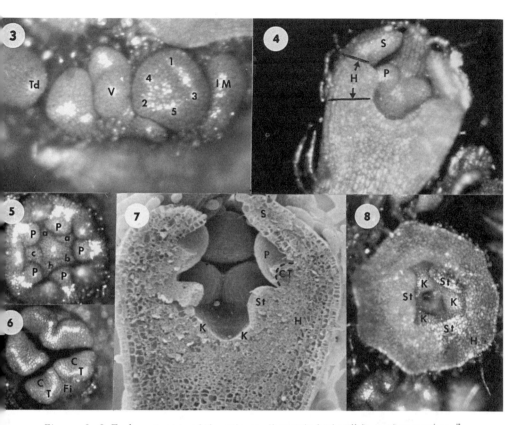

Figures 3–8. Early ontogeny of the primary flower in leaf axil four of cucumber. 3. The axillary complex, viewed from above, with tendril, vegetative branch, primary flower showing order of sepal initiation, and inflorescence meristem. Two days after initiation of primary flower. ×114. 4. A primary flower, longitudinal section, showing the origin of petals and hypanthium, 3.5 to four days. ×131. 5. A flower, from above, during the origin and early fusion of four stamens into pairs (a-a, b-b, ca. four days.). Stamen c remains isolated. ×81. 6. A simple and two compound stamens, from above, ca. five days. Connectives, thecae, and filaments differentiated. ×86. 7. A flower, longitudinal section, at 5.5 days, showing the origin of carpels and corolla tube. The hypanthium supports all floral whorls. ×151. 8. A flower cut across stamens at six days. The carpel margins fused in stamen radii. ×89. C, connective; CT, corolla tube; Fi, filament; H, hypanthium; IM, inflorescence meristem; K, carpel; P, petal; S, sepal; St, stamen; T, theca; Td, tendril; V, vegetative branch.

initiation, when the most basal region of fused carpel margins began to elongate. By seven days in most flowers, the ovary wall and fused carpel margins (septal ridges) thickened radially as the ovary elongated in the region below the "floral cup" (Figure 9). At this point other flowers of the nodal inflorescence were already inhibited in their development.

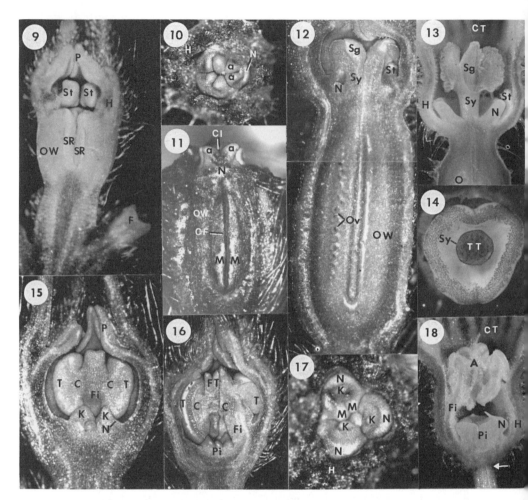

Figures 9–18. Late ontogeny of the primary female and male flowers in leaf axil four in cucumber. 9. A female flower, longitudinal section, seven days after initiation of primary flowers. The ovary wall and septal ridges thicken and lengthen. The stamens are inhibited, as are flowers of nearby inflorescence. ×45. 10. A dissected eight-day female flower, viewed from above. Tips of each carpel become bilobed and swell (a-a), initiating the stigma. The nectary arises as three discrete swellings of carpel tissues at the base of the stigma-style complex. ×45. 11. A ten-day female flower with the "floral cup" removed. A sector of ovary wall has been removed in the radius of the carpel surmounted by stigmatic lobes a-a. The two carpel margins expand outward and will form placental tissues. The cleft between a-a can be traced into the ovary as a carpellary fissure. ×27. 12. A female flower at 12 days, showing the young ovules, large ovary, and stigma-style complex, weak staminodia, and the young nectary. ×23. 13. A female flower at anthesis, 22 days, showing strong development of the nectary and female components and weak growth of the staminodia. ×4.5. 14. The nectary of a female flower at anthesis with the "floral

Marginal fusion of the carpel apices did not occur; rather, by eight days each apex became bilobed, signalling stigma differentiation (Figure 10). The two stigmatic lobes of each carpel began to swell from the tips downward. The floral nectary was also initiated at eight days, but not as a ring of tissue surrounding the style, as stated by Judson (19). Instead, the nectary began as three independent eruptions at the base of the style, one in each of the three carpel radii (Figure 10). Marginal fusion of these separate nectary tissues later occurred to form the single ringlike nectary encircling the base of the mature style.

By ten days the two stigmatic lobes of each carpel had spread tangentially as they enlarged (Figure 11). The cleft separating the lobes of each carpel could be followed basipetally along the inside surface of the carpel, through the ovarian region, as a carpellary fissure (Figure 11). Thus, the original septal ridges still had not fused, but rather, they kept pace with ovary enlargement such that both the ovary wall and the three septa grew as a unit. The tissues labelled M in Figure 11 are each the centrifugally enlarging margins of only one carpel, which is surmounted by stigmatic lobes, labelled a-a. Each original septal ridge (cf. Figure 9) therefore had to bifurcate along its inner edge, then expand outward, i.e., toward the reader's eye in Figure 11. After expanding to the ovary wall, the two margins in each carpel spread tangentially, initiated ovules, and so formed one of the three T-shaped placentae seen in mature ovary transections (see Figure 2B). This process is illustrated quite well by Leins and Galle (27) and was described to some extent by Judson (19). Placentation is parietal (31).

In the largest ten-day flowers, ovule initiation had begun. By 12 days ovules were easily seen with the naked eye (Figure 12), although integuments were not yet present. The stigma-style complex occupied most of the "floral cup" in 12-day flowers. From that time until anthesis, development

---

cup" and stigma-style complex removed. Nectar-secreting pores stain with aqueous toluidine blue, as does stylar transmitting tissue. ×5.7. 15. A male flower at seven days, showing a well-developed compound stamen but inhibition of carpels. The nectary tissue arises at the outer base of each carpel. ×38. 16. A male flower at nine days, with massive anthers and a small pistillodium. ×24. 17. The pistillodium of a male flower, viewed from above at ten days with the "floral cup" removed. The carpel tips remain free but the carpel margins (M-M) will fuse with the margins of adjacent carpels. The nectary tissues are obvious, but they are yet discrete. ×44. 18. A male flower at anthesis, 23 days, showing strong development of stamens and engorgement of the nectary tissues of the pistillodium. The abscission zone forms at base of the "floral cup" (arrow). ×4. A, anther; C, connective; CF, carpellary fissure; Cl, stigmatic cleft; CT, corolla tube; F, flower of inflorescence: Fi, filament; FT, filament tip; H, hypanthium; K, carpel; M, carpel margin; N, nectary tissue; O, ovary; Ov, ovule; OW, ovary wall; P, petal; Pi, pistillodium; Sg, stigma; SR, septal ridge; St, stamen; Sy, style; T, theca; TT, transmitting tissue.

consisted primarily of tissue differentiation, megasporogenesis, and organ enlargement, except for the staminodia.

The nectary attained its greatest size at anthesis. Nectar secretion began in the morning on the day of anthesis through nectariferous pores that covered the upper surface of the nectary (Figure 14). Collison and Martin (5) described these cells as stomate-like guard cells, which remained capable of opening and closing under vacuum stress during SEM observation.

At anthesis the "floral cup" is narrowest at stigma level, where it is also quite hairy (Figure 13). To reach the nectary, insect visitors must pass between the stigmatic and "floral cup" surfaces, often depositing pollen in these regions. The stigma is apparently of the "dry type," having unicelled papillae (14). Ovules are reached by pollen tubes that grow downward from the stigma either through the transmitting tissue of the style (Figure 14) or through the epidermal layers of the three carpellary fissues in the ovary (24). Fuller and Leopold (7) found that pollination in cucumbers stimulated ovary development within 24 hours, although syngamy occurred at 30–36 hours after pollen deposition.

Male Flower Development

The first obvious evidence that the primary flower would be male was the abrupt enlargement of the stamens at six to seven days after flower initiation and the weak growth of the three carpel primordia (Figure 15). As in female flowers, the nectary of male flowers began as three discrete eruptions, each on the basal abaxial flank of a carpel lobe, above the carpel-hypanthium juncture (Figure 15; cf. Figure 10). Ovary development in the male flower was insignificant. In contrast to floral abortion in the female nodal complex, the male node-four inflorescence continued to develop as the primary flower differentiated.

By 9–11 days the primary flower had enlarged greatly (Figure 16). The filaments thickened and became heavily vascularized in order to support the massive anthers, the connectives expanded centripetally to the floral axis, and the filament tips renewed elongation. By this time two microsporangia had differentiated in each theca, and single strands of microsporocytes could be found. A full account of microsporogenesis is given by Heimlich (13). The pistillodium began to swell at ten days. Because its three carpels were fused along their tangential margins (Figure 17), the nectary expanded as a single cushionlike mass, and so pushed the three carpel tips (the vestigial stigmas) upward into the substaminal space.

Between ten days and anthesis the major change was that of cell expansion within the enlarging organs and completion of microsporogenesis. In contrast to female flowers, the "floral cup" of males at anthesis was most constricted at the level of the anthers, the connectives of which were conjoined at the floral axis (Figure 18). Large staminal hairs and hairs on the inner wall of the "floral cup," probably aid in insect pollination by trapping pollen. To attain food, a nectar-feeding insect (and perhaps a pollen-feeding one as well) must penetrate this labyrinthine network of hairs and winding thecae. The pistillodium was greatly swollen and heavily vascularized; it also had numerous nectariferous pores on its upper surface that appeared identical to such pores in female flowers. The pedicel was delimited from the "floral cup" by a slightly constricted band encircling the pedicel (Figure 18, arrow). An abscission zone formed in this band before or during flower fade. It had no counterpart in the female flower. By the end of the day of anthesis all anthers in the male flower had dehisced, and the flowers showed signs of dehydration and wilting. By the second day the flower had abscised or would do so at a touch.

Quantitative Aspects of Flower Development

It was extremely difficult to determine on ontogenetic grounds alone when and at what critical stage the sexes first diverged in development; thus quantitative analysis became necessary. Perianth members could be excluded from such comparison because they showed no measureable differences between the sexes or cultivars at any developmental stage. There were, however, considerable differences in elongation rates between the stamens and the stigma-style complex of male and female flowers (Figure 19).

At five days after primary flower initiation, male stamens extended at a much greater rate than female staminodia, and ultimately were four to five times the staminodium length (Figure 19A). After seven to eight days the female stigma-style complex elongated much more rapidly than the male pistillodium, so that, at anthesis it was four times longer (Figure 19B). In males the pistillodium elongated at only a very low rate but increased in size after 14–15 days, when nectary differentiation became active. Total carpel length (not figured), which includes ovary length, first differed in the two sexes at about six days. The carpels of the female flowers thereafter greatly elongated to anthesis, and such growth continued afterward in the case of the ovary.

The growth rate relationships between functional and nonfunctional sexual structures are readily analyzed by allometry, which is the study of the change in proportions of two structures with increase in size, through log transformation of the data. Time need not be accounted for in the method, since the ratio of the two respective growth rates allows the common time factor to be cancelled. The attainment of form can then be viewed as a function of absolute size already achieved rather than of calendar time. Classical and excellent treatments of the use of allometry can be found in Clark and Medawar (4), Huxley (16), Richards (34), and Simpson and Roe (37).

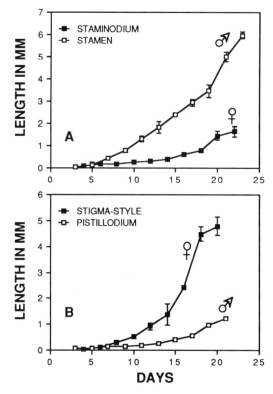

Figure 19. Elongation of the stamen vs. the staminodium (A) and elongation of the stigma-style complex vs. the pistillodium (B), for homologous primary male and female flowers of cucumber from day of flower initiation to anthesis. Vertical bars show standard deviations of means.

Values for the stamen and stigma-style lengths of each sex plotted on double log coordinates, from the time these organs were first measureable to anthesis, are shown in Figure 20. Each line is a best-fit linear regression, taking the form, $Y = bX^k$. The exponent k is, in each instance, the slope of the regression line and the ratio between the specific growth rates of the stigma-style and the stamen. The initial slopes in Figure 20 are very different between the sexes so that, with regard to the sexual structures, there is little doubt that male and female flowers differed in emphasis from the outset. The female flower initially showed a substantial increase in the rate of stigma-style elongation compared with the rate of staminodium elongation ($k = 3.80$). In contrast, the male flower initially showed a pronounced lag in the stigma-style elongation rate compared with the stamen elongation rate ($k = 0.63$). Slopes later changed markedly between sexes when the staminodia were abruptly carried upward by the lengthening corolla tube in the female and when the nectary began to swell in the male. The k values for the upper regression lines (between sexes) could not be separated on a statistical basis, so that the allometric growth constants for stigma-style length vs. stamen length could be considered identical for both sexes late in development.

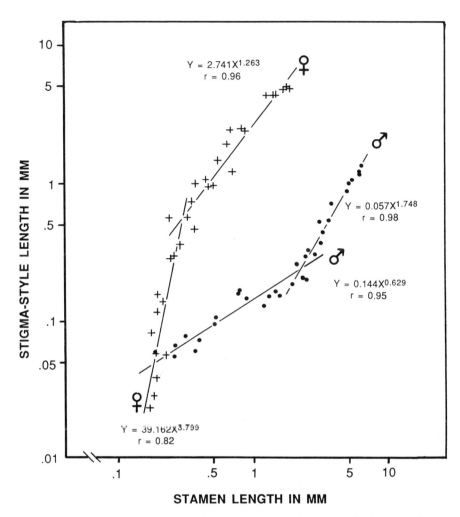

Figure 20. Allometric relationship between stigma-style (or pistillodium) and stamen (or staminodium) lengths in homologous female and male flowers. Each point represents the dimensions in one flower. Least-squares linear regressions take the form $Y = bX^k$. For each regression the value r is the correlation coefficient.

Discussion

In terms of gross morphology no differences could be found between primary male and female flowers immediately after initiation. These findings agree with those of Galun et al. (10) and Leins and Galle (27). Atsmon et al. (2) reported that in the monoecious types tested, anthesis of the first female flower on the main axis was earlier than that for male flowers at nodes immediately above or below. They concluded that there is faster development of the female bud. I have already noted that node-four female flowers of 'GY 3' bloomed one day before homologous male flowers of 'SMR 58', if counted from time of flower initiation.

At six days after flower initiation, the growth rate of the stamens in the male flower began an abrupt and permanent increase over the growth rate of the staminodium in the female flower. A small but significant difference in carpel elongation also occured between the sexes at that time. This initial difference apparently results from a lack of ovary development in the male and because of the pronounced development of a functional stigma and style in the female. A high growth rate in female carpels of cucumbers (15) and squash (39, 40) continues to fruit maturity. In the male flower there is late development of the stigma-style (pistillodium) that corresponds to the engorgement of nectary tissue.

When the rate of elongation of the stigma-style complex is compared with that of the stamens on an allometric basis, the data show dissimilarity in the allometry constants between the sexes prior to the time that sex can be determined by eye. In this regard Huxley's (16, p.235) statement is pertinent: "The line of biological least resistance would be to produce adult vestigiality of an organ by reducing its growth coefficient," i.e., k in the equation $Y = bX^k$. In cucumbers this occurs very early in development, in fact at the outset, and long before k changes in the plotted data. The data thus agree with the argument of Reeve and Huxley (32) that the above equation is only applicable at the "small end" of the growth of the two structures, since they are then in their exponential phase. Indeed, on a statistical basis, male and female regression lines have the same growth coefficients after k changes in the plotted data, although b differs.

Huxley (16) felt that the chief genetic factor controlling k is a system of rate genes. Hutchins (15) found that "shape correlations" may be made between selected embryological structures of cucumber and their mature counterparts. The responsible genes apparently affect all the organs within a phenotype in the same way and control shape as a whole by correlating length-width ratios from anthesis to fruit maturity. Sinnott (38) also proposed that genetic "shape-controls" in some *Cucurbita* ovaries may exist almost at the inception of ovary development.

The pattern of allometric growth of stigma-style length vs. stamen length in the female flower is nearly a mirror image of the pattern in the male (Figure 20). The two patterns demonstrate the interplay between enhancement and suppression in the attainment of sexual dimorphism in this species. It would be of interest in this regard to see whether a hermaphroditic cultivar, i.e., one with perfect flowers, would have an allometric growth relationship intermediate to those of male and female flowers, with k approximating the average slope for both sexes. A single unbroken regression line should be expected. In considering the typical monoecious pattern, a "promotion-suppression" mechanism would agree with what is known about hormonal control of sex in cucumbers—a control that itself may be thought of as a promotion-suppression phenomenon. The sex expression of the whole plant can be envisioned as a plastic system under a genotypic control that is quite easily manipulated, either naturally or artificially.

Galun and Atsmon (9) visualized that the sexual development of cucumber flower buds may depend on the developmental state of the bud and its subtending leaf, and therefore on gradients of growth promoters and inhibitors. Atsmon et al. (2) proposed that there is first the bisexuality of the individual bud, then the bisexuality of the plant as a whole, and that the sexual tendency of a bud is affected by environmental factors from the early primordial stage.

Ito et al. (17) in their study of sex reversions in cucumber flowers concluded that sex determination was induced before appreciable morphological expression. Lvova (28) reported that sex in cucumber flowers was determined before corolla formation. In a histochemical study of cucumber flower development Yamamoto et al. (40) could distinguish male and female flower buds two to three plastochrons earlier than sex could be identified morphologically. We therefore cannot exclude the possibility that morphogenetic "shifts" affecting maleness or femaleness are established even earlier, at or before the time the primary flower is initiated in the leaf axil, in which case the "sex" of the young axillary complex establishes the sex of the flower, and not the reverse.

The term *bisexual* should be viewed more critically when describing young, sexually undifferentiated flower primordia in the cucumber and other species. The term *presexual* would perhaps be more fitting. Bisexual misleads us by its general failure to recognize that flowers committed from initiation as male, female, or perfect may nonetheless have identical early developmental histories. True bisexuality would therefore be lacking. In contrast, the term presexual makes no distinction between flowers that are truly bisexual early in development (whether this potential is retained to anthesis in a perfect flower, or later lost in producing a unisexual flower) and flowers with unisexual fates from inception (whether or not they appear to have an early bisexual phase). In any case, the sexual phase of floral development begins only when carpels, stamens, or both start to enlarge and differentiate as functional organs.

Acknowledgments

I thank Ernst C. Abbe for stimulating discussions and advice during the course of this investigation and David W. Davis for granting field space and helping in cultivar selection. Northrup, King and Co., Minneapolis, generously provided seed. Research funds were provided by Anderson Summer Fellowships in Botany, University of Minnesota, and by a grant from the Department of Botany.

Literature Cited

1. Atsmon, D., and E. Galun. 1960. A morphogenetic study of staminate, pistillate, and hermaphroditic flowers in *Cucumis sativus* L. Phytomorphology 10: 110–115.
2. Atsmon, D., E. Galun, and K. M. Jakob. 1965. Relative time of anthesis in pistillate and staminate cucumber flowers. Ann. Bot. 29: 277–282.
3. Chakravarty, H. L. 1958. Morphology of the staminate flowers in the Cucurbitaceae with special reference to the evolution of the stamen. Lloydia 21: 49–87.
4. Clark, W. E. le Gros, and P. B. Medawar, eds. 1947. Essays on Growth and Form. Oxford Univ. Press, London.
5. Collison, C. H., and E. C. Martin. 1975. A scanning electron microscope study of cucumber nectaries, *Cucumis sativus*. J. Apic. Res. 14: 79–84.
6. Eichler, A. W. 1875. Blüthendiagramme. Vol. 1. Engelmann, Leipzig.
7. Fuller, G. L., and A. C. Leopold. 1975. Pollination and the timing of fruit-set in cucumbers. HortScience 10: 617–619.
8. Galun, E. 1961. Study of the inheritance of sex expression in the cucumber. The interaction of major genes with modifying genetic and non-genetic factors. Genetica 32: 134–163.
9. Galun, E., and D. Atsmon. 1960. The leaf-floral bud relationship of genetic sexual types in the cucumber plant. Bull. Res. Counc. Israel, Sect. D Bot. 9: 43–50.
10. Galun, E., Y. Jung, and A. Lang. 1963. Morphogenesis of floral buds cultured in vitro. Dev. Biol. 6: 370–387.
11. Goffinet, M. C. 1978. Comparative ontogeny of male and female flowers of *Cucumis sativus* L.—floral inception to anthesis. Ph.D. dissertation, Univ. Minnesota, St. Paul.
12. Heimlich, L. F. 1927. The development and anatomy of the staminate flower of the cucumber. Amer. J. Bot. 14: 227–237.
13. Heimlich, L. F. 1929. Microsporogenesis in *Cucumis sativus*. La Cellule 39: 7–24.
14. Heslop-Harrison, Y., and K. R. Shivanna. 1977. The receptive surface of the angiosperm stigma. Ann. Bot. 41: 1233–1258.
15. Hutchins, A. E. 1934. Morphological relationships in the ontogeny of the cultivated cucumber, *Cucumis sativus* L. Minnesota Agric. Exp. Sta. Tech. Bull. 96.
16. Huxley, J. S. 1932. Problems of Relative Growth. Dial Press, New York.
17. Ito, H., T. Kato, K. Hashimoto, and T. Saito. 1954. Factors responsible for the sex expression of Japanese cucumber. II. Anatomical studies on the sex expression and the transformation of the cucumber flowers induced by pinching (in Japanese, English summary). J. Jap. Soc. Hort. Sci. 23: 65–70.
18. Judson, J. E. 1928. The floral development of the staminate flower of the cucumber. Pap. Mich. Acad. Sci. 9: 163–168.

19. Judson, J. E. 1929. The morphology and vascular anatomy of the pistillate flower of the cucumber. Amer. J. Bot. 16: 69–86.

20. Kubicki, B. 1969. Investigation on sex determination in cucumber (*Cucumis sativus* L.). III. Variability of sex expression in the monoecious and gynoecious lines. Genet. Polon. 10: 3–22.

21. Kubicki, B. 1969. Investigations on sex determination in cucumber (*Cucumis sativus* L.). IV. Multiple alleles of locus ACR. Genet. Polon. 10: 23–67.

22. Kubicki, B. 1969. Investigations on sex determination in cucumber (*Cucumis sativus* L.). V. Genes controlling intensity of femaleness. Genet. Polon. 10: 69–85.

23. Kubicki, B. 1969. Investigations on sex determination in cucumber (*Cucumis sativus* L.). VI. Androecism. Genet. Polon. 10: 87–99.

24. Kubicki, B. 1969. Investigations on sex determination in cucumber (*Cucumis sativus* L.). VII. Andromonoecism and hermaphroditism. Genet. Polon. 10: 101–120.

25. Kubicki, B. 1969. Investigations on sex determination in cucumbers (*Cucumis sativus* L.). VIII. Trimonoecism. Genet. Polon. 10: 123–142.

26. Kubicki, B. 1972. The mechanism of sex determination in flowering plants. Genet. Polon. 13: 53–64.

27. Leins, P., and P. Galle. 1971. Entwicklungs geschichtliche Untersuchungen an Cucurbitaceen-blüten. Osterr. Bot. Z. 119: 531–548.

28. Lvova, I. N. 1977. Morphogenesis of cucumber plants of different sexual types under the influence of plant growth regulators. *In* T. Kudrev, I. Ivanova, and E. Kranov, eds., Plant Growth Regulators: Proc. Second Intl. Symp. Pl. Growth Regulators, Sofia, Bulgaria, 1975. Bulgarian Acad. Sci., Sofia.

29. Matienko, B. T. 1969. Sravnitel 'naya Anatomiya i Ul 'trastruktura Plodov Tykvennykh (Comparative Anatomy and Ultrastructure of Cucurbitaceous Fruits). Kartya Moldovenyaske, Kishinev.

30. Porath, D., and E. Galun. 1967. In vitro culture of hermaphrodite floral buds of *Cucumis melo* L.: Microsporogenesis and ovary formation. Ann. Bot. 31: 283–290.

31. Puri, V. 1954. Studies in floral anatomy. VII. On placentation in the Cucurbitaceae. Phytomorphology 4: 278–299.

32. Reeve, E. C. R., and J. S. Huxley. 1947. Some problems in the study of allometric growth. *In* W. E. le Gros Clark and P. B. Medawar, eds., Essays on Growth and Form. Oxford Univ. Press, London.

33. Reuther, E. 1876. Beiträge zur Entwickelungs geschichte der Blüthe. Bot. Zeit. 25: 385–395; 26: 401–416; 27: 417–431; 28: 433–447.

34. Richards, O. W. 1935. Analysis of the constant differential growth ratio. Carnegie Institution of Washington. Tortugas Lab. Pap. 29: 171–183.

35. Sattler, R. 1968. A technique for the study of floral development. Can. J. Bot. 46: 720–722.

36. Shifriss, O. 1961. Sex control in cucumbers. J. Hered. 52: 1–12.

37. Simpson, G. G., and A. Roe. 1939. Quantitative Zoology. McGraw-Hill, New York.

38. Sinnott, E. W. 1936. A developmental analysis of inherited shape difference in cucurbit fruits. Amer. J. Bot. 70: 245–254.

39. Sinnott, E. W. 1945. The relation of cell division to growth rate in cucurbit fruits. Growth 9: 189–194.

40. Sinnott, E. W., and G. B. Durham. 1929. Developmental history of the fruit in lines of *Cucurbita pepo* differing in fruit shape. Bot. Gaz. 87: 411–421.
41. Tillman, O. I. 1906. The embryo sac and embryo of *Cucumis sativus*. Ohio Naturalist 6: 423–430.
42. Whitaker, T. W. 1931. Sex ratio and sex expression in the cultivated cucurbits. Amer. J. Bot. 18: 359–366.
43. Yamamoto, M., T. Matsui, and H. Eguchi. 1974. Histochemical observations of flower buds in cucumber with special reference to sex differentiation. Proc. Int. Hort. Congr. 19: 343 (Abstract).

PART IV

Utilization

24 | Biodynamic Cucurbits in the New World Tropics

Richard Evans Schultes

ABSTRACT. The Cucurbitaceae provide mankind with a rich store of biodymanic plants. Many are part of the pharmacopoeias of primitive and rural peoples and some may prove to be important in the future of medicine and related fields. A preliminary survey of biodynamic cucurbits of New World tropical regions is presented. Although knowledge of the ethnopharmacological properties of cucurbits remains incomplete, the present sample documents its diversity and suggests promising avenues for future studies.

The large and diverse family Cucurbitaceae obviously has attracted man's attention as a source of useful plants. The great variety of foods and sundry kinds of oils obtained from this family are well known. Perhaps less widely recognized, and for the most part unstudied, is the wealth of ethnopharmacological applications of numerous members of the family in the pharmacopoeias of primitive societies.

There is an appreciably greater number of genera and species in the Old World than in the New, and the folk medicinal uses of cucurbits in Asia and Africa are far more extensive than in the Americas. Yet many of the ethnopharmacological uses in the tropics of the Americas are of extreme interest from both the academic and practical points of view. This brief summary of the biodynamic cucurbits of the New World is preliminary. Many local uses undoubtedly have been overlooked. It is to be hoped, however, that the following notes, gleaned from the sparse literature, herbarium records, and my own field work in the Amazon, may serve as a catalyst for future phytochemical and pharmacological investigations.

In primitive and rural societies, the inhabitants necessarily have had to depend largely, sometimes exclusively, on plants as the source of their medicines and poisons. Hundreds of years of experimentation have led to the discovery of plants with some biological activity. Whether these biodynamic plants can effect cures or act merely as palliatives or stimulants or in other alleviatory capacities, the fact that they have any biological activity indicates that they have one or more physiologically active constituents. Consequently, they are worthy of phytochemical and probably pharmacological study. Such studies frequently lead to the discovery of compounds of interest to synthetic chemists or to constituents, which, in themselves, may have real value as sources of new medicinally valuable agents.

Biodynamic Neotropical Cucurbitaceae

In the Neotropics, members of the following genera have been used biodynamically: *Anguria* (= *Psiguria*), *Calycophysum*, *Cayaponia*, *Ceratosanthes*, *Citrullus*, *Cucumis*, *Cucurbita*, *Cyclanthera*, *Elaterium* (= *Rytidostylis*), *Fevillea*, *Gurania*, *Ibervillea*, *Lagenaria*, *Luffa*, *Melothria*, *Momordica*, *Sechium*, *Sicana*, and *Sicyos*. In the following enumeration and discussion, species are arranged alphabetically, using the names that appeared in the cited reports. If different, currently accepted names, as reported by Jeffrey (21, pers. comm., 1987), are also indicated.

Anguria vogliana Suesseng. (= *Psiguria triphylla* (Miq.) C. Jeffr.). In Venezuela medicine men "crush the leaves for curing" (53). The rhizome is considered to be toxic in Venezuela, where the plant is called *pasana* (45).

Anguria umbrosa Kunth. (= *Psiguria umbrosa* (Kunth) C. Jeffr.). In Venezuela the rhizome is considered to be toxic (53).

Calycophysum brevipes Pitt. In Venezuela the fruit is said to have a toxic latex (1).

Cayaponia attenuata (Hook. & Arn.) Cogn. In El Salvador this plant is believed to be a remedy for the bite of *tamapaz* (53).

Cayaponia cabocla (Vell.) Mart. In Brazil the oil of this plant is said to be purgative and drastic. The oil is employed to treat skin eruptions (32).

Cayaponia citrullifolia (Griesb.) Cogn. ex Griseb. In Paraguay, where this species is called *tayuyá*, the roots are employed as a "blood purifier" (1).

Cayaponia cordifolia Cogn. In Brazil the fruit is considered to be a drastic purgative (32).

Cayaponia diffusa Manso. In Brazil the fruit and root are believed to be purgatives and also are used in treating snakebite (32).

Cayaponia espelina (Manso) Cogn. In Brazil the root of this perennial is considered diuretic, tonic, antisyphilitic, antidiarrhetic, and strongly purgative (51).

Cayaponia fluminense (Vell.) Cogn. The fruit is considered to be a drastic purgative in Brazil (32).

Cayaponia glandulosa (Poepp. & Endl.) Cogn. Among the Tikunas of the Colombian Amazon, the leaves and young stems are dried and pulverized and employed in hammocks and clothes as an insect repellent. The fruit is chopped up and boiled in water; the resulting infusion is thought to be effective in treating "liver complaints" (46).

Cayaponia globosa (Manso) Cogn. In Brazil the seeds are taken as drastic purgatives and in high doses for treating hemorrhages (32).

Cayaponia kathematophora R. E. Schult. This extensive vine is cultivated by the Indians of the middle course of the Río Apaporis of Colombia for its unusually large, glossy brown seeds that, when hollowed out, are used in making anklets and necklaces. Among the Makunas, the plant is called *ka'-moo-ka*; the Taiwanos know it as *pa-moo'-pa*; the Kabuyarí name is *wa-cha'*; in the Puinave language, it is *way-yot'*; and the Matapies refer to it as *wa'yaw*. The seeds are believed to be toxic if eaten (44).

Cayaponia ophthalmica R. E. Schult. The soft green bark is employed in preparing a soothing wash for treating conjunctivitis, which is a wide-spread disease in the Colombian Amazon. This stout heliophile is cultivated by the Indians of the Río Apaporis as a medicinal plant. A spot test for alkaloids, using Dragendorff reagent, was negative on fresh material. The Makunas call the vine *mun-te'-ka*; the Puinave name is *tsun-jo'* (44, 46).

Cayaponia pedata Cogn. In Brazil the fruit is used as a purgative (51).

Cayaponia racemosa (Mill.) Cogn. In El Salvador, this vine is said to be toxic, especially to cattle (1).

Cayaponia ruizii Cogn. The Waorani of Ecuador eat the seed of this liana, and hunters know that it is a preferred food of the toucan, other birds, and the woolly monkey (11). Among the Kofáns of Colombia and Ecuador, the seed, when toasted for five minutes, is edible; these Indians call the plant *sau-ra'-kit-sa* and *kan-bi'-fa-cho* (45).

Cayaponia triangularis (Cogn.) Cogn. In the Brazilian Amazon the fruits and roots are known as *purga de gentia* and are valued as a strong purgative (23).

Cayaponia sp. The stems are reportedly burned, and among the Kofán, the ashes are applied to external sores on the ankles. These Indians know the plant as *cho-rok-o-pi-sĕ-hĕ-pa* (35, 46).

Ceratosanthes palmata (L.) Urb. In Curaçao a decoction of the root is used as a gargle for throat infections; the root soaked in rum relieves flatulence and grated in gin is said to relieve flatulence following childbirth. It is reported to be abortifacient. A poultice of the crushed leaves combined with oil and nutmeg is placed on the navel to relieve flatulence. A decoction of the leaf is drunk to treat the discharge of white mucus from the vagina. In Venezuela the root is considered to be emetic (52).

Citrullus lanatus (Thunb.) Matsum. & Nakai. This Old World species, the watermelon, has acquired sundry medicinal uses in the New World. The

Mexicans employ a leaf decoction as an antimalarial (33). In Puerto Rico the ripe flesh of the fruit is used as a diuretic and tonic and to alleviate bronchitis and pulmonary ailments (31). In Curaçao it is believed that the rind placed on the forehead will reduce a headache (6), and in Venezuela the same material is used as a poultic to treat "liver troubles" (8). The inhabitants of the Bahamas boil the seeds to prepare a diuretic drink; while in the Caicos Islands the dried, parched, and ground up seeds are steeped in water that is then drunk to treat urinary burning (29). It is interesting and possibly significant that the Chickasaw Indians of North America believe that the seeds are effective in treating urinary problems (50).

Cucumis anguria L. In Mexico a decoction of the root is valued as a remedy for "stomach trouble" (1). The inhabitants of Colombia treat kidney problems with a decoction and believe that the fruit, taken in salad form, dissolves kidney stones (33). In Curaçao the fresh fruit is eaten to treat jaundice (6). The Cubans apply the juice of the leaves to freckles, and after being steeped in vinegar, the leaves are considered valuable in treating ringworm. The fruit is applied to hemorrhoids (9). The root is taken in decoction in Cuba to reduce edema (41).

Cucumis melo L. In Colombia the seeds are considered to be taenifuges (17). The root is esteemed in Mexico (25) and Colombia (17) as an emetic.

Cucurbita maxima Duch. ex Lam. The winter squash, native to the New World and widely cultivated as a food plant, is not lacking in medicinal applications in the American tropics. In Colombia the seeds are toasted and eaten as an apparently effective vermifuge (17, 33). Similarly, in Mexico the seeds are used in a water extract against taenia and worms (25).

Cucurbita moschata (Duch. ex Lam.) Duch. ex Poir. and *Cucurbita pepo* L. The medicinal uses of these two food plants are so similar that they can be enumerated together. Both species, of American origin, have found places in native pharmacopoeias from Mexico to South America. In Mexico the seeds, given with castor oil, are anthelmintic (25). The seed oil is applied to persistent ulcers in Venezuela, and a decoction is given to reduce eruptive fevers (8). Jamaicans employ a seed decoction as a diuretic (39). In Trinidad the leaves are poulticed on sprains, and a decoction of the flowers has been used to treat measles and smallpox (54). Colombians report that a root infusion is febrifugal and also is applied to syphilitic ulcers (34). The seeds are said to be excellent vermifuges and taenifuges and the fruits are diuretic (16). *Cucurbita pepo* fruit is considered an antidiabetic in Yucatán (22); while in Curaçao, a decoction of the flowers or fruits is used in treating jaundice (10). The uses in Brazil are similar: the seeds are active against taenia, especially in children, when given in coconut milk (26).

Cyclanthera explodens Naud. The Kamsá Indians of Sibundoy, Colombia, call this vine *semarron'-shajush* and consider it medicinal, but the precise use is not known (7).

Elaterium gracile (Hook. & Arn.) Cogn. (= *Rytidostylis gracilis* Hook. & Arn.). The plant is reported as a cattle poison in San Salvador (1).

Fevillea amazonica (Cogn.) C. Jeffr. Among the Tikunas of the Colombian Amazon, oil from the seed is reputed to hasten the healing of serious burns when applied three or four times a day over a period of ten days (46).

Fevillea cordifolia L. This species is employed in South America as a laxative; the seeds are the source of sekua fat and mandiro fat, both of which were formerly used in the candle and soap industries (32). In the Brazilian Amazon the seeds are used against jaundice and are considered tonic, stomachic, and purgative; the fat from the seeds is applied with friction for erysipelas (23). In Colombia the seeds, taken orally and applied externally as a poultice are used in cases of snakebite (16), and the oil is believed to prevent loss of hair (33). In Costa Rica the seeds are considered valuable in treating yellow fever (38). The Puerto Ricans grate the seeds and take them in alcohol for stomachache and colic (41); in Jamaica a similar solution is taken for snakebite and is rubbed on rheumatic pains (24). The Jamaicans also believe that *Fevillea cordifolia* is an antidote for "poison," and the grated seeds are poulticed on sprains and wounds (2). The seed oil is employed in treating dermatitis caused by *Hippomane mancinella* L. and was formerly considered valuable in treating leprosy in Venezuela (38). In Cuba the oil was given to horses and other large animals in cases of fever and abdominal troubles (41). The seeds in decoction are widely considered to be strongly emetic and purgative (30).

Fevillea trilobata L. In Brazil the oil extracted from the seeds is used for skin diseases and rheumatism (51). The seeds are considered to be toxic and stomachic and are used in treating liver complaints (32)

Gurania acuminata Cogn. In the lower region of the Río Caquetá, Colombia, the inhabitants believe that a tea of the leaves is one of the most effective vermifuges (46).

Gurania bignoniacea (Poepp. & Endl.) C. Jeffr. In the Amazon regions of Colombia, extensive medicinal use is made of this vine. The Makuna crush the leaves and flowers and apply them to infected sores and wounds that refuse to heal; they call the plant *hě'-ně-gaw*. The Tukanos, who know the vine as *mee'-chee*, prepare the leaves and roots in decoction as an anthelmintic. The *colonos* (settlers who come from the mountainous sections of Colombia to the Amazon) rub the leaves on fungal infections of the skin, a use that possibly was adopted from the Indians (46).

Gurania eriantha (Poepp. & Endl.) Cogn. The Witotos, who live along the Putumayo of Colombia, call this species *usiya'-o*. No medicinal use is reported (46).

Gurania guentheri Harms. The Kofáns of Colombia and Ecuador take an infusion of the leaves as a strong vermifuge; their name for the vine is *yama-cho'-ro* (46).

Gurania insolita Cogn. The Tikunas of Colombia prepare the crushed flowers as a poultice and apply it to boils and other external infections (46).

Gurania pachypoda Harms. The Tikunas of the Río Loretoyacu, Colombia, employ the crushed leaves as a poultice to relieve headache. In the Iquitos region of Peru, the vine is known as *mashu-mikuna* (46).

Gurania rhizantha (Poepp. & Endl.) C. Jeffr. Tikuna women in the Río Loretoyacu region prepare a tea from the roots and woody stem for a condition that seems to be irregular menstruation. The Kofáns dry the leaves of this common plant, burn them, and apply the ashes to sores of the skin; they know the vine as *akie-ka-kie-sě'-hě-pa* and *cho-rok-o-pi'* (46).

Gurania rufipila Cogn. Along the Río Apaporis, Colombia, the Tanimuka Indians report that the stem and roots are toxic and that care must be taken not to confuse it with other species; they call the plant *mee-ree-fee'-ka-no-ma-ka* (46).

Gurania speciosa (Poepp. & Endl.) Cogn. Fresh material of this vine was questionably positive when submitted to a spot test for alkaloids with Dragendorff reagent (46). In Colombia the seeds are toasted and ground to prepare a vermifuge, and the Inganos of the Putumayo region use them against snakebite (17).

Gurania spinulosa (Poepp. & Endl.) Cogn. The Tikunas of the Río Loretoyacu use a tea of the roots for treating faulty menstruation (46). A decoction of the leaf is considered a remedy for constipation in Trinidad (55). In the Ecuadorean Amazon the leaves are boiled and "applied to scrapes" (Herb. Coll. *Shemluck 320*).

Gurania ulei Cogn. The Witotos of the Colombian Amazon know this vine as *ma-ru-chao'*. They apparently do not use the plant (46).

Ibervillea lindheimeri (A. Gray) Greene. The fruits are very poisonous (1).

Lagenaria siceraria (Mol.) Standl. The bottle gourd, the source of gourds provided by the woody outer pericarp, has several medicinal uses in tropical American folk-medicine. In the Brazilian Amazon the pulp of the ripe fruit is considered to have emollient properties; it is also employed as a laxative, and a tea of the seeds is considered to be efficaceous against nephritis (23, 46). The pulp of the ripened fruit is "dangerously purgative," and the leaves have steroidal sapogenins (30). In Curaçao, a leaf decoction is used to reduce flatulence, and a decoction of the leaves together with *Rivina humilis* L., the fruits of which are the source of a reddish dye, is used to "relieve gas" (6). In many parts of tropical America, the leaves are mixed with an oil and are applied as a poultice to the skin for dermatological problems and tumors (39). In Colombia the powdered seeds are valued as a vermifuge (34), while in Panama the seeds are eaten as a remedy for "headache" (14).

Luffa cylindrica (L.) M. J. Roem. The source of a vegetable sponge, this Old World species has acquired a few local medicinal uses in the American tropics. In Colombia the fruit is considered emetic, cathartic, and purgative

(17). The Brazilians value an alcoholic preparation of the seeds as a vermifuge (16), and the stems and leaves are said to be used in treating hepatic afflictions, ammenorrhea, and anemia (32). In Guatemala the oily seeds are reported to be emetic-cathartic (13), and the juice of the unripe fruit serves as a purgative. In San Salvador the mashed leaves are applied to the forehead for headache (19). A decoction of the unripe fruit is prepared in Cuba for treating "asthma associated with venereal diseases" (41). In the marketplace in Belém, Pará, Brazil, the fruit, mixed with *Jatropha curcas* L., is employed by the Afro-Brazilians in culture rites, and the "latex" is used to treat sinusitis (5).

Luffa operculata (L.) Cogn. Like *L. cylindrica*, this New World species is also the source of a vegetable sponge and is similarly valued in tropical America as a medicinal plant. In the Brazilian Amazon, a water decoction of the fruit makes a drastic diarrhea remedy; the dried fruit, when powdered and taken with a weak extract of the leaves of *Passiflora laurifolia* L., is said to have abortifacient properties (23). The Peruvians use an infusion for treating syphilis and apply it externally to cauterize wounds (43). In Colombia the fruit is valued as a strong purgative, emetic, and sudorific, but it must be used with extreme caution. It also is applied externally to treat ringworm, chronic ophthalmia resulting from conjunctivitis, and sinusitis (34, 36). The seeds, soaked in rum, are esteemed in Trinidad for treating snakebite (55). when steeped in alcohol and rubbed on as a liniment, they are recommended in some parts of South America for relieving sciatica (39).

Melothria fluminensis Gardn. In Brazil the seeds of this vine are used in the treatment of colic in animals (51), and the flowers and leaves in decoction are applied in clusters for uterine afflictions (32).

Melothria pendula L. This plant is a folk medicine in Yucatán for gonorrhea, swellings, and inflammations (47). The fruit is a drastic purgative and is used in Brazil for treating horses and cattle (16, 32).

Momordica charantia L. The parboiled leaves of this perennial vine are served as a vegetable, and the unripe fruits, boiled or fried, are consumed in salads in Peru. Sap from the leaves and fruit is taken for colic and as a vermifuge (51). It is the most common medicine for diabetes among the populations of the Iquitos region of Peru; an infusion of the leaves taken "before breakfast tends to decrease the blood sugar level for a short time." It is locally known as *papailla* (4). In Colombia the seeds are considered to be toxic (33). In Brazil the plant is applied in baths to cure eczemas and herpes. A tincture of the stems is considered febrifugal, the roots are believed to be purgative and emetic, and the leaves are used for rheumatic and menstrual pains (32).

In the Caribbean *M. charantia* is one of the most common remedies. It is used to treat colds and fevers, rheumatism and arthritis. In the Caicas Islands a decoction of the vine is taken with salt before and after childbirth

(28, 30), and in Aruba and Curaçao a decoction with sugar is drunk to reduce blood pressure (6). The natives of Barbados take a decoction to treat influenza (18). In the Coro Islands the plant is considered an effective remedy for kidney stones (27). The Jamaicans drink a decoction of the fruit and diced leaves for colds, fever, and stomachache and value it as a relief for constipation in small children. It is also a general tonic and emmenogogue and is widely believed to be a "cancer cure" in the Caribbean (27, 39). In Cuba it is reported to be used to improve the appetite and as a treatment for colic, hepatic problems, fever, and kidney stones. A lotion is applied to reduce inflammation from skin diseases (41). In Yucatán a root decoction is said to have anthelmintic and aphrodisiac properties (22, 39).

Sechium edule (Jacq.) Sw. The chayote, cultivated for its edible fruit, has numerous medicinal uses in Central America and the West Indies. The young shoots and roots are likewise consumed as food. In Yucatán a decoction of the leaves is esteemed as a treatment for stones in the bladder, as a remedy for arteriosclerosis, and to lower blood pressure (22). In other parts of Mexico the leaves are boiled with those of *Casimiroa edulis* Llave to prepare a decoction for reducing hypertension (25). The raw fruit is used to cauterize wounds in Guatemala (12). The Jamaicans drink the juice from the grated fruit to reduce hypertension (3), whereas in Cuba the root and fruit are used as a strong diuretic and are considered to be effective in treating pulmonary troubles. The fresh fruit is poulticed on skin inflammations, and an emulsion of the seeds is taken to relieve intestinal inflammation (41).

Sicana odorifera (Vell.) Naud. The edible fruit is very fragrant and is employed throughout the American tropics to scent clothes (51). In Puerto Rico the juice is thought to be a remedy for angina (41). In Brazil the seeds in infusion are reported to be anthelmintic, purgative, febrifugal, and effective in promoting menstrual discharge. A leaf decoction is a valuable treatment for a variety of uterine complaints, and an infusion of the rind of the fruit and seeds is reputedly a cure for uterine hemorrhages (9, 16). The leaves and flowers in Yucatan are considered a laxative, vermifuge, and emmenogogue (22).

Sicyos polyacanthus Cogn. In Brazilian folk medicine, the fruit, when taken internally, is considered to act as a drastic purgative, and when used externally, to have properties as a resolutive (32).

Conclusions

The extensive and diverse medicinal and toxic uses that have been characteristic of the members of the Cucurbitaceae in tropical America call for much more detailed chemical and pharmacological investigation. While a majority of the folklore uses refer to the vermifugal, purgative, or toxic properties of the numerous species, some of the other "medicinal" values of the several species deserve investigation.

The most comprehensive summary of our chemical knowledge of the family was presented by Hegnauer (20) in 1964, although, since that time, other investigators have dealt with individual groups of compounds.

Triterpenes have been found in *Cucurbita, Citrullus, Ecballium, Lagenaria,* and *Momordica.* Saponins have been found in the seeds of *Citrullus, Lagenaria,* and *Momordica charantia,* which has a very potent fish-poison constituent; seed oils from a number of genera have been isolated; and phenolic compounds have been found in *Citrullus* and *Cucurbita* (20). Only *Momordica* is known to contain alkaloids (41), although *Gurania speciosa* was questionably positive for alkaloids when tested with Dragendorff reagent (46).

This preliminary survey of only a few ethnobotanical sources indicates the extraordinary potentialities that may be present in a chemicopharmacological study of the cucurbits that are used, and have been used for centuries, for biodynamic properties recognized in folk medicine. The most important physiological uses of the cucurbits of the New World tropics are as 1) purgatives, 2) emetics, 3) anthelmintics, and 4) poisons. All of the uses of the members of the Cucurbitaceae in aboriginal or rural societies, however, deserve modern scientific examination. It is a family that phytochemically and pharmacologically has been neglected.

Literature Cited

1. Altschul, S. von R. 1973. Drugs and Foods from Little Known Plants. Notes in Harvard University Herbarium. Harvard Univ. Press, Cambridge, MA.
2. Asprey, G. F., and P. Thornton. 1953. Medicinal Plants of Jamaica. Parts I and II. West Indian Med. J. 2: 233–252; 3: 17–41.
3. Asprey, G. F., and P. Thornton. 1955. Medicinal Plants of Jamaica. Parts III and IV. West Indian Med. J. 4: 69–82, 145–168.
4. Ayala-Flores, F. 1984. Notes on some medicinal and poisonous plants of Amazonian Peru. Adv. Econ. Bot. 1: 1–8.
5. Berg, M. E. van den. 1984. Ver-o-Peso: The ethnobotany of an Amazonian market. Adv. Econ. Bot. 1: 140–149.
6. Brenneker, P. 1961. Jerba-Kruiden van Curaçao en hun gebruik. Boekhandel St. Augustinus, Curaçao.
7. Bristol, M. L. 1965. Sibundoy Ethnobotany. Ph.D. dissertation, Harvard Univ., Cambridge, MA.
8. Chiossone, V. C. 1938. Flora Médica del Estado Lara. Coop. de Artes Gráficas, Caracas.

9. Cruz, G. L. 1965. Livro Verde das Plantas Medicinais Industriais do Brasil. Velloso, Belo Horizonte.

10. Daal, M. de veer. 1912. I. De geneesmeddelen van Groot-Nederland. II. De geneesmiddelen van Curaçao (Cited in Morton, 29.)

11. Davis, E. W., and J. A. Yost. 1983. The ethnobotany of the Waorani of eastern Ecuador. Bot. Mus. Leafl. Harvard Univ. 29: 159–217.

12. Dieseldorff, E. P. 1940. Las Plantas Medicinales del Departamento de Alto Verapáz. Published by the Author, Guatemala.

13. Dieterle, J. V. A., 1976. Cucurbitaceae. In D. L. Nash, ed., Flora of Guatemala. Fieldiana, Bot. 24 (XI, 4): 306–395.

14. Duke, J. A. 1968. Darien Ethnobotanical Dictionary. Battelle Memorial Institute, Columbus, OH.

15. Eldridge, J. 1975. Bush medicine in the Exumus and Long Island, Bahamas: a field study. Econ. Bot. 29: 307–332.

16. Friese, F. W. 1934. Plantas Medicinais Brasileiras. Inst. Agron. Estado, São Paulo.

17. García-Barriga, H. 1975. Flora Medicinal de Colombia. Vol. 3. Univ. Nacional, Bogotá.

18. Gooding, E. G. B. 1942. Facts and beliefs about Barbadian plants. J. Barbados Mus. Hist. Soc. 9: 192–194; 10: 3–6.

19. Guzmán, D. J. 1947. Especies útiles de la Flora Salvadoreña. Imprenta Nacional, San Salvador.

20. Hegnauer, R. 1964. Chemotaxonomie der Pflanzen. Birkhauser Verlag, Basle.

21. Jeffrey, C. 1978. Further notes on Cucurbitaceae. IV. Some New-World taxa. Kew Bull. 33: 347–380.

22. Lavadores, V., G. 1969. Estudio de las 119 plantas Medicinales más conocidas en Yucatán, Mexico. Published by the Author, Mérida.

23. LeCointe, P. 1934. Arvores e Plantas Uteis. Livraria Classica, Belém do Pará.

24. Manfred, L. 1947. 7000 Recitas Botánicas a Base de 1300 Plantas Medicinales Americanas. Editorial Kerr, Buenos Aires.

25. Martínez, M. 1959. Las Plantas Medicinales de México, 4th ed. Ediciones Botas, México.

26. Mors, W. B., and C. T. Rizzini. 1966. Useful Plants of Brazil. Holden-Day, San Francisco, CA.

27. Morton, J. F. 1974. Folk-remedy plants and esophageal cancer in Coro, Venezuela. Morris Cerb. Bull. 25: 24–34.

28. Morton, J. F. 1975. Current folk remedies of northern Venezuela. Quart. J. Crude Drug Res. 13: 97–122.

29. Morton, J. F. 1977. Medicinal and other plants used by people on North Caicos (Turks and Caicos Islands, West Indies). Quart. J. Crude Drugs Res. 15: 1–24.

30. Morton, J. F. 1981. Atlas of Medicinal Plants of Middle America. Charles C. Thomas, Springfield, IL.

31. Nuñez-Melendez, E. 1975. Plantas Medicinales de Puerto Rico. Univ. Puerto Rico Est. Exper. Agrícola. Río Piedras, Bol. 176.

32. Penna, M. 1946. Dicionario Brasileiro de Plantas Medicinais, 3rd ed. Levraria Kosmos Editora, Rio de Janeiro.

33. Pérez-Arbeláez, E. 1956. Plantas Utiles de Colombia, 3rd ed. Libreria Colombiana Camacho Roldán, Bogotá.

34. Pérez-Arbeláez, E. 1975. Plantas Medicinales y Venenosas de Colombia. Hernando Salazar, Medellín.

35. Pinkley, H. V. 1973. The Ethnobotany of the Kofán Indians of Ecuador. Ph.D. dissertation, Harvard Univ., Cambridge, MA.

36. Piovano, G. 1960. *Luffa operculata* (L.) Cogn. puo servire a curare la sinusite? Nuovo G. Bot. Ital. (n.s.) 67: 557–560.

37. Pittier, H. 1926. Manual de las Plantas Usuales de Venezuela. Published by the Author, Caracas.

38. Pittier, H. 1957. Ensayo sobre Plantas Usuales de Costa Rica, 2nd ed. Univ. Costa Rica, San José.

39. Pompa, G. 1974. Medicamentos Indígenas. Editorial America, Miami.

40. Raffauf, R. F. 1970. A Handbook of Alkaloids and Alkaloid-Containing Plants. Wiley-Interscience, New York.

41. Roíg y Mesa, J. T. 1945. Plantas Medicinales, Aromáticas o Venenosas de Cuba. Cultural, Havana.

42. Rose, J. N. 1899. Notes on Useful Plants of Mexico. Contrib. U.S. Natl. Herb. 5: 209–259.

43. Sagástegui, A. A. 1973. Manual de las Malezas de las Costas Norperuanas. Talleres Gráficos, Univ. Nac. de Trujillo, Trujillo.

44. Schultes, R. E. 1964. Plantae Colombianae XVIII. Bot. Mus. Leafl. Harvard Univ. 20: 321–324.

45. Schultes, R. E. 1977. De plantis toxicariis e Mundo Novo tropicale XVI. Bot. Mus. Leafl. Harvard Univ. 25: 126.

46. Schultes, R. E. 1986. De plantis toxicariis e Mundo Novo tropicale commentationes XXVIII. Ethnobotanical notes on cucurbits in the northwest Amazon. Bot. Mus. Leafl. Harvard Univ. 30: 239–245.

47. Standley, P. C. 1930. Flora of Yucatán. Field Mus. Nat. Hist., Bot. Ser. 3 (3), Publ. 279.

48. Standley, P. C. 1931. Flora of the Lancetilla Valley, Honduras. Field Mus. Nat. Hist., Bot. Ser. 10, Publ. 283.

49. Steggerda, M. 1929. Plants of Jamaica used by natives for medicinal purposes. Amer. Anthropol. (n.s.) 31: 431–434.

50. Taylor, L. A. 1940. Plants Used as Curatives by Certain Southeastern Tribes. Botanical Museum Harvard Univ., Cambridge, MA.

51. Uphof, J. C. T. 1968. Dictionary of Economic Plants, 2nd ed. Cramer, Lehre.

52. Van Meeteren, N. 1947. Volkskunde von Curaçao. Drukkerij Schepenheuvel, Willenstad, Curaçao.

53. Von Reis, S., and F. J. Lipp, Jr. 1982. New Plant Sources for Drugs and Foods from the New York Botanical Garden Herbarium. Harvard Univ. Press, Cambridge, MA.

54. Willis, J. C. [H. K. Airy-Shaw]. 1966. A Dictionary of the Flowering Plants and Ferns, 7th ed. Cambridge Univ. Press, Cambridge.

55. Wong, W. 1976. Some folk medicinal plants from Trinidad. Econ. Bot. 30: 103–142.

25 | Cucurbits of Potential Economic Importance

Thomas W. Whitaker

ABSTRACT. *Cucurbita ficifolia, Cyclanthera pedata, Praecitrullus fistulous, Cucumis anguria, Cucumis metuliferus, Cucumeropsis mannii, Telfaria pedata,* and *Hodgsonia macrocarpa* are identified as having great potential as foods sources, primarily in tropical regions. Descriptive information concerning their uses and culture is given for each. Other species, less promising for the immediate future, are mentioned.

There is cause to wonder why the Cucurbitaceae, considering their importance as a source of food, have been largely ignored by crop-reporting services, crop planners, nutritionists, dieticians, and others responsible for the quantity and quality of our food supply. This attitude appears shortsighted, and a satisfactory explanation is not readily apparent. It is true that none of the cucurbits are produced in prodigious monocultures, as are cereals such as rice, wheat, and corn; vegetables, e.g., potatoes and tomatoes; the sugar crops, including beets and cane; or fruit crops, notably grapes and citrus. This fact probably accounts for the woeful lack of acreage and harvest data for cucurbits worldwide. Nevertheless, the cucurbits are a potentially important, although largely unexploited and unrecorded, food resource.

An outstanding example of a well-devised program calculated to make more and better use of cucurbits is that of the late W. P. Bemis and his colleagues at the University of Arizona (1, Gathman and Bemis, this volume). This program was designed to exploit the food potential of a single species, *Cucurbita foetidissima* HBK, the buffalo gourd, as a source of edible oil and protein cake. As population pressures increase in the tropical and subtropical regions of the world, it is probable that the Cucurbitaceae

318

will become increasingly important as a source of food because the family is primarily adapted to these climatic regions. The buffalo gourd program could well serve as a model to harness other species for food for humans and domesticated animals.

With this thought in mind I suggest a list of cucurbit species with known food or medicinal properties that have not yet been exploited or are exploited only to a limited degree. Following are species that are worthy of consideration as additions to our food arsenal in tropical and subtropical areas throughout the world.

Cucurbita ficifolia (Malabar gourd)

Cucurbita ficifolia Bouché (see Andres, this volume) has not been cultivated as widely as it deserves. It could probably be successfully cultured in the highland, arid areas of southern Arizona, southern New Mexico, northwestern Mexico, and perhaps other tropical and subtropical areas with altitudes ranging from 1200 to 2600 m.

The immature fruits of *C. ficifolia* can be used as summer squash. The greenish white flesh makes delicately flavored soups. Impregnated with sugar, the flesh is transformed into candy, the so-called "dulces" of Latin America. My Bolivian friends tell me the flesh can be fermented to produce a passable beer. Furthermore, the fruits can be stored for over a year with no special refrigeration, an added bonus where the food budget is tight and refrigeration absent.

Cyclanthera pedata (wild cucumber, *korilla*, *caygua*, *achoccha*, *caihua*)

Cyclanthera pedata (L.) Schrad. produces a vigorous vine as long as 5 m, with pale greenish, flattened fruits that resemble cucumbers but are softer because of the spongy, partly hollow seed cavity. The glabrous foliage has a strong odor reminiscent of cucumbers when crushed. The fruits, about 15 cm long, are produced in relative abundance. They are used prior to maturity much as summer squash is used in the United States. Raw, the fruits are tasty ingredients of salads. Stuffed with meat, fish, or cheese and baked, they are delicious. The shoots are also edible. The fruits contain numerous black seeds, each having the appearance of a small mud turtle.

Cyclanthera pedata is native to the New World. It is cultivated from Mexico to Peru and Ecuador, and it frequently occurs as an escape, being found in thickets and disturbed sites near habitations. It should be easy to cultivate it in the tropics and subtropics of both hemispheres and deserves to be more widely grown in these areas. It is more cold tolerant than most cucurbits, and continues to be productive in the fall after cold weather curtails the production of summer squash and cucumbers (16). Because of its cold tolerance, it is cultivated in mountainous valleys up to 2000 m elevation (19).

A second species of *Cyclanthera*, *C. explodens* Naud., with a distribution and temperature tolerance similar to that of *C. pedata*, is also cultivated for its edible fruits (18).

Praecitrullus fistulosus (squash melon, round melon, *tinda*)

Praecitrullus fistulosus (Stocks) Pang. was described in *Citrullus* as *C. lanatus* var. *fistulosus* Stocks, but it differs from the watermelon and other species of *Citrullus* by having 12 instead of 11 pairs of chromosomes. The squash melon is a popular summer vegetable in the northwestern states of India (17). The entire fruit is used as a cooked vegetable, usually when immature, before the seeds become hard. It is generally cooked with other vegetables, chiefly legumes (2). According to Choudhury (4), its nutritive value is substantially greater than that of other cucurbits. Its seeds are also roasted and eaten (Chakravarty, this volume). The squash melon is not widely disseminated throughout the tropics, although it should be adapted wherever watermelons are grown.

Cucumis anguria (gherkin, West Indian gherkin, bur gherkin)

The gherkin, *Cucumis anguria* L., a native of Africa brought to the Americas when the slave trade was active, has spread from the West Indies to many countries in Latin America. It was once thought to be native to the West Indies, hence its name, but it has been shown to be closely related to the African species, *C. longipes* Hook. f., which is now considered to be a botanical variety of *C. anguria* (8). In some areas, *C. anguria* has become established in the New World as an escape from cultivation, and appears to be an element of the indigenous flora. Vines of this species are extraordinarily prolific, but the fruits have been little utilized up to the present. The spiny fruits, the size and shape of an egg and having long peduncles, are produced in prodigious numbers. They are mainly used for pickles when immature. The fruits can also be used in soups and stews.

The young, pickled fruits of *C. anguria* are frequently marketed as gherkins, but so are small, immature cucumbers, *C. sativus* L.; this labeling often creates confusion. *Cucumis anguria* is a source of food that could be more widely utilized by populations tied to the starchy diets of tropical areas.

Cucumis metuliferus (jelly melon, African horned cucumber, kiwano)

Fruits of *Cucumis metuliferus* C. H. Mey. ex Schrad., a trailing annual herb, have recently appeared in specialty markets in the United States and Europe, suggesting it may have a future in some areas as a substitute for slicing cucumbers. The fruits are ovate-ellipsoid and about 15 cm long. Immature fruits are mottled grayish green, but at maturity they turn an attractive reddish orange. They are covered with stout, scattered, fleshy spines, each tipped with a stiff bristle. The spines, when injured, ooze dark bloodlike fluid. The seeds are imbedded in an emerald green, jellylike matrix. To some the flesh has a pleasant taste, much like a cucumber, but others report it to have an unpleasant aftertaste (9). Some vines produce fruits that are extremely bitter, hence growers will have to be careful of seed sources.

The jelly melon is native to Africa, but was introduced to Australia more than 60 years ago and has become a weed there. The fruits have been exported in recent years from Africa, i.e., Kenya, to Europe, and from New Zealand to the United States. The name *kiwano* was given to the fruit by New Zealand growers, apparently with the hope of duplicating the commercial success of the kiwifruit, *Actinidia chinensis* Planch. (9). The fruit ships well and can be stored for six months without cold storage (9).

Cucumis metuliferus is resistant to viruses (15), nematodes (5), powdery mildew, and white flies (den Nijs and Custers, this volume), but attempts to transfer these resistances to cultivated species of *Cucumis* have not been successful. *Cucumis metuliferus* cannot be crossed with the cucumber (10). Embryos have been formed in crosses between C. *melo* and C. *metuliferus*, but these aborted (den Nijs and Custers, this volume).

Cucumeropsis mannii (white-seeded melon)

An herbaceous annual climber, *Cucumeropsis mannii* Naud. is native to tropical Africa. The vine is large and often grown on trellises or planted around trees for support when cultivated. The ellipsoidal, cylindrical fruits are about the size of a small watermelon. The numerous, smooth, white seeds are edible. They are up to 2 cm long and are embedded in white, jellylike pulp. In Nigeria the seeds are processed to make a soup thickener or flavoring. They also can be roasted and ground with peanuts (*Arachis hypogaea* L.), chili peppers (*Capsicum* spp.), and sometimes shrimp to make a fine, oily, edible paste (12). The seeds are also processed to produce a cooking oil for domestic purposes, but none is exported (14). The species should be adapted for culture in the humid tropics of both hemispheres and deserves to be more widely cultivated.

Telfairia pedata (oyster nut, Zanzibar oil vine)

Telfairia pedata (Sims) Hook. is a large, vigorous, dioecious, perennial climber native to tropical Africa and is cultivated sparingly in East and Central Africa. It grows from sea level to elevations of 2000 m (6).

The vine is huge and according to one account climbs in trees to a height of 14–18 m. A single plant may cover an area as large as a tennis court. The ribbed fruit is large, weighing up to about 13 kg. At maturity and while still attached to the vine, the gourd splits and the seeds drop to the ground. They are edible after the seed coat is removed. The bitterness of the seeds is mostly in the seed coats (6). The seeds have about the same flavor as almonds or Brazil nuts and are a good substitute in confections. The kernel has about 60% edible oils and is esteemed by native Africans. The seeds are eaten raw, roasted, pickled, or cooked in soup. The leaves and vine tips are also eaten. Cultivation of this gourd should be attempted in Southeast Asia and perhaps in tropical America.

A related species, *T. occidentalis* Hook. f., fluted gourd, occurs in West Africa. It is cultivated extensively by the Ibos of southern Nigeria, primarily as a leafy vegetable but also for its seeds, which are roasted or boiled and eaten or dried, powdered, and used to thicken soups (11). A cooking oil can be extracted from the cotyledons. The seeds are collected just before the fruits are fully mature; otherwise they are bitter (13). The fluted gourd is partly drought-resistant and offers potential for cultivation in dry tropical regions (11).

Hodgsonia macrocarpa (Chinese lard fruit)

Not much is known about *Hodgsonia macrocarpa* (Blume) Cogn. It is an immense climbing vine with fruit as large as watermelons. The Chinese have attempted to cultivate it for the very large seeds. The vines may attain a length of 25–35 m and in the wild often climb trees. The seeds, up to eight per fruit, are 5–7.5 cm long and up to 3.8 cm in diameter. They yield 70 to 80% of an excellent edible oil, high in vitamins A and D. The oil tastes much like lard, hence the common name. It has been estimated that only 15 plants of *Hodgsonia* produce as much oil as a hectare of rape, the principal oil seed crop of China (7).

According to Chien (3), plants of *H. macrocarpa* are highly resistant to drought and waterlogging, are relatively easy to cultivate, produce 40–80 fruits per plant annually, and have a life span of more than 70 years. Parts of this statement must be over-enthusiastic. I find it hard to believe the plants are both resistant to drought and waterlogging. Nevertheless, the culture and propagation of the species appears to be relatively simple.

Hodgsonia macrocarpa is native to the tropical forests of Sikkim, the eastern parts of Bengal and Assam, and the Chinese province of Yunan. My experience with this species has been limited. In New Delhi, I examined a single, vigorous, young plant growing over a lath house at the Indian Agricultural Research Institute. At the South China Botanical Gardens, Guangzhou, there was a single male plant climbing over trees and bushes. There is reason to believe this species might do well under cultivation in parts of Southeast Asia, such as Malaysia and Indonesia.

Other Cucurbits

There are many other species that are eaten occasionally, but the prospects for their domestication and increased use are less promising. For example, two species of *Corallocarpus* are used for food by the San of the Kalahari Desert in southern Africa. The leaves of *C. bainesii* (Hook. f.) A. Meeuse and the roots of *C. welwitschii* (Naud.) Hook. f. ex Welw. are part of their diet. Fruits and seeds of the spiny desert shrub, *Acanthosicyos horridus* Welw. ex Hook f. (Sandelowsky, this volume) are also eaten in season.

Literature Cited

1. Bemis, W. P., J. W. Berry, C. W. Weber, and T. W. Whitaker. 1978. The buffalo gourd: a new potential horticultural crop. HortScience 13: 235–240.
2. Burkill, I. H. 1966. A Dictionary of the Economic Products of the Malay Peninsula, 2nd ed. Ministry Agric. Cooperatives, Kuala Lumpur.
3. Chien, H. 1963. "Lard fruit" domesticated in China. Euphytica 12: 261–262.
4. Choudhury, B. 1967. Vegetables, 2nd ed. Natl. Book Trust, New Delhi.
5. Fassuliotis, G. 1967. Species of *Cucumis* resistant to the root-knot nematode, *Meloidogyne incognita acrito*. Pl. Dis. Reporter 51: 720–723.
6. Herklots, G. A. C. 1972. Vegetables of Southeast Asia. George Allen and Unwin, London.
7. Hu, S.-Y. 1964. The economic botany of *Hodgsonia*. Econ. Bot. 18: 167–169.
8. Meeuse, A. D. J. 1962. The Cucurbitaceae of Southern Africa. Bothalia 8: 1–111.
9. Morton, J. F. 1987. The horned cucumber, alias "Kiwano" (*Cucumis metuliferus*, Cucurbitaceae). Econ. Bot. 41: 325–327.
10. Norton, J. D. and D. M. Granberry. 1980. Characteristics of progeny from an interspecific cross of *Cucumis melo* with *C. metuliferus*. J. Amer. Soc. Hort. Sci. 105: 174–180.
11. Okoli, B. E. 1983. Fluted pumpkin, *Telfairia occidentalis*: West African vegetable crop. Econ. Bot. 37: 145–149.
12. Okoli, B. E. 1984. Wild and cultivated cucurbits in Nigeria. Econ. Bot. 38: 350–357.
13. Omidiji, M. O. 1977. Tropical cucurbitaceous oil plants of Nigeria. Veg. Hot Humid Tropics Newslett. 2: 37–39.

14. Omidiji, M. O. 1978. Tropical leafy and fruit vegetables in Nigeria. Veg. Hot Humid Tropics Newslett. 3: 6–18.
15. Provvidenti, R., and R. W. Robinson. 1974. Resistance to squash mosaic virus and watermelon mosaic virus in *Cucumis metuliferus.* Pl. Dis. Reporter 58: 735–738.
16. Robinson, R. W. 1987. Cold tolerance in the Cucurbitaceae. Cucurbit Genet. Coop. Rep. 10: 104.
17. Whitaker, T. W. 1973. The cucurbits of India. Proc. Amer. Soc. Hort. Sci. Trop. Reg. 17: 255–259.
18. Wunderlin, R. P. 1978. Cucurbitaceae. *In* R. E. Woodson, Jr., and R. W. Schery, eds., Flora of Panama. Part IX. Ann. Missouri Bot. Gard. 65: 285–368.
19. Yamaguchi, M. 1983. World Vegetables. Avi, Eastport, CT.

26 | Cucurbits of India and Their Role in the Development of Vegetable Crops

H. L. Chakravarty

ABSTRACT. The common Indian cucurbits are discussed, with particular attention given to their cultivation and economic uses. Short botanical descriptions and accounts of the culture and use of 16 species in nine genera are provided.

The cucurbits have been major vegetable crops in India since ancient times. They are easily cultivated, ubiquitous components of household gardens. To the people of the country, especially villagers, they are important sources of nutrition. For some cultivars of *Cucurbita*, *Lagenaria*, or *Benincasa*, extensive field space may be unnecessary, since these plants climb well over any standing structure or support, thriving on bamboo platforms, small trees, or even the roofs of houses. This attribute is especially useful in the monsoonal areas in eastern India and on the lower Gangetic plain, where small, elevated patches of land may be surrounded by water.

The noteworthy cucurbit vegetables of India include 16 species in nine genera: *Lagenaria*, *Luffa*, *Benincasa*, *Momordica*, *Cucumis*, *Citrullus*, *Praecitrullus*, *Cucurbita*, and *Sechium*. All of these genera except *Cucurbita* and *Sechium* include species indigenous to India. Among the cultivated species are an array of landraces and cultivars; indeed, the cultivated forms in a country as vast as India are uncountable. In addition to their value as food plants, these vegetable cucubits may be important elements in traditional medical practice (3–5).

Not included in the following enumeration are two species, *Coccinia grandis* (L.) Voigt and *Cucurbita ficifolia* Bouché. The young shoots of the indigenous *Coccinia grandis* are sometimes used as pot herbs, while the fruits of the introduced *Cucurbita ficifolia* are eaten or used as animal feed (Andres, this volume).

Major Cucurbit Crops of India

Lagenaria siceraria (Mol.) Standl. (bottle gourd, calabash gourd, white-flowered gourd, *lauki, kadu* [Hindi], *lau, ladu* [Bengali], *dudhiva, ghiba*)

Bottle gourds are large, softly pubescent, monoecious or dioecious climbing or trailing herbs with angular stems and bifid tendrils. The leaves are long-petioled and five-lobed. The flowers are solitary and the corollas white. The mature fruits are large, up to 2 m long, usually bottle- or dumbbell-shaped, cylindrical or spherical, with a nearly woody rind or shell. The numerous seeds are ovate-oblong, 9–20 mm long, smooth, elegantly marbled, and white-margined.

Bottle gourds are popular vegetables throughout India and are grown almost year round, but winter is the best season. The vines are allowed to trail on the ground, but the fruits, usually 10–15 per plant, develop better when the vines climb trees or other supports from which the fruits may hang.

Bottle gourds are gathered while tender; this is indicated by the presence of soft hairs all over the body of the fruit, at which time they weigh from 0.5 to 1.3 kg. The young shoots are also consumed as a vegetable. The shells of some massive fruits of bottle gourds are utilized in India for the manufacture of stringed musical instruments like the *bina, sitar,* and *tanpura.* Cultivars or selections from West Africa, bearing almost perfectly round, large fruits, are excellent for such purposes. Otherwise the fruits are used for many kinds of utensils, ranging from bottles to ladles, horns, and snuff boxes (6).

Luffa cylindrica (L.) M. J. Roem. (sponge gourd, vegetable sponge, *ghia-torui* [Hindi], *dhundul* [Bengali], *ghiyatoti-bhol* [Assamese], *ghosale* [Marathi])

Sponge gourds are extensive, glabrous, monoecious climbers with trifid tendrils, large palmately lobed leaves, and a fleshy, glandular bract borne at the base of the petiole. The male flowers, in elongate clusters, and the solitary female flowers are often borne in the same leaf axils. The sepals are glandular and the corollas yellow. The fruits are fusiform, cylindrical, or clavate, not prominently veined, and 25–50 cm long or longer. The seeds are ovate, usually black, inconspicuously marginate but slightly winged all round, and about 12 mm long.

Sponge gourds are indigenous to the Indian subcontinent but are widely grown throughout the tropics. In India they are suitable for cultivation during spring and summer. While the tender fruits of this species are used as a vegetable, they are far inferior to those of *L. acutangula*. The sponge gourd is grown principally for the fibrovascular bundles of the mature fruits, which are used as bath sponges.

Luffa acutangula L. (angular luffa, ribbed luffa, ribbed gourd, *ara-torui*, *jhinga* [Hindi], *jhinga* [Bengali and Assamese])

The angular luffa is distinguished most readily from the preceding species by its hispid vestiture, five- to seven-angled leaves, pale yellow corollas, ten-angled or ten-ridged fruits, and wingless seeds.

Angular luffas are cultivated widely in India. Two crops are raised during the year, one early in summer and the other late in autumn (monsoon crop). The summer crop is ready for harvesting from May and the monsoon crop from September. The fruits are collected when still tender, for with age they become fibrous and unpalatable. The average yield is 15–20 fruits per plant. A Bihar cultivar, 'Satputiya' (syn. *L. hermaphrodita* Singh & Bhandari) bears bisexual flowers. A cross between 'Satputiya' and common 'Tori' may produce about five times the yield of either.

Trichosanthes cucumerina L. var. *anguina* (L.) Haines. (snake gourd, *chachinda* [Hindi], *chichinga* [Bengali], *chichendara* [Oriya], *padwal* [Marathi], *padavali* [Gujarati], *lingapotla* [Telugu], *pudal* [Tamil], *galartori* [Punjabi])

The snake gourd is an annual, monoecious creeper, with slender, angular stems, bi- or trifid tendrils, and generally deeply five-lobed, softly pubescent leaves. The male flowers are borne in racemes and have connate anthers; the female flowers are solitary. The fruits are up to 1 m long or longer, very narrowly cylindrical, often twisted but smooth and green with seven or eight longitudinal white stripes. When mature they turn a brilliant orange. The finely rugulose seeds are mostly about 15 mm long.

Snake gourds are grown extensively in southern India and throughout the lower Gangetic plain. They are a warm season crop, but can be cultivated throughout the year in the plains. The principal harvest season is from June to October. The fruits are used primarily as a vegetable or in curries. They are harvested when tender but are very succulent and do not keep well. Fifty fruits per plant is the average yield, with each fruit averaging about 1 kg.

Trichosanthes dioica Roxb. (pointed gourd, *palwal* [Hindi], *potol* [Bengali], *patal* [Oriya], *putulika* [Sanskrit], *kommupotla* [Telugu], *kombu-pudalai* [Tamil], *kaadu-padavala* [Kan.], *patolam* [Mal.])

The pointed gourd is a dioecious climber with a perennial rootstock. It differs further from the snake gourd in having simple or bifid tendrils;

cordate, ovate-oblong, sparsely scabrous leaves; male flowers that are solitary or in axillary pairs, with free anthers; and oblong fruits, to about 12 cm long, with globose seeds, 5–6 mm long.

Extensively cultivated throughout the warmer parts of India, particularly in Uttar Pradesh, West Bengal, and Assam, the pointed gourd also occurs in a wild state on the plains of northern India. The fruits and leaves are commonly used as vegetables. They also may be pickled and used in confectionary. A few agricultural types, differing mainly in the size, shape, and skin color of the fruits, i.e., variously green-striped, especially when young, are grown in India. Parthenocarpic fruit formation has been induced by growth regulators (9).

A humid and hot climate is preferred. Early plantings are usually made during February to April and late plantings from May to July. In upper India, the pointed gourd can be grown during the rainy season. Sandy to loamy soil with good drainage favors growth. The crop cannot withstand waterlogging.

Pointed gourds are mainly propagated by root or stem cuttings, since plants raised from seed are usually weak and the fruits produced are useless for commercial purposes. Stems and branches from one-year-old vines, with leaves removed, are made into coils known as *guchchi* and buried 1.2–2.5 cm deep. The connection of *guchchi* with the parent plant is not broken until the roots develop from the nodes buried in the soil. Both male and female cuttings are needed; 2% of male cuttings planted at random between the female plants is considered sufficient for pollination, but growers usually plant about 25%. The plants produce two crops in two consecutive years, the first year's crop being less than that of the second year. After the last harvest, the old crop is uprooted except for plants needed for cuttings.

Benincasa hispida (Thunb.) Cogn. (ash gourd, white gourd, wax gourd, *gol-kaddu* [Hindi], *chal-kumra* [Bengali], *kushmanda* [Sanskrit])

This cultigen is a pubescent, monoecious annual, with bifid or trifid tendrils, and reniform-orbicular, deeply cordate, five- to seven-lobed leaves. The male and female flowers are solitary, with large yellow corollas. The stamens are free. The fruits are cylindrical or spheroidal, hairy when young, but with age covered with a waxy white bloom. The seeds are light yellow, compressed-ovoid, distinctly marginate, smooth, and 10–12 mm long.

Apparently of Indo-Malayan origin, the ash gourd is cultivated extensively in India. The vines are trained on the roofs of huts in the villages. The young fruit is popular as a vegetable and is made into curries. The ripe fruits can also be consumed. In northern and western India, the ripe fruits are used in the preparation of a delicious transparent sweet called *pithe*. The sliced pulp is dried in the sun and is used in some parts of India during the off-season. The bloom from the fruits is collected for making candles. The fruits are especially valued for their medicinal and cooling properties and also for their beneficial effect in treating nervous disorders.

Momordica charantia L. (bitter gourd, carilla gourd, karela [Hindi], karena [Oriya], karala [Bengali], pava-kai [Tamil])

The bitter gourd is a monoecious climber, with slender, subglabrous stems, simple tendrils, and suborbicular, five- to seven-lobed leaves. The flowers are solitary on peduncles with foliaceous bracts; the corollas are yellow. The fruits are oblong, 8–20 cm long, with a muriculate-tuberculate surface. When mature, the three valves of the fruit dehisce and recurve from the apex to expose the yellowish brown seeds attached to orange-red, pulpy flesh. The seeds, 13–16 mm long, are ovate-oblong, sculptured, and subtridentate at the base and apex.

Bitter gourds are apparently of Indo-Malayan origin, occurring in the wild in India, but are best known as a tropical crop. In India, two primary varieties of *M. charantia* are grown, both with muriculate-tuberculate fruits in many forms (1). Variety *charantia* bears large, fusiform fruits, not tapering at both ends, with numerous triangular tubercles, giving it the appearance of a crocodile's back. Variety *muricata* (Willd.) Chakravarty bears small, rounded fruits, more or less tapering at each end, also with tubercles. The seeds are sculptured in each variety, but are smaller in var. *muricata*.

Bitter gourds are commonly cultivated during the hot season. In the hills an early summer crop is sown during April to July. The hot season crop is sown in June to July. The fruit, although bitter, is wholesome and much liked when young. Various curry preparations are made out of it, and it is also pickled. *Momordica charantia* is a good source of calcium. Abundant calcium carbonate crystals in the form of cystoliths and pure crystals as calcium oxalate are found in every part of the plant (2). The fruits are widely used as medicinals.

Momordica dioica Roxb. ex Willd. (kaksa [Hindi], kakrol, ban-kerela [Bengali], kanchan-arak [Santal], kartoli [Maharashtra], Agakral [Teluga], karlikai [Kan.], tholoopavail [Tamil])

This species is a perennial, dioecious climber, with tuberous roots, slender stems, and simple, filiform tendrils. The leaves are broadly ovate, entire or deeply or shallowly three- to five-lobed. The flowers are solitary on bracteate (male) or ebracteate (female) peduncles, with yellow corollas. The fruits are ovoid or ellipsoid, 2.5–6.5 cm long, with a densely echinate surface. The seeds are pale yellow, ovoid, slightly compressed, and 6–7 mm long.

Momordica dioica is found throughout India, mostly in the wild state, but partly cultivated in the Gangetic plain. The species is grown for its fruits, which are quite palatable, free from bitterness, and rather sweet. The large tuberous roots are astringent, and traces of alkaloids have been detected in them. They are used in indigenous medicine for bowel complaints and urinary troubles.

Cucumis melo L. (melon, muskmelon, sweet melon, *kharbuz* [Hindi], *kharmuj, futi*)

The melons are silky pubescent, monoecious, scandent annuals, with simple tendrils and moderately three-, five-, or seven-lobed leaves. The male flowers are borne in fascicles, the females are solitary. The corollas are yellow. The three stamens are free. The fruits of the cultivated forms are highly variable in size, shape, color, and markings; those of the wild variety are oblong to turbinate and green. The seeds are white, oblong to elliptic, highly compressed, smooth, and 10–12 mm long.

Two botanical varieties of *C. melo* are currently recognized: var. *agrestis* Naud. includes the wild populations, which are found throughout India; var. *melo* includes all of the cultivated forms, although in early classifications many of these were recognized as botanical varieties. Within var. *melo*, cultivars may be placed in about seven principal groups, e.g., the Reticulatus Group, in which cultivars have a more or less netted fruit surface.

Melons are cultivated throughout India, particularly in the drier northwestern areas. Numerous cultivars and races are known that differ in their fruit characteristics, i.e., size and shape, thickness, texture, color, and flavor of the pulp, and character of the rind, being soft or hard, yellow, green, cream, or orange. They also differ in their cultural reqirements, although melons grow well in sandy river beds, which are less suitable for most other crops.

The melons are primarily eaten as a dessert fruit. In an immature state they may used in salads like a cucumber or in curries. The flesh may be canned or used to prepare a syrup or jam. The seeds, slightly roasted, are used like almonds or pistachios.

Cucumis sativus L. (cucumber, gherkin, *khira* [Hindi], *sasa* or *khira* [Bengali], *kaknai* [Oriya])

This species is distinguished from *C. melo* primarily by its harsher pubescence, less extensively lobed leaves, solitary or clustered female flowers, and prickly or tuberculate fruits, at least when immature, that usually are oblong and yellowish green to green with yellowish bands.

Young cucumbers are often pickled, and the older ones are used for salads or in curries. The large, ripe fruits are sometimes used to prepare chutney. Kernels of the seeds are used in confectionary. The presence of proteolytic enzymes and malic acid and succinic acid has been reported (6). The odorous principle is extractable and is sometimes used in the preparation of perfumes.

The cultivars are broadly divided into two classes, a hot weather crop and a rainy season crop. The former comprises creeping plants, which yield small egg-shaped fruits known as gherkins. The rainy season crop is more common in India and bears much larger fruits. Eight to ten thousand fruits (1000–3000 kg) per acre have been recorded.

Citrullus lanatus (Thunb.) Matsum. & Nakai (watermelon, *tarbuz* [Hindi], *tarmuj* [Bengali])

Watermelons are monoecious, trailing annual herbs with bifid tendrils and scabrous, variously pinnately lobed leaves. The male and female flowers are extra-axillary and solitary, with yellow corollas. The fruits are large, smooth, and glossy, with many forms distinguished by their size, shape, and skin color, which varies from light to deep green or striped light and deep green. The pulp is dull red to bright red or yellow, and the seeds are black, red or reddish, oblong, inconspicuously marginate, and 12–15 mm long.

The watermelon is a native of southwestern Africa. Numerous cultivars, differing mainly in fruit characters, are grown in India, where the flesh provides a sweet, cooling repast. The seeds of watermelons are used as food and are sometimes powdered and baked like bread. A bitter cultivar known as *kirtut* in India is used as a purgative like *C. colocythis* (L.) Schrad. Plants are propagated from seeds sown in January to February, with a second sowing possible in June and July.

Praecitrullus fistulosus (Stocks) Pang. (squash melon, round melon, *tinda*)

Praecitrullus fistulosus is a diffuse, monoecious annual, with hollow, banded stems, which are villous at first, then scabrous. The tendrils are 3- to 5-fid. The leaves are cordate-ovate in outline and cut into five rounded lobes. The male flowers are clustered, the female solitary. The corolla is rotate and sulfur yellow. The fruit is spherical, hispid at first, then smooth and bright green. The seeds are black and ridged along the margin.

Squash melons, native to northern India and Pakistan, are cultivated largely in Uttar Pradesh, Punjab, and Bombay, and is much used as a summer vegetable. The fruits are picked somewhat before maturity, when about 7–9 cm in diameter. With the seeds removed, the fruit is cooked with spices or together with meat in stews or curries. It also is fried in ghee with split gram peas and curry (12), pickled, or made into preserves. The seeds are roasted and eaten like those of *Cucurbita*, muskmelons, and watermelons.

Cucurbita maxima Duch. ex. Lam. (squash, red gourd, *lal kumra* [Hindi], *metha kumra* [Bengali], *parangikayi* [Tamil], *lal-dudiya*, *lal-bhopli* [Mar.])

This cultigen is a trailing, monoecious annual, with slightly prickly stems, 2- to 6-fid tendrils, and generally reniform-orbicular, unlobed or five-lobate rather than deeply lobed leaves. The male and female flowers are solitary. The corollas are yellow, with outwardly spreading, obtuse lobes. The fruits are of various shapes, i.e., spherical to oblong to turban-shaped, and are generally large. The peduncle is striate, not ridged or prominently expanded at the apex. The seeds are white or pale yellow, 20–24 mm long, and obscurely marginate.

The species, apparently of South American origin, is cultivated in several forms throughout India, usually as a rainy season vegetable. All parts of the plant are eaten, generally after cooking. In some parts of India, tender shoots and leaves are eaten as a salad or in vegetable curry. The flowers are usually fried. The seeds are often eaten after they are roasted and the seed coats removed, or are used to prepare sweet dishes. The kernels yield a dark, brownish red oil having a bland fatty taste.

Cucurbita moschata (Duch. ex Lam.) Duch. ex Poir. (crookneck squash, *mitha-kadu* and *sitaphal* [Hindi], *kumra* [Bengali], *lal-kumra* and *mitha-lau* [Assam], *kali-dudhi* [Mar.])

The crookneck squash is distinguished from *C. maxima* by its soft pubescence, often white-blotched leaves, usually foliaceous calyx, and wide-spreading but acute corolla lobes. The peduncle is angled and much expanded at the apex, while the seeds are 16–20 mm long, marginate, grayish white to pale brown with dark marginal bands.

This squash is widely cultivated in India, but requires a warmer climate than *C. maxima*. There are two primary forms, one with smooth, oblong fruits and the other with fluted fruits, which are flattened or spheroidal in shape. The yellow flesh of the fruit is cooked and eaten as a vegetable.

Cucurbita pepo L. (vegetable marrow, pumpkin, squash, *kadmiah*, *kumra*, *keala* [Hindi], *kumra* [Bengali], *kumbala* [Kan.], *kohala* [Mar.])

Vegetable marrow is distinguished from *C. maxima* and *C. moschata* by its more hispid to prickly vestiture, angulately and often deeply lobed leaves, nonfoliaceous calyx, and erect, acute corolla lobes. The fruiting peduncle is strongly angled and enlarges gradually toward the apex. The seeds are whitish yellow, 7–25 mm long, and marginate.

Cucurbita pepo, North American in origin, includes a wide range of cultivated forms in two subspecies, i.e, subsp. *pepo* and subsp. *ovifera* (L.) Decker var. *ovifera* (7, Decker, this volume). Cultivars differ widely in the coloration and shape of the fruits, some being grown as ornamental gourds (8), as well as in other characters. The tender shoots may be eaten as a vegetable, but the species is grown principally for the immature fruits, which are comsumed as a vegetable or in curries. The seeds are edible, as are the oils derived from them, the lower grades of which are used as illuminants.

Usually the tender or enlarging fruits of *Cucurbita*, irrespective of species, are called squash in western countries. In India all three species, *C. maxima*, *C. moschata*, and *C. pepo*, are called *kumra* or *mitha kadu*.

Sechium edule (Jacq.) Sw. (chayote, *chow-chow, quash* [Bengali], *seema-kattirikkai* [Tamil], *seeme-badane* [Kan.])

Chayote is a vigorous, monoecious, tuberous perennial with climbing annual stems, 2- to 5-fid tendrils, and broadly ovate to triangular, but scarcely lobed leaves. The flowers are small with yellow corollas; the males are clustered, the females solitary. The fruits are irregularly pear-shaped to ovoid, 7–20 cm long, green or white, fleshy with a single large flat seed about 5 cm long, which often is viviparous.

A native of Central America and Mexico, where many cultivars are found (Newstrom, this volume), chayote is extensively cultivated in southern India and the northern hills at elevations up to 1700 m. The fruits may vary in color from dark green to ivory white, and weigh from 100 g to 1 kg. The surface is deeply wrinkled or corrugated and prickly.

Chayote prefers moderate temperature. Sprouted fruits are usually utilized for propagation, but cuttings from the stem also can be used. The vines are allowed to spread over netted platforms, which they quickly cover, with fruits hanging down. Except for February and March, harvest occurs throughout the year in southern India, although there are two principal fruiting seasons and the vines are pruned at the end of each season (10, 11). A single plant may produce 75–300 fruits in a year.

The unripe fruits are used as a vegetable and are variously prepared, either cooked with spices to make a curry or simply pan-fried. They are also sometimes used raw as salad. The large tuberous roots are a source of an arrowrootlike starch, the seeds are cooked in butter as a delicacy in southern India, and the tender shoots are used as a vegetable. The vines and the tubers both serve as excellent fodder for the cattle.

Future Developments

The further development of cucurbit crops in India, where the problems of feeding a large human population are formidable, is essential. Cucurbits, if properly nurtured, can assist greatly in alleviating this problem because of their ease of culture and abundant production in short periods of time. Indian soils, in general, are quite suitable for cucurbits. Of those described here, the cultivation of muskmelons and watermelons can be attempted in the drier, arid regions of India, where experience (3, 4) suggests they will thrive under irrigation. Along with the development of cereal crops, vegetable production on a large scale must be taken up without delay. Easily grown, with massive or sufficiently bulky fruits, short periods of production, and good food value, the cucurbits have unique advantages favoring of their cultivation.

Acknowledgments

It is a pleasure to acknowledge, with thanks, the help of Gour Gopal Maity in reading and commenting on the manuscript.

Literature Cited

1. Aiydaurai, S. G. 1951. Preliminary studies on bitter gourd (*Momordica charantia*). Madras Agric. J. 38: 245–248.
2. Chakravarty, H. L. 1937. Physiological anatomy of the leaves of Cucurbitaceae. Philipp. J. Sci. 63: 409–431.
3. Chakravarty, H.L. 1966. Monograph on the Cucurbitaceae of Iraq. Tech. Bull. No. 133. Ministry of Agriculture, Baghdad.
4. Chakravarty, H.L. 1976. Plant Wealth of Iraq. I. Ministry of Agriculture and Agrarian Reform, Baghdad.
5. Chakravarty, H. L. 1982. Cucurbitaceae. *In* K. Thothathri, ed., Fascicles of Flora of India. Fasc. 11. Botanical Survey of India, Howrah.
6. Chopra, R. N., and A. C. Roy. 1933. A proteolytic enzyme in cucumber (*Cucumis sativus*). Indian J. Med. Res. 21: 17–23.
7. Decker, D. 1988. Origin(s), evolution, and systematics of *Cucurbita pepo* (Cucurbitaceae). Econ. Bot 42: 4–15.
8. Firminger, T. A. 1947. Manual of Gardening for India. Thaker Spink, Calcutta.
9. Ghosh, P. C. 1963. Parthenocarpic fruit development in *Trichosanthes dioica* Roxb. by using growth regulators. Sci. Cult. 29: 146–148.
10. Kolhe, A. K. 1962. Chau-Chau does well in Maharashtra. Indian Hort. 6: 8–10.
11. Seemanthani, B. 1964. A study of practices and problems in the cultivation of some perennial vegetables in Madras State. South Indian Hort. 12: 1–15.
12. Stocks, J. E. 1851. An account of the dilpasand, a kind of vegetable marrow. Hooker's J. Bot. Kew Gard. Misc. 3: 74–77.

27 | Domestication of Buffalo Gourd, Cucurbita foetidissima

A. C. Gathman and W. P. Bemis

ABSTRACT. Buffalo gourd, *Cucurbita foetidissima*, is a perennial cucurbit native to the semiarid lands of the western United States and northern Mexico. Seeds of buffalo gourd are high in edible oil and protein. The storage root, although bitter, can be processed to yield a high-quality food or industrial starch. The potential of this species as a new domesticate was recognized in the early 1940's, when it was suggested as a possible oilseed crop. Summarized is the large body of accumulated knowledge about the cultivation and utilization of buffalo gourd. Commercial development of this species awaits appropriate pilot-scale research.

Cucurbita foetidissima HBK, known commonly as buffalo gourd, has recognized potential as a crop for arid lands. For over fifteen years at the University of Arizona, Tucson, and elsewhere, buffalo gourd has been the subject of a wide range of research dealing with its biology, agronomic requirements, and potential use. In this chapter the results of that research are summarized.

Buffalo gourd (Figure 1) is a xerophytic, perennial cucurbit, ranging from Guanajuato in central Mexico in the south through Nebraska in the north and from Missouri to the Rocky Mountains, with some probably recent introductions in California and Illinois (6). It occurs primarily as a weed in and around agricultural fields and along roads and railroad embankments, as well as on the banks of arroyos and in low areas in open country (6, 19).

Buffalo gourd is less drought tolerant than *Cucurbita digitata* A. Gray and related species, and seems to lie between the mesophytic cucurbits and the true xerophytes in its environmental adaptations (8). Although it is not an extreme xerophyte, the species is drought resistant by virtue of its large, fleshy storage root, which is covered with an impermeable suberized per-

Figure 1. The habit of *Cucurbita foetidissima*, showing the relative proportions of vegetative structures, flowers, and fruits.

iderm (19). The vines are frost sensitive and die back in the winter, but the roots can survive air temperatures to −25°C, particularly when the soil is insulated by snow cover.

Vines develop in early spring from nodes around the crown of the root, commonly as many as 20 from a two-year-old plant, but may die back under stress during the summer. New vines continue to sprout from the perennial root when conditions are favorable. Any developing node of the vine can produce adventitious roots if it remains in contact with moist soil, and such new roots will sprout vines. Large clonal populations are frequently produced in this manner, and asexual reproduction is probably more important than seed propagation in the wild (7). Under optimal conditions this species is capable of prodigious growth. Dittmer and Talley (19) reported that a wild plant produced 360 annual shoots covering an area 12 m in diameter with a total vine length of over 2000 m, all developed in a single growing season. The central tap root of this plant alone weighed 72 kg.

Buffalo gourd is generally described as monoecious; however, Curtis and Rebeiz (16) reported that about 25% of seeds collected from a large wild colony produced gynoecious plants. In such plants, development of staminate primordia ceases when staminate tissues reach 2–3 mm in length. Lack of stamen development usually induces the subsequent abortion of stamenless flower buds (46). Application of aminoethoxyvinylglycine (AVG) has been shown to promote staminate bud development in gynoecious plants (45). Dossey et al. (20) reported monoecy to be due to a homozygous recessive allele and gynoecious plants to be heterozygous. The dominant allele for gynoecy, symbolized G, has been found to be widespread in natural populations. Of 47 buffalo gourd germplasm accessions collected by Bemis et al. (3), 22 were segregating for gynoecious and monoecious plants. Gynoecious plants have been found to outperform monoecious plants in fruit yield (16), a fact that may prove important agronomically.

Domestication

Until World War II, buffalo gourd was a weed of value only to certain Indian tribes (33). The disruption of vegetable oil supplies that occurred at that time stimulated the consideration of wild *Cucurbita* as potential oilseed producers. Curtis (15) summarized the advantages of some xerophytic cucurbits, particularly buffalo gourd, with the following four points: 1) the plants are perennial, 2) they can be grown on waste lands in regions of low rainfall, 3) they produce an abundant crop of fruit that contains seed rich in oil and protein, and 4) the hard fruit lends itself to mechanical harvesting. There was a flurry of interest in buffalo gourd as an oilseed producer in the 1940's and early 1950's. Reports were published on the use and composition of the seed (12, 47) and on the agronomic properties of the plant (36).

The first organized effort to modify the germplasm of *C. foetidissima* to produce a crop plant was that of Curtis in Lebanon (16). Using germplasm from a single collection site in Texas, he found tremendous variability in yield, ranging from no fruits up to 283 fruits per plant in the first year. Other characteristics, such as vine habit and sex expression, were also highly variable.

While buffalo gourd has been grown since 1963 at the University of Arizona, it was at first only one of several species used to study genetic barriers to interspecific hybridization. In 1973, at the urging of Curtis and others, a research project on the domestication and utilization of buffalo gourd was established. The first step was to create a germplasm base in order that maximum use of the species' wide genetic variability could be made in breeding. Three collections trips were made from 1975 to 1977, and seed from the collections was planted in germplasm nurseries at Tucson. The 145 accessions planted, including some sent by interested cooperators, were assumed to represent a cross-section of the genetic diversity of *C. foetidissima* in its native range (29).

Plant Breeding and Agronomic Studies

The initial phase of the breeding and genetics portion of this project centered on screening for agronomically superior plants and production of relatively homogeneous seed for use in field trials. Phenotypic variation among accessions was extensive, particularly for gross yield components. For instance, seed weight per fruit had a coefficient of variation (C.V.) of 36.9% (41). Of approximately 700 plants in the 1976 germplasm nursery, 32 were selected on the basis of first season fruit yield. After the seed from these plants was analyzed for crude fat and protein content, two superior plants were chosen to establish seed production plots. Isolation plots were established with seeds of the two selections, 142-1 and 158-2. Because the two plants were gynoecious, their seeds segregated 1:1 for sex type, and monoecious plants were rogued from the seed parent line. Reciprocal hybrid production plots, 158-2 × 142-1 and 142-1 × 158-2, were thus established. The larger plot (158-2 × 142-1) produced 3500 fruit and 23 kg of seed in the first season from 114 gynoecious plants. This seed stock, known as Arizona Hybrid #1, was made available to interested persons and used to establish cultural plots (29).

Later selections were made, and plants from eight accessions were chosen for further evaluation. Seven experimental hybrids and three open-pollinated lines were tested in a replicated yield trial in 1979. The yield test was seeded at the Agricultural Experiment Station, Marana, Arizona (elevation 610 m), in plots 4 × 21 m. Each plot contained approximately 120 plants. The first season yields ranged from 20.4 to 908.4 kg/ha. Typically, higher yields are obtained in the second season of growth of buffalo gourd, so these data do not reflect the full potential of buffalo gourd seed yield. They do, however, demonstrate the great genetic variability of this species. It should also be noted that all of the hybrids outyielded the three selected open-pollinated lines (29).

In another study in 1981, plots at Marana yielded 933 kg/ha of seed in the first season when irrigated at 50% soil water depletion. Plots that were not irrigated after plant establishment yielded 792 kg/ha of seed. Total consumptive water use for the irrigated plots was 871 mm, and for the unirrigated plots 620 mm (35). Last, a planting in Maricopa, Arizona (elevation 360 m), in 1983 yielded 226 kg/ha of seed in the first season. This low yield was probably due to the extremely high temperatures at this location. The second season yield of this planting was 748 kg/ha (35).

Diseases and Pests

Early workers reported no disease or pest problems for buffalo gourd (15, 36). However, once cultural plots were established in Arizona, certain pathogens became apparent.

A 1981 planting of buffalo gourd in Marana suffered severe losses from Texas root rot, caused by *Phymatotrichum omnivorum* Shear (40). Avoiding infested soils is the only control measure for this disease, which is limited to alluvial soils in the southwestern United States. In 1978, numerous dead buffalo gourd plants were observed in field plots in southern Arizona. A bacterium (*Erwinia carotovora* Jones) and a fungus (*Fusarium solani* Mart.) were isolated from the decayed stems and roots. Loss due to this condition may range as high as 80%. The only control measure that has been effective is crop rotation.

Viral disease symptoms, including leaf mosaic and distortion, have been noted in field plantings in Arizona, particularly in older plantings. Provvidenti et al. (37) reported that *C. foetidissima* was resistant to cucumber mosaic virus, tobacco ringspot virus, bean yellow virus, and watermelon mosaic viruses 1 and 2. Tomato ringspot virus produced systemic symptoms. Rosemeyer et al. (39), using different viral strains, found it to be susceptible to cucumber mosaic virus, watermelon mosaic viruses 1 and 2, squash mosaic virus 2, and lettuce infectious yellow virus. Squash mosaic virus and cucumber mosaic virus were identified as the cause of some mild leaf mosaic symptoms in the field. The cause of the most important viral disease of buffalo gourd that has been observed, which is characterized by severe foliar distortion, was identified as watermelon curly mottle virus. Control of these virus conditions has been accomplished mainly by growing this crop as an annual or biennial, since disease buildup reduces yields of older plantings. Resistance has not yet been found (38).

Buffalo gourd is susceptible to root-knot nematode, *Meloidogyne* spp., which causes galling of roots but no foliar symptoms. The first year starch yield of buffalo gourd roots decreased 20% when one to two *Meloidogyne javanica* Treub larvae per milliliter of soil were present. While no immunity to *M. javanica*, *M. incognita* Kaford & White, or *M. arenaria* Neal was found in a greenhouse test of ten *C. foetidissima* accessions, a wide variation in the ability of the nematodes to reproduce was noted in the different lines. Breeding for nematode tolerance in buffalo gourd should be possible (26).

Insect pests remain insignificant on buffalo gourd plants in Arizona. The spotted and striped cucumber beetles (*Diabrotica undecimpunctata* Mannerheim and *Acalymma vittatum* Fabricus) occasionally cause some damage to seedlings, but this is rarely severe. One reason for the low damage levels in older plants as a result of these insects may be the relatively low cucurbitacin levels in the mature leaves of C. *foetidissima* (32). A gall-forming relative of the squash vine borer, *Melittia snowi* Edwards, is found in buffalo gourd plants, but causes little damage. In a study in Illinois, buffalo gourd was not attractive to the squash bug, *Anasa tristis* Defeer, the squash vine borer, *Melittia cucurbitae* Harris, or the potato leaf-hopper, *Empoasca fabae* Harris. The potato aphid, *Macrosiphum euphorbiae* Thomas, and the melon aphid, *Aphis gossypii* Glor, were somewhat attracted to buffalo gourd plants (30). Insect pests to date seem unlikely to limit the feasibility of cultivating C. *foetidissima*, except through their role as vectors of viral disease. Further studies are necessary to find disease-resistant and tolerant lines of buffalo gourd and to develop cultural practices to minimize disease damage.

Utilization and Products

In general, the potential products of this species fall into four groups: 1) seed oil, 2) seed protein, 3) root starch, and 4) byproducts. The food products of buffalo gourd have been reviewed recently (42). Efforts have been made to improve the quality and quantity of specific products of buffalo gourd through plant breeding. Variation in a number of these products was reported by Scheerens et al. (41). Crude protein content of defatted embryos was relatively constant (C.V. = 5%); however, the percentage of embryo in seeds had a C.V. of 9%, indicating that it may be possible to increase both protein and oil yields by selecting for lower seed coat weight.

Seed Oil

Oil was considered the most promising product of buffalo gourd in early investigations, and it still shows potential in certain areas (18). The oil extracted from C. *foetidissima* seed has been characterized by several authors. Shahani et al. (47) reported an extremely dark color in the crude oil and were unable to obtain an acceptable color even after refining and bleaching. They suggested that this was caused by field-drying of the fruit and was similar to the effects of field-weathering in cottonseed. Vasconcellos et al. (51), using seed extracted from fruit harvested at maturity, reported no difficulty with oil color. The carotenoid content was similar to that of cottonseed oil. *Cucurbita foetidissima* seed oil refined by the process of Vasconcellos and Berry (52) is comparable with other processed commercial edible oils in color and flavor.

Scheerens et al. (41) reported the fatty acid composition of buffalo gourd oil to be rather variable, with a range of 49.5 to 68.8% for linoleic acid in 15 randomly chosen accessions. Oleate (18:1) and linoleate (18:2) were found to be highly negatively correlated, as would be expected if they represent successive steps in a biosynthetic pathway. The authors concluded that selection for either polyunsaturation or monounsaturation of buffalo gourd seed oil was possible; such selections were later carried out and high- and low-linoleate lines were produced. Individual seeds have been obtained with linoleic acid contents ranging from 27.4 to 82%. Heritability of linoleate content was determined by regression of progeny on midparent to be 0.84 under relatively constant environmental conditions (24), although long days and high growing temperatures decreased the level of unsaturation (23). It should be relatively easy to breed a desired form of unsaturation into agronomically superior stocks as they are developed.

Conjugated fatty acids are highly prone to oxidation and thus promote rancidity in vegetable oils. Unlike some other xerophytic cucurbits, buffalo gourd produces only small amounts of conjugated fatty acids in seed oil (1). Although the conjugated diene content of 2.3% is higher than that of cottonseed oil (51), the level drops to 0.48% after processing (52). This small residual amount does not affect oil quality, since oxidative stability of the processed oil was found acceptable even in the absence of antioxidants (52). Feeding studies with mice (4) and chicks (31) showed *C. foetidissima* seed oil to be nutritionally comparable with commercial oils. No toxicity has been observed.

Seed Protein

The seed of buffalo gourd contains approximately 30% protein by weight. The seed coats have a high protein content (17%), but this material is not digestible by monogastric animals. The predicted seed coat digestibility in ruminants is 23%, which compares favorably with those obtained for sunflower seed coats (20%) and cottonseed hulls (26%), both of which are currently used as ruminant roughages (21). The embryo is 37.5% protein, and defatting raises the protein content to 75%.

The amino acid composition of seed protein in buffalo gourd accessions is relatively constant, with few amino acids showing a C.V. higher than 5% (41). On the basis of composition alone, *C. foetidissima* seed protein is comparable with that of most other oilseeds with the exception of soybean, which is substantially higher in lysine. Feeding experiments with rats (28, 50), mice (2, 48), and chicks (7, 17) all found buffalo gourd seed meal inferior to casein or whole egg controls as a protein source. It is deficient mainly in threonine, methionine, and lysine (55%, 57%, and 70%, respectively, of FAO/WHO requirements). Glutamic acid, arginine, and aspartic acid predominate. Like most oilseed proteins, buffalo gourd seed protein needs amino acid supplementation or blending with complementary protein for complete nutrition in monogastric animals.

Daghir et al. (17) found a nervous disorder and enteritis in chicks fed a diet containing 30 percent buffalo gourd whole seed meal. The authors noted a bitter flavor in the buffalo gourd meal used in the study and suggested that cucurbitacins caused the observed toxic effects. Since no such bitterness was noted previously in clean buffalo gourd seed (4), contamination of the seed meal used by Daghir et al., either with fruit pulp or extraneous material, seems likely. Defatted, decorticated buffalo gourd seed meal has been analyzed for the presence of antinutritional factors. The meal has 20.21 trypsin inhibitor units (TIU) per milligram protein, compared with 120.83 TIU/mg protein in soybean flour. Heat treatment reduces both of these to an insignificant level. Lectins are essentially absent from *C. foetidissima* seed (28).

In organoleptic evaluation panels, African students rated buffalo gourd seed meal highly acceptable as a substitute for cereals in a number of traditional African dishes. Because the dishes prepared with *C. foetidissima* seed meal were significantly higher in protein than their traditional grain-containing equivalents, the use of buffalo gourd seed as a food could help alleviate protein deficiencies in some African countries (49).

Root Starch

While early interest in buffalo gourd centered on its seed products, its potential as a starch source has become increasingly apparent. Berry et al. (9) reported a dry weight starch content of 55% in *C. foetidissima* roots. Starch content has since been shown to increase with plant age during the growing season, except for a sharp decrease when and if fruit begin to set (10, 43).

It has been proposed that perennial buffalo gourd plantings could be managed to produce both fruit and roots by harvesting roots in a portion of the field each year (3). Maximum root yields, however, are obtained in high density plantings in which pistillate flowering is completely suppressed. Nelson et al. (34) achieved yields of fresh roots as high as 34,500 kg/ha in high density plantings, and found a quadratic relationship between plant density and yield with a theoretical maximum yield at 550,000 plants/ha.

Although the root itself is extremely bitter because of its cucurbitacin content, starch extracted from the root by dispersal in water, filtering, and centrifugation is essentially free of cucurbitacins (5). Puddings made with buffalo gourd root starch were indistinguishable from those made with corn starch in organoleptic evaluations (22). Buffalo gourd root starch was compared with potato, tapioca, and corn starches in a feeding trial using mice (21). Raw buffalo gourd starch had a digestibility of 41.1%, between that of potato starch (26.5%) and tapioca (97.2%) or corn starch (96.5%). Cooking improved its digestibility to 95.6%, a level not significantly different from cooked corn and tapioca starches.

The physical and rheological properties of buffalo gourd root starch are promising. The granules average 6 μm in diameter, slightly smaller than those of corn (9). Like corn starch, buffalo gourd root starch maintains a stable level of viscosity over time at 90°C; however, its viscosity is considerably higher than that of corn starch at the same concentration. The initial pasting temperature and gelatinization temperature are similar to those of corn starch (22). Hydroxypropylation, a modification commonly applied to food starches, improves the freeze-thaw stability and pasting characteristics of buffalo gourd starch (13). Modified buffalo gourd starch should be suitable for a wide range of food industry applications.

Buffalo gourd roots can also be ground and used directly as a substrate for alcoholic fermentation. Scheerens et al. (44) performed yeast fermentation of buffalo gourd root slurries and found conversion efficiencies of 82 to 87%. Because of the high fiber concentration and foaming of buffalo gourd roots, they had to add 50% water to their slurries. Therefore, the final concentration of alcohol was low (3.3 to 4.5%, compared with 6 to 7% in potato controls). Although first-year roots served as an acceptable fermentation substrate, the authors reported inhibition of yeast growth in slurries made from perennial roots. Such inhibition did not occur in an earlier study (25). Further research is needed to determine how the conditions under which roots are grown and stored affect their fermentability.

Forage and Cellulose

The prodigious vine growth of buffalo gourd has prompted speculations that the leaves and stems remaining after root or seed harvest could be used as a forage for ruminants (3, 11). Although high protein content has been reported for the vines of this plant (11), other factors limit their use as feed. Numerous observations have suggested that the fresh vines are unacceptable to sheep or cattle for grazing. Dried leaves and stems have been used in feeding trails with sheep; however, the animals accepted them only after the addition of 7.5% molasses. The apparent digestibility coefficient for total dry matter was 60% (14). Although no toxicity was noted in this study, the leaves and stems contain cucurbitacins, which are known to be toxic to mammals.

Cucurbitacins

Another potential use of buffalo gourd currently being considered is as an attractant for *Diabrotica* beetles. The potential of cucurbitacins as *Diabrotica* phagostimulants is reviewed by Metcalf and Rhodes (this volume). Fruit and roots of buffalo gourd have fairly high cucurbitacin levels (0.3 and 0.34%, respectively). The main form present in leaves, cucurbitacin E, is detected by diabroticites at about ten times the limit of response to cucurbitacin B (32). However, the primary cucurbitacins of buffalo gourd roots have recently been tentatively identified as cucurbitacins B and D and their glycosides. The large quantity of buffalo gourd byproducts produced during fruit and root processing would make them a good source of diabroticite bait.

Commercialization

The potential of buffalo gourd as a crop plant has been extensively investigated, but pilot plant work is still needed before large plantings will be economically justified. Admittedly, bridging the gap between basic research and commercialization is the most difficult step in the process of plant domestication. The greatest barrier is the disparity between the time scales of the biological processes involved in commercialization and the fiscal concerns of corporate and private enterprise. The size of the investment required and the length of time before a return is realized frequently make commercial development of new crops impractical for private industry. Aspects of the economic development of buffalo gourd have been reviewed (18).

Several attempts have been made to develop buffalo gourd as a commercial crop, but none has come to fruition. The largest area yet planted was 80 acres, established in 1981 in Queensland, Australia by Primary Energy Australia, Ltd. The success of this planting was limited, mainly because of weed competition and to poor adaptation of the Arizona Hybrid #1 germplasm to Australian conditions. A planting of 32 acres was made in 1983 in Eloy, Arizona by Agrocenter Real Estate. This project was abandoned after one year. Presently the only production-scale planting of buffalo gourd in existence belongs to the New Mexico Energy R&D Institute (New Mexico State University, Las Cruces), where the roots are used as a feedstock in a fuel alcohol project.

Perhaps the most important potential use of buffalo gourd is as a source of edible oil and protein for subsistence farmers in semi-arid Third World areas such as the Sahel region of Africa. Seeds have been provided to researchers in numerous countries, but successful development of buffalo gourd as a food crop in these regions would require on-site breeding work to select locally adapted germplasm.

Conclusions

The unique position of buffalo gourd as an arid-adapted plant capable of producing economic quantities of starch, oil, and protein makes it an important potential addition to the crops of the world. As water supplies for agriculture dwindle in many parts of the world, new plants adapted to arid and semi-arid conditions will be in great demand. The potential of buffalo gourd as a food, feed, and fuel source should not be ignored.

Literature Cited

1. Bemis, W. P., J. W. Berry, M. J. Kennedy, D. Woods, M. Moran, and A. J. Deutschman, Jr. 1967. Oil composition of *Cucurbita*. J. Amer. Oil Chem. Soc. 44: 429–430.
2. Bemis, W. P., J. W. Berry, and C. W. Weber. 1977. Breeding, domestication, and utilization of the buffalo gourd. *In* Non-Conventional Proteins and Foods, Proceedings of NSF Grantee-Users Conference. Univ. Wisconsin, Madison.
3. Bemis, W. P., J. W. Berry, and C. W. Weber. 1978. The buffalo gourd: a new potential horticultural crop. HortScience 13: 235–240.
4. Bemis, W. P., J. W. Berry, and C. W. Weber. 1979. The buffalo gourd: a potential arid land crop. *In* G. A. Ritchie, ed., New Agricultural Crops. Westview Press, Boulder, CO.
5. Bemis, W. P., J. W. Berry, and C. W. Weber. 1979. Domestication studies with the feral buffalo gourd. *In* J. R. Goodin and D. K. Northington, eds., Arid Land Plant Resources. International Center for Arid and Semi-Arid Land Studies, Lubbock, TX.
6. Bemis, W. P., L. C. Curtis, C. W. Weber, and J. W. Berry. 1978. The feral buffalo gourd, *Cucurbita foetidissima*. Econ. Bot. 32: 87–95.
7. Bemis, W. P., L. C. Curtis, C. W. Weber, J. W. Berry, and J. M. Nelson. 1975. The buffalo gourd (*Cucurbita foetidissima* HBK): a potential crop for the production of protein, oil, and starch on arid lands. USAID Tech. Series Bull. 15. U.S. Gov. Printing Office, Washington, DC.
8. Bemis, W. P., and T. W. Whitaker. 1969. The xerophytic *Cucurbita* of northwestern Mexico and southwestern United States. Madrono 20: 33–41.
9. Berry, J. W., W. P. Bemis, C. W. Weber, and T. Philip. 1975. Cucurbit root starches: Isolation and some properties of starches from *Cucurbita foetidissima* HBK and *Cucurbita digitata* Gray. J. Agric. Food Chem. 23: 825–826.
10. Berry, J. W., J. C. Scheerens, and W. P. Bemis. 1978. Buffalo gourd roots: chemical composition and seasonal changes in starch content. J. Agric. Food Chem. 26: 364–366.
11. Berry, J. W., C. W. Weber, M. L. Dreher, and W. P. Bemis. 1976. Chemical composition of buffalo gourd, a potential food source. J. Food Sci. 41: 465–466.
12. Bolley, D. S., R. H. McCormack, and L. C. Curtis. 1950. The utilization of the seeds of the wild perennial gourds. J. Amer. Oil Chem. Soc. 27: 571–574.
13. Butler, L. E., D. D. Christianson, J. C. Scheerens, and J. W. Berry. 1986. Buffalo gourd root starch, Part IV. Properties of hydroxypropyl derivatives. Starch/Starke 38: 156–159.
14. Cossack, Z., L. B. Waymack, C. W. Weber, and J. C. Scheerens. 1979. Nutritional availability of buffalo gourd (*Cucurbita foetidissima*) residue for sheep. Amer. Soc. Anim. Sci. Proc., W. S. 30: 156–158.
15. Curtis, L. C. 1946. The possibilities of using species of perennial cucurbits as source of vegetable fats and protein. Chemurgic Dig. 5: 221–224.

16. Curtis, L. C., and N. Rebeiz. 1974. The domestication of a wild, perennial, xerophytic gourd: *Cucurbita foetidissima*, the buffalo gourd. Report of the Arid Lands Agricultural Development Program, Ford Foundation, New York.

17. Daghir, N. J., H. K. Mahmoud, and A. El-Zein. 1980. Buffalo gourd (*Cucurbita foetidissima*) meal: nutritive value and detoxification. Nut. Rep. Int. 21: 837–848.

18. DeVeaux, J., and E. B. Schultz, Jr. 1985. Development of buffalo gourd as a semiarid land starch and oil crop. Econ. Bot. 39: 454–472.

19. Dittmer, H. J., and B. P. Talley. 1964. Gross morphology of tap roots of desert cucurbits. Bot. Gaz. 125: 121–126.

20. Dossey, B. F., W. P. Bemis, and J. C. Scheerens. 1981. Genetic control of gynoecy in the buffalo gourd. J. Hered. 72: 355–356.

21. Dreher, M. L., J. C. Scheerens, C. W. Weber, and J. W. Berry. 1981. Nutritional evaluation of buffalo gourd root starch. Nut. Rep. Int. 23: 1–8.

22. Dreher, M. L., A. M. Tinsley, J. C. Scheerens, and J. W. Berry. 1983. Buffalo gourd root starch, Part II: Rheologic behavior, freeze-thaw stability, and suitability for use in food products. Starch/Starke 35: 157–162.

23. Gathman, A. C. 1983. Inheritance of fatty acid composition of seed oil in the buffalo gourd, *Cucurbita foetidissima* HBK. Ph.D. dissertation, Univ. Arizona, Tucson.

24. Gathman, A. C., and W. P. Bemis. 1983. Heritability of fatty acid composition of buffalo gourd seed oil. J. Hered. 74: 199–200.

25. Gathman, A. C., D. Grygiel, A. Kartman, N. McDonald, D. Marburger, and J. W. Berry. 1979. A solar energy fermentation process for fuel alcohol production from xerophytic starches. Report to the National Science Foundation. Washington, DC.

26. Heard, B. L. 1981. Root-knot nematode on buffalo gourd. M.S. thesis, Univ. Arizona, Tucson.

27. Henderson, C. W., J. C. Scheerens, and J. W. Berry. 1986. Antinutritional factors in *Cucurbita* seed meals. J. Agric. Food Chem. 34: 434–436.

28. Hensarling, T. P., T. J. Jacks, and A. N. Booth. 1973. Cucurbit seeds. II. Nutritive value of storage protein isolated from *Cucurbita foetidissima* (buffalo gourd). J. Agric. Food Chem. 21: 986–988.

29. Hogan, L., and W. P. Bemis. 1983. Buffalo gourd and jojoba; potential new crops for arid lands. Adv. Agron. 36: 317–349.

30. Howe, W. L., and A. M. Rhodes. 1976. Phytophagous insect associations with *Cucurbita* in Illinois. Environ. Entomol. 5: 747–751.

31. Khoury, N. N., S. Dagher, and W. Sawaya. 1982. Chemical and physical characteristics, fatty acid composition and toxicity of buffalo gourd oil. J. Food Tech. 17: 19–26.

32. Metcalf, R. L., A. M Rhodes, R. A. Metcalf, J. Ferguson, E. R. Metcalf, and P.-Y. Lu. 1982. Cucurbitacin contents and diabroticite feeding on *Cucurbita* spp. Environ. Entomol. 11: 931–937.

33. Nabhan, G., and J. Thompson. 1985. Wild *Cucurbita* in Arid America: Ethnic Uses, Chemistry and Geography; An Annotated Bibliography. Native Seeds/SEARCH, Tucson.

34. Nelson, J. M., J. C. Scheerens, J. W. Berry, and W. P. Bemis. 1983. Effect of

plant density and planting date on root and starch production of buffalo gourd grown as an annual. J. Amer. Soc. Hort. Sci. 108: 198–201.
35. Nelson, J. M., J. C. Scheerens, T. L. McGriff, and A. C. Gathman. 1988. Irrigation and plant spacing effects on seed production of buffalo and coyote gourds. Agron. J. 80: 60–65.
36. Paur, S. 1952. Four Native New Mexico Plants of Promise as Oilseed Crops. New Mexico Agric. Exp. Sta., Press Bulletin 1064, Las Cruces.
37. Provvidenti, R., R. W. Robinson, and H. M. Munger. 1978. Resistance in feral species to six viruses infecting *Cucurbita*. Pl. Dis. Reporter 62: 326–329.
38. Rosemeyer, M. E. 1981. Isolation and characterization of two viruses from *Cucurbita foetidissima* HBK, buffalo gourd. M.S. thesis, Univ. Arizona, Tucson.
39. Rosemeyer, M. E., J. K. Brown, and M. R. Nelson. 1986. Five viruses isolated from field-grown buffalo gourd (*Cucurbita foetidissima*), a potential crop for semiarid lands. Pl. Dis. Reporter 70: 405–409.
40. Rosemeyer, M. E., B. H. Wells, and A. Zaid. 1982. Diseases of the buffalo gourd, *Cucurbita foetidissima*, in Arizona. Phytopathology 72: 955. (Abstract).
41. Scheerens, J. C., W. P. Bemis, M. L. Dreher, and J. W. Berry. 1978. Phenotypic variation in fruit and seed characteristics of buffalo gourd. J. Amer. Oil Chem. Soc. 55: 523–525.
42. Scheerens, J. C., and J. W. Berry. 1986. Buffalo gourd: composition and functionality of potential food ingredients. Cereal Foods World 31: 183–192.
43. Scheerens, J. C., M. L. Dreher, W. P. Bemis, and J. W. Berry. Buffalo gourd root starch, Part III. Effects of plant age and spacing on the physicochemical and rheological properties. Starch/Starke 35: 193–198.
44. Scheerens, J. C., M. J. Kopplin, I. R. Abbas, J. M. Nelson, A. C. Gathman, and J. W. Berry. 1987. Feasibility of enzymatic hydrolysis and alcoholic fermentation of starch contained in buffalo gourd (*Cucurbita foetidissima*) roots. Biotechnol. Bioeng. 29: 436–444.
45. Scheerens, J. C., H. M. Scheerens, A. E. Ralowicz, T. L. McGriff, M. J. Kopplin, and A. C. Gathman. 1988. Staminate floral induction on gynoecious buffalo gourd following application of aminoethoxyvinylglycine. HortScience 23: 138–140.
46. Scheerens, J. C., Y. M. R. Yousef, A. E. Ralowicz, A. C. Gathman, and H. M. Scheerens. 1987. Floral Development, flowering patterns and growth rate of monoecious and gynoecious buffalo gourd. J. Amer. Soc. Hort. Sci. 112: 574–578.
47. Shahani, H. S., F. G. Dollear, K. S. Markley, and J. R. Quinby. 1951. The buffalo gourd, a potential oilseed crop of the southwestern drylands. J. Amer. Oil Chem. Soc. 28: 90–95.
48. Thompson, S. A., C. W. Weber, J. W. Berry, and W. P. Bemis. 1978. Protein quality of buffalo gourd seed and seed fractions. Nut. Rep. Int. 18: 515–519.
49. Tinsley, A. M., J. W. Berry, and C. W. Weber. 1982. Development of acceptable African foods from the buffalo gourd (*Cucurbita foetidissima*). Paper presented to the 42nd Annual Institute of Food Technologists Meeting. Las Vegas.
50. Tu, M., W. D. Eustace, and C. W. Deyoe. 1978. Nutritive value of buffalo gourd seed protein. Cereal Chem. 55: 766–772.

51. Vasconcellos, J. A., W. P. Bemis, J. W. Berry, and C. W. Weber. 1981. The buffalo gourd, *Cucurbita foetidissima* HBK, as a source of edible oil. Amer. Oil. Chem. Soc. Monograph 9.
52. Vasconcellos, J. A., and J. W. Berry. 1982. Characteristics of laboratory processed *Cucurbita foetidissima* seed oil. J. Amer. Oil Chem. Soc. 59: 79–84.

28 | *Acanthosicyos horridus*, a Multipurpose Plant of the Namib Desert in Southwestern Africa

B. H. Sandelowsky

ABSTRACT. *Acanthosicyos horridus*, known as !nara, is a sand-binding shrub endemic to the Namib Desert of southwestern Africa. The flesh of the fruit and the seeds are edible. Its use among the Topnaar Hottentots is reviewed. Archaeological studies indicate that !nara has been a food for humans during the past 8000 years. It is suggested that !nara has potential for domestication in arid climates similar to those that favor the buffalo gourd, *Cucurbita foetidissima*.

The Namib Desert stretches along the Atlantic Coast from southern Angola to near the Cape of Good Hope in a band approximately 100 km across. Eastward it reaches the edge of the central highlands of Namibia. Gravelly plains with Inselberge alternate with dune fields, which contain the world's highest sand dunes. River valleys, usually carrying surface water for only a few days a year, resemble longitudinal oases and commonly represent striking boundaries between the plains and the dune fields. While surface flow has retreated from the coast, fresh subsurface water continues to maintain insular patches of vegetation and associated animal life in this vast desert region. *Acanthosicyos horridus* Welw. ex Hook. f., a cucurbit endemic to the Namib, is a common constituent of these patches and may also be found on dunes flanking river beds.

Acanthosicyos horridus, commonly known as !nara (pronounced with an initial dental click) among desert peoples, and narra, butterpips, or botterrpitte among those of European origin, is a food resource for humans and animals and has been so since antiquity. In this chapter consideration is given to some of its botanical, archaeological, and ethnobotanical aspects, and attention is drawn to its potential as a crop for arid lands.

349

Botanical Aspects

Acanthosicyos horridus was discovered in the mid-1800's by Friedrich Welwitsch, who is better known for another curious plant of the Namib Desert, *Welwitschia mirabilis* Hook. f. (3, 6). *Acanthosicyos horridus* is one of two species of the genus, the other being *A. naudinianus* (Sond.) C. Jeffr., which is widely distributed in tropical and southern Africa. !Nara is a perennial, thorny, dioecious shrub (Figure 1). It appears leafless, the leaves being reduced to stiff scales. Photosynthesis takes place in its stems and sturdy, paired stipular spines, which are 2–3 cm long. The ovary and fruit are beset with spine-tipped protuberances, and the fruits, about 20 cm in diameter, are macelike in appearance. Their inner pulp is yellow and the seeds, which are swollen, are about 15 mm long. The cucurbitacins of !nara have been considered by Rehm et al. (11) and Hylands and Magd (9). Seed proteins have been studied by Joubert and Cooper (10), and an analysis of amino acid content of seed proteins was presented by Schwartz and Burke

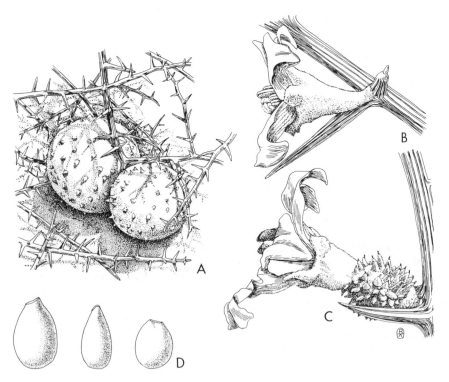

Figure 1. Acanthosicyos horridus. A. Fruit and stems. B. Male flower. ×1.2. C. Female flower. ×1.2. D. Seeds. ×1.2.

Figure 2. Acanthosicyos horridus. Sand excavated to show the size of the taproot.

(12). Arnold et al. (1) summarized the nutritional composition of the flesh and seeds of the fruit.

A special feature of !nara is its adaptation to dune habitats. Functionally, the dense intergrowth of the stems of plants serve to stabilize dunes. Considering the height of dunes topped by some plants and the assumption that ground water is being tapped, the length of some roots must be quite remarkable. An excavation of !nara in a dune population is suggestive of the size of the taproot (Figure 2). In another instance the Kuiseb River in flash flood largely excavated the taproot of a stunted plant at the edge of a population of !nara plants. Here the root crept along a crevice in rock for several meters, before disappearing into the earth.

Herre (8) found that the seeds of !nara germinated fairly easily in a hothouse, but the plants languished and eventually perished, presumably because an appropriate soil/water relationship could not be provided. In the Namib Desert rainfall is sparse, episodic, and local. It averages about 20 mm a year, but is complemented by fog, which occurs at least 100 days annually. Established plants of !nara survive rainless years well, but local inhabitants report that germination only follows rain. Despite these observations, most aspects of !nara biology remain unknown.

Uses of !Nara

Current human interest in !nara centers largely on its fruits, which are used primarily as food. Both the flesh and the seeds are eaten. The Topnaar Hottentots, who are also known as !Naranin (people of the !Nara), are identified with the plants. For those living along the Kuiseb River, which is to the southeast of Walvis Bay and Swakopmund, it is a staple food. Dentlinger (4) described the harvest and use of !nara by the Topnaar, and also has documented the trade of the seeds with bakeries in Cape Town and with dealers on the Grand Parade, a market in that city.

When the melons are ripe, Topnaar working in Walvis Bay and Swakopmund will leave their jobs for a week or two and move to !nara patches that have been owned traditionally for generations. The melons are harvested and processed, and the flesh and seeds stored or distributed to the villages on the banks of the Kuiseb River.

The flesh of ripe melons may be eaten raw. In one of the earliest accounts of !nara, Dinter (5) mentioned an extremely unpleasant, burning sensation in the mouth sometime after eating the flesh. When told this, local people found it amusing, for they knew that immature fruits are bitter, whereas ripe fruits are not. The bitterness is due to the presence of cucurbitacins (7). My personal experience was of the extreme sweetness of the melon's yellow, fibrous flesh, and of a certain strong taste, which is difficult to counter, even with spices such as cloves and cinnamon.

Besides eating the melon raw, the boiled flesh may be mixed with wheat flour, used as an admixture to corn porridge, or dried. A kind of fruit roll, '=goa-garibeb (= lateral click), made from the cooked flesh of the fruit also keeps well, is easy to carry, and is tasty (4). Leftovers of boiled fruit flesh are served to dogs and chickens in small dishlike hollows in the sand. Donkeys, goats, and cattle seem to like the skins of the melons that are fed to them after the flesh is removed.

The seeds are the most important and common food. A nutritional analysis of the seeds reported by Arnold et al. (1), showed them to be particularly rich in protein (30.7 g/100 g) and energetically rich (2709 kJ/100 g), as well as being significant sources of several minerals. Indigenous people commonly carry and eat the pips, and they may be used as snack foods. As noted above, the seeds are traded as far away as Cape Town. During the first part of this century, bakers in Namibia used them as an almond substitute, until a story was circulated that said that the Topnaar process the seeds for sale by passing them through the alimentary canal. This misunderstanding may have arisen because the Topnaar do process the kernels for their own use in a different way from those they sell. Seeds that are used locally are cracked inside the mouth. The more of the sweet flesh that adheres to the seed coat the better. Dried, however, this flesh does appear dark and dirty. Consequently, the seeds for sale are cleaned by boiling.

The Himba Herero living in the Kaokoveld are also familiar with the plant, although they do not consider it a staple, but rather an emergency food to be used only during time of drought. One Himba informant was adamant about the curative value of the root. The medicinal value of !nara was noted by Versfeld and Britten (13).

In addition to the value of its fruit, !nara also is a forage. Both wild and domestic herbivores find the new growth attractive. Older growth is not heavily browsed, which is fortunate, for the role of !nara in dune stablization would otherwise be greatly diminished.

Antiquity of Use

My interest in !nara originally developed in conjunction with archaeological studies being conducted at Mirabib Hill Shelter in the central Namib Desert (12). Here !nara seed coats were uncovered in most layers of the excavation. In fact, seed coats and even kernels are a common component of archaeological digs in the Namib and pro-Namib. Radiocarbon dating of the layers at Mirabib Hill Shelter placed the oldest seed coats of !nara at 8000 years.

To determine whether or not the seeds had been used as human food, the markings on archaeologically recovered seed coats were compared with those derived from extant animal and human feedings. In the wild whole ripe melons, including seeds, may be eaten. In other instances, birds may remove the seeds, or insect larvae may consume them. Small mammals, such as gerbils, rats, or mice, may also eat the seeds. In doing so they leave gnaw marks on the seed coats, and often the seed coats are nibbled transversely to the seeds margins.

Seed coats collected from the yards of Topnaar houses illustrate the effects of human feeding. Most often these were half seed coats that had been split from the other half along the margins. Occasionally there would be a jagged edge where the seed coat was broken or ripped open in order to remove the kernel. These patterns of breakage closely resembled those observed in the seed coats from the archaeological site, and it was therefore concluded that they were indeed the remains of meals taken during the past 8000 years.

Potential for Utilization and Domestication

A century of modern farming in the deserts of Namibia has proved to be relatively unsuccessful. In part, this could be due to the fact that the crops on which most of this farming relies are derived from more temperate regions. Until recently, indigenous species have largely been ignored. Interest in utilizing these resources, especially in low input, sustainable agricultural systems is now increasing (14).

The potential for arid zone plants has recently been considered from a number of perspectives. Arnold et al. (1) suggested that the highly specific habitat requirements of *A. horridus* make it an unlikely candidate for domestication, and suggest that it be encouraged as a wild adjunct to agricultural species. On the other hand, the high protein content and energetic value of the seeds, general usefulness of the fruits, including the presence of interesting chemical compounds, and even the uniqueness of its adaptation and potential as a sand-binding plant indicate that domestication should not be discounted. In fact, parallels exist with the buffalo gourd, *Cucurbita foetidissima* HBK, a species of xeric North America, which has been the subject of intensive investigation by Bemis and his coworkers (2, Gathman and Bemis, this volume).

Literature Cited

1. Arnold, T. H., M. J. Wells, and A. S. Wehmeyer. 1985. Khosian food plants: Taxa with potential for future economic exploitation. *In* G. E. Wilkins, J. R. Goodin, and D. V. Field, eds., Plants for Arid Lands. George Allen & Unwin, London.
2. Bemis, W. P., J. W. Berry, and C. W. Weber. 1979. The buffalo gourd: A potential arid land crop. *In* G. A. Richie, ed., New Agricultural Crops. Westview Press, Boulder, CO.
3. Bornman, C. H. 1978. *Welwitschia*: Paradox of a Parched Paradise. C. Struik, Cape Town.
4. Dentlinger, U. 1977. The !nara plant in the Topnaar Hottentot culture of Namibia. Munger Africana Libr. Notes 7(38): 1–39.
5. Dinter, K. 1912. Vegetabilische Veldkost in Südwestafrika. Selbstverlag, Okahandja.
6. Dolezal, H. 1959. Friedrich Welwitsch Leben und Werk. Portugaliae Acta Biol., Ser. B, Sist. 6: 257–332.
7. Enslin, P. R., F. J. Joubert, and S. Rehm. 1956. Bitter principles in the Cucurbitaceae. II. Elaterase, an active enzyme for the hydrolysis of the bitter principle glycosides. J. Sci. Food Agric. 7: 646–655.
8. Herre, H. 1974. Die Narapflanze. Namib und Meer. Ges. Wiss. Entwicklung Mus. 5/6: 27–31.
9. Hylands, P. J., and M. S. Magd. 1986. Cucurbitacins from *Acanthosicyos horridus*. Phytochemistry 25: 1681–1684.
10. Joubert, F. J., and D. C. Cooper. 1954. Chemistry of naras seed (*Acanthosicyos horridus* Hook.). Part 1—A physico-chemical study of the protein from naras seed. J. S. Afr. Chem. Inst. 7: 99.
11. Rehm, S., P. R. Enslin, A. D. J. Meeuse, and J. H. Wessels. 1957. Bitter principles of the Cucurbitaceae. VII. The distribution of bitter principles in this family. J. Sci. Food Agric. 8: 679–686.

12. Sandelowsky, B. H. 1977. Mirabib—An archaeological study in the Namib. Madoqua 10(4): 221–283.
13. Schwartz, H. M., and R. P. Burke. 1958. The chemistry of naras seed (*Acanthosicyos horridus* Hook.). III. The amino acid composition of the protein. J. Sci. Food Agric. 9: 159–162.
14. Versfeld, W., and G. F Britten. 1915. Notes on the chemistry of !naras. S. African J. Sci. 12: 232–239.
15. Wickens, G. E., J. R. Goodin, and D. B. Field. 1985. Plants for Arid Lands. George Allen & Unwin, London.

29 | Cucurbit Seeds: Cytological, Physiochemical, and Nutritional Characterizations

T. J. Jacks

ABSTRACT. Seed yields among and within cucurbit species vary widely but are comparable with those of commercial oilseeds. The cytological features of cucurbit seeds are similar in all respects to those of typical oilseeds. The bulk of the cytoplasm consists of two organelles: spherosomes, the sites of lipid storage; and protein bodies (aleurone grains), the sites of storage protein. Embedded within the protein bodies are globoids that contain metallic salts of phytic acid and crystalloids that contain storage globulin. Starch grains are absent. Decorticated seeds contain by weight 50% oil and 35% protein. The oil is unsaturated and generally edible, although certain species contain conjugated trienoic fatty acids that preclude edibility but are valuable industrially as drying oils. Albumins and globulins constitute the protein portion. Albumins are of low molecular weight and are composed of many water-soluble species. Globulins (cucurbitins) are salt-soluble oligomers consisting of sub-units that have molecular masses of about 54,000 daltons. Like other oilseed proteins, cucurbit proteins are rich in arginine, aspartic acid, and glutamic acid, and are deficient in lysine and amino acids containing sulfur. The nutritional value of the protein is similar in magnitude to other oilseeds. Cellulosic cell wall materials and phytic acid constitute the carbohydrates of cucurbit seeds; starch is absent and free sugars are sparse.

In 1972 the first general review of research concerning the oil and protein of cucurbit seeds was published (14). Since then, much research has been conducted on the characteristics and properties of cucurbit seed oil and protein, particularly of wild, xerophytic cucurbits such as the buffalo gourd. Indeed, even review articles summarizing this research have appeared. These review articles either featured cucurbits (7, 12, 19) or included cucurbits with other underexploited arid plants (10, 11). This chapter therefore at-

356

tempts to summarize the most recent and pertinent information on cucurbit seeds.

Seed Characteristics

Yield

Since the carbohydrate-rich flesh of the fruits is the common article of commerce, seed yields of cucurbits seldom have been investigated. Wide variation in seed yield has been noted, even within a species, however, wild cucurbits growing in desert areas produced estimated yields of 500 to 3000 pounds of seeds per acre (5, 6, 20). These yields are comparable with yields of commercial oilseeds.

Cytological Structure

Mature cucurbit seeds lack endosperm. The embryo fills the seed almost completely. The cotyledons provide food reserves for germination and are the predominant tissue of the seed. Epidermal cells are subtended by palisade cells and an abundance of parenchyma cells that contain the food reserves. Vascular tissues are also present.

The cytological features of the cotyledons are shown in Figure 1, which is a composite electron micrograph portraying typical parenchyma cells that compose the cotyledonary storage tissue of several cucurbits (9). The bulk of the cytoplasm consists of two organelles: spherosomes (lipid bodies) and protein bodies (aleurone grains). Starch grains are absent. Spherosomes are about 1 μm in diameter, are surrounded by half-unit membranes (22), and contain the reserve oil (23). Protein bodies are 5–20 μm in diameter, are enclosed in unit membranes, contain storage protein (15, 18), and harbor two inclusions: crystalloids and globoids. Crystalloids are crystalline deposits of storage protein (cucurbitin) and are generally abundant and large in cucurbits (15, 18). Globoids are composed mostly of metallic salts of phytic acid (16, 21). In addition to phosphorus due to phytic acid, potassium, magnesium, and sometimes calcium are located in the globoids (15). These cations constitute the metallic salts of phytin in the globoid; they are absent in the proteinaceous matrix of the protein body in which the globoids are embedded (15, 16).

Figure 2 shows scanning electron micrographs (SEM) of seeds of *Cucurbita digitata* A. Gray. The low magnification used in Figure 2A allows the entire sectioned seed to be seen. The sample was sliced with a razor blade, treated with hexane, and sputter-coated (2). The outer boundary of the

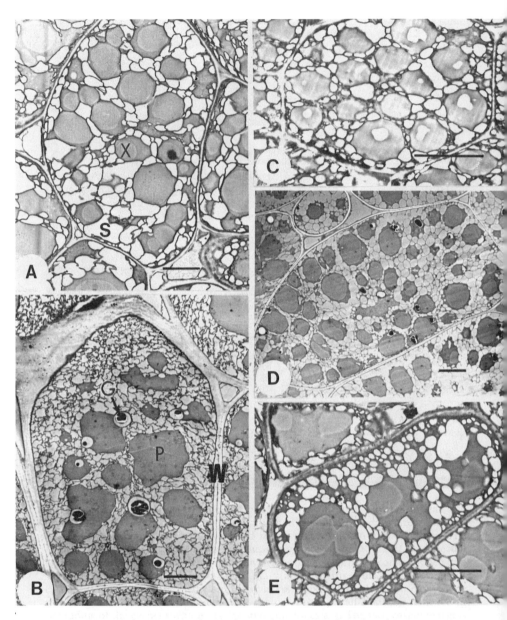

Figure 1. Cotyledon tissues of cucurbit seeds. A. *Cucurbita foetidissima* HBK. B. *Cucurbita pepo.* C. *Cucurbita palmata* S. Wats. D. *Cucurbita digitata* A. Gray. E. *Apodanthera undulata* A. Gray. W, cell wall; P, protein body; S, spherosome; G, globoid; X, crystalloid. Bars represent 5 μm. From Hensarling et al. (9), reprinted with permission of the American Oil Chemists' Society.

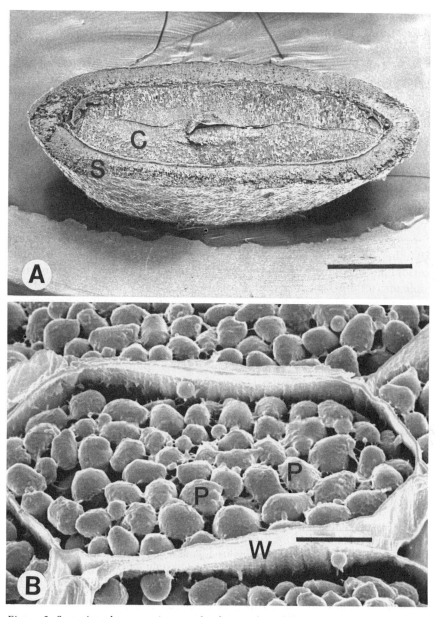

Figure 2. Scanning electron micrograph of *Cucurbita digitata*. A. Cross-section of a dormant seed showing the outer seed coat (S), two cotyledons (C), and the cleft between the cotyledons that contains the first pair of embryonic leaves. Bar represents 1 mm. B. Mesophyll cell of the cotyledon showing the cell wall (W) and the spherical protein bodies (P) inside of the cell. Bar represents 10 μm.

section is the seed coat; the remainder is composed of two cotyledons separated by the "true" leaves of the embryo.

A view of the sample in Figure 2A is shown at a higher magnification in Figure 2B. This SEM shows the intracellular structure of a cotyledonary cell. The spherical objects within the cell wall are the protein bodies shown in cross-section in Figure 1. The spherosomal network shown in Figure 1 was destroyed by the preparative procedures used to obtain this SEM, but the intracellular abundance of the protein bodies is plainly visible.

Composition

The amount of seed coat per seed varies considerably (14), from 18% in *Cucurbita pepo* L. to 60% in *Lagenaria siceraria* (Mol.) Standl. Indeed, some lines of *C. pepo* have little or no coat. The amounts of oil and protein in decorticated seeds are somewhat less varied. Decorticated seeds contain, by weight, 49.5 ± 2.3% oil and 35.0 ± 2.4% protein (12, 14).

Oil. Unsaturated fatty acids preponderate in cucurbit oils, and in some seeds conjugated triene is the form of one-third of this unsaturation (12, 14). Cucurbit seeds are important sources of edible oil, and digestibility in rats supports this (12). Conjugated trienes may be present. In that case, the oil is valuable as a drying oil, and studies of conjugated, triene-rich cucurbit oils illustrate their value as protective coatings (5). It is also of interest that the sterol composition of 12 genera of cucurbits has recently been described (1).

Protein. Decorticated cucurbit seeds contain by weight about 35% protein. Traditionally, seed proteins are classified as globulins and albumins according to their solubility in certain solvents. Globulins are insoluble in water but soluble in concentrated salt solutions, whereas albumins are soluble in water or dilute salt solutions. Both solubility properties depend on the pH values of the solvents. Biochemically, oilseed globulins are generally considered storage proteins, while albumins are believed to be metabolic (catalytic) proteins.

Albumins have not been as thoroughly investigated as globulins. A recent, thorough study of *Cucumis sativus* L. albumin (24) showed that about one-fourth of its protein is water soluble and has a low molecular weight. From its contents of cysteine and nitrogen-rich amino acids, a large portion of the albumins was thought to act biochemically as storage protein for sulfur in addition to nitrogen.

Cucurbit seed globulins, which account for about 70–90% of the total protein content, contain about 18% nitrogen, are soluble in 10% salt solutions from which they readily crystallize upon dilution, and are also soluble in both acidic and basic solutions of low ionic strength. Since the classical isolation of crystalline cucurbitin in 1892 (17), many chemical and physiochemical studies of cucurbitin have been conducted. Varied results from determinations of the molecular weight of the native oligomeric globulins and the number and molecular weights of its subunits have been obtained. In addition to slight natural variations among species, the variability appears due to preparative procedures as well as conditions and modes of analyses. Earlier studies of the molecular weight of the oligomeric globulin, the number of subunits, and their electrophoretic heterogeneity have been reviewed (12, 14).

A recent, thorough study (4) showed that cucurbitins from pumpkin and other cucurbits had values of only 12 S (90% of the protein) and 18 S (sedimentation constants in Svedberg units). By electrophoretic analyses and chromatography combined with ultracentrifugation, native cucurbitin was found to be a hexamer of about 325,000 daltons (12 S form), composed of six monomeric subunits of about 54,000 daltons each, forming a dimer of 630,000 daltons (18 S form). The large and small polypeptide subunits of the monomer ranged from 33,000 to 36,000 daltons and 22,000 to 25,000 daltons, respectively. Studies of the secondary structure of cucurbitin have shown its conformational modes consist of 5% alpha-helical, 32% pleated sheet, and 62% unordered structures (13). These values are similar in distribution to those of other oilseed globulins.

Analyses of the amino acid compositions of total cucurbit seed protein (meal) and purified globulin (12, 14) indicate that cucurbit seeds, as oilseeds in general, are rich in glutamic acid (and glutamine), arginine, and aspartic acid (and asparagine). The abundance of these nitrogen-rich amino acids accounts for the 18% of the nitrogen content of cucurbit protein.

From amino acid compositions, evaluations of the nutritional potentials of cucurbit meals and globulins were calculated according to FAO/WHO standards (8). The A:E ratios, which are the amounts of each essential amino acid relative to the total amount of essential amino acids, indicate that, like most other oilseeds, cucurbit seeds are deficient in lysine and sulfur-containing amino acids (12, 14).

A protein that is unduly rich in the ten essential amino acids would not provide sufficient nitrogen for other metabolic processes without obligatory catabolism of the essential amino acids. Thus, the proportion of the total nitrogen intake that essential amino acids form indicates how a given protein fulfills nutritional requirements for proteins. This proportion, the E/T ratio (8), is indicative of the amount of protein nitrogen supplied by essential amino acids per gram of nitrogen; it is 2.18 for cucurbit meal and 2.67 for cucurbit globulin. The value for meal is similar in magnitude to those for other seeds and the value for globulin is similar to those for fish and pork tenderloin (8).

Reviews (12, 14, 19) of nutritional evaluations of cucurbit seeds in feeding studies with mice and rats have shown the following. According to protein efficiency ratios and net protein retention values, meals and isolated proteins were inferior to animal sources of protein but were typical for oilseed products. Generally, the digestibility of cucurbit seed products was good. Supplementation of cucurbit meals with certain amino acids increased the nutritive values of the meals.

Carbohydrate. Cellulosic cell wall materials and phytic acid compose the bulk of the carbohydrate in cucurbit seeds. Free sugars are sparse and starch is absent.

Literature Cited

1. Akihisa, T., P . Ghosh, S. Thakur, F. U. Rosenstein, and T. Matsumoto. 1986. Sterol compositions of seeds and mature plants of family Cucurbitaceae. J. Amer. Oil Chem. Soc. 63: 653–658.
2. Arnott, H. J., and M. A. Webb. 1983. Scanning electron microscopy and histochemistry of oilseeds. *In* D. B. Bechtel, ed., New Frontiers in Food Microstructure. Amer. Assoc. Cereal Chem., St. Paul, MN.
3. Asenjo, C. F., and J. A. Goyco. 1951. Preliminary note on the nutritional value of pumpkin seeds (from Puerto Rico). Bol. Colegioquim. Puerto Rico 8: 14–16.
4. Blagrove, R. J., and G. G. Lilley. 1980. Characterization of cucurbitin from various species of the Cucurbitaceae. Eur. J. Biochem. 103: 577–584.
5. Bolley, D. S., R. H. McCormack, and L. C. Curtis. 1950. The utilization of the seeds of the wild perennial gourds. J. Amer. Oil Chem. Soc. 27: 571–574.
6. Curtis, L. C. 1946. The possibilities of using species of perennial cucurbits as source of vegetable fats and protein. Chemurgic Digest. 13: 221, 223–224.
7. DeVeaux, J. S., and E. B. Shultz. 1985. Development of buffalo gourd (*Cucurbita foetidissima*) as a semiarid land starch and oil crop. Econ. Bot. 39: 454–472.
8. FAO/WHO Expert Group. 1965. Protein requirements. FAO Meetings Report Series 37: 1–71.
9. Hensarling, T. P., T. J. Jacks, and L. Y. Yatsu. 1974. Cucurbit seeds: III. Ultrastructure of quiescent storage tissues. J. Amer. Oil Chem. Soc. 51: 474–475.
10. Hinman, C. W. 1984. New crops for arid lands. Science 225: 1445–1448.
11. Hinman, C. W. 1986. Potential new crops. Sci. Amer. 255: 32–37.
12. Jacks, T. J. 1986. Cucurbit seed protein and oil. *In* R. L. Ory, ed., Plant Proteins: Applications, Biological Effects, and Chemistry. Am. Chem. Soc. Symposium Series No. 312. Amer. Chem. Soc., Washington, DC.
13. Jacks, T. J., R. H. Barker, and O. E. Weigand, Jr. 1973. Conformations of oilseed storage proteins (globulins) determined by circular dichroism. Int. J. Peptide Protein Res. 5: 289–291.
14. Jacks, T. J, T. P. Hensarling, and L. Y. Yatsu. 1972. Cucurbit seeds. I. Characterizations and uses of oils and proteins. A review. Econ. Bot. 26: 135–141.

15. Lott, J. N. A. 1980. Protein bodies. *In* N. E. Tolbert, ed., The Biochemistry of Plants. A Comprehensive Treatise. Vol. 1: The Plant Cell. Academic Press, New York.

16. Lui, N. S. T., and A. M. Altschul. 1967. Isolation of globoids from cottonseed aleurone grains. Arch. Biochem. Biophys. 121: 678–684.

17. Osborne, T. B. 1892. Crystallized vegetable proteins. Amer. Chem. J. 14: 662–689.

18. Pernollet, J.-C. 1978. Protein bodies of seeds: ultrastructure, biochemistry, biosynthesis and degradation. Phytochemistry 17: 1473–1480.

19. Scheerens, J. C., and J. W. Berry. 1986. Buffalo gourd: composition and functionality of potential food ingredients. Cereal Foods World 31: 183–192.

20. Shahani, H. S., F. G. Dollear, K. S. Markley, and J. R. Quimby. 1951. The buffalo gourd, a potential oil seed crop of the southwestern drylands. J. Amer. Oil Chem. Soc. 28: 90–95.

21. Wada, T., and E. Maeda. 1980. Morphology of globoids in protein bodies and phosphorus content in various crop seeds. Jap. J. Crop Sci. 49: 51–57.

22. Yatsu, L. Y., and T. J. Jacks. 1972. Spherosome membranes. Half unit-membranes. Pl. Physiol. 49: 937–943.

23. Yatsu, L. Y., T. J. Jacks, and T. P. Hensarling. 1971. Isolation of spherosomes (oleosomes) from onion, cabbage, and cottonseed tissues. Pl. Physiol. 48: 675–682.

24. Youle, R. J., and A. H. C. Huang. 1981. Occurrence of low molecular weight and high systeine containing albumin storage proteins in oilseeds of diverse species. Amer. J. Bot. 68: 44–48.

PART V

Crop Improvement
and Protection

30 | Cell, Tissue, and Organ Culture Techniques for Genetic Improvement of Cucurbits

Todd C. Wehner, Rebecca M. Cade, and Robert D. Locy

ABSTRACT. Cucurbit tissue, cell, and organ culture is an area of research that is growing rapidly. Several important advances have been made within the past several years. Embryo culture has been used to rescue the progeny of interspecific crosses of species of *Cucumis*. Axillary buds have been induced to produce multiple shoots in cucumbers (*Cucumis sativus*) and watermelons (*Citrullus lanatus*), and both organogenesis and embryogenesis have been accomplished from several tissue sources of cucurbits. Cucumber and muskmelon (*Cucumis melo*) protoplasts have been isolated from leaf mesophyll cells and grown into callus. Protoplast-derived callus has been induced to form embryos in cucumbers and muskmelons. No successful protoplast fusions have yet been reported. In anther culture experiments with squash (*Cucurbita pepo*), only roots have been induced to form. However, cucumber anthers formed callus that later produced mature embryos. These breakthroughs may soon make genetic manipulations in culture a useful tool in cucurbit breeding programs.

Eight decades after the first report of an attempt to grow plant tissue in culture (27), successful techniques have been developed for many crop species, especially solanaceous species like tobacco, *Nicotiana tabacum* L. Progress with other crop species, however, including the cucurbits, has been much slower. Several reviewers have covered the application of in vitro techniques to crop plant improvement (25, 34, 44, 45, 54, 55). For that reason, we will review the status of tissue culture systems only as they relate to cucurbits.

Research activity has increased recently in the area of cell, tissue, and organ culture of cucurbits. Much of the basic work has been done in other crop species, so that often it is necessary only to adapt the available techniques to the cucurbit species of interest. We know of no reports in which in vitro culture of cucurbits has been used for selection. However, progress has been made in the areas of embryo culture, propagation, embryogenesis, organogenesis, protoplast fusion, and anther and pollen culture (Table 1).

Propagation

In vitro propagation via shoot tip culture has been used successfully in several cucurbit species. The advantage of in vitro methods over conventional vegetative propagation is that selected clones can be multiplied in a short period of time. Shoot tip culture has an advantage over regeneration from callus since it usually induces less genetic variability in the resulting clones. Thus, plants that are regenerated from callus tend to be less uniform than plants produced from axillary or apical buds.

Handley and Chambliss (28) cultured the axillary buds of cucumber, *Cucumis sativus* L., on a Murashige-Skoog (MS) medium (44) with 0.1 mg/l each of naphthalene acetic acid (NAA) and 6-furylamino purine (kinetin). They were able to produce both shoots and roots with little callus formation, and the plants were successfully transferred to soil. A fivefold increase in the number of plants obtained was estimated for bud culture over the traditional method of rooted cuttings.

Fortunato and Mancini (21) were able to increase that efficiency ten times by using enhanced axillary branching. With that technique many shoots can be produced from a single bud. Shoot proliferation in cucumber occurred on a medium containing 20 to 30 ppm of 2-dimethylallyl-amino-purine (2-iP), while elongation and rooting were achieved by transferring the shoots to a medium with 1.0 ppm indole-3-acetic acid (IAA), 1.0 ppm 2-iP, and 0.025 ppm gibberellic acid-3 (GA_3). More recently, Aziz and McCown (2) obtained single shoots from axillary buds of cucumber by culturing them on a medium containing benzyladenine (BAP).

Similarly, Barnes (4) stimulated axillary bud development in watermelon, *Citrullus lanatus* (Thunb.) Matsum. & Nakai, shoot tips with a high kinetin and low IAA medium. Those shoots likewise were induced to elongate and root on a medium containing IAA. Alternatively, axillary shoots from the first medium could be transferred back to the same medium, i.e., high kinetin and low IAA, where an average of 10.3 axillary shoots were obtained per shoot per cycle. A similar protocol has also been used (37) in the propagation of buffalo gourd, *Cucurbita foetidissima* HBK. The use of a high cytokinin/low auxin medium is fairly routine for shoot tip culture in most plant species.

Other components of the culture medium may also influence the multiplication of buds in culture. In particular, ammonium and nitrate sources of nitrogen in the commonly used MS medium may not be optimum for cucumber shoot tip propagation (38).

Table 1.   Summary of research in cell and tissue culture of cucurbits

Species	Research findings	Reference
Propagation		
Cucumis sativus	Asparagine and glutamine stimulated shoot tip growth in culture better than the Murashige-Skoog nitrogen sources (ammonium + nitrate)	38
C. sativus	Axillary buds cultured to produce one shoot/bud	2, 28
C. sativus	Enhanced axillary branching used to produce a mean of 10 shoots/bud	21
Citrullus lanatus	Axillary buds from shoot tips developed into shoots in culture (mean of 4.5/shoot tip) that could then be subcultured into plantlets and grown into whole plants	4, 5
Cucurbita foetidissima	Apical shoot tips and axillary buds produced 4–9 multiple shoots in 4 weeks	37
Embryo and ovule culture		
Cucumis metuliferus × *C. africanus*	Embryos from interspecific cross cultured into plantlets	13, 16
C. metuliferus × *C. melo*	Embryos from interspecific cross cultured but failed to form growing points	32
C. metuliferus × *C. melo*	Embryos from interspecific cross cultured into plantlets and grown into whole plants	46
C. metuliferus × *C. melo*	Culture of immature seeds used to obtain an extra generation of plants per year	24
Cucurbita pepo × *C. moschata*	Embryos from flat seeds cultured into mature plants using sterile media and soil; obtained 45 plants from 65 embryos	58
C. pepo	Unfertilized ovules cultured; some haploid plants obtained	9
Floral bud culture		
Cucumis sativus and *C. melo*	Floral buds induced to develop on a modified White's medium. Hormonal control of sex expression was studied	22, 49
Organogenesis		
Cucumis sativus var. *hardwickii*	Cotyledon explants made from cotyledons 2.5–4 mm long formed buds in culture	14
C. sativus	Callus and roots (but not shoots) were obtained by culturing stem, root, tendril, petiole, leaf, flower, and seedling hypocotyl and cotyledon tissue	1
C. sativus	Callus, but not shoots, obtained from field-grown leaf tissue	60
C. sativus	Apical stem explants formed shoots in culture	39
C. sativus	Hypocotyl and cotyledon explants formed callus in culture; subcultured callus formed shoots	53
C. sativus	Hypocotyl and cotyledon explants formed buds in culture	15

(continued)

Table 1. (*Continued*)

Species	Research findings	Reference
C. sativus	Best growth of hypocotyl explants in culture occurred on medium containing glucose or sucrose plus galactose; subcultures grew fastest on medium containing stachyose or raffinose	26
C. sativus	Cotyledon but not hypocotyl explants from 28 lines formed shoots in culture	8, 59
C. sativus	Cotyledons formed best callus with NAA + BAP; root and shoot formation occurred at low NAA + BAP concentration	47
C. sativus	Callus initiated from cucumber fruit tissue	3
C. melo	Seedling root callus developed chorophyll best on a modifed White's medium with 4% sucrose added, but grew best if only 2% sucrose was added	7
C. melo	Cotyledon callus induced to form well-developed shoots at good level of efficiency	42
C. anguria var. longipes	Calli from stem segments produced roots at higher frequencies than leaf-derived calli; leaf-derived calli produced more shoots than stem-derived calli	23
Cucurbita pepo and C. maxima	Mature cotyledons produced callus and shoots while immature embryo cotyledons produced only callus	35
Embryogenesis		
Cucumis sativus	Embryoids produced from anthers developed into whole plants	36
C. sativus	Embryos developed from leaf callus grown in dark for 6 weeks	40
C. sativus	Embryos obtained from hypocotyl callus	51
C. sativus	Embryos obtained from protoplast-derived callus	31, 48
C. melo	Somatic embryos obtained from callus on hypocotyl explants	6, 7, 42
C. melo	Embryos obtained from protoplast-derived callus	43
Cucurbita pepo	Fruit tissue explants formed callus in culture; subcultured callus formed embryonic outgrowths that formed plantlets when subcultured	52
C. pepo	Occasional hypocotyl and cotyledon explants formed callus only after 3 months in culture; callus formed embryos that were grown into whole plants	29, 30
Protoplast culture		
Cucumis sativus	Leaf mesophyll protoplasts isolated and cultured into callus, but not into plantlets	11, 12

Table 1.   (Continued)

Species	Research findings	Reference
C. sativus	Glycine found to have stabilizing effect on leaf protoplasts; embryos developed form protoplast derived calli on solid medium	48
C. sativus	Protoplasts isolated from cotyledons; embryos formed on a low auxin/high cytokinin solid medium	31, 56
C. melo	Protoplasts isolated from leaf mesophyll and grown into callus	41
C. melo	Protoplast derived calli form embryos	43
Anther and pollen culture		
Cucumis sativus	Anthers formed callus in culture; subcultured callus formed embryoids, then plantlets; the origin of embryoids (from either sporophytic or gametophytic tissues) was not checked	36
Cucurbita pepo	Stamens cultured at the uninucleate stage formed roots but not shoots	10

Embryo and Ovule Culture

Techniques for in vitro culture of embryos are potentially useful in two important areas of cucurbit breeding. Culture of immature embryos could be used to reduce the generation time of crops to permit more cycles of selection of generation advances. For example, an extra generation of cucumbers could be grown each year in the greenhouse using the technique (24). In addition, embryo culture might be used to rescue the normally inviable F_1 hybrids of interspecific crosses before the embryos abort. Wall (58) obtained 45 mature plants by culturing 65 embryos from flat seeds of the cross Cucurbita pepo L. × C. moschata (Duch. ex Lam.) Duch. ex Poir. He used a solid medium to grow the embryos for 7 to 12 days before transferring them to soil. Embryo culture permitted more plants of the interspecific hybrid to be obtained than if the usual process of using the rare, plump seeds were used (of 7000 seeds from interspecific crosses, only 4 were plump).

Viable embryos from the cross of Cucumis metuliferus Naud. × C. africanus L. f. were obtained by removing those longer than 3.0 mm from the fruits approximately 30 days after pollination and culturing them into plantlets (16). The reciprocal cross did not produce embryos larger than 3.0 mm, and those embryos could not be regenerated into plantlets. Additional research on embryo culture using the same cross identified the early heart stage as the optimum time for removing the embryos from ovules that might abort in advanced stages. Early heart stage embryos were 0.1 to 0.8 mm long and were excised 17 to 22 days after pollination. The embryos were cultured into plantlets in 32% of the cases, whereas only 15% of the embryos at the late globular stage produced plantlets (13).

Embryo culture has also been used in an effort to rescue embryos from the cross of *Cucumis metuliferus* with the muskmelon, *C. melo* L. Embryos in culture developed cotyledons but not growing points (32). In another case, success has been reported in making the above cross using embryo culture (46), but the plants were not checked to verify that they were interspecific hybrids.

Fassuliotis (20) attempted to transfer nematode resistance from *C. metuliferus* into *C. melo* using the above technique, but only one of the many embryos cultured was raised into a fruit-bearing plant. It was, however, considered to be apomictic, not an interspecific hybrid.

Most recently, unfertilized ovules of *Cucurbita pepo* have been cultured in vitro to obtain haploid plants (9). Ovules excised one to two days before flowering were best for regenerating plants. Under optimum conditions, four to seven plantlets were obtained per 100 cultured ovules. The ovules then grew into plants after transfer to a hormone-free medium. Although most of the resulting plants were diploid, a few were haploid-diploid chimeras, or polyploids. Eventually, it may be possible to adapt these techniques for rescue of interspecific hybrids before the embryos abort, but that has not yet been attempted.

Organogenesis

A basic requirement for nearly all uses of cell and tissue culture techniques in crop improvement is the regeneration of whole plants from cultured protoplasts, cells, or tissues. Regeneration from callus is accomplished in one of two ways: embryogenesis or organogenesis. Organogenesis involves the adventitious regeneration of any plant organ, including leaves, shoots, flowers, and roots.

Most of the work on cucumber organogenesis has been difficult to repeat in other laboratories, and the results have often been unpredictable. There is no clear-cut protocol available yet, although work is proceeding in several locations that may solve that problem (Zamir Punja, pers. comm., 1986). Maciejewska-Potapczykowa et al. (39) were the first to report organogenesis from callus produced by cucumber stem pieces in culture, but they did not describe the methods for obtaining shoots, since that was not the objective of their experiments.

Many studies have been carried out on cucumber callus. Alsop et al. (1) obtained callus only on several organ explants with various concentrations of NAA and BAP. Some budlike knobs were observed in callus grown at 0.1

mg/l of NAA and 0.1 mg/l of BAP. Aziz and McCown (2) also described budlike nodules on callus from cucumber internode pieces but could only induce roots to form. Callus growth of muskmelon seedling explants was best on modified White's medium with 0.1 to 1.0 mg/l of NAA and 2% sucrose, although chlorophyll development was best if 40% sucrose was added (17). In another study, callus subcultures grew faster when stachyose or raffinose was used as a carbon source instead of sucrose, galactose, or glucose (21). These results can be understood in light of the fact that the cucumber is one of the few species that translocates carbohydrates primarily as stachyose (50).

Others working with cucumbers have been able to produce buds, shoots, or both from callus formed on either hypocotyls or cotyledons (15, 47, 53, 59). Cotyledons presently appear to be the better explant for use in organogenesis experiments (8, 41, 59). Cotyledon callus, however, charac- teristically proliferates fibrous roots, whereas hypocotyl callus does not (47, 53). The proliferation may be caused by higher endogenous levels of auxin in the cotyledon tissue. Besides stems, cotyledons, and hypocotyls, other organs, including leaves (23, 60), fruit tissues (2), and embryos (14), have been used as explants for organogenesis in Cucurbitaceae.

Other factors are involved in organogenesis. Wehner and Locy (59) found a great deal of variation in the percentage of shoots produced among differ- ent cucumber genotypes. Some accessions from the United States Depart- ment of Agriculture germplasm collection had as many as 53% of cotyledon pieces forming shoots; while commercial cultivars in the United States never exceeded 13%. Other researchers have also observed differences among genotypes (8). The age of the fruit from which seed is obtained may also affect shoot production. Mature cotyledons that were explanted displayed a greater ability to form callus and differentiate into shoots and roots than cotyledons in formative stages (35).

Embryogenesis

Some of the earliest work in cucurbit tissue culture dealt with the forma- tion of somatic embryos in culture (see Figure 1). Schroeder (52) was the first to recognize this phenomenon in cucurbits. Zucchini (*Cucurbita pepo*) pericarp tissue was grown on a Nitsch and Nitsch medium for 18 months. After that, spherical cell masses of friable callus were discovered that even- tually produced torpedo-shaped embryoids. Finally, root hairs and leaf pri- mordia developed on some of the embryoids, but none of the embryos developed further.

In 1972 Jelaska (29) observed somatic embryos in callus cultures from pumpkin (*Cucurbita pepo*) hypocotyls and cotyledons. No correlation was

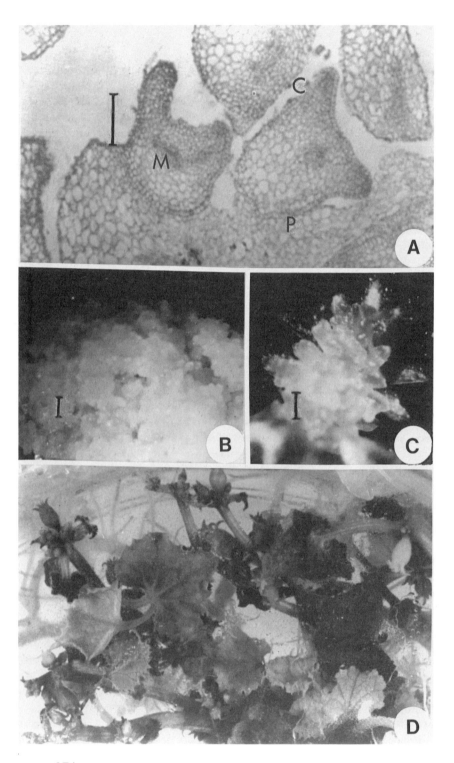

found between the hormones used and the number of embryos produced. This is not surprising, since the media used were undefined, i.e., nutrient concentrations were not known. She attributed the ability of cotyledons to form embryos to physiological factors within the plant. Later Jelaska (30) reported organogenesis from the same cultures.

In the 1980's additional research on embryogenesis was reported for cucurbits. Blackmon et al. (7) and Blackmon and Reynolds (6) obtained somatic embryos from callus hypocotyl explants of muskmelon after two to three months on a modified Nitsch and Nitsch medium. No additional development of the embryos into plants was reported.

Lazarte and Sasser (36) were able to induce embryoid formation from anthers of cucumber on a modified Nitsch and Nitsch medium having raffinose as the carbon source. Callus was initiated with 0.1 mg/l BAP and 0.1 mg/l GA. Pieces of callus were then placed on a Nitsch and Nitsch medium with or without kinetin. Dark green areas of the callus differentiated into embryoids, which developed into embryos when all growth regulators were removed.

Malepszy et al. (40) were able to induce embryogenesis from leaf explants of cucumber with high frequency. They used a primary medium high in auxin and cytokinin followed by a secondary medium with the same level of cytokinin and no auxin. The highest number of embryos (220 from one explant) was obtained on a primary medium containing 1.2 mg/l 2,4,5-trichlorophenoxy acetic acid and 0.8 mg/l BAP. The paper did not mention whether the embryos were grown into mature plants.

Moreno et al. (42) could obtain only embryos from muskmelon hypocotyls, while only cotyledons underwent organogenesis. The embryos did not undergo further development. Rajasekaran et al. (51) were also able to obtain a few embryos from hypocotyl explants by using a medium high in 2,4-D medium. Both Orczyk and Malepszy (48) and Moreno et al. (41, 42) were able to adapt their embryogenesis techniques to later work on protoplast cultures, which will be discussed in the next section. Jia et al. (31) also obtained embryos from protoplast-derived cucumber callus. Cotyledon tissue, however, was used as the explant source instead of leaf tissue. A low auxin level and a high cytokinin level were used instead of the high auxin/low cytokinin ratio that is generally used to obtain somatic embryos. Embryos developed fully when placed on a half-strength MS medium de-

---

Figure 1. A. Heart-stage cucumber embryos developing on edges of callus from cotyledon explants. C, cotyledon, P, protoderm, and M, meristem tissue. Bar is 0.1 mm long. B. Embryogenic cucumber callus on Murashige-Skoog medium containing 2 mg/1 2, 4-D and 0.5 mg/l kinetin. Bar is 1 mm long. C. Cluster of late-stage cucumber embryos. Bar is 1 mm long. D. Cucumber plantlets rooting on Murashige-Skoog medium with no growth regulators.

void of exogenous hormones. Similar work has been done at the ARCO Plant Cell Research Institute by Trulson and Shahin (56). Embryos were initiated from cotyledon and root explants on MS medium containing 1 mg/l 2,4-D, 1 mg/l NM, and 0.5 mg/l BAP. After three weeks on the above medium, the callus was transferred to the same medium but without 2,4-D for two weeks of maturation. Plants were regenerated with high frequency when the callus was transferred to the same medium with no growth regulators.

Selection in Culture

Several potential uses exist for selecting cucurbits in in vitro culture, especially if single-cell cultures could be used. Plants could be selected for resistance to herbicides (e.g., atrazine and trifluralin), to high or low temperatures, to high salt (e.g., NaCl) concentrations, or resistance to diseases, if the techniques were available. Additional research is needed, however, before selection of single cells in culture will be a useful technique for cucurbit breeding. Until those advances have been made, shoot tip cultures could be used. Twenty or more shoot tips could be grown and tested in a single Petri plate, and the procedure for growing resistant shoot tips into whole plants has been demonstrated for both cucumbers (28) and watermelons (5).

Protoplast Culture

Recently, there has been increased interest in the use of plant protoplasts for genetic manipulation in culture. Protoplast fusion, i. e., the fusion of two cells to form a hybrid daughter cell, is one attractive use for the system. Crosses that are difficult or impossible to make by sexual methods because of incompatibility barriers may be possible through protoplast fusion if the techniques are available for the crop of interest.

It has been possible to isolate and grow cucumber protoplasts from leaf mesophyll cells (11) and to grow cucumber mesophyll cells into callus (12). These techniques have also been adapted for muskmelon (43), but until recently, the only morphogenic response has been root production in cucumber (12).

Embryos recently were obtained from callus grown from leaf protoplasts of cucumbers and muskmelons (31, 41, 48). Orczyk and Malepszy (48) found that glycine had a stabilizing effect on leaf protoplasts. They were able to transfer mini-calli to solid medium containing 0.8 mg/l 2,4-D and 2-iP, and later obtain embryos. The embryos grew into plants when transferred to a hormone-free medium. The efficiency of embryo and plantlet production was not given.

Moreno et al. (42) achieved both organogenesis and embryogenesis from the protoplast-derived calli of muskmelon. They found that an agitated liquid medium was better than a solid medium for obtaining embryos, and, under the best conditions, 1308.3 ± 322.7 embryos per replicate were obtained. Few of the embryos developed further into plantlets. Jia et. al. (31) and Trulson and Shahin (56) used a different approach, as mentioned previously, in that cotyledons were used as a protoplast source instead of leaf mesophyll. Both papers reported successful regeneration of embryos, several of which developed into mature plants. The use of cotyledons instead of leaf mesophyll was probably the key to their success, since cotyledon tissue is younger and more active metabolically.

There are no reports of the use of protoplast fusion for the interspecific hybridization of cucurbits. However, interspecific hybridization of cucumber with other *Cucumis* species would be useful to plant breeders, and attempts have been made using more conventional methods (32, 33). Possible applications include incorporation of root-knot nematode resistance from other *Cucumis* species into cucumbers and muskmelons. *Cucumis ficifolius* A. Rich. and *C. metuliferus* have the highest levels of resistance of 14 *Cucumis* species tested by Fassuliotis (18, 19). Thus far, however, plants have not been recovered from the interspecific crosses that have been attempted.

Anther and Pollen Culture

Successful culture of anthers or pollen grains and subsequent regeneration into whole plants would provide a method for production of haploid cells for selection, dihaploids (i.e., chromosome-doubled haploids to be used as inbred lines) for breeding programs, and as haploid plants for genetic studies (57). Little work has been done on anther culture in cucurbits. However, stamen cultures of squash (*Cucurbita pepo*) formed roots but not shoots if cultured when the microsporocytes were at the uninucleate stages (10). Also, a recent report has demonstrated successful embryogenesis from cultured anthers of cucumber (36). Anthers from plants of 'SMR 58' were cultured on several different media to produce callus, then embryoids, and finally plantlets. It was not determined whether the embryoids originated from sporophytic or gametophytic tissue.

In studies of hormonal control of sex expression in cucumbers (22) and muskmelons (49), floral buds have been induced to develop on a modified White's medium. That procedure has been used to provide a source of "clean" pollen for use in tissue culture studies.

Conclusions

There is much room for improvement in the area of in vitro culture of cucurbits. Embryo culture and anther and pollen culture have not been developed as fully as needed. However, where research has been done, success has been fairly rapid. Embryogenesis and regeneration of whole plants from protoplasts are now becoming routine. With those tools available, selection in culture and protoplast fusion may not be far off.

Literature Cited

1. Alsop, W. R., W. W. Cure, G. F. Evans, and R. L. Mott. 1978. Preliminary report on in vitro propagation of cucumber. Cucurbit Genet. Coop. Rep. 1: 1–2.
2. Aziz, H. A., and B. H. McCown. 1985. Hormonal response curves of shoot and callus cultures of cucumber (Cucumis sativus L.). HortScience 20: 540 (Abstract).
3. Aziz, H. A., B. H. McCown, and R. L. Lower. 1986. Callus initiation from cucumber (Cucumis sativus L.) fruits. Cucurbit Genet. Coop. Rep. 9: 3.
4. Barnes, L. R. 1979. In vitro propagation of watermelon. Scientia Hort. 11: 223–227.
5. Barnes, L. R., F. D. Cochran, R. L. Mott, and W. R. Henderson. 1978. Potential uses of micropropagation for cucurbits. Cucurbit Genet. Coop. Rep. 1: 21–22.
6. Blackmon, W. J., and B. D. Reynolds. 1982. In vitro shoot regeneration of Hibiscus acetosella, muskmelon, watermelon, and winged bean. HortScience 17: 588–589.
7. Blackmon, W. J., B. D. Reynolds, and C. E. Postek. 1981. Production of somatic embryos from callused cantaloupe hypocotyl explants. HortScience 16: 451 (Abstract).
8. Bouabdallah, L., and M. Branchard. 1986. Regeneration of plants from callus cultures of Cucumis melo L. Z. Pflanzenzucht. 96: 82–85.
9. Chambonnet, D., and R. Dumas de Vaulx. 1985. Obtention of embryos and plants from in vitro culture of underfertilized ovules of Cucurbita pepo. Cucurbit Genet. Coop. Rep. 8: 66.
10. Chang, Y.-F., and J. E. Lazarte. 1981. Anther culture of squash (Cucurbita pepo L.). HortScience 16: 451 (Abstract).
11. Coutts, R. H. A., and K. R. Wood. 1975. The isolation and culture of cucumber mesophyll protoplasts. Pl. Sci. Lett. 4: 189–193.
12. Coutts, R. H. A., and K. R. Wood. 1977. Improved isolation and culture methods for cucumber mesophyll protoplasts. Pl. Sci. Lett. 9: 45–51.
13. Custers, J. B. M. 1981. Heart shape stage embryos of Cucumis species more successful in embryo culture than advanced stage embryos. Cucurbit Genet. Coop. Rep. 4: 48–49.
14. Custers, J. B. M., and J. H. W. Bergovoet. 1980. In vitro adventitious bud formation on seedling and embryo explants of Cucumis sativus L. Cucurbit Genet. Coop. Rep. 3: 2–4.
15. Custers, J. B. M., and L. C. Buijs. 1979. The effects of illumination, explant

position, and explant polarity on adventitious bud formation in vitro of seedling explants of *Cucumis sativus* L. cv. 'Hokus'. Cucurbit Genet. Coop. Rep. 2: 2–4.

16. Custers, J. B. M., and G. van Ee. 1980. Reciprocal crosses between *Cucumis africanus* L. and *C. metuliferus* Naud. II. Embryo development in vivo and in vitro. Cucurbit Genet. Coop. Rep. 3: 50–51.

17. Fadia, V. P., and A. R. Mehta. 1976. Tissue culture studies on cucurbits: chlorophyll development in *Cucumis* callus cultures. Phytomorphology 26: 170–175.

18. Fassuliotis, G. 1967. Species of *Cucumis* resistant to the root-knot nematode, *Meloidogyne incognita acrita*. Pl. Dis. Reporter 51: 720–723.

19. Fassuliotis, G. 1970. Resistance of *Cucumis* spp. to the root-knot nematode, *Meloidogyne incognita acrita*. J. Nematol. 2: 174–178.

20. Fassuliotis, G. 1977. Embryo culture of *Cucumis metuliferus* and the interspecific hybrid with *C. melo*. 4th Ann. Colloquium Plant Cell Tissue Culture. Ohio State Univ., Columbus.

21. Fortunato, I. M., and L. Mancini. 1985. Tecnica sulla propagazione vegetativa in vitro di centriolo (*Cucumis sativus* L.). Annali della Facotadi Agraria, Universita de Bari 32: 651–659.

22. Galun, E., Y. Jung, and A. Lang. 1963. Morphogenesis of floral buds of cucumber cultured in vitro. Dev. Biol. 6: 70–387.

23. Garcia-Sogo, M., I. Granelli, B. Garcia-Sogo, L. A. Roig, and V. Moreno. 1986. Plant regeneration from explant-derived calli of *Cucumis anguria* L. var. *longipes*. Cucurbit Genet. Coop. Rep. 9: 108–110.

24. George, B. F., and L. V. Crowder. 1970. The use of a modified embryo culture technique in cucumber breeding. HortScience 5: 329 (Abstract).

25. Green, C. E. 1977. Prospects for crop improvement in the field of cell culture. HortScience 12: 131–134.

26. Gross, K. C., D. M. Pharr, and R. D. Locy. 1981. Growth of callus initiated from cucumber hypocotyls on galactose and galactose-containing oligosaccharides. Pl. Sci. Lett. 20: 333–341.

27. Haberlandt, G. 1902. Culturversuche mit isolierten Pflanzenzellen. Sitz-Ber. Mat.-Nat. Kl. Kais. Akad. Wiss. Wien 111: 69–92.

28. Handley, L. W., and O. L. Chambliss. 1979. In vitro propagation of *Cucumis sativus* L. HortScience 14: 22–23.

29. Jelaska, S. 1972. Embryoid formation by fragments of cotyledons and hypocotyls in *Cucurbita pepo*. Planta 103: 278–280.

30. Jelaska, S. 1974. Embryogenesis and organogenesis in pumpkin explants. Physiol. Pl. 31: 257–261.

31. Jia, S., Y. Fu, and Y. Lin. 1986. Embryogenesis and plant regeneration from cotyledon protoplast culture of cucumber (*Cucumis sativus* L.). J. Pl. Physiol. 124: 393–398.

32. Kho, Y. O., J. Franken, and A. P. M. den Nijs. 1981. Species crosses under controlled temperature conditions. Cucurbit Genet. Coop. Rep. 4: 56.

33. Kho, Y. O., A. P. M. den Nijs, and J. Franken. 1980. Interspecific hybridization in *Cucumis* L. II. The crossability of species, and investigation of in vivo pollen tube growth and seed set. Euphytica 29: 661–672.

34. Kleinhofs, A., and R. Behki. 1977. Prospects for plant genome modification by nonconventional methods. Ann. Rev. Genet. 11: 79–101.

35. Lange, N. E., and J. A. Juvik. 1986. Age dependence for organogenesis of seed explants from four *Cucurbita* accessions. Cucurbit Genet. Coop. Rep. 9: 93.

36. Lazarte, J. E., and C. C. Sasser. 1982. Asexual embryogenesis and plantlet development in anther culture of *Cucumis sativus* L. HortScience 17: 88.

37. Lee, C. W., and J. C. Thomas. 1985. Tissue culture propagation of buffalo gourd. HortScience 20: 218–219.

38. Locy, R. D., and T. C. Wehner. 1982. Cucumber shoot tip growth on 9 nitrogen sources in in vitro culture. Cucurbit Genet. Coop. Rep. 4: 20–22.

39. Maciejewska-Potapczykowa, W., A. Rennert, and E. Milewska. 1972. Callus induction and growth of tissue cultures derived from cucumber plant organs of four different sex types. Acta Soc. Bot. Pol. 41: 329–339.

40. Malepszy, S., and A. Nadolsky-Orczyk. 1983. In vitro culture of *Cucumis sativus*. I. Regeneration of plantlets from callus formed by leaf explants. Z. Pflanzenphysiol. 111: 273–276.

41. Moreno, V., M. Garcia-Sogo, I. Granelli, B. Garcia-Sogo, and L. A. Roig. 1985. Plant regeneration from calli of melon (*Cucumis melo* L. cv. 'Amarillo Oro'). Pl. Cell Tiss. Organ Cult. 5: 139–146.

42. Moreno, V., L. Zubeldia, B. Garcia-Sogo, F. Nuez, and L. A. Roig. 1986. Somatic embryogenesis in protoplast-derived cells of *Cucumis melo* L. *In* W. Horn, C. J. Jensen, W. Odenbach, and O. J. Schieder, eds., Genetic Manipulation in Plant Breeding. Proc. Int. Symp., Berlin, 1985. W. de Gruyter, Berlin.

43. Moreno, V., L. Zubeldia, and L. A. Roig. 1984. A method for obtaining callus cultures from mesophyll protoplasts of melon (*Cucumis melo* L.). Pl. Sci. Lett. 34: 195–201.

44. Murashige, T. 1974. Plant propagation through tissue cultures. Ann. Rev. Pl. Physiol. 25: 136–166.

45. Narayanaswamy, S. 1977. Regeneration of plants from tissue cultures. *In* J. Reinert and Y. P. S. Bajaj, eds., Applied and Fundamental Aspects of Plant Cell, Tissue and Organ Culture. Springer-Verlag, New York.

46. Norton, J. D. 1980. Embryo culture of *Cucumis* species. Cucurbit Genet. Coop. Rep. 3: 34.

47. Novak, F. J., and M. Dolezelova. 1982. Hormone control of growth and differentiation in the in vitro cultured tissue of cucumber (*Cucumis sativus* L.). Biologia 37: 283–290.

48. Orczyk, W., and S. Malepszy. 1985. In vitro culture of *Cucumis sativus* L. V. Stabilizing effect of glycine on leaf protoplasts. Pl. Cell Rep. 4: 269–273.

49. Porath, D., and E. Galun. 1967. In vitro culture of hermaphrodite floral buds of *Cucumis melo* L.: microsporogenesis and ovary formation. Ann. Bot. 31: 283–290.

50. Pristupa, N. A. 1957. The transport form of carbohydrates in pumpkin plants. Fizcol. Rast. 6: 26–32.

51. Rajasekaran, K., M. G. Mullins, and Y. Nair. 1983. Flower formation in vitro by hypocotyl explants of cucumber (*Cucumis sativus* L.). Ann. Bot. 52: 417–420.

52. Schroeder, C. A. 1968. Adventive embryogenesis in fruit pericarp tissue in vitro. Bot. Gaz. 129: 374–376.

53. Sekioka, T. T., and J. S. Tanaka. 1981. Differentiation in callus cultures of cucumber (*Cucumis sativus* L.). HortScience 16: 451 (Abstract).

54. Skirvin, R. M. 1978. Natural and induced variation in tissue culture. Euphytica 27: 241–266.

55. Thomas, E., P. J. King, and I. Potrykus. 1979. Improvement of crop plants via single cells in vitro—an assessment. Z. Pflanzenzucht. 82: 1–30.

56. Trulson, A. J. and E. A. Shahin. 1986. In vitro plant regeneration in the genus *Cucumis*. Pl. Sci. 47: 35–43.

57. Vasil, I. K., and C. Nitsch. 1975. Experimental production of pollen haploids and their uses. Z. Pflanzenphysiol. 76: 191–212.

58. Wall, J. R. 1954. Interspecific hybrids of *Cucurbita* obtained by embryo culture. Proc. Amer. Soc. Hort. Sci. 63: 427–430.

59. Wehner, T. C., and R. D. Locy. 1981. In vitro adventitious shoot and root formation of cultivars and lines of *Cucumis sativus* L. HortScience 16: 759–760.

60. Wehner, T. C., and R. D. Locy. 1981. Tissue culture propagation of field-grown cucumber selections. Cucurbit Genet. Coop. Rep. 4: 20–22.

31 | Introducing Resistances into Cucumbers by Interspecific Hybridization

A. P. M. den Nijs and J. B. M. Custers

ABSTRACT. The need for resistances to several major diseases and pests in the greenhouse-grown cucumber, *Cucumis sativus*, prompted a thorough interspecific hybridization program in *Cucumis*. A broad collection of mostly African species was studied to gain insight into variability for resistance, crossability, and phylogeny of the genus. Serious classification problems were encountered with several variable species. Crossability barriers were evident either by the absence of pollen tubes or their slow growth through the style or by embryo abortion. Patterns of crossability largely agreed with the division of the genus into the subgenera *Cucumis* (with *C. sativus* as the sole well-documented species) and *Melo*. Meiotic analysis of F_1 hybrids between African wild species of subgenus *Melo* revealed a further division of this crossability group into two subgroups. Embryo rescue techniques yielded viable hybrids in certain combinations of species in subgenus *Melo*; however, crosses between species of that subgenus and the cucumber did not produce viable hybrids, although significant progress was made. Following pollinations by *C. zeyheri* and *C. metuliferus*, a selected accession of *C. sativus* var. *hardwickii* yielded embryos, but these could not be rescued. The same accession showed improved crossability with *C. melo* because hybrid embryos developed much further than usual in this combination. A useful variation with respect to prefertilization barriers in crosses with *C. sativus* was also found in *C. zeyheri*.

Despite great progress in breeding the cucumber, *Cucumis sativus* L., the available variation within the species now appears insufficient. Resistance to several important diseases and pests in the greenhouse culture of cucumbers in the Netherlands could not be detected in a large collection of cucumber cultivars and landraces from all over the world (17). Cucumber green mottle mosaic virus, for which strict phytosanitary measures are necessary, remains a serious threat to the industry. Root knot nematodes severely attack cucumber roots, necessitating yearly fumigation or steam sterilization. Although there is a trend toward more rockwool culture, the nematode problem is likely to remain important. Integrated control of the greenhouse whitefly is still fraught with difficulties, so that resistance is an attractive alternative to frequent spraying. Powdery mildew resistance is being employed in the cucumber, but at the low light intensity of early spring cultivation the resistant plants appear susceptible to a physiological disorder leading to leaf chlorosis and necrosis (22). Other sources of powdery mildew resistance, such as the dominant form in the melon, may prove to be attractive alternatives.

Breeders would welcome sources of resistance, if available, in species that can be successfully hybridized with the cucumber. In addition, the novel variation generated by introduction of foreign genetic material would help to broaden the genetic base for selection in the crop. Resistances have been reported in African species of *Cucumis*, which are more or less related to the cultivated cucumber (10, 17, 25). Interspecific hybridization is commonly used for the introduction of valuable characters into crop plants, but in the case of cucumbers this approach has been unsuccessful (7, 8, 20). The urgent need for resistant cultivars prompted a new, more thorough investigation, which was started in 1978 by a group of researchers at the Institute for Horticultural Plant Breeding (17). This chapter presents the most important results from taxonomic studies, resistance screening, crossability analyses, and attempts to overcome the crossability barriers.

Collection and Taxonomic Studies

Wild species often vary for important characters; thus, we attempted to assemble a large and varied collection of many different species of *Cucumis*. Accessions from the center of origin of the cultivated cucumber, i.e., the Indo-Asiatic subcontinent, proved very difficult to obtain, but through personal contacts with researchers and the cooperation of the Vavilov Institute of Plant Industry, Leningrad, our collection now contains over 35 accessions of primitive and feral cucumber types. Wild species, all originating from the African continent, constitute the bulk of the collection. Many accessions were obtained from the collection of the USDA Plant Introduction Stations at Experiment, Georgia, and Ames, Iowa, and from several botanical gardens. An effort was made to trace the origins of contributed accessions through documentation and morphological comparison, and as a result, several identical samples were combined. On the other hand, a number of accessions proved to contain more than one idiotype; these were

Figure 1. Cucumis africanus. A pre-Linnaean illustration, representative of the African species of *Cucumis* subgenus *Melo*. From Hermann, P. 1698. Plate 134. *Cucumis echinatus Colocynthidis folio.* Paradisus batavus. Lugduni-Batavorum.

separated into their components. Plants of each accession were self-pollinated and reciprocally crossed for seed increase. Several accessions contained hybrids between allied species, since they segregated upon inbreeding. A few intermediates proved to be true to seed, and their taxonomic status remains uncertain.

The systematics of *Cucumis* has it origins in pre-Linnaean literature (as an example see Figure 1), and classifications of the genus have varied widely (6, 14, 15, 17, 18). At one point over 150 species were recognized, but the recent review of Jeffrey (15) distinguished some 25 species. In his treatment, which is followed here, two subgenera, *Cucumis* ($x = 7$) and *Melo* ($x = 12$), are recognized. The difference in base chromosome numbers and intergeneric sterility are the basis of the subdivision and testify to the evolutionary divergence of the cucumber and the melon and its relatives. Subgenus *Cucumis* contains *C. sativus* and perhaps two or three other species; subgenus *Melo* contains the remainder, including all the African species, distributed through four essentially cross-sterile groups: Metaliferus Group, Anguria Group, Melo Group, and Hirsutus Group.

For identification of accessions, plants were grown during the summer season in a heated greenhouse with temperatures set at 23°C during the day and 18°C at night, in containers with drip irrigation to restrain root development. This procedure induced more profuse flowering and less vegetative growth, and resulted in a plant habit that resembled that seen in natural growing conditions. Herbarium sheets were prepared and fruits were photographed and preserved. Root systems were classified as either annual or perennial. All plants were compared with original descriptions. For some species reference collections, including type specimens, were consulted. The chromosome number of doubtful accessions and suspected polyploids was established. This aided identification.

In total 176 accessions were positively identified. The accessions represented 14 species, excluding the cultivated cucumbers and melons (Table 1). In keeping with earlier suggestions (6, 8, 15) and the full crossability between *C. sativus* and *C. hardwickii* Royle, the latter is listed as *C. sativus* var. *hardwickii* (Royle) Alef. (It appears that the original form of *C. hardwickii* found by Royle in 1839 has never been rediscovered [18].) Both varieties of *C. anguria* L. are listed separately, as are two forms of *C. zeyheri* Sond., one a diploid, the other a tetraploid.

During our study, we observed that for several species the existing descriptions only cover part of the variation within such a species. There appears to be a need to treat the existing, partly discordant taxonomic literature together with reference specimens in herbaria and observations of living plants. A variety of fruit shapes of different species is assembled in Figure 2.

Table 1. Wild and primitive *Cucumis* species of the Institute for Horticultural Plant Breeding (IVT) collection with their numbers of chromosomes and accessions

Species	2*n* chromosome no.[1]	No. of accessions
C. africanus	24	4
C. anguria var. anguria	24	12
C. anguria var. longpipes	24	14
C. dipsaceus	24	16
C. ficifolius	48	8
C. figarei	72	10
C. heptadactylus	48	1
C. meeusei	48	1
C. melo var. agrestis	24	27
C. metuliferus	24	18
C. myriocarpus	24	24
C. prophetarum	24	2
C. sagittatus	24	4
C. sativus var. hardwickii	14	9
C. zeyheri (diploid)	24	22
C. zeyheri (tetraploid)	48	4
Total		176

[1]According to Kroon (pers. comm., 1980), Dane (6), or Ramachandran and Narayan (27).

Screening for Disease Resistance

Following the initial seed increase and taxonomic identification, accessions of the different species were tested for resistance to cucumber green mottle mosaic virus (CGMV), to root knot nematodes (*Meloidogyne incognita* Chitwood), and to powdery mildew (*Sphaerotheca fuliginea* (Slecht. ex Fr.) Poll.). In addition, several species were evaluated for resistance to the glasshouse whitefly (*Trialeurodes vaporariorum* Westwood).

To test for CGMV resistance, cotyledons were inoculated with sap from frozen infected leaves. Infection induced mottling and distorted growth of true leaves within two to three weeks. Symptomless plants were checked for the presence of the virus by inoculating a sensitive cucumber line. Resistance to root knot nematodes was tested by planting seeds in small plastic tubes filled with silver sand, which were inoculated a few days after germination of the seeds with approximately 25 larvae per tube. After five weeks at 23°C, the roots of the plantlets were washed free of sand and examined. The rating of this test correlated well with data from soil-grown plants (2). Lastly, resistance to powdery mildew was examined on fully grown plants in the greenhouse after natural infection. In further tests, accessions of promising species were screened as young plants following inoculation with the mildew. Infestation by greenhouse whitefly was assessed on mature plants following a natural epidemic.

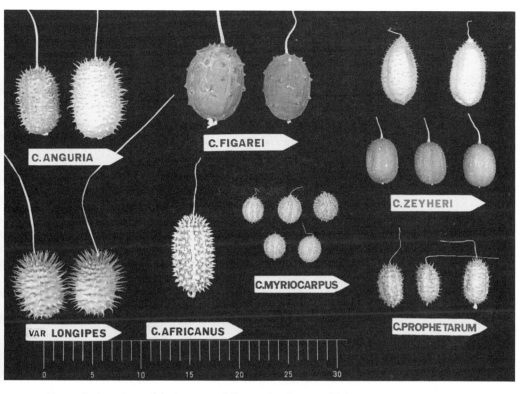

Figure 2. A variety of fruit types of *Cucumis* subgenus *Melo*.

Results of the screenings are summarized in Table 2. Among the wild species, C. *anguria* and C. *zeyheri* stood out because of their total immunity to CGMV. Segregation data of interspecific crosses and repeated backcrosses with C. *anguria* indicated that a single dominant gene confers CGMV resistance in this species (21). As is common in C. *sativus*, accessions with symptomless carriers were found in many other species.

Resistance to root knot nematodes occurred in C. *metuliferus* E. H. Mey. ex Schrad. and diploid C. *zeyheri*. Earlier, good levels of nematode resistance were reported in C. *metuliferus*, C. *ficifolius* A. Rich., and C. *anguria* (10). Among accessions of C. *metuliferus*, however, we observed large differences in nematode resistance: the mean number of galls in the seedling test ranging from 2.5 to 9 per plant, whereas C. *sativus* averaged 25 galls per plant. Thus, no total resistance was obtained. Prebreeding and selection is advisable in such cases. More detailed tests subsequently showed the resistance of diploid C. *zeyheri* to be the most valuable (2).

Powdery mildew resistance was rather common in the mature plant test, but subsequent tests using artificial inoculation of potted plants indicated

Table 2. Distributions of tested accessions of *Cucumis* species at Institute for Horticultural Plant Breeding (IVT) over classes of resistance to four pathogens

Species	CGMV		Root knot nematodes			Powdery mildew			Greenhouse whitefly		
	R	S	0	1	2	0	1	2	0	1	2
C. africanus	1							1			
C. anguira var. anguria	7			2	3	3			1	3	2
C. anguria var. longpipes	7				3	4	1		1	1	2
C. dipsaceus		12			6	6			2	2	1
C. ficifolius	3	2		2	1	1					
C. figarei					1		1		1		
C. heptadactylus				1							
C. meeusei					1			1			
C. melo var. agrestis		2			2		1	8		1	4
C. metuliferus		6	3	11	1	5	2		6		
C. myriocarpus		18			2	5	5	1	5	2	5
C. prophetarum		1				1			1		
C. sagittatus		1									
C. sativus var. hardwickii		5			2		1	4	1		
C. zeyheri (diploid)	16		1	3	10	6				3	2
C. zeyheri (tetraploid)	4				3	2					

R, resistant; S, susceptible; 0, no or hardly any infection; 1, slight infection; 2, heavy infection. Blanks, species not tested.

that *C. anguria* is especially promising in this respect. Although none of the primitive melons (*C. melo* var. *agrestis* Naud.) proved to be resistant to powdery mildew resistant, we had several resistant *C. melo* L. cultivars at our disposal. The mechanism of this resistance may differ from that in the cucumber (22), but detailed studies have not been made. The greenhouse whitefly infestations clearly showed that *C. metuliferus* was resistant, while other species gave varied reactions.

Detailed data on reactions of individual accessions, insofar not revealed in earlier publications, can be obtained from the Institute for Horticultural Plant Breeding on request.

Analysis of Interspecific Crossability

The presence of resistances in only the African species necessitated an investigation of the crossability of those species with the cucumber. In fact, this had already been attempted as a part of earlier interspecific hybridization programs in *Cucumis* with the emphasis on improving the cultivated melons (6, 8). The only positive result at the end of the 1970's, however, was the development of embryos in the cross *C. sativus* × *C. melo* (20), but these embryos ceased growth at an early stage. To improve on these results we undertook a new and broader analysis of the interspecific crossability in the genus *Cucumis*, using the species as identified by our group. Apart from evaluating the crossability of the resistant species with *C. sativus*, it was our aim to add to the understanding of the phylogenetic relations between the different species involved. Four species were not included, namely *C. africanus* L. f., *C. heptadactylus* Naud., *C. meeusei* C. Jeffr., and *C. sagittatus* Peyr., because they lacked apparent potential.

We first concentrated on identifying and localizing prefertilization barriers by examining in vivo pollen germination, tube growth, and penetration in the ovules with ultraviolet microscopy (16). In several combinations of African species fertilization proved to be no problem, but this was usually not so with pollen from these species on *C. sativus*. Only the pollen of *C. melo* grew through the style and into the ovules of *C. sativus* (see Niemirowicz-Szczytt and Kubicki [20]). Occasionally, pollen tubes of diploid and tetraploid *C. zeyheri* grew into the ovaries of *C. sativus*, indicating a potential for fertilization. In the reciprocal crosses, with the African species as the female parent, pollen growth of *C. sativus* was generally arrested in the stigma or the upper part of the style. We attempted to evaluate more than one accession per species because of possible crossability variation, but generally these accessions exhibited similar pollen tube growth and similar acceptance of the pollen.

A number of the African cross combinations yielded seed that germinated well. Twenty-seven crosses produced vigorous F_1 hybrids (24). The offspring of all crosses were intermediate in morphology between the parents, especially with respect to fruit characteristics. Pollen fertility of the hybrids ranged from 0–87% stainable pollen. Self-pollinations and backcrosses by both parents yielded many well-filled fruits with several hybrids, but were unsuccessful with some others. Together, all these data resulted in a crossability polygon, as presented in Figure 3. The two varieties of *C. anguria* are combined on the polygon, as offspring from crosses with either *C. anguria* var. *anguria* or var. *longipes* (Hook. f.) Meeuse were very similar in fertility. This reduces the number of hybrid offspring presented to 19.

In the polygon three fertility groups are distinguished, ranging from full fertility to nearly complete sterility. A fourth group contains the inviable seeds and seedlings. A fifth category is formed by the absence of a line between species. This indicates that following pollination no seeded fruits were obtained. From both latter categories it appears that *C. sativus*, *C. melo*, and *C. metuliferus* are isolated species.

In general, our data confirm Jeffrey's (15) subdivision of subgenus *Melo* into four crossability groups of which one, containing only *C. hirsutus* Sond., is not shown in the polygon. Nevertheless, in the large intercrossable Anguria Group different degrees of crossability exist, and many species are not as closely related as suggested by his classification. The presence of polyploid species in the Anguria Group complicates the analysis, because

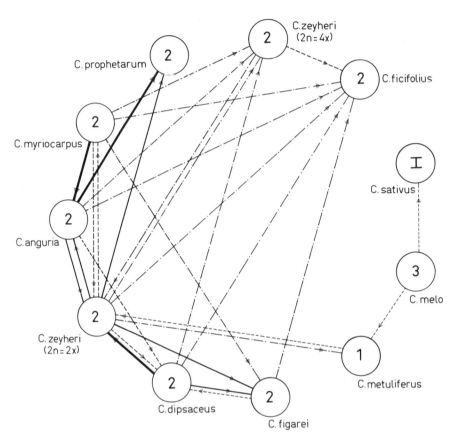

Figure 3. Crossability polygon of *Cucumis* species. I. *Cucumis* subgenus *Cucumis.* 1. *Cucumis* subgenus *Melo,* Metuliferus Group. 2. *Cucumis* subgenus *Melo,* Anguria Group. 3. *Cucumis* subgenus *Melo,* Melo Group. Arrows point in the direction of the female parent. Moderately to strongly self-fertile and cross-fertile hybrids ————; sparingly self-fertile and moderately cross-fertile hybrids ————; self-sterile, usually not cross-fertile hybrids — · — · — · —; inviable seeds or seedlings — — — —; absence of a line indicates that seeded fruits were not obtained. From den Nijs and Visser (24), reprinted with permission of *Euphytica.*

the complete sterility of triploid hybrids from crosses between the tetraploids *C. ficifolius* and *C. zeyheri* and several diploid species does not itself preclude a certain degree of phylogenetic relationship among the species.

To further define the relationships in the Anguria Group, meiosis was studied in the hybrids produced (26), and these results were compared with those of similar and related hybrids studied by other investigators (6, 7, 28). From these cytological data and several morphological lines of evidence it was concluded that the crossing relationships between species of the An-guria Group should be refined and two subgroups should be distinguished: the Myriocarpus Subgroup with C. *africanus* as the only additional species and the Anguria Subgroup containing C. *anguria*, C. *dipsaceus* Spach, C. *ficifolius*, C. *prophetarum* L., and C. *zeyheri*. Most polyploids belong to the Anguria Subgroup. It should be noted that C. *africanus* was unfortunately not included in our own study, so that its phylogenetic status had to be determined on the basis of data in the literature only.

We conclude that the species with the interesting resistances, i.e., C. *metuliferus*, diploid C. *zeyheri*, C. *anguria*, and C. *melo*, appear well iso-lated from the cucumber. With the exception of diploid C. *zeyheri* and C. *anguria*, they are also phylogenetically isolated from each other. Thus, bridge crosses are unlikely to succeed, and special techniques to overcome the barriers to crossability were investigated.

Improvement of Crossability

Several early attempts to overcome crossing barriers in *Cucumis* were unsuccessful (8, 10, 17) or yielded haploids rather than hybrids (9). Em-bryos were found in the cross between C. *metuliferus* and C. *melo* 'Gulfstream' (11), but these aborted. Prefertilization barriers occurred in all the other combinations. Recently, partly fertile hybrids from the cross C. *metuliferus* × C. *anguria* were obtained through embryo culture and soma-tic embryogenesis (12).

In selected cross combinations we evaluated the merits of mentor pollen (29), growing the plants in containers instead of in open soil, temperature treatments, and application of a growth regulator, while simultaneously searching for genetic variation for crossability throughout the experiments. Most of the results have been described earlier (3, 4, 5, 13, 16, 23, 24) and are only summarized here. The experiments concentrated on C. *sativus*, C. *metuliferus*, diploid C. *zeyheri*, and C. *melo* in all cross combinations. In a few experiments, C. *ficifolius*, which generally permitted good growth of foreign pollen, was used as the female parent. Most of the cultivated musk-melons had hermaphrodite flowers, and these were emasculated. Since emasculation easily damaged the ovary, the melons were used mainly as male parents.

Mentor pollen techniques involved pollinations with a mixture of fresh pollen of the foreign male parent and maternal pollen rendered "genetically dead" by irradiation (23). Initially an irradiation dose of 1 kGray (100 krad) appeared to be suitable, as the pollen effectively induced fruit set in all attempted crosses. Most ovules, however, were stimulated into a par-thenocarpic growth pattern, developing into large but empty seeds. In the diploid crosses *C. sativus* × *C. zeyheri* and its reciprocal cross, *C. zeyheri* × *C. metuliferus*, and *C. metuliferus* × *C. sativus*, only very few ovules contained putative hybrid embryos. These, however, did not grow larger than 0.02–0.06 mm (23). The low number of embryos probably was due to competition between the two types of pollen, which could be reduced by a higher irradiation dose (1).

Growing the plants in 25 l containers instead of in open soil not only increased flowering intensity, but also provided a means of overcoming crossing barriers, especially in the diploid cross *C. metuliferus* × *C. zeyheri* (5). The female flowers of the *C. metuliferus* plants in open soil never set fruit upon pollination with *C. zeyheri*, but in the containers they readily developed seeded fruits. The seeds germinated well in soil, and hybrid plants were produced, which, however, were completely sterile. Growth in containers also favored crossability of *C. metuliferus* × *C. melo*, but in this case it did not improve embryo growth. In crosses with *C. sativus* and *C. zeyheri* as female parents this special growing condition did not have any effect (5).

To evaluate temperature treatments the plants were grown at 17, 20, 23, and 26°C in phytotron greenhouses. In several crosses pollen tube growth improved at higher temperatures, but a positive effect on fertilization success could not be reproducibly demonstrated (13). For the cross *C. metuliferus* × *C. melo* a temperature of 26°C seemed to be required for fruit set, while 23°C was required for the cross *C. ficifolius* × *C. melo*. In both crosses, however, the percentage of fruit set was very low at these temperatures, and the embryos did not grow larger than 0.05–0.08 mm. Fruit set in the diploid cross *C. metuliferus* × *C. zeyheri* appeared to decrease with higher temperature (13).

Aminoethoxyvinylglycine (AVG) was applied to flowers at pollination time in order to delay fruit abscission. Aminoethoxyvinylglycine, which inhibits ethylene production and activity, slightly increased fruit and seed set in hybrid muskmelon seed production (19). In our experiments the AVG treatments initially seemed to improve fruit formation in certain crosses, but this effect could not be reproduced (5).

Genetic variation for crossability was found in diploid *C. zeyheri*. Rare plants of accession IVT Genebank number (Gbn) 0181 yielded numerous fruits with embryos after pollination with *C. metuliferus* (embryo size 0.9–1.5 mm) and with *C. sativus* (embryo size 0.08–0.1 mm). Vegetatively propagated daughter plants also exhibited this exceptional crossability as female parents, but not as male parents. In addition, genetic variation for crossability was found in the cucumber itself. *Cucumis sativus*, IVT Gbn 1811A, a *hardwickii* type, gave exceptionally large embryos (0.25–0.4 mm) in the cross with *C. melo* and appeared to accept pollen of both *C. metuliferus* and diploid *C. zeyheri*, which regularly resulted in fruits containing embryos 0.03–0.09 mm long (5).

By using a number of techniques we have significantly improved the crossability in our most interesting interspecific diploid crosses in *Cucumis*. In the cross *C. metuliferus* × *C. zeyheri* the crossing barriers were almost completely overcome, resulting in vigorous, although sterile, hybrid plants (see Figure 3). In the crosses *C. sativus* × *C. metuliferus*, *C. sativus* × *C. zeyheri*, *C. zeyheri* × *C. sativus*, and *C. zeyheri* × *C. metuliferus* the prefertilization barriers were overcome to such extent that the production of a large number of embryos, unfortunately still abortive, can readily be achieved. In these cases embryo rescue techniques may be helpful.

Embryo Culture

The embryos obtained in the crosses described above did not grow further than an early developmental stage. Postfertilization barriers also occurred in certain successful crosses between species within the Anguria Group (Figure 3). In order to rescue the abortive hybrid embryos, an embryo culture system that first was developed using vital nonhybrid embryos was applied.

Embryo rescue succeeded with heart-shaped to mature embryos in certain cross combinations (3, 4). For instance, the cross *C. anguria* var. *longipes* × diploid *C. zeyheri* yielded seeds with almost mature embryos that failed to germinate in soil. Through embryo rescue, hybrid plants were obtained that were sparingly self-fertile (Figure 3). Almost mature embryos out of inviable seeds from the reciprocal crosses between *C. myriocarpus* and *C. zeyheri* also germinated in vitro, but the seedlings died soon afterward from collapse of the root collar (24). Embryos of the cross diploid *C. zeyheri* × *C. metuliferus* reached an intermediate developmental stage in situ, and through embryo rescue ten hybrid plants were obtained. These plants were completely sterile.

In contrast to these successes with embryos already differentiated from crosses between species within the subgenus *Melo*, our embryo culture procedures appeared unsuccessful with embryos from the crosses between species of subgenera *Cucumis* and *Melo*. These embryos, which in situ were always globular-shaped, in vitro did not start differentiation, usually dying within a month. We hypothesize that unbalanced chromosome complements rather than disturbed interaction between embryo and endosperm or maternal tissue limit embryo growth and development in these intersubgeneric combinations. Therefore, as an alternative approach, we attempted to induce callus formation from these hybrid embryos in order to promote chromosome elimination and chromosome rearrangements. To date, however, the calli obtained have grown slowly, and we have not yet succeeded in regenerating plants from them. In addition, a program of somatic fusion of isolated protoplasts of the species has been started. It may have potential for introducing the desired characteristics into the cucumber.

Conclusions

The following conclusions can be drawn from our investigations.

There is a need for a modern revision of *Cucumis*. Jeffrey's (15) study provides a good reference point for such a taxonomic review.

Valuable resistances to various important pathogens attacking the cultivated cucumber have been found in the African group of species, especially in diploid *C. zeyheri*, *C. metuliferus*, *C. anguria*, and *C. melo*. There is still a need for further collection and evaluation of taxa more closely related to *C. sativus*.

Analysis of crossability of the African species *C. melo*, and *C. sativus* confirmed, by and large, the subdivision of the genus into several crossability groups (8, 15, 28). The large African crossability group can be further divided into two subgroups (26).

Mentor pollen induced high fruit set in several crosses between isolated species, but the yield of hybrid embryos was low. The application of AVG did not significantly enhance fruit set.

Growing the plants in containers generally improved flowering, but it also proved to be a means of overcoming the crossing barriers between *C. metuliferus* × diploid *C. zeyheri*. This confirms the importance of the physiological condition of the plant for crossing results.

Significant variation for crossability was discovered in diploid *C. zeyheri*. Selected clones of this species yielded fruits with small embryos following pollination by *C. sativus* and by *C. metuliferus*. An exceptional accession of *C. sativus* var. *hardwickii* was found to accept pollen of both *C. zeyheri* and *C. metuliferus*, resulting in small, abortive embryos.

Embryo culture techniques have been developed that can rescue hybrid embryos that have reached at least the heart-shaped stage, but our methods were not suited for the still undifferentiated embryos that always occur in crosses between the subgenera *Cucumis* ($x = 7$) and *Melo* ($x = 12$). Also, induction of callus from the embryos of combinations between both subgenera may facilitate chromosomal rearrangements. This phenomenon may be essential in overcoming the somatic incompatibility between the two subgenera. Alternatively, somatic fusion of protoplasts may provide a means of accomplishing genetic interchange.

Acknowledgments

We acknowledge the cooperation of the other members of the working group on interspecific hybridization in *Cucumis* and of the Department of Plant Taxonomy of the Agricultural University, Wageningen. We appreciate the help of Charles Jeffrey, Royal Botanic Gardens, Kew, in identifying part of our collection, and we thank Hoffman-La Roche, Inc., New York, for providing AVG.

Literature Cited

1. Boom, J. M. A. van den, and A. P. M. den Nijs. 1983. Effects of γ-radiation on vitality and competitive ability of *Cucumis* pollen. Euphytica 32: 677–684.
2. Boukema, I. W., G. T. M. Reuling, and K. Hofman. 1984. The reliability of a seedling test for resistance to root-knot nematodes in cucurbits. Cucurbit Genet. Coop. Rep. 7: 92–93.
3. Custers, J. B. M. 1980. Overcoming incongruity in interspecific crosses in Cucumis. Proc. Eucarpia Conference on Breeding of Cucumbers and Melons, Wageningen. Pp. 50–55.
4. Custers, J. B. M., and G. van Ee. 1980. Reciprocal crosses between *Cucumis africanus* L. f. and *C. metuliferus* Naud. II. Embryo development in vivo and in vitro. Cucurbit Genet. Coop. Rep. 3: 50–51.
5. Custers, J. B. M., and A. P. M. den Nijs. 1986. Effects of aminoethoxyvinylglycine (AVG), environment and genotype in overcoming hybridization barriers between *Cucumis* species. Euphytica 35: 639–647.
6. Dane, F. 1976. Evolutionary studies in the genus *Cucumis*. Ph.D. dissertation, Colorado State Univ. Fort Collins.
7. Dane, F., D. W. Denna, and T. Tsuchiya. 1980. Evolutionary studies of wild species in the genus *Cucumis*. Z. Pflanzenzucht. 85: 89–109.
8. Deakin, J. R., G. W. Bohn, and T. W. Whitaker. 1971. Interspecific hybridization in *Cucumis*. Econ. Bot. 25: 195–211.
9. Dumas de Vaulx, R., and D. Chambonnet. 1980. Haploid induction in muskmelon (*Cucumis melo* L.) by pollination with *Cucumis ficifolius* A. Rich. PI193967. Proc. Eucarpia Conference on Breeding of Cucumbers and Melons, Wageningen. Pp. 39–43.
10. Fassulliotis, G. 1967. Species of *Cucumis* resistant to the root-knot nematode *Meloidogyne incognita acrita*. Pl. Dis. Reporter 51: 720–723.
11. Fassuliotis, G. 1977. Self-fertilization of *Cucumis metuliferus* Naud. and its cross-compatibility with *C. melo* L. J. Amer. Soc. Hort. Sci. 102: 336–339.
12. Fassuliotis, G., and B. V. Nelson. 1988. Interspecific hybrids of *Cucumis metuliferus* × *C. anguria* obtained through embryo culture and somatic embryogenesis. Euphytica 37: 53–60.
13. Franken, J., J. B. M. Custers, and R. J. Bino. 1989. Effects of temperature on pollen tube growth and fruit set in reciprocal crosses between *Cucumis sativus* and *C. metuliferus*. Pl. Breeding 100: 150–153.
14. Gabaev, S. 1933. Systematische Untersuchungen an Gurkenarten und Varietaeten. Angew. Botanik 15: 290–307.
15. Jeffrey, C. 1980. A review of the Cucurbitaceae. J. Linn. Soc. Bot. 81: 233–247.
16. Kho, Y. O., A. P. M. den Nijs, and J. Franken. 1980. Interspecific hybridization in *Cucumis* L. II. The crossability of species, an investigation of in vivo pollen tube growth and seed set. Euphytica 29: 661–672.

17. Kroon, G. H., J. B. M. Custers, Y. O. Kho, A. P. M. den Nijs, and H. Q. Varekamp. 1979. Interspecific hybridization in *Cucumis* L. I. Need for genetic variation, biosystematic relations and possibilities to overcome crossability barriers. Euphytica 28: 723–728.

18. Leeuwen, L. van, and A. P. M. den Nijs. 1980. Problems with the identification of *Cucumis* L. taxa. Cucurbit Genet. Coop. Rep. 3: 55–59.

19. Natti, T. A., and J. B. Loy. 1978. Role of wound ethylene in fruit set of hand pollinated muskmelons. J. Amer. Soc. Hort. Sci. 103: 834–836.

20. Niemirowicz-Szczytt, K. A. W., and B. Kubicki. 1979. Cross-fertilization between cultivated species of genera *Cucumis* L. and *Cucurbita* L. Genet. Polon. 20: 117–124.

21. Nijs, A. P. M. den. 1982. Inheritance of resistance to cucumber green mottle mosaic virus (CGMV) in *Cucumis anguria* L.. Cucurbit Genet. Coop. Rep. 5: 57–58.

22. Nijs, A. P. M. den. 1985. Rootstock-scion interactions in the cucumber: implications for cultivation and breeding. Acta Hort. 156: 53–60.

23. Nijs, A. P. M. den, and E. H. Oost. 1980. Effect of mentor pollen on pistil-pollen incongruities among species of *Cucumis* L. Euphytica 29: 267–271.

24. Nijs, A. P. M. den, and D. Visser. 1985. Relationships between African species of the genus *Cucumis* L. estimated by the production, vigour and fertility of F_1 hybrids. Euphytica 34: 279–290.

25. Norton, J. D., and D. M. Granberry. 1980. Characteristics of progeny from an interspecific cross of *Cucumis melo* with *C. metuliferus*. J. Amer. Soc. Hort. Sci. 105: 174–180.

26. Raamsdonk, L. W. D. van, A. P. M. den Nijs, and M. C. Jongerius. 1989. Meiotic analyses of *Cucumis* hybrids and an evolutionary evaluation of the genus *Cucumis*. Pl. Syst. Evol. 163:133–146.

27. Ramachandran, C., and R. K. J. Narayan. 1985. Chromosomal DNA-variation in *Cucumis*. Theoret. Appl. Genet. 69: 497–502.

28. Singh, A. K., and K. S. Yadava. 1984. An analysis of interspecific hybrids and phylogenetic relationships in *Cucumis* (Cucurbitaceae). Pl. Syst. Evol. 147: 237–252.

29. Stettler, R. F. 1968. Irradiated mentor pollen: its use in remote hybridization of black cottonwood. Nature 219: 746–747.

32 | Prospects for Increasing Yields of Cucumbers via *Cucumis sativus* var. *hardwickii* Germplasm

R. L. Lower and J. Nienhuis

ABSTRACT. The increasing trend toward mechanical harvesting of pickling cucumbers has focused attention on the problem of low yield. Previous studies suggested that response to selection for yield within existing populations would be slow and expensive. The low heritability and variance associated with fruit number in existing populations might be increased by incorporating multiple fruiting genotypes into the germplasm pool. A possible source of multiple fruiting is *Cucumis sativus* var. *hardwickii*, an annual monoecious taxon thought to be either a feral form or a progenitor of the cultivated cucumber, *C. sativus*. One of the potentially useful characteristics of var. *hardwickii* is its ability to set sequentially a large number of seeded fruits. The incorporation of var. *hardwickii* germplasm into an adapted population provides a new source of variation in which recurrent selection can be initiated.

Over the past 60 years yields of pickling cucumber (*Cucumis sativus* L.) have increased more than threefold (29). This increase in productivity has been due primarily to improved cultural practices and the development of disease-resistant cultivars, rather than to direct selection for increased yield. Numerous plant introductions were exploited in the development of disease-resistant cultivars (23).

The Korean cucumber 'Shogoin' (PI220860), the source of completely or predominantly pistillate plants (7), perhaps has played the most significant role in cucumber breeding. Backcrossing to adapted cultivars and selection led to the release of gynoecious inbred lines, which triggered the development of F_1 hybrids (23). These hybrids were more uniform, earlier, and in some instances higher yielding than their monoecious progenitors. They tended to concentrate fruit set and were instrumental in the shift towards mechanical harvesting of the processing cucumber industry.

The principal current problem in selection for increased yield is that cultivars appear to plateau at approximately one fruit per plant at populations of between 170,000 and 250,000 plants per hectare (19). The cause of this limitation is not known. It may be that fruit set in cucumbers is 1) controlled by a translocated inhibitor produced by the developing fruit, 2) limited because seed development is such a strong sink for metabolites that they are unavailable for additional fruit set, or 3) controlled by other factors (5, 17, 28). In this chapter means of increasing yields are discussed, but attention is focused on increasing the number of fruits per plant by introducing germplasm from *Cucumis sativus* var. *hardwickii* (Royle) Alef. This wild or feral form characteristically sets a large number of seeded fruits.

General Approaches to Increasing Yield

One possible solution to low fruit yield is to increase plant density per unit area by changing plant architecture, thus maximizing fruit number per hectare (12, 15). Two dwarf plant types, compact (*cp*) and determinate (*de*), capitalize on reduced plant size and have the potential for more concentrated fruit set and increased yield in high density plantings (15, 25).

A second approach to increasing yield is the development of seedless cultivars. Parthenocarpic fruit set can be promoted either genetically or by growth regulators (2, 5, 6, 8, 24). Gynoecious parthenocarpic cultivars may lead to higher yields when compared with conventional seeded cultivars (1, 3, 8). Parthenocarpic slicing cultivars have been very successful in greenhouse culture, but pickling types have met with very limited success because of unstable sex expression, erratic fruit setting ability, or poor processing quality.

A more traditional approach to improving yield is through population improvement and direct selection for yield components. Whereas disease resistance, plant architecture, and parthenocarpy are governed primarily by few genes with major effects, yield components are likely controlled quantitatively by many genes, each with small effects. Therefore, the primary objective of plant breeders in improving yield is to manipulate populations and choose breeding techniques that increase the frequency of favorable genotypes within populations. The prerequisites to successful selection for increased yield in pickling cucumbers are the development of methods for efficiently measuring yields, and the existence of genetic variability within the source populations.

Marketable yield in pickling cucumbers is expressed as dollar value, which is the sum of products of value and weight of various grade sizes. Smith (26) found that a simpler evaluation of yield based only on fruit number could be very efficient because of its high correlation (.84) with dollar value in once-over mechanically harvested operations.

The existence of significant additive and dominant genetic variance for various yield components was recently shown by Ghaderi and Lower (9). Smith et al. (27) suggested that the low genetic variance for yield in existing populations might be increased by identifying multiple fruiting genotypes among plant introductions and incorporating them into the germplasm pool.

Increasing Yield via var. *hardwickii*

One promising source of multiple fruiting genotypes for pickling cucumbers is *Cucumis sativus* var. *hardwickii*, an annual, monoecious cucurbit, which may be either a feral form or the progenitor of *Cucumis sativus* (4, 31). The var. *hardwickii* line currently used in our breeding program, 'LJ 90430' obtained from G. W. Bohn, differs from *C. sativus* cultivars in several morphological and flowering characteristics (13). Typically, 'LJ 90430' plants are larger than *C. sativus* cultivars, seem to lack apical dominance, and have more and larger laterals. Seeds of 'LJ 90430' are one-third the size of seeds of *C. sativus* cultivars. The fruits are bitter, ellipsoidal in shape, and weigh 25–35 g when mature. 'LJ 90430' is a short-day plant that produces flowers only when the photoperiod is less than 12 hours at 30°C day/20°C night temperatures (13).

One of the most potentially useful characteristics of 'LJ 90430' is its ability to set sequentially large numbers of seeded fruits per plant. Fruits with developing seeds do not inhibit later fertilized fruits, as is common in *C. sativus* cultivars (13). Horst and Lower (14) reported that 'LJ 90430' averaged 80 fruits per plant under North Carolina conditions.

A recent genetic study examined heterosis and the types of gene action involved in fruit yield per plant, as well as other characteristics in a cross between a gynoecious inbred, 'Gy 14', and 'LJ 90430' (16, 21). Generation means were significantly different for all the characteristics measured (Table 1). Heterosis above both mid- and high-parent was observed for fruit weight per plant and main stem length. An additive dominance model accounted for most of the variance among generation means for fruit weight per plant, main stem length, number of lateral branches, and length-diameter ratio of the mature fruit. In addition, additive times additive epistasis was involved in variation among generation means for fruit per plant yield. Lower et al. (16) suggested that if gene effects reflect the types of genetic variances within this population, then some form of intrapopulational recurrent selection should be effective in improving performance for fruit per plant yield.

Because of the narrow germplasm base and low genetic variance for yield in currently available cucumber populations, our emphasis is on the development of two populations in which recurrent selection can be initiated. Recurrent selection is a useful breeding technique for the evaluation and

Table 1. Generation means and heterosis estimates for five characteristics in a cross between *Cucumis sativus* gynoecious inbred 'Gy 14' and *Cucumis sativus* var. *hardwickii* line 'LJ 90430'

Generation	Fruit no. per plant	Fruit weight per plant (kg)	No. of lateral branches	Main stem length (cm)	Fruit length-diameter ratio
P_1(Gy 14)	4.03	2.21	4.10	121.31	2.24
P_2(LJ 90430)	93.91*	2.20	39.91	137.15	1.47[3]
F_1	53.12	6.70	16.50	290.32	1.99[4]
F_2	35.24	4.18	12.85	214.76	1.93[4]
BC_{P1}	14.27	3.51	6.72	179.21	2.11
BC_{P2}	75.66	4.31	25.30	253.61	1.73
H_{MP}[1]	4.15	4.49**	−5.46*	161.09**	0.14
H_{HP}[2]	—	4.49**	—	153.17**	—

Source: After Lower et al. (16).
[1] Heterosis above mid-parent.
[2] Heterosis above high-parent.
[3] Estimate from appropriate gene effects model.
[4] Generation means not significantly different for trait (0.5 level).
*Statistically significant deviation at $P < 0.05$.
**Statistically significant deviation at $P < 0.01$.

incorporation of exotic germplasm, since random recombination during each cycle of selection will serve to break linkages and to foster new gene combinations. The introduction of exotic germplasm into a breeding population may often produce greater short-term results than if locally adapted materials are recycled in long-term breeding projects (20). However, caution should be exercised in the introduction of exotic germplasm into horticultural crops, since undesirable quality characteristics may be closely associated with desirable traits, either because of linkage or pleiotropy. Hence, cucumber populations that incorporate 'LJ 90430' germplasm may result in higher yields, but this increase may be accompanied by losses of disease resistance or quality attributes necessary for the processing industry. For this reason, we developed two divergent populations; one an exotic with enhanced genetic variance for yield, but with limited horticultural quality; and the other a broad-based population with limited genetic variance for yield but with a desirable range of horticultural quality and disease resistances. The exotic population, referred to as "hardwickii semi-exotic," is a random-mated F_2 generation derived from a cross between 'Gy 14' and 'LJ 90430'. The other broad-based population, referred to as the "gynoecious synthetic," is a random-mated population derived from over 50 breeding lines and cultivars, all of which had been previously classified as gynoecious or predominately pistillate in sex expression.

The breeding procedures most appropriate for population development depend to a large extent on whether the breeder is seeking an improved inbred line or an F_1 hybrid (20). The development of both the "hardwickii semi-exotic" and the "gynoecious synthetic" populations provides an opportunity to improve population performance per se (intrapopulational improvement), from which superior inbred lines can be extracted. Test crosses between diverse improved populations (interpopulation improvement) provide an opportunity to maximize heterosis in F_1 hybrids.

A comparison of intrapopulational and interpopulational recurrent selection methods provides information about which is most effective and allows inferences to be drawn about the types of gene action involved in response to selection. Cucumbers are well-adapted to test different selection schemes, as both self-pollinations and cross-pollinations can be made on the same plant. The objective of current research is to compare rates of gain in the "hardwickii semi-exotic" and "gynoecious synthetic" populations subjected to two recurrent selection schemes, S1 line (S1) and reciprocal full-sib (RFS).

S1 is principally an intrapopulational improvement scheme in which selection is based on selfed progeny performance (Figure 1). RFS is an interpopulational improvement scheme, in which selection is based on full-sib progeny performance between individuals in both populations (Figure 2). In S1, selection is for increased performance in each population separately; therefore, improved performance of the population hybrid would be an indirect effect. Conversely, in RFS direct selection is for performance of the population hybrid, and indirect effects of selection would be for performance of the populations per se (19). The relative magnitude of direct and indirect responses to selection are governed by the magnitude of the correlation between effects of genes in populations per se and in population hybrids (19). In both selection methods the three generations required to complete a cycle of selection can be completed in one year.

The types of genetic variances involved in fruit yield per plant in these two populations will influence the rate of gain in the two recurrent selection schemes. RFS selection will capitalize on both additive and nonadditive gene effects (overdominance and epistasis involving dominance interactions); whereas S1 selection will emphasize response due to additive gene effects (10, 11). Previous genetic studies have indicated that most of the genetic variances for fruit yield per plant in both adapted and exotic populations are additive in nature (9, 26, 27). Hence, S1 selection might be expected to result in the highest rate of gain. However, other considerations in a breeding program might make RFS more attractive. One of the advantages of RFS over S1 selection is that only half as many progeny rows have to be evaluated to maintain the same selection intensity (Figure 2). In addition, where the primary goal of a breeding program is the development of single-cross hybrids, RFS selection provides an efficient means of developing parents with improved combining ability (30). Gynoecious inbreds could be easily isolated from the "gynoecious synthetic" population to make F_1 hybrid production feasible.

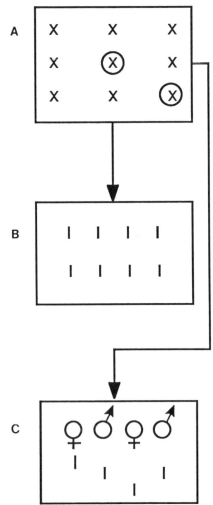

Figure 1. Outline of S1 line selection for fruit number in the cucumber populations "hardwickii semi-exotic" and "gynoecious synthetic." A. Generation of family structure (winter, greenhouse). Self-pollinate 100 individuals in each population. B. Progeny test (early summer). Progeny test S1 families for fruit/plant yield. C. Recombination block (late summer). Plant approximately two to three weeks after progeny test planting, with each family represented in a random male and female row. Based on progeny test, random mate the top 15% of S1 families, with other families rogued out before flowering. Harvest seed only from female rows.

We feel strongly that public breeders should develop and maintain breeding populations with sufficient flexibility to meet both short-term and long-term goals. Clearly, the incorporation of exotic hardwickii germplasm into adapted populations is an attempt to meet a specific long-term goal of increased yield. However, the utility of this approach and other broadly based populational studies should outlive the tenure of any one breeder. The existence of experimental populations may someday provide a reservoir of variability from which future breeders may extract solutions to problems that we cannot yet anticipate.

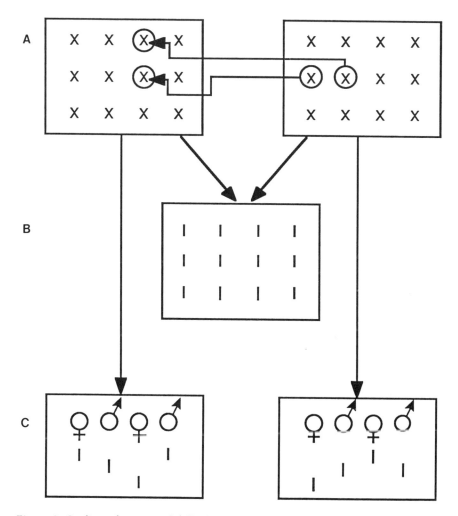

Figure 2. Outline of reciprocal full-sib recurrent selection for fruit number in two cucumber populations. Left "hardwickii semi-exotic," right "gynoecious synthetic." A. Generation of family structures (winter greenhouse). Self-pollinate and produce full-sib crosses between 100 individuals in both populations. B. Progeny test (early summer). Progeny test full-sib families for fruit/plant yield. C. Recombination block (late summer). Plant approximately two to three weeks after progeny test planting, with each selfed family represented in a random male and female row. Based on full-sib progeny test, random mate the top 15% selfed families, with other families rogued out before flowering. Harvest seed only from female rows.

Literature Cited

1. Baker, L. R., J. W. Scott, and J. E. Wilson. 1973. Seedless pickles—a new concept. Mich. State Univ. Res. Rep. 277.
2. Cantliffe, D. J. 1974. Promotion of fruit set and reduction of seed number in pollinated fruit of cucumbers by chlorflurenol. HortScience 9: 577–578.
3. Colwell, H. T. M., and J. O'Sullivan. 1981. Economics of harvest timing for once-over harvesting of cucumbers. J. Amer. Soc. Hort. Sci. 106: 163–167.
4. Deakin, J. R., G. W. Bohn, and T. W. Whitaker. 1971. Interspecific hybridization in *Cucumis*. Econ. Bot. 25: 195–211.
5. Denna, D. W. 1973. Effects of genetic parthenocarpy and gynoecious flowering habit on fruit production and growth of cucumber *Cucumis sativus* L. J. Amer. Soc. Hort. Sci. 98: 602–604.
6. de Ponti, O. M. B., and F. Garretsen. 1976. Inheritance of parthenocarpy in pickling cucumber (*Cucumis sativus* L.) and linkage with other characters. Euphytica 25: 633–642.
7. Dolan, D. D., and S. W. Braverman. 1971. Plant Germplasm for the Northeast. New York Res. Rep. No. 6. Geneva, NY.
8. El-Shawaf, I. I. S., and L. R. Baker. 1981. Performance of hermaphroditic pollen parents in crosses with gynoecious lines for parthenocarpic yield in gynoecious pickling cucumber for once-over mechanical harvest. J. Amer. Soc. Hort. Sci. 106: 356–359.
9. Ghaderi, A., and R. L. Lower. 1981. Estimates of genetic variance for yield in pickling cucumbers. J. Amer. Soc. Hort. Sci. 106: 237–239.
10. Hallauer, A. R., and S. A. Eberhart. 1970. Reciprocal full-sib selection. Crop Sci. 10: 315–316.
11. Hallauer, A. R., and J. B. Miranda. 1981. Quantitative genetics in maize breeding. Iowa State Univ. Press, Ames.
12. Hogue, E. J., and H. B. Henney. 1974. Ethephon and high density plantings increased yield of pickling cucumbers. HortScience 9: 72–73.
13. Horst, E. K. 1977. Vegetative and reproductive behavior of *Cucumis sativus* as influenced by photoperiod, temperature and planting density. M.S. thesis, North Carolina State Univ., Raleigh.
14. Horst, E. K., and R. L. Lower. 1979. A source of germplasm for the cucumber breeder. Cucurbit Genet. Coop. Rep. 1: 5.
15. Kaufmann, C. S., and R. L. Lower. 1976. Inheritance of an extreme dwarf plant type in the cucumber. J. Amer. Soc. Hort. Sci. 101: 150–151.
16. Lower, R. L., J. Nienhuis, and C. H. Miller. 1982. Gene action and heterosis for yield and vegetative characteristics in a cross between a gynoecious pickling cucumber inbred and a *Cucumis sativus* var. *hardwickii* line. J. Amer. Soc. Hort. Sci. 107: 75–78.
17. McCollum, J. P. 1934. Vegetative and reproductive responses associated with fruit development in the cucumber. Cornell Univ. Agric. Exp. Sta. Memoir 163, Ithaca, NY.
18. Miller, C. H., and G. R. Hughes. 1969. Harvest indices for pickling cucumbers in once-over harvested systems. J. Amer. Soc. Hort. Sci. 94: 485–487.
19. Moll, R. H., and C. W. Stuber. 1971. Comparisons of response to alternative

selection procedures with two populations of maize (*Zea mays* L.). Crop Sci. 11: 706–710.

20. Moll, R. H., and C. W. Stuber. 1974. Quantitative genetics—empirical results relevant to plant breeders. *In* N. C. Brady, ed., Advances in Agronomy. Academic Press, New York.

21. Nienhuis, J., R. L. Lower, and R. R. Horton. 1980. Heterosis estimates for several characteristics in a cross between a gynoecious inbred of *Cucumis sativus* L. and *C. hardwickii* Royle. Cucurbit Genet. Coop. Rep. 3: 20–21.

22. O'Sullivan, J., and H. T. M. Colwell. 1980. Effect of harvest date on yield and grade distribution relationships for pickling cucumbers harvested once-over. J. Amer. Soc. Hort. Sci. 105: 408–412.

23. Peterson, C. E. 1975. Plant introductions in the improvement of vegetable cultivars. HortScience 10: 575–599.

24. Pike, L. M., and C. E. Peterson. 1969. Inheritance of parthenocarpy in cucumber (*Cucumis sativus* L.). Euphytica 18: 101–105.

25. Prend, J., and C. A. John. 1976. Improvement of pickling cucumber with the determinate (*de*) gene. HortScience 11: 427–428.

26. Smith, O. S. 1977. Estimation of heritabilities and variance components for several traits in a random mating population of pickling cumbers. Ph.D. dissertation, North Carolina State Univ., Raleigh.

27. Smith, O. S., R. L. Lower, and R. H. Moll. 1978. Estimation of heritabilities and variance components in pickling cucumbers. J. Amer. Soc. Hort. Sci. 103: 222–225.

28. Tiedjens, A. A. 1928. Sex ratios in cucumber flowers as affected by different conditions of soil and light. J. Agr. Res. 36: 720–746.

29. USDA Agricultural Statistics. 1918–78. U.S. Govt. Printing Office, Washington, DC.

30. West, D. R., W. A. Compton, and M. A. Thomas. 1980. A comparison of replicated S1 per se vs. reciprocal full-sib index selection in corn. I. Indirect response to population densities. Crop Sci. 20: 35–42.

31. Whitaker, T. W., and W. P. Bemis. 1976. Cucurbits. *In* N. W. Simmonds, ed., Evolution of Crop Plants. Longman, New York.

33 | Developmental Aspects of the *B* Genes in *Cucurbita*

Oved Shifriss

ABSTRACT. The *B* genes are nuclear elements that bring about precocious depletion of chlorophyll. In fruits, they are expressed in all known genetic backgrounds. In vegetative organs, they may or may not be expressed. If they are expressed, some or all vegetative organs are affected, depending upon the genetic background. Moreover, productive cultivars of high fruit quality evolved through certain combinations of *B* and other genes. These observations support two hypotheses. First, the capacity of plants to synthesize chlorophyll throughout life is sustained in part by a genetic regulatory mechanism. It is suggested that the variations in precocious depletion of chlorophyll in *Cucurbita* represent a promising model for investigations of this mechanism in relation to plastid development. Second, a potentially deleterious gene can become a beneficial gene exclusively if 1) in addition to its deleterious effects this gene exhibits some beneficial effects, and 2) there exist in the gene pool elements that selectively suppress the deleterious effects.

Immature angiosperm fruits that are not enclosed by bracts synthesize chlorophyll in the outer layer of their cells and appear green. As fruits mature, they often turn to a bright color during ripening. This change in fruit color is preceded by chlorophyll depletion unmasking the presence of other pigments.

It can be argued that chlorophyll synthesis in immature fruits has a selective advantage for two reasons. First, accessible photosynthates contribute to fruit and seed development. Second, the green color of immature fruits, being indistinguishable from foliage color, serves as camouflage that protects these tender organs against predators. The change to bright color during ripening may also have a selective advantage because it promotes seed dispersal by predators.

If the preceding points are valid, any genetic variant whose fruits fail to

406

synthesize chlorophyll early in ontogeny would be less adapted than the wild type. Indeed, I am not aware of any successful variant of this kind among natural populations of the angiosperms. However, such variants exist in two cultivated species of *Cucurbita*: *C. pepo* L. and *C. maxima* Duch. ex Lam. Chlorophyll depletion in the fruits of each of these variants is conditioned by a major nuclear gene, designated *B*. The *B* genes of the two species are similar in expression, but their relationship is not yet understood. Breeding experiments showed that different combinations of *B* and other genes produce a wide range of variation in fruit and plant development. Most of these variations are deleterious, but some are nondeleterious or beneficial.

The objectives of the present chapter are to describe the origin of the *B* genes as well as the variations in their developmental expression, and to discuss the significance of these and other genetic variations that block chlorophyll synthesis.

The *B* Genes: Their Origin and Relation to Present-Day Cultivars

Cucurbita pepo

The fruits of most cultivars of *C. pepo* are green prior to anthesis. Cultivars bearing such fruits exclusively represent the standard mode of pigmentation. As standard fruits grow they either remain green up to maturity or turn to other colors (orange, yellow, tan, white) sometime between anthesis and maturity.

The situation is different in the bicolor-fruited cultivars of the ornamental gourds, *C. pepo* var. *ovifera* (L.) Alef. A bicolor fruit is made up of green and yellow portions, usually in polar orientation (Figure 1A). A bicolor-fruited plant can produce both green and bicolor fruits. The bicolor fruits appear irregularly during plant development, and they often vary considerably in extent and design of pigmentation (Figure 1B–F). More important, yellow pigmentation in a bicolor fruit occurs in the ovary sometime before anthesis. This deviation from the standard timing of fruit color differentiation represents the precocious mode of pigmentation (1, 10–12).

Starting with a single bicolor-fruited plant of the 'Pear' gourd (Figure 2), a program of inbreeding and selection was initiated in 1948 in an attempt to develop genetically stable lines that differ in extent of precocious fruit pigmentation. As a result, two true-breeding precocious lines were established. The fruits of one line (line 11) were highly variable phenotypically, ranging from green to spotted to distinctly bicolor in design. The fruits of the other line, known as precocious 'Pear', were either completely yellow or almost

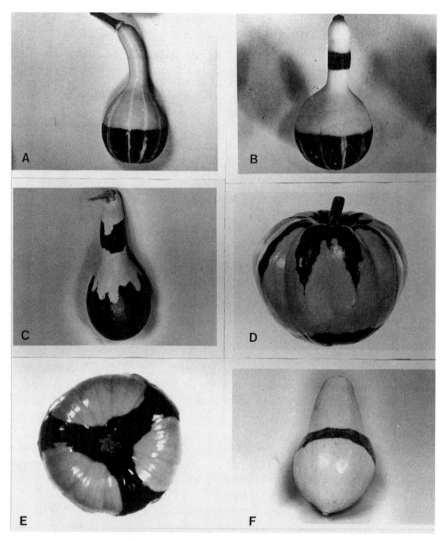

Figure 1. Variations in the bicolor fruit design in C. *pepo*. Fruit size is not to scale.

completely yellow, i.e., only the very tip of their distal end was green, depending on the environment.

Results of breeding experiments demonstrated that the standard and precocious modes of pigmentation are conditioned by alleles at the major locus B: B^+ for standard and B for potentially bicolor fruit. Furthermore, there are at least two precocious alleles, B (as in precocious 'Pear', BB) and B^w (a weak allele of B, as in line 11, B^wB^w). The results also suggested that the B alleles are generated by nuclear instability, and that the expression of these

Figure 2. A bicolor-fruited plant of the ornamental 'Pear' gourd. Gene *B* originated as a mutation in one of the inbred generations obtained from this plant.

alleles, particularly the weak allele, is highly sensitive to environmental conditions. Further studies showed that the expression of both B^+ and *B* is affected by other genes. Therefore, in reality, the two modes of pigmentation are under the control of two genetic systems: the B^+ system and the *B* system.

Gene *B* was transferred from precocious 'Pear' to nine standard squash cultivars. Significantly, the fruits of the resulting *BB* lines were not only precociously pigmented, but also completely yellow (Figure 3) in different environments. Some of the *BB* breeding lines were used either as parents or germplasm for the development of 15 *B* cultivars in squash. Among the 15, the following six represent different groups of summer and winter squash: 'Golden Zucchini' (introduced in 1973), 'Multipik' (1981), 'Jersey Golden Acorn' (1982), 'Sunburst' (1985), 'Orangetti' (1986) and 'Autumn Gold' (1987).

There is little doubt that the mutation from B^+ to *B* (or B^w) occurred occasionally in the past and that some of the old squash cultivars (21) carried one of the *B* alleles, most probably B^w. But the *B* gene has not been utilized effectively in squash breeding until recent times.

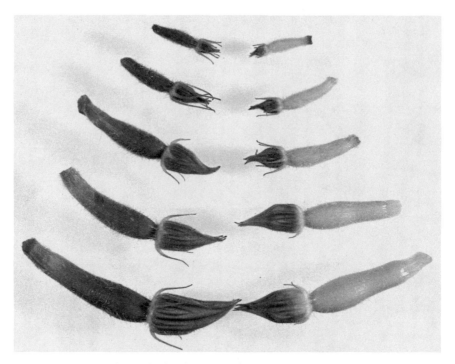

Figure 3. Pistillate flower buds of B^+B^+ (left, green) and BB (right, yellow) inbreds of *C. pepo* 'Early Prolific Straightneck' background prior to anthesis.

Cucurbita maxima

My first acquaintance with a precocious cultivar in this species was an introduction from India, PI165558 (11). Analysis of a cross between PI165558, bearing precociously pigmented fruits, and 'New Hampshire Buttercup', bearing green fruits, demonstrated that precocious fruit pigmentation is conditioned by a dominant gene. This gene was also designated by symbol B.

A subsequent survey suggested that cultivars which bear golden, yellow, pink, white, or variegated fruits belong to the B system, that this system includes leading commercial cultivars such as 'Boston Marrow', which probably originated over 150 years ago (21), and that the B^+ system includes only cultivars that bear persistently green fruits.

If this division of cultivars is correct, it implies that nongreen mutations within the B^+ system are either extremely rare or do not occur in *C. maxima*, that mutations from B^+ to B are relatively common, and that in selecting nongreen fruits humans unconsciously fixed only B alleles.

Cucurbita moschata

I am not aware of any precocious cultivar or mutant in *Cucurbita moschata* (Duch. ex Lam.) Duch. ex Poir. However, two different *BB* breeding lines of *C. moschata* were established in 1980 through interspecific transfers of the two *B* genes from *C. maxima* and *C. pepo*. The *BB* line that carries the *B* of *C. maxima* is known as IL-B (developed by A. M. Rhodes of the Illinois Agricultural Experiment Station), and the *BB* line that carries the *B* of *C. pepo* is known as NJ-B (developed in the New Jersey Agricultural Experiment Station). The donor of *C. maxima B* was PI165558 and the donor of *C. pepo B* was 'Jersey Golden Acorn' (16).

Cucurbita maxima and *C. pepo* are reproductively isolated from one another by strong genetic barriers (22). These barriers represent an insurmountable obstacle to a study of the relationship between the two *B* genes through breeding experiments. However, the transfer of these genes to *C. moschata* opened the opportunity for such a study.

The analysis of the cross IL-B × NJ-B has not been completed. The F_1 hybrid of the two *BB* inbreds was precocious, vigorous, and fertile. The F_2 generation was highly variable with respect to precocious chlorophyll depletion in different organs during plant development, and critical classification of many plant phenotypes was uncertain (16). Nevertheless, I believe that future studies of this cross, at the organismal and molecular levels, will clarify not only the relationship between the *B* genes but also the important developmental consequences associated with them.

Mutability at the *B* Locus

According to a previous suggestion (12), nuclear instability—the source of *B* genes—occurs largely during gametogenesis, and fruit variation on the same plant is due to phenotypic plasticity of *B*.

Following the substitution of *B* for B^+ in squash cultivars of different backgrounds, it became evident that while most *BB* lines are stable genetically, others are somewhat labile, spontaneously generating B^w or B^+. For example, the *BB* lines of 'Early Prolific Straightneck' and 'Fordhook Zucchini' are highly stable genetically. In contrast, 'Jersey Golden Acorn' (JGA) and NJ260 (17) are slightly unstable. Some data were obtained on mutability of *B* in the cross JGA, *BB* × NJ260, *BB*. The F_1 generation of this cross was stable phenotypically. The F_2 generation consisted of 3231 plants, among which 32 (1%) were distinct deviants. Of these 32, 28 bicolor-fruited individuals proved to be BB^+ and BB^w, 2 were chimeric, and 2 other plants produced fruits of unusual bicolor pattern (Figure 1F), the genotype of which has not been determined. Each of the chimeric plants was made of two persistently different regions: one bearing completely yellow fruits (BB) and the other bearing bicolor and green fruits (BB^+). While it is not unreasonable to assume that back-mutations from B to B^+ or B^w are generated during gametogenesis, the finding of two chimeric plants shows that mutations affecting B can occur in the soma.

Some puzzling data were obtained from backcross progenies, $BB^+ \times B^+B^+$, in the ornamental gourds. These backcross data showed consistent deficiency in the B^+B^+ class (Table II, $Bb \times bb$, in reference 11). The cause of this deficiency is not known. Different interpretations can be entertained, and among them is the high rate of mutation from B^+ to B in some ornamental gourds.

Timing of Gene Expression and Control of the Bicolor Design

The terms 'standard' and 'precocious' were used in the first section of this chapter to distinguish between two groups of cultivars. In the standard group the onset of chlorophyll depletion occurs at different times during post-anthesis stages. In the precocious group the onset of chlorophyll depletion occurs at different times during the early pre-anthesis stages. This restricted use of the term precocious is operationally valid in *C. pepo* and in several cultivars of *C. maxima*. In a broader sense, however, the term applies to all variants in which chlorophyll synthesis is blocked prior to fruit maturity or prior to natural senescence of any vegetative organ. The main usefulness of this term is that it focuses attention on deviations in timing of gene expression.

Figure 4 illustrates diagrammatically the timing of fruit pigmentation during development among the cultivars of the two genetic systems in both *C. pepo* and *C. maxima*. It also illustrates the relationship between timing and extent of pigmentation. The following conclusions are drawn on the basis of present knowledge.

First, in *C. pepo*, the cultivars of the B^+ system differ widely in onset of fruit color, and their fruits are uniformly pigmented, i.e., they do not exhibit polarized regions of color. The variation in timing ranges from a few days before anthesis (Figure 4, *C. pepo*, number 3, e.g., 'Early Prolific Straightneck') to about 20 days after anthesis or just before full maturity (Figure 4, *C. pepo*, number 2, e.g., 'Small Sugar'). In both *C. pepo* and *C. maxima*, the green-fruited cultivars may turn to another color sometime after maturity or in storage. In these cases, pigmentation is often incomplete, but it is not polarized. The timers in the B^+ system have not been identified. They may be different alleles of B^+ or different modifiers.

Second, in both species, the cultivars of the B system differ in onset of fruit color, but these cultivars exhibit an inverse relationship between timing

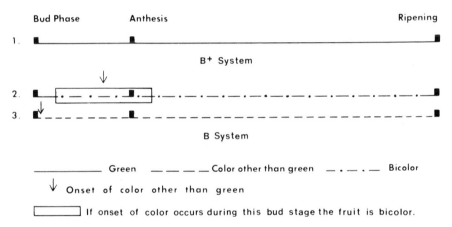

Figure 4. Fruit color development in cultivars of the *B*⁺ and *B* systems in *C. pepo* and *C. maxima*.

and extent of fruit pigmentation. Thus, unlike the *B*⁺ system, the *B* system is capable of producing bicolor fruits (Figure 4: *C. pepo*, number 4; *C. maxima*, number 2) of polar orientation. This difference between the two systems is clearly evident under certain environmental conditions in which their onset of fruit color overlaps. The timers in the *B* system are the *B* alleles and their modifiers (see below).

Third, the *B* system of *C. pepo* differs from the *B* system of *C. maxima* in two respects: 1) The expression of *B* in *C. pepo* is strictly confined to pre-anthesis stages. If B is not expressed during these stages the fruit will be indistinguishable from a B^+B^+ fruit. In contrast, the expression of some B genes of *C. maxima* occurs beyond the pre-anthesis stages (Figure 4, *C. maxima*, number 2). The most extreme cases are some BB^+ hybrids of *C. maxima*. The onset of pigmentation in these hybrids occurs about 20 days after anthesis, but the fruits are bicolor. 2) Unlike the bicolor fruits of *C. pepo*, the bicolor fruits of some backgrounds of *C. maxima* are prone to exhibit partial longitudinal striping.

The factors that control the predisposition of a plant to produce bicolor fruits actually regulate the timing of B expression. These factors include the strength of the B gene involved, its dosage, the kind and number of modifier genes present, and the environment (12, 19). Elements known as Ep modifiers extend the potential boundaries of pigmentation set by the B genes, while Ep^+ modifiers have the opposite effect.

A BB plant that carries a high number of Ep modifiers produces completely pigmented fruits in different environments. A B^wB^+ plant that carries a high number of Ep^+ modifiers produces numerous green fruits and some bicolor ones. In between these extremes, a B^wB^w (or BB^+) plant with a balance between Ep and Ep^+ modifiers is likely to produce bicolor fruits of perfect design with respect to polar orientation (Figure 1A), if this plant grows in a favorable environment. Future studies of the genetic-environment control of the bicolor design may contribute to our understanding of polarity in the design of multicellular organisms.

Finally, a bicolor design can persist throughout development only in the presence of genotype Y^+Y^+ W^+W^+. The green tissue of a bicolor fruit behaves during development as the green tissue of young fruits of the B^+ system. Gene Y for yellow color, and gene W for white color (as well as other pigment-inhibiting genes) act at post-anthesis stages. Therefore, in the presence of either Y or W or both, a bicolor design fades gradually during post-anthesis stages.

Effects of the B Genes on Fruit Growth and Quality

Fruit growth

In *C. pepo*, BB lines of many genetic backgrounds exhibit a lower rate of fruit growth than their near-isogenic B^+B^+ inbreds. A lower growth rate is often associated with smaller, more slender fruit. A decrease in fruit size is not evident in BB lines of other genetic backgrounds, particularly under favorable growing conditions. Moreover, the decrease in fruit size, when it exists, is less drastic in BB^+ than in BB plants of comparable background. In *C. maxima*, the B of PI165558 lowers the rate of fruit growth in some genetic backgrounds (e.g., 'Green Delicious'), but many BB lines of this species do not exhibit lower rates of fruit growth. Indeed, the most rapidly growing and largest fruits of C. *maxima* are produced by *BB* cultivars in which chlorophyll depletion occurs early in pre-anthesis stages (e.g., 'Big Max' and 'Exhibition').

If a lower rate of fruit growth in *BB* lines of some backgrounds is due to deficiency in accessible photosynthates, then a normal rate in *BB* lines of other backgrounds must be due to an alternative pathway that provides an adequate supply of essential metabolites.

The *BB* line of 'Fordhook Zucchini' (C. *pepo*) bears poorly developed fruits with excessively firm placental tissue and few seeds. In addition, these fruits tend to crack prematurely under certain conditions. Obviously, such a line is unacceptable as a seed producer. Crossing this line with another cultivar (B^+B^+) of the zucchini group, followed by selection, resulted in a new *BB* line that bears normal fruits and retains the desirable characteristics of the original *BB* parent. Abnormally developed *BB* fruits similar to those of the *BB* line of 'Fordhook Zucchini' have not been observed in C. *maxima*.

Fruit quality

One of the most favorable effects of the *B* gene on fruit quality is its increase of flesh carotenoids in some genetic backgrounds. Indirect evidence of this effect accumulated gradually from 1958 to 1962. It was initially based on observations of relative flesh color intensity. The evidence showed that the interaction of *B* and genes for heavy fruit pigmentation, such as *L* (L^+ being for light fruit pigmentation), in the absence of pigment inhibiting genes, greatly intensifies the golden color of the flesh. This evidence was subsequently supported by chemical determinations of flesh carotenoids in several different pairs of near-isogenic inbreds, B^+B^+ and *BB* (8, 18).

The *B* and *L* interaction is regarded here as the enhancer system of flesh carotenoids in *Cucurbita*. There are probably several different *L* genes, and Paris and Nerson (5) successfully identified two of them, L_1 and L_2. But other genes may also play a role in the enhancer system, particularly in relation to different carotenoids. The potential increase in total flesh carotenoids by this system varies from less than two to over ten orders of magnitude, depending on the *L* genes involved and the initial carotenoid level in the comparable B^+B^+ background. Differences in reported levels of carotenoids may be due to differences in chemical procedures (8). In addition, it is necessary to take into consideration differences in size between the isogenic fruits (and the size of their cells), B^+B^+ and *BB*. From a genetic point of view, differences in carotenoid levels must be based on an equal number of cells.

One of the striking differences in flesh carotenoids is between B^+B^+ and

BB fruits of 'Fordhook Zucchini' background. Externally B^+B^+, L_1L_1, L_2L_2 fruits are dark green, gradually becoming "black" during development. Internally, these fruits are inconspicuously pigmented (almost white). In contrast, *BB*, L_1L_1, L_2L_2 fruits are intensely golden, externally and internally. Thus, the B^+ and L system is distinguished by intense green pigmentation, and the B and L system is distinguished by intense yellow pigmentation in the flesh. Schaffer et al. (7) showed that while B can increase the level of carotenoids in the flesh, it does not increase the level of carotenoids in the skin. This is an interesting observation because the different expressions of B in different tissues of the fruit are probably related to a single effect of B that occurs earlier in ontogeny (20).

Other reported effects of B on fruit are related to flavor, texture, and relative tolerance to damage caused by some viruses (12, 18). As to each of these reports, one can say that B is either "good," "neutral," or "bad," depending upon the genetic background. But it is often difficult to assess the role of B in the absence of critical genetic information.

Expression of the *B* Genes in Vegetative Organs

The fruit is the consistent target of the B genes. In vegetative organs, these genes can be either expressed or suppressed. With respect to precocious chlorophyll depletion in vegetative organs, the *BB* cultivars, breeding lines, and clones are divided into six groups.

1. All aerial vegetative organs are green. Plants are vigorous and productive. Examples: some inbreds of 'Pink Banana' (*C. maxima*) and a breeding line JGA-II (*C. pepo*).

2. All aerial vegetative organs are light green. Plants are low yielders. Example: some inbreds of 'Golden Delicious' (*C. maxima*).

3. Leaf blades are prone to yellowing at seedling stages, particularly in response to certain environmental conditions, including low temperatures. The leaf blades are green at later stages of plant development. The turning from yellow to green leaf blades may be abrupt or gradual. If gradual, the transitional stages include variegated leaf blades in which yellowing is often confined to veins. Aerial vegetative organs other than leaf blades are green throughout development. Examples: 'Jersey Golden Acorn' (*C. pepo*) and a breeding line known as precocious 'Small Sugar' or PSS (*C. pepo* [12]). 'Jersey Golden Acorn' is less sensitive to leaf yellowing than PSS.

4. All aerial vegetative organs (stems, tendrils, petioles, and leaf blades) are prone to yellowing in response to some environmental conditions, including low temperatures. The extent of yellowing is highly variable and generally inconspicuous under field conditions. Plants are vigorous and productive in tropical environments of high solar radiation and high tem-

peratures, conditions that suppress yellowing. Example: PI165558 (*C. maxima*, [11]). In addition to the *B* genes, PI165558 may carry a genetic element for precocious depletion of chlorophyll in the stems.

5. Stems, tendrils, and petioles are golden, but leaf blades are green under field conditions. Although the shoot apices are green, the onset of chlorophyll depletion occurs early, and this is associated with a gradual increase in intensity of golden color (and high carotenoid levels), externally and internally. Example: IL-B (a breeding line of *C. moschata* that carries the precocious genes of PI165558, [16]).

6. Semi-lethals and lethals. This group includes difficult to reproduce inbreds of variegated plants (green/yellow mosaicism) that often turn into completely yellow individuals. These inbreds evolved in *C. maxima* from crosses between *BB* and *B⁺B⁺* inbreds, as well as between different *BB* inbreds in which one of the parents was PI165558. In addition, the group includes difficult to maintain albino and yellow seedlings. The latter are lethals except when grown under special conditions. These lethals are segregates of NJ-B × IL-B (*C. moschata* lines with *B* genes derived from *C. pepo* and *C. maxima*, respectively, [16]).

In *BB* cultivars of group 3 (*C. pepo*) the propensity for yellow leaf blades can be overcome by substituting gene *Ses-B* for *Ses-B⁺*. Gene *Ses-B* selectively suppresses the expression of *B* in leaf blades but not in fruit (13). The breeding line JGA-II of group 1 and 'Jersey Golden Acorn' of group 3 are near-isogenic, the former being *Ses-B Ses-B* and the latter, *Ses-B⁺ Ses-B⁺*. The *B⁺B⁺* cultivars of *C. pepo* are divided into three classes: 1) those that carry *Ses-B⁺*, 2) those that carry a strong *Ses-B*, and 3) those that carry selective suppressors of varying degrees of suppression.

Line IL-B of group 5 (*C. moschata*) was developed from the cross *C. maxima* (PI165558) × *C. moschata*, followed by six backcrosses to *C. moschata* and selection for golden color. The recurrent parent of *C. moschata* used in the last few backcrosses had dark green fruits and dark green stems. Therefore, the intensely golden stems of IL-B could have resulted from interaction between the precocious genes of PI165558 and the *L* genes for heavy pigmentation contributed by the *C. moschata* parent. Thus there are two enhancer systems of carotenoids in *Cucurbita*, one in fruits and the other in stems.

In some cases the effect of *B* on vegetative organs is particularly striking. Two cases in which shoots are affected are described below.

When plants of a *BB* breeding line (closely related to PSS of group 3) are grown perennially under greenhouse conditions, they often behave in an unusual manner in winter. With the passage of time, the basal portions of their main stem and roots become intensely golden. Shoots that originate in basal leaf axils in winter are yellow, being entirely devoid of chlorophyll. Some of these shoots turn quickly to green, the yellow stem being 5 to 10 cm in height before reversion. Other shoots remain completely yellow up to a height of about 75 cm before reversion, and the transitional phase of these shoots is relatively long (Figure 5). Shoots that originate in upper leaf axils of the main stem are green.

An extraordinary *BB* inbred was developed from a segregate of the interspecific cross of *C. pepo* 'Jersey Golden Acorn' *BB*, × *C. moschata* 'Burpee Butterbush', B^+B^+. Under field conditions, the normal stems and leaves of this inbred are green, but most fruits are fused with shoots. The fused shoots are invariably yellow, but when they grow beyond the fruit they turn into green either abruptly or gradually, as in Figure 5.

A radically new precocious variant of *C. pepo* was found in an F_2 generation of the cross 'Bicolor Spoon', B^wB^w, × 'Table King', B^+B^+. The fruits and stems of this variant are green, but the color of its leaf blades changes during plant development. Leaf blades that are differentiated early in development are green, whereas those that are differentiated late in development

Figure 5. Gradual reversion of a shoot from yellow to green. This shoot originated in one of the basal leaf axils of a *BB* plant (of 'Small Sugar' background) late in life. The shoot was completely yellow up to a height of 75 cm before reversion began. All shoots that originated in upper leaf axils were completely green.

are yellow or partially yellow. Inbreeding resulted in a pure line of late leaf yellowing. The genotype for late yellowing has not been identified, but it is clear that its target is the leaf at a particular time during plant development, not the fruit.

Overview

There is a vast literature on chlorophyll-deficient mutants in plants, and numerous variegated cultivars are known in which chlorophyll synthesis is adversely affected by different genetic systems. Many chlorophyll-deficient mutants are lethal or semi-lethal. Similar variants were uncovered in *Cucurbita* through inbreeding of open-pollinated cultivars (9) as well as through interspecific crosses (6). There are also similarities between *Cucurbita* and other taxa with respect to some interesting developmental variations.

As to precocious depletion of chlorophyll in *Cucurbita*, it appears that different groups of genes target different organs at specific times during ontogeny. Further investigations of this phenonmenon may reveal the existence of a genetic regulatory mechanism that sustains the capacity of plants to synthesize chlorophyll throughout development.

Although the *B* genes in *Cucurbita* are potentially deleterious, they can become beneficial. Glass (3) speculated that some deleterious genes are maintained in natural populations of *Drosophila* because they also have some beneficial effects. Mayr (4) elaborated on this idea in the broader context of "gene coadaptation," a concept originally proposed by Dobzhansky (2). Unlike the selective suppressors in *Cucurbita*, however, the suppressors that Glass discovered in *Drosophila* completely mask the phenotypes of the deleterious genes.

While our understanding of the *B* genes and those that affect their expression has increased markedly over the years, intriguing questions remain to be answered, including the following: Do the *B* genes reflect the operation of mobile elements? What function, if any, does *Ses-B* have in B^+B^+ cultivars? Can an interaction of the *B* gene and other genes enhance female expression (14, 15, 17)? Is the signal that the *B* gene sends to plastids a diffusible substance (10–12)?

Acknowledgments

This chapter forms a paper of the Journal Series, New Jersey Agricultural Experiment Station, Project 12002. I thank Cyril R. Funk for support and encouragement. I am grateful to Vincent A. Abbatiello for preparing Figures 1, 3, and 5 and to Toby Scriba for preparing Figure 2.

Literature Cited

1. Bailey, L. H. 1958. The Garden of Gourds. Gourd Society of America. Boston. Reprinted from the original 1937 edition, Macmillan, New York.
2. Dobzhansky, T. 1951. Genetics and the Origin of Species, 3rd ed. Columbia Univ. Press, New York.
3. Glass, B. 1957. In pursuit of a gene. Science 126: 683–689.
4. Mayr, E. 1970. Population, Species and Evolution. Harvard Univ. Press, Cambridge, MA.
5. Paris, H. S., and H. Nerson. 1986. Genes for intense fruit pigmentation of squash. J. Heredity 77: 403–409.
6. Robinson, R. W. 1987. Novel variation in an interspecific cross. Cucurbit Genet. Coop. Rep. 10: 85–86.
7. Schaffer, A. A., C. D. Boyer, and T. Gianfagna. 1984. Genetic control of plastid carotenoids and transformation in skin of *Cucurbita pepo* L. Theor. Appl. Genet. 68: 493–501.
8. Schaffer, A. A., H. S. Paris, and I. M. Ascarelli. 1986. Carotenoid and starch content of near-isogenic B^+B^+ and BB genotypes of *Cucurbita*. J. Amer. Soc. Hort. Sci. 111: 780–783.
9. Shifriss, O. 1945. Male sterilities and albino seedlings in cucurbits. J. Heredity 36: 47–52.
10. Shifriss, O. 1955. Genetics and origin of the bicolor gourds. J. Heredity 46: 213–222.
11. Shifriss, O. 1965. The unpredictable gourds. Amer. Hort. Mag. 44: 184–201.
12. Shifriss, O. 1981. Origin, expression and significance of gene B in *Cucurbita pepo* L. J. Amer. Soc. Hort. Sci. 106: 220–232.
13. Shifriss, O. 1982. Identification of a selective suppressor gene in *Cucurbita pepo* L. HortScience 17: 637–638.
14. Shifriss, O. 1985. Origin of gynoecism in squash. HortScience 20: 889–891. (See Corrigenda, HortScience 20: 990, 1985.)
15. Shifriss, O. 1986. A gynoecious line of B^+B^+ genotype in *Cucurbita pepo*. HortScience 21: 319.
16. Shifriss, O. 1986. Relationship between the B genes of two *Cucurbita* species. Cucurbit Genet. Coop. Rep. 9: 97–99.
17. Shifriss, O. 1987. Synthesis of genetic females and their use in hybrid seed production. U.S. Patent 4686319.
18. Shifriss, O. 1987. Notes on squash breeding. Cucurbit Genet. Coop. Rep. 10: 93–99.
19. Shifriss, O., and H. S. Paris. 1981. Identification of modifier genes affecting precocious fruit pigmentation in *Cucurbita pepo* L. J. Amer. Soc. Hort. Sci. 106: 653–660.
20. Shifriss, O., and T. H. Superak. 1987. Precocious fruits pigmentation in B^wB^+ plants of *Cucurbita pepo* L. Cucurbit Genet. Coop. Rep. 10: 100–102.
21. Tapley, W. T., W. D. Enzie, and G. P. Van Eseltine. 1937. The Vegetables of New York. Vol. 1, part 4. The Cucurbits. New York Agric. Exp. Sta., Geneva.
22. Whitaker, T. W., and G. N. Davis. 1962. Cucurbits: Botany, Cultivation, and Utilization. Interscience, New York.

34 | Interspecific Trisomics of *Cucurbita moschata*

J. D. Graham and W. P. Bemis

ABSTRACT. Phenotypic effects and transmission rates of the extra chromosome in interspecific trisomics of *Cucurbita moschata* with one chromosome from *C. palmata* were compared with those of a primary trisomic of *C. moschata*. Based on gross morphological similarities, 17 interspecific trisomic lines were placed in six phenotypic groups, suggesting that six different *C. palmata* chromosomes were recovered. Fruit from one of the interspecific trisomics exhibited the hard rind of *C. palmata*, indicating that this is a dominant trait carried on one chromosome. Some phenotypic effects of the extra chromosome were similar in both the interspecific and primary trisomics, showing a chromosomal effect due to genic imbalance. Transmission of the extra chromosome through the female was less than the expected 50%. None of the extra chromosomes was transmitted through the male parent.

The addition of a single alien chromosome to a species is one way to overcome barriers to introgression and to introduce agronomically useful traits into the recipient species. Since the univalent would be expected to be included in a maximum of only 50% of the gametes, the agronomic usefulness of a monosomic alien addition line would be limited unless it could be vegetatively propagated. Addition lines, however, may be used to obtain genetic information about individual chromosomes of the donor species. If phenotypic differences between the addition lines and the recipient represent specific genes from the donor species, these genes can be relegated to particular chromosomes. Phenotypic differences due to the effect of the added chromosome itself are caused by a genic imbalance in the trisomic. This trisomic effect can be investigated by comparing the phenotypic effects of the extra chromosome in addition lines with the effects of the extra chromosome in primary trisomics of the recipient. Transmission rates of the added chromosome through the male and female gametes would give an indication of the genic imbalance caused by the addition as well as the possibility of obtaining the disomic addition through selfing. Theoretically, the disomic addition could be chromosomally stable, which would increase the agronomic potential of the respective monosomic addition line.

The first report of a fertile interspecific trisomic of *Cucurbita moschata* (Duch. ex Lam.) Duch. ex Poir. was by Bemis (4), who added a single chromosome of *C. palmata* S. Wats. to *C. moschata* 'Butternut'. After backcrossing the amphidiploid of the two species to the diploid *C. moschata* parent, the resulting triploid was backcrossed to *C. moschata*, which produced interspecific aneuploids, including fertile interspecific trisomics. Although the addition of a single alien chromosome to the normal complement of a species is usually referred to as a monosomic alien addition, the term *interspecific trisomic* is more descriptive when the added chromosome is from within the genus of the recipient species.

The foregoing procedure produced 17 fertile interspecific trisomics, which could be identified by their phenotypic differences from diploid 'Butternut'. These trisomic lines were grouped according to similarities of gross morphologically distinctive traits. Transmission rates of the *C. palmata* chromosome through the male and female gametes were determined, using the segregation ratios in the progeny of reciprocal crosses between the interspecific trisomics and diploid *C. moschata*. One primary trisomic of 'Butternut', obtained by crossing the triploid with the diploid, was used to compare the morphological effects and transmission rates of the extra *C. moschata* chromosome with those of the extra *C. palmata* chromosome in the *C. moschata* background.

Morphological and Taxonomic Relationships

Some of the features of the wild, xerophytic *C. palmata* that distinguish it from *C. moschata* are its perennial roots, deeply lobed leaves, short tendrils, peduncles not expanded at their attachment to the fruit, and nearly round five-carpelled pepos (2). Its fruits are extremely bitter (3). The runners root readily at the nodes, producing colonies with little genic variation (7). In contrast, the domesticated, mesophytic *C. moschata* has a five-angled peduncle that expands, then curves inward at the attachment to its tricarpellate fruit. Its root is fibrous and annual (1, 2). The cultivar 'Butternut' originated from 'Canada Crookneck' (14), the fruit shape of which is occasionally produced by 'Butternut' plants (10).

Since no natural polyploidy has been found in *Cucurbita* and all species have 20 pairs of similar chromosomes, species differentiation is largely genic or cytoplasmic and not due to changes in chromosome number or gross rearrangements (13). Phenetic studies (11), interspecific hybridization (5), and evidence from the squash and gourd bees (9) place *C. moschata* and *C. palmata* in separate groups of the genus. Although *C. moschata* is the mesophytic species of *Cucurbita* most closely related to all of the xerophytic species (13), *C. palmata* is comparatively distantly related to it (8, 9). In hybrids between *C. moschata* 'Butternut' and *C. palmata*, the frequency of bivalents at metaphase I ranged from six to nine, showing a lack of chromosome homology between the two taxa (4). In a study of the interspecific hybrid, Bemis (3) suggested that both fruit size and the perennial root of *C. palmata* were dominant characters, and that the flesh thickness and carpel number of *C. moschata* fruit were dominant. Most characters, including leaf shape, tendril length, skin color, flesh color, and bitterness of the fruit, were intermediate between the two parents. From phenetic studies (6), it was concluded that the hybrid had an intermediate phenotype because of the extreme genetic divergence of the parental species.

Morphology of the Trisomics

The most noticeable influence of single *C. palmata* chromosomes on the diploid *C. moschata* complement was on fruit shape (Figure 1). Overall, the trisomics were surprisingly similar to diploid *C. moschata*, considering the distant relationship between the two species. This may be due to a loss of the *C. palmata* chromosomes with a more marked effect in the cross between the triploid and the diploid. Although 17 fertile interspecific trisomics were obtained, they were classified into six phenotypic groups, indicating that only 6 of the 20 *C. palmata* chromosomes may have been recovered. This number corresponds to the six to nine bivalents observed in metaphase I in the interspecific hybrid. Some of the infertile aneuploids from the above cross may have been trisomic, but these were not studied.

Interspecific Trisomics

The phenotypic group P1 had fruits with a restricted neck, which distinguish it from the other trisomics. This trisomic could be identified in the seedling stage by the narrow angle between the outer leaf veins where they attach to the petiole. The fruits of P2 were parthenocarpic, of variable size, and had a hard thick rind. The vines of this group exhibited considerable fasciated growth. The shape of P3 fruits was not as consistent as in the other trisomics, and the fruits were often cracked at maturity. The most distinguishing character of this trisomic was the droplets of a sticky exudate that covered the surface of the leaves. The cotyledons of most P3 seedlings were borne at an acute angle. Groups P4 and P5 had fruits of similar shape with a

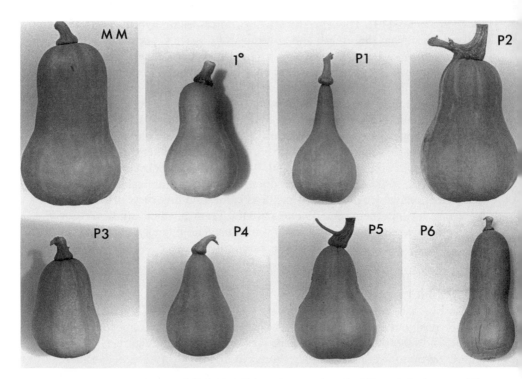

Figure 1. Fruit shapes of diploid (MM), primary trisomic (1°), and six interspecific trisomics (P1–P6) of *C. moschata.*

slightly restricted neck. Only P5, however, had pale yellow staminate flowers. Group P6 had club-shaped fruits, and the vines were less vigorous than the disomic. Otherwise, the vines of the interspecific trisomics were as vigorous as those of the disomic. The carpel number, flesh color, lack of bitterness of the fruit, and fibrous root system of all the trisomics were similar to those of the disomic. However, the hard rind of P2 indicates that this dominant trait is carried on one *C. palmata* chromosome. A study of the isozyme phenotypes (12) revealed that the loci coding fumarase and hard rind were both situated on the alien chromosome of P2.

The Primary Trisomic and the Trisomic Effect

The fruit of the primary trisomic, although smaller, had the shape of disomic 'Butternut'. The leaves had droplets of sticky exudate, and the cotyledons of most seedlings were at an acute angle. Since both of these traits were found in P3, they were probably due to a trisomic effect and not to specific genes. The occurrence of male flower buds at the branch of the early tendrils was also due to the trisomic effect. This character was found in most plants of the primary and interspecific trisomics.

Transmission of the Extra Chromosome

Transmission of the *C. palmata* chromosome through the female was highest (32%) in P3, the interspecific trisomic which most closely resembled the primary trisomic, and was lowest (15%) in P6, the least vigorous trisomic. It might be expected that the most frequent *C. palmata* chromosome obtained in the trisomics would have the highest transmission rate. Although 8 of the 17 interspecific lines were included in P1, this trisomic had an intermediate transmission rate of 26%. It is possible that different *C. palmata* chromosomes have similar effects on the morphology of the disomic, and that P1 includes different trisomics. Each phenotypic group was cytologically confirmed as trisomic by using the iron-aceto-carmine squash technique, but the *C. palmata* univalents could not be differentiated.

Some of the reduction in transmission frequency was due to the loss of the univalent through lagging during meiosis, which was cytologically observed. Since pollen fertility was high, competition between normal pollen and pollen carrying the extra chromosome may have accounted for the lack of transmission through the male. Perhaps by doubling the chromosomes of the trisomics, producing tetraploids with the disomic additions ($4n + 2$), some transmission through the male would be achieved.

Literature Cited

1. Bailey, L. H. 1930. The domesticated cucurbitas. Gentes Herb. 2: 59–115.
2. Bailey, L. H. 1943. Species of *Cucurbita*. Gentes Herb. 6: 267–322.
3. Bemis, W. P. 1963. Interspecific hybridization in *Cucurbita*. II. *C. moschata* Poir. × xerophytic species of *Cucurbita*. J. Heredity 54: 285–289.
4. Bemis, W. P. 1973. Interspecific aneuploidy in *Cucurbita*. Genet. Res. 21: 221–228.
5. Bemis, W. P., and J. M. Nelson. 1963. Interspecific hybridization within the genus *Cucurbita*. I. Fruit set, seed and embryo development. J. Ariz. Acad. Sci. 2: 104–107.
6. Bemis, W. P., A. M. Rhodes, T. W. Whitaker, and S. G. Carmer. 1970. Numerical taxonomy applied to *Cucurbita* relationships. Amer. J. Bot. 57: 404–412.
7. Bemis, W. P., and T. W. Whitaker. 1969. The xerophytic *Cucurbita* of northwestern Mexico and southwestern United States. Madrono 20: 33–41.
8. Goldberg, R. B., W. P. Bemis, and A. Seigel. 1972. Nucleic acid hybridization studies within the genus *Cucurbita*. Genetics 72: 253–266.
9. Hurd, P. D. Jr., E. G. Linsley, and T. W. Whitaker. 1971. Squash and gourd bees (*Peponapis*, *Xenoglossa*) and the origin of the cultivated *Cucurbita*. Evolution 25: 218–234.
10. Pearson, O. H. 1968. Unstable gene systems in vegetable crops and implications for selection. HortScience 3: 271–274.
11. Rhodes, A. M., W. P. Bemis, T. W. Whitaker, and S. G. Carmer. 1968. A numerical taxonomic study of *Cucurbita*. Brittonia 20: 251–266.

12. Weeden, N. F., J. D. Graham, and R. W. Robinson. 1987. Identification of two linkage groups in C. *palmata* using alien addition lines. HortScience 21: 1431–1433.
13. Whitaker, T. W., and W. P. Bemis. 1965. Evolution in the genus *Cucurbita*. Evolution 18: 553–559.
14. Whitaker, T. W., and G. N. Davis. 1962. Cucurbits. Botany, Cultivation and Utilization. Interscience, New York.

35 | Viral Diseases and Genetic Sources of Resistance in *Cucurbita* Species

R. Provvidenti

ABSTRACT. A number of viruses have been reported to infect *Cucurbita* species. Attention is focused on those of economic importance in the United States, although many are important elsewhere. Only a few sources of viral resistance have been found in cultivated species of *Cucurbita*. Wild species have proved to be valuable sources of germplasm for effective control of these pathogens. An enumeration of viruses and sources of resistance to them is presented and is organized according to vectors: aphids—clover yellow vein virus, cucumber mosaic virus, watermelon strain of papaya ringspot virus, watermelon mosaic virus, zucchini yellow fleck virus, and zucchini yellow mosaic virus; beetles—squash mosaic virus; nematodes—tomato ringspot virus; leafhoppers—beet curly top virus; and whiteflies—squash leaf curl virus.

Viral diseases are a significant threat to squash grown in the United States (16). Since no effective control of the vectors has been achieved with insecticides, and eradication of overwintering weed hosts is impractical, the development of resistant cultivars represents the best approach to control viral diseases (14). For years the search for resistance to viral diseases was confined primarily to the domestic and foreign cultivars of *Cucurbita pepo* L., *C. maxima* Duch. ex Lam., *C. moschata* (Duch. ex Lam.) Duch. ex Poir., and *C. mixta* Pang.; however, the results were mostly unsatisfactory. No high level of resistance was ever found, and the few purportedly tolerant lines discovered were never used successfully in squash breeding programs (7, 14).

In 1976 Munger (10) reported a high level of resistance to cucumber mosaic virus in *C. martinezii* L. H. Bailey from Mexico. The search for resistance was expanded by Provvidenti et al. (16) to include 20 wild species

427

grown from seeds collected in the subtropical and tropical regions of the American continents.

In this chapter the principal viruses affecting *Cucurbita* species are discussed and grouped according to their vectors, and known sources of resistance for each are listed.

Viruses Transmitted by Aphids

The most important viruses transmitted by aphids infecting squash are clover yellow vein (CYVV), cucumber mosaic (CMV), watermelon strain of papaya ringspot (PRSV-W), watermelon mosaic (WMV), zucchini yellow fleck (ZYFV), and zucchini yellow mosaic (ZYMV). They are efficiently spread by a number of aphid species, notably *Aphis citricola* Van Der Goot, *A. craccivora* Koch, *A. gossypii* Glover, *Aulacorthum solani* Kaltenbach, *Macrosiphum euphorbiae* Thomas, *Myzus persicae* Sulzer, *Toxoptera citricidus* Kirk, and others. These viruses are transmitted in a stylet-borne manner (nonpersistent), but their acquisition and retention may vary with the aphid species, biotypes, hosts, and strains of the viruses (14).

Clover Yellow Vein Virus

A potyvirus, formerly known as the severe strain of bean yellow mosaic virus (BYMV-S), CYVV is characterized by flexuous rod particles about 750 nm long, which contain a single strand of RNA. CYVV has a very extensive host range within the Leguminosae but also can infect several cucurbit species (17). It can be mechanically transmitted, but there is no evidence of spread via seed. The most striking symptoms usually occur in cultivars of yellow summer squash (*C. pepo*), and consist of small, roundish, bright yellow spots that remain distinct throughout the season. The number of lesions per leaf may vary from a few inconspicuous specks to numerous spots covering the entire leaf area. Temperature and perhaps other environmental factors appear to be critical in the development of symptoms. Diseased plants are somewhat stunted, and the number of viable seeds is somewhat reduced, but the appearance of the fruits is not affected (17).

A high level of resistance to CYVV was found in *C. andreana* Naud., *C. cordata* S. Wats, *C. ficifolia* Bouché, *C. foetidissima* HBK, *C. gracilior* L. H. Bailey, *C. martinezii*, *C. okeechobeensis* (Small) L. H. Bailey, *C. pedatifolia* L. H. Bailey, and *C. scabridifolia* L. H. Bailey. Plants of these species failed to be infected locally or systemically by the virus (14). Also found to be resistant were *C. palmata* S. Wats., *C. palmeri* L. H. Bailey, *C. sororia* L. H. Bailey, and many cultivars of *C. maxima*, *C. moschata*, and *C. mixta*, which usually respond with only localized infection (16). No studies have been conducted regarding the inheritance of resistance.

Cucumber Mosaic Virus

CMV is a member of the cucumovirus group, characterized by three functional pieces of a single strand of RNA encapsulated in three types of similar particles about 28 nm in diameter. The virus has been reported to be transmitted by more than 60 aphid species, but its transmission efficiency depends upon several factors. Generally, the virus is acquired by aphid instars within one minute, but their ability to transmit the virus declines and is lost within two hours. CMV is easily transmitted mechanically, and although there is no evidence that it is seedborne in cucurbits, it can be carried in the seeds of 19 other plant species. This is a virus of worldwide distribution with a host range of about 800 plant species. Unquestionably, CMV is one of the most destructive viral agents affecting squash, causing severe plant stunting, prominent foliar mosaic, malformation, and reduction of leaf size. Infected fruits usually remain small, distorted, and often discolored. Seed production is negligible in severely affected fruits (4).

Through the years many attempts were made to find sources of resistance to CMV in cultivated squash, particularly in accessions of *C. pepo*; however, results were mostly negative, and in some cases the available tolerance was too marginal to be fully exploited (7). Conversely, resistance was found to be common among wild species, including *C. cordata*, *C. cylindrata* L. H. Bailey, *C. digitata* A. Gray, *C. ecuadorensis* Cutler & Whitaker, *C. ficifolia*, *C. foetidissima*, *C. galeotii* Cogn., *C. gracilior*, *C. lundelliana* L. H. Bailey, *C. martinezii*, *C. okeechobeensis*, *C. palmata*, *C. palmeri*, *C. pedatifolia*, and *C. scabridifolia* (16). Inoculated plants of these species responded with localized symptoms but without systemic spread of the virus.

Resistance to CMV from *C. martinezii* has been successfully transferred into *C. pepo*, directly with the help of embryo culture, and indirectly using *C. moschata* as a bridge species. This resistance appears to be partially dominant (23). More recently, an accession of *C. moschata* from Nigeria and some cultivars of *C. maxima* from Africa and Australia were reported to be resistant to this virus (13, 15).

Papaya Ringspot Virus

There are two pathotypes of papaya ringspot virus: 1) PRSV-P (papaya strain), which infects *Carica papaya* L. and most of the Cucurbitaceae, and 2) PRSV-W (watermelon strain), formerly known as watermelon mosaic virus 1 (WMV-1). Of these two pathotypes, PRSV-W is of great economic importance in squash as well as in other cucurbits. Virus particles consist of flexuous rods, about 780 × 12 nm, containing a single strand of RNA.

PRSV-W is efficiently spread by 20 aphid species. It also is easily transmitted mechanically, but is not spread by seed. PRSV-W is widespread in the warm regions of the world, but occasionally it occurs in temperate zones. The leaf blades of infected plants are usually strongly mosaiced, distorted, and narrowed. Fruits are malformed, exhibiting color break and knobby overgrowths (18).

Only a few species seem to be resistant to this destructive virus. A high level of resistance was found in *C. ecuadorensis*, in some accessions of *C. ficifolia* and *C. foetidissima*, and in an accession of *C. moschata* from Nigeria (15, 16). Resistance in the wild species appears to be polygenic, whereas in the Nigerian squash it appears to be monogenically recessive. A low level of tolerance was also detected in 'Zapallito Redondo', a cultivar of *C. maxima* from Uruguay (13).

Watermelon Mosaic Virus

Formerly known as watermelon mosaic virus 2 (WMV-2), this potyvirus consists of flexuous particles about 760 nm long, which contain a single strand of RNA. WMV usually is spread by the same vectors as PRSV (about 20 or more aphid species), but also is easily transmitted by mechanical means. No seed transmission has been reported for this virus. The natural host range of WMV includes most of the Cucurbitaceae and many leguminous species. It is present in the tropics as well as in temperate and cool regions of the world. Generally, the symptoms caused by WMV are less severe than those caused by other cucurbit viruses; however, they may vary considerably depending upon the species and viral strain involved. Symptoms include green mosaic, leaf rugosity, green vein banding, and chlorotic rings. The foliage remains essentially of normal size. Fruits are not distorted, but their colors may be adversely affected, particularly those that are yellow or red (19).

Resistance has been located in a few species, including *C. ecuadorensis*, *C. ficifolia*, *C. foetidissima*, *C. pedatifolia*, and in the Nigerian squash (*C. moschata*) (14). A selection from a Chinese cultivar of *C. maxima* (PI419081) also possesses a good level of resistance (13). Resistance appears to be polygenic in wild species and monogenic recessive in cultivated species.

Zucchini Yellow Fleck Virus

ZYFV, a potyvirus with particles about 750 nm long and a single strand of RNA, was first recognized in 1981. It occurs mainly in the Mediterranean basin, where it infects mostly squash and cucumbers. It is spread by several species of aphids and by mechanical means. Symptoms incited by this virus consist of pinpoint yellow flecks on the foliage and fruit malformation (22). No information is available regarding resistance or tolerance in squash or in other cucurbits.

Zucchini Yellow Mosaic Virus

Like ZYFV, ZYMV was first recognized in 1981. Since then, it has assumed great economic importance. ZYMV was reported almost simultaneously from Italy and France, where it was named muskmelon yellow stunt virus. Presently, it is known to occur in 20 countries on five continents. Similar to other potyviruses, ZYMV particles are flexuous, about 750 nm long, and contain a single strand of RNA. It is spread by a number of aphid species, but also is easily transmissible mechanically. Circumstantial evidence has suggested seed transmission of this virus, but it has been difficult to demonstrate this type of dispersal (6). ZYMV is one of the most destructive viruses occurring in squash, where it incites foliage mosaic, severe malformation, and plant stunting. Fruit symptoms occur as knobby areas and strongly resemble those caused by PRSV-W. Thus, it is often very difficult to differentiate the symptoms caused by these two viruses. In the tropics ZYMV is often associated with PRSV-W. A few strains and pathotypes of ZYMV have already been identified (6).

Cucurbita ecuadorensis and the Nigerian *C. moschata* appear to be the only sources of resistance to this destructive virus. Resistance from the first species tends to be polygenic and very difficult to transfer into *C. pepo*. Better results have been obtained in crosses between *C. ecuadorensis* and *C. maxima*. Resistance from the Nigerian squash is valuable for interspecific crosses with cultivars of *C. pepo* and with other cultivars of *C. moschata*, particularly 'Butternut'. This resistance appears to be partially dominant (11, 15).

Viruses Transmitted by Beetles

Only one virus of this group, squash mosaic (SqMV), is of economic importance in cucurbits, particularly in squash and melons (12).

SqMV is a comovirus with isometric particles about 30 nm in diameter. They contain a single strand of RNA. The major vectors are the striped and spotted cucumber beetles (*Acalymma trivittatum* Mannerhein and *Diabrotica undecimpunctata* Mannerhein, respectively). These species acquire the virus within five minutes and can retain it for about 20 days. SqMV does not multiply in the vector but can be recovered from regurgitation fluid, feces, and hemolymph. In nature the host range of this virus is limited to cucurbits, particularly squash and melons, in which most of the infection can be traced to infected seeds. Two pathotypes have been characterized: strain I, which causes prominent symptoms in melon but mild symptoms in squash; and strain II, which causes symptoms in the reverse. These two pathotypes are also serotypes and thus can be easily distinguished in immunodiffusion tests. Plants infected with SqMV may show a variety of symptoms, including mosaic, ringspots, green vein banding, and protrusion of veins at the leaf margins. Under certain environmental conditions, infected leaves may develop prominent enation (2, 12).

Attempts to find sources of resistance in cultivated species have been unsuccessful (14). However, *Cucurbita ecuadorensis*, *C. martinezii*, and *C. okeechobeensis* seem to possess some tolerance to this virus. Plants of these wild species react with mild and transient symptoms, and virus titer is low (16). No information is available regarding the inheritance of this tolerance.

Viruses Transmitted by Nematodes

The most common viruses transmitted by nematodes affecting cucurbits are tobacco ringspot (TRSV) and tomato ringspot (TomRSV). TRSV causes severe symptoms in cucumbers and melons but mild symptoms in squash; conversely, TomRSV is more prevalent and devastating in squash (14).

TomRSV, like other members of the nepovirus group, is characterized by isometric particles about 28 nm in diameter, which encapsulate single-stranded RNA with a bipartite genome. It is easily transmitted by mechanical means, seeds, and pollen, but the main vectors are nematodes (*Xiphinema americanum* Cobb and the closely related *Xiphinema* spp.). After acquiring the virus these nematodes are capable of retaining and spreading it for months; however, the virus does not multiply in the vector, and there is no transmission through eggs. TomRSV infects woody and herbaceous species in more than 35 plant families. It can cause severe damage in summer and winter squash. Particularly striking are the symptoms in some cultivars of *C. maxima*. Infected plants exhibit mosaic, leaf reduction, short internodes, proliferation of flower buds, and prominent ringspots and discoloration on the fruits (20).

The following wild species offer resistance to this virus: *C. cylindrata*, *C. digitata*, *C. ficifolia*, *C. gracilior*, *C. palmata*, *C. palmeri*, *C. pedatifolia*, and *C. sororia*. Plants of these species develop local infection, often as chlorotic ringspots, but the virus does not spread systemically (16). No information is available on the inheritance of the resistance.

Viruses Transmitted by Leafhoppers

There are very few viruses transmitted by leafhoppers that can infect squash, the most important being beet curly top virus (BCTV) (14).

BCTV is a geminivirus with isometric particles of about 20 nm in diameter. They occur singly or in pairs, and each contains a single strand of RNA.

The virus is restricted to phloem tissue and is transmitted mainly by *Circulifer tenellus* Baker, in which it circulates without multiplying. BCTV can infect a wide range of hosts, causing prominent leaf curling and distortion. Squash plants are generally severely stunted and show puckering and outward cupping of leaves, which may assume globular shape. Flowers tend to abort prematurely. BCTV is usually found in arid and semiarid regions of the Americas and the eastern Mediterranean basin, where it apparently originated (21). No information is available regarding sources of resistance in *Cucurbita* species or in other cucurbits.

Viruses Transmitted by Whiteflies

Several viruses transmitted by whiteflies have been reported to occur in cucurbits, including squash leaf curl (SLCV), melon leaf curl, (MLCV), cucumber vein yellowing (CVYV), and cucumber yellows (CYV). Squash leaf curl virus seems to be the most prevalent in squash (14).

First recognized in 1981, SLCV is a geminivirus with particles of about 22 nm in diameter. Each contains a single strand of RNA. It is spread by *Bemisia tabaci* Gennadius, in which it is circulative with a relatively long latent period. SLCV can be transmitted mechanically, provided that an appropriate buffer is used. In squash the virus causes reduction of leaf size, thickening of veins, enations, upward curling of the blades, and plant stunting. Serologically, it is related to bean golden mosaic virus and cassava latent virus. SLCV has been found in southern California and Mexico and may be present in other semitropical and tropical areas of the world, where it can cause severe losses (3).

Resistance was reported to occur in the wild species *Cucurbita ecuadorensis*, *C. lundelliana*, and *C. martinezii* (9). Tolerance was detected in some cultivars of *C. moschata* and *C. pepo* (8), but whether it can be exploited for breeding remains to be determined.

Conclusions

Substantial progress has been made in recent years in locating sources of resistance to viral diseases affecting *Cucurbita* species (8–11, 15, 16), but additional sources are still needed for several viruses and their pathotypes. It is well known that resistance can be of various kinds, such as hypersensitivity, tolerance, resistance to virus spread, and immunity. Most of these sources of resistance have proved to be rather stable. It is also evident that resistance can be viral strain-specific, that is, a specific gene for each pathotype. Consequently, additional studies are needed to determine the occurrence of pathotypes, which, in turn, can be recognized and differentiated solely by resistance genes (14).

Many of the known sources of resistance have been found in wild species (14), but due to the paucity of collections available, only a limited number of accessions of each species has been analyzed. Because each wild species includes many genotypes, it is likely that a reservoir of genetic diversity remains untapped. Our work (15, 16) has revealed that some species, e.g., *C. ecuadorensis*, *C. ficifolia*, *C. foetidissima*, and a Nigerian accession of *C. moschata*, are multiresistant. They represent very valuable germplasm, offering resistance to the most important viruses affecting squash: CMV, PRSV-W, WMV, and ZYMV. Yet, because of genetic incompatibility, only *C. ecuadorensis* and the Nigerian squash are at present widely used as sources of multiresistance in crosses with cultivars of summer and winter squash.

The relationships among *Cucurbita* species have been extensively investigated and a large number of interspecific hybridizations have been made (1, 5, 23, 24); however, genetic incompatibility among species remains a formidable barrier. As our resistance studies continue, traditional breeding methods are being augmented by other novel techniques, including early embryo rescue, tissue culture, protoplast regeneration, and cellular fusion. Further advances in biotechnology undoubtedly will offer enormous benefits, making possible the transfer of resistance genes among *Cucurbita* species as well as among diverse species of the same family.

Literature Cited

1. Bemis, W. P. 1963. Interspecific hybridization in *Cucurbita*. II. *C. moschata* × xerophytic species of *Cucurbita*. J. Heredity 54: 285–289.
2. Campbell, R. N. 1971. Squash mosaic virus. Descriptions of Plant Viruses No. 43. Commonw. Mycol. Inst./Assn. Appl. Biol., Kew.
3. Cohen, S., J. E. Duffus, R. C. Larsen, H. Y. Liu, and R. A. Flock. 1983. Purification, serology, and vector relationships of squash leaf curl virus, a whitefly-transmitted geminivirus. Phytopathology 73: 1669–1673.
4. Francki, R. I. B., D. W. Mossop, and T. Hatta. 1979. Cucumber mosaic virus. Descriptions of Plant Viruses No. 213. Commonw. Mycol. Inst./Assn. Appl. Biol., Kew.
5. Greber, R. S., and M. E. Herrington. 1980. Reaction of interspecific hybrids between *Cucurbita ecuadorensis*, *C. maxima*, and *C. moschata* to inoculation with cucumber mosaic virus and watermelon mosaic viruses 1 and 2. Austral. Pl. Path. 9: 1–2.
6. Lisa, V., and H. Lecoq. 1984. Zucchini yellow mosaic virus. Descriptions of Plant Viruses No. 282. Commonw. Mycol. Inst./Assn. Appl. Biol., Kew.
7. Martin, W. W. 1960. Inheritance and nature of cucumber mosaic virus resistance in squash. Diss. Abstr. 20: 3462.

8. McCreight, J. D. 1984. Tolerance of *Cucurbita* spp. to squash leaf curl virus. Cucurbit Genet. Coop. Rep. 7: 71–72.
9. McCreight, J. D. 1986. Reaction of *Cucurbita* species to squash leaf curl. Hort Science 21: 873 (Abstract).
10. Munger, H. M. 1976. *Cucurbita martinezii* as a source of resistance. Veg. Improv. Newslett. 18: 4.
11. Munger, H. M., and R. Provvidenti. 1987. Inheritance of resistance to zucchini yellow mosaic virus in *Cucurbita moschata*. Cucurbit Genet. Coop. Rep. 10: 80–81.
12. Nelson, M. R., and H. K. Knuhtsen. 1973. Squash mosaic virus variability: Review and serological comparisons of six biotypes. Phytopathology 63: 920–926.
13. Provvidenti, R. 1982. Sources of resistance and tolerance to viruses in accessions of *Cucurbita maxima*. Cucurbit Genet. Coop. Rep. 5: 46–47.
14. Provvidenti, R. 1986. Viral diseases of cucurbits and sources of resistance. Tech. Bull. No. 93. Food Fertiliz. Technol. Center, Taipei.
15. Provvidenti, R., D. Gonsalves, and H. S. Humaydan. 1984. Occurrence of zucchini yellow mosaic virus in cucurbits from Connecticut, New York, Florida, and California. Pl. Dis. Reporter 68: 443–446.
16. Provvidenti, R., R. W. Robinson, and H. M. Munger. 1978. Resistance in feral species to six viruses infecting *Cucurbita*. Pl. Dis. Reporter 62: 326–329.
17. Provvidenti, R., and J. K. Uyemoto. 1973. Chlorotic leaf spotting of yellow summer squash caused by the severe strain of bean yellow mosaic virus. Pl. Dis. Reporter 57: 280–282.
18. Purcifull, D., and D. Gonsalves. 1984. Papaya ringspot virus. Descriptions of Plant Viruses No. 292. Commonw. Mycol. Inst./Assn. Appl. Biol., Kew.
19. Purcifull, D., H. Hiebert, and J. Edwardson. 1984. Watermelon mosaic virus 2. Descriptions of Plant Viruses No. 293. Commonw. Mycol. Inst./Assn. Appl. Biol., Kew.
20. Stace-Smith, R. 1970. Tomato ringspot virus. Descriptions of Plant Viruses No. 18. Commonw. Mycol. Inst./Assn. Appl. Biol., Kew.
21. Thomas, P. E., and G. I. Mink. 1979. Beet curly top virus. Descriptions of Plant Viruses No. 210. Commonw. Mycol. Inst./Assn. Appl. Biol., Kew.
22. Vovlas, C., E. Hiebert, and M. Russo. 1981. Zucchini yellow fleck virus, a new potyvirus of zucchini squash. Phytopath. Medit. 20: 123–128.
23. Washek, R. L., and H. M. Munger. 1983. Hybridization of *Cucurbita pepo* with disease resistant *Cucurbita* species. Cucurbit Genet. Coop. Rep. 6: 92.
24. Whitaker, T. W., and G. N. Davis. 1962. Cucurbits: Botany, Cultivation, and Utilization. Interscience, New York.

36 | Growth, Assimilate Partitioning, and Productivity of Bush and Vine Cultivars of *Cucurbita maxima*

J. Brent Loy and Cyril E. Broderick

ABSTRACT. Bush strains of *Cucurbita maxima* offer the potential for easier crop management and increased yields of squash because of their adaptability to high-density planting and conventional row crop culture. Bush strains exhibit a more uniform pattern of growth than vine strains, and achieve full leaf canopy development rapidly at high density plantings. The bush cultivar 'Autumn Pride' has both a higher net assimilation rate and a higher harvest index than the vine cultivar 'Blue Hubbard'. Fresh weight yields of 'Autumn Pride' increased as plant density increased from 5600 to 22,000 plants/ha. At the highest plant density both dry and fresh weight yields of 'Autumn Pride' were significantly higher than those of 'Blue Hubbard' grown at a near optimum plant density of 5600 plants/ha. Use of bush genotypes for increasing productivity and production efficiency can be exploited best in short growing seasons because of earlier fruit maturity associated with the bush habit of growth.

The pattern of development of a crop plant in part determines growth potential, the proportion of assimilate diverted into the harvested portion, and the manner in which the plant exploits a particular growing season (8, 15). A key strategy for maximizing crop photosynthesis is to increase the rate of leaf canopy development (7). Morphological and developmental attributes that contribute to leaf canopy development, such as rate of leaf initiation and expansion, leaf spacing, final leaf size, and number of active shoot meristems (15), can be genetically manipulated to achieve faster canopy development. Altering plant population densities can also drastically affect canopy development and concomitant crop yields (18). The degree to which increased photosynthetic capacity contributes to higher

436

crop yields depends upon the ability of the plant to partition assimilates into the economically harvested portion (6, 7).

Based on yield per acre, production labor, and overall nutrient content per acre, squash is recognized as one of the more efficient nutrient producers in comparison with several other vegetable crops (13). However, there is little fundamental information on the potential productivity of winter squash, *Cucurbita maxima* Duch. ex Lam., and reliable data on optimum spacing for maximizing yields are lacking. Commercially important cultivars of *C. maxima* have a prostrate, vining growth habit. However, bush cultivars adapted for high density planting and standard row crop culture have been and are now being developed. For continued cultivar improvement, basic knowledge is needed about the comparative growth patterns of bush and vine squash and the manner in which these patterns relate to productivity at different planting densities.

Results reported herein represent several comparative studies on growth, development, and productivity of bush or short internode strains and vine cultivars of *C. maxima*. These experiments have been conducted since 1973 at the New Hampshire Agricultural Experiment Station. The primary objective of these studies has been to ascertain the potential productivity of the bush phenotype compared with the vine phenotype of squash.

Materials and Methods

Leaf Production of Bush and Vine Strains

Comparative studies of leaf number and area of bush and vine strains of *C. maxima* were conducted in 1974. Two F_3 bush breeding lines were used, NH265 and NH457. Plants of NH265 usually produced one fruit per plant at high density spacing. Plants of NH457 produced 6 to 12 small 'Gold Nugget' type fruits per plant. Although these strains segregated for fruit traits other than size, they both exhibited uniform bush habits of growth. Vine cultivars were 'Blue Hubbard' and 'Buttercup'. 'Blue Hubbard' plants are robust and have large leaves and fruit. 'Buttercup' plants have smaller leaves and produce small, turban-shaped fruit with a high dry matter content.

Plant spacings (within row and between rows) were 0.9 × 1.2 m and 0.9 × 1.5 m for bush lines, and 1.2 × 1.8 m and 1.8 × 1.8 m for vine cultivars. Spacing studies in 1973 established these plant spacings as near optimum for maximizing yields of these strains. A randomized complete block design was used with three replications and eight plants per plot for yield determination. Four plants per plot in two of the three blocks were randomly selected for leaf counts and measurements. Data on leaf numbers and size were taken at weekly intervals for five weeks, beginning 12 days after transplanting into the field. Leaf areas were determined by multiplying leaf width by leaf length by a correction factor to account for leaf shape. Correction factors were determined by using planimeter traces of subsamples of leaves from each strain.

Light Interception

Experiments on light interception were conducted in 1973, using two F_4 bush strains, NH57-2-17 and NH1838-20-54, and the vine cultivars 'Blue Hubbard' and 'Buttercup'. Plants for light interception measurements were from plots similar to those of the 1974 spacing trial, but with plant spacings of 0.9 × 1.2 and 1.2 × 1.2 m for bush plants and 1.2 × 2.4 and 1.8 × 2.4 m for vine plants. Between 10 A.M. and 2 P.M. on sunny days, photosynthetically active radiation (PAR) was measured with a LICOR LI185 light meter and a silicon photocell sensor covered with a Schott glass filter having a cutoff point at 695 nm. Measurements were made at the top of the canopy and at ground level at 6 cm intervals along the X and Y axes intersecting each plant. Measurements were replicated twice for each strain at each spacing over three different time periods. Percent absorbed PAR was calculated as follows:

$$PAR = (incident - reflected\ PAR) - ground\ level\ PAR \times 100/incident$$

Plant Density and Yield

Yield of a new bush cultivar 'Autumn Pride' was compared with that of 'Blue Hubbard' in 1980 and 1981. 'Autumn Pride' produces a fairly robust bush plant with large leaves and usually one large fruit per plant at high density spacing (Figure 1). In both years the experimental design was a split block between 'Blue Hubbard' and 'Autumn Pride' as the main treatments. 'Autumn Pride' subplots consisted of four planting densities, with rows 1.5 m apart and within row spacings of 0.3, 0.6, 0.9, and 1.2 m. 'Blue Hubbard' was planted at one spacing, 1.2 × 1.5 m, near optimum for maximum yields. All subplots were 37 m². Two sets of guard rows were used around the 'Blue Hubbard' plots, and one row around each 'Autumn Pride' subplot. Treatments were replicated four times in 1980 and three times in 1981.

Dry Matter Partitioning

Dry matter accumulation and partitioning in 'Autumn Pride' and 'Blue Hubbard' were investigated in 1981, with six sampling times and five replications in a randomized complete block design. Plant spacing was initially

Figure 1. A plant of 'Autumn Pride' showing the compact, uniform growth pattern of the bush phenotype, the upright habit of growth, and a single immature fruit.

1.8 × 1.8 m, but with continued sampling over time the area per plant increased so that interplant competition was minimized. Plants were sampled biweekly to obtain specific leaf weights and fresh and dry weights of stems (including petioles), fruits, and leaves. Leaf areas during the first two sampling periods were estimated by multiplying the specific leaf weight (SLW) of the last fully expanded leaf on the main stem by the total leaf dry weight per plant. Following fruit set at about six weeks from transplanting, SLW values per plant were determined from the average of four samples taken from a fully expanded young leaf, a leaf located mid-vine, a leaf subtending a developing fruit, and an older, fully expanded leaf.

Results and Discussion

Growth and Leaf Canopy Development in Bush and Vine Strains

Rates of leaf production were not appreciably different among the bush and vine strains during the first four weeks of growth (Figure 2). After four weeks, leaf production in vine cultivars increased more rapidly than in bush strains because of the development of extensive secondary and tertiary lateral shoots. Bush plants had fewer vigorous secondary shoots and no tertiary

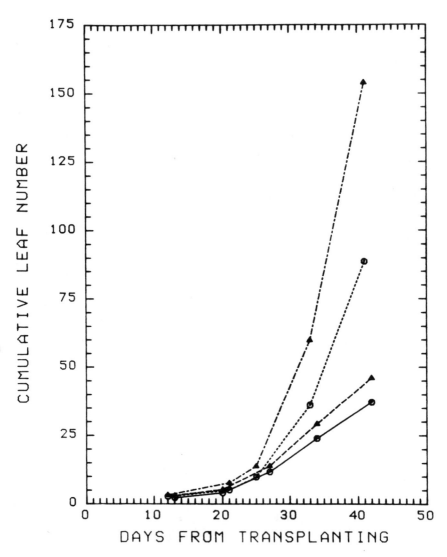

Figure 2. Cumulative leaf production in two bush lines of squash (NH265 and NH457) at 0.9 × 1.2 m and 0.9 × 1.5 m spacings, and in two vine cultivars ('Blue Hubbard' and 'Buttercup') at 1.2 × 1.8 m and 1.8 × 1.8 m spacings. 'Blue Hubbard' ○ — — — —, 'Buttercup' ▲ · — · — · —, NH265 ○ —————, NH457 ▲ — — —.

laterals. Cumulative leaf areas developed similarly to leaf production, but with a notable exception. 'Blue Hubbard' produced fewer leaves than the other vine strain, 'Buttercup', but had a larger leaf area due to larger leaves. Because of their closer spacing, bush strains had a higher leaf area index (LAI) than the vine cultivars during the first five weeks of growth from transplanting (Figure 3). In a more recent study with the bush cultivar

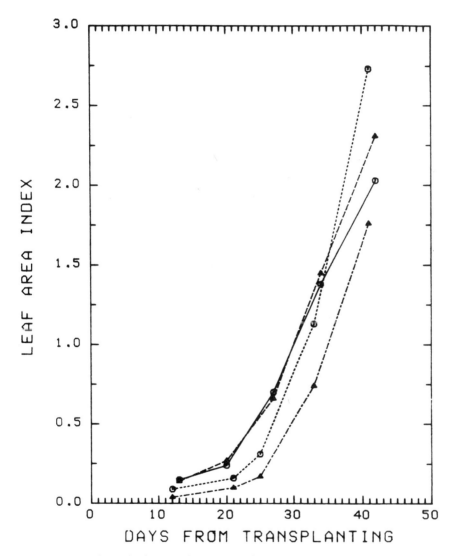

Figure 3. Cumulative leaf area index. Data as for Figure 2. 'Blue Hubbard' ○ —— ——, 'Buttercup' ▲ · — · — · —, NH265 ○ ————, NH457 ▲ ————————.

'Autumn Pride' (1), LAIs of between 3 and 6 were reached at six weeks of growth in the field and then declined. The highest LAI was obtained at the highest planting density of 22,222 plants/ha. Following fruit set and the development of a strong reproductive sink in large-fruited bush strains, vegetative growth is suppressed (17); this coupled with leaf senescence can lead to a decline in LAI prior to fruit maturity. Vine strains, on the other hand, exhibit an exponential phase of leaf production and leaf area beginning at about five to six weeks from transplanting (Figure 2, 3). Under conditions of relatively high density planting, LAIs as high as 9 may be reached late in the growing season (1). This is at the upper LAI range of 4 to 9, which has been cited as optimum for many crop plants (2, 4, 14, 16, 18), and probably represents excessive vegetative growth as a compromise for faster canopy cover in vine strains grown at relatively close spacing.

In comparison with vine strains that exhibit a random orientation of trailing vines, the uniform and compact growth pattern of bush strains is well suited to high density planting, which facilitates rapid development of a full leaf canopy. This is supported by results obtained in 1973 on interception of photosynthetically active radiation (PAR) by leaf canopies of bush and vine strains. Bush plants absorbed more PAR than vine plants during the first six weeks of growth. At the highest spacing for bush plants (0.9 × 1.2 m), the leaf canopy at six weeks from transplanting absorbed 90% of incoming PAR. The vine strains, 'Blue Hubbard' and 'Buttercup', absorbed only 63% and 68%, respectively, of incoming PAR at a plant density of 1.2 × 1.8 m. Plant densities in this study were less than optimum for maximum productivity in both bush and vine strains, but probably closer to optimum for the vine strains. Thus, the results reflect the greater efficiency of bush plants at high density plantings in utilizing light during the early part of the growing season.

Growth and Partitioning of Dry Matter in Bush and Vine Squash

Total biomass (exclusive of roots) accumulation in bush 'Autumn Pride' and vine 'Blue Hubbard' at low density spacing was not statistically different through the first eight weeks of development in the field (Table 1). By six weeks, however, 'Autumn Pride' had partitioned 18% of its biomass into fruit, compared with only 3% for 'Blue Hubbard'. This difference in partitioning was both a reflection of earlier flowering and fruit set (three to five days) in the bush strain, and more extensive vegetative growth in the vine strain. At week ten, fruit constituted 65% of the biomass in 'Autumn Pride' and 56% in 'Blue Hubbard'. The harvest indices (proportion of fruit biomass at maturity) were 70% and 53%, respectively, for bush and vine plants. These values may be slightly inflated because of irretrievable loss of leaves through senescence.

Table 1. Distribution of dry weight in stems, leaves, and fruits of 'Autumn Pride' (bush) and 'Blue Hubbard' (vine) plants

Weeks after transplanting	Organ	Dry wt (g)[1]	
		Bush	Vine
2	Leaves	1.8	3.0
	Stems	0.7	0.9
	Fruits	0.0	0.0
	Total	2.5 ± 0.3	3.9 ± 0.3
4	Leaves	41.4	43.4
	Stems	17.2	23.2
	Fruits	0.0	0.0
	Total	58.6 ± 3.3	66.6 ± 5.3
6	Leaves	331	504
	Stems	265	437
	Fruits	129	31
	Total	725 ± 97	972 ± 54
8	Leaves	616	732
	Stems	689	912
	Fruits	1250	870
	Total	2555 ± 155	2514 ± 441
10	Leaves	584	1886
	Stems	500	1688
	Fruits	2040	4600
	Total	3124 ± 286	8174 ± 683
12	Leaves	368	1880
	Stems	414	2191
	Fruits	1790	5400
	Total	2572 ± 161	9471 ± 626

Source: From Broderick (1).
[1]Values (±SE for total weights) represent means of five replications.

Leaf area ratios (LARs) for 'Autumn Pride' were consistently lower than those for 'Blue Hubbard' throughout the latter part of the growing season (Table 2), indicating that bush plants were more efficient than vine plants in the production of assimilates per unit of leaf area. Similar results were found in a greenhouse study comparing the growth of these same cultivars (1). Likewise, cumulative net assimilation rates (NAR) were higher in 'Autumn Pride' than in 'Blue Hubbard' during the last six weeks of data collection. The highest NAR values in bush plants coincided with the period of rapid reproductive development, suggesting that rates of photosynthesis were enhanced during this period. Enhancement of rates of photosynthesis by increasing sink demand has been noted in other crops (7). Leaf thickness is similar in 'Autumn Pride' and 'Blue Hubbard', (ca. 200 μm) but the palisade

Table 2. Specific leaf weights, leaf area ratios, and cumulative net assimilation rates of 'Autumn Pride' (bush) and 'Blue Hubbard' (vine) squash plants grown in the field in 1981

Treatment	Weeks after transplanting				
	4	6	8	10	12
Specific leaf weight (mg/cm^2)					
Bush	6.8	7.3	5.9	5.9	8.0
Vine	5.5	8.3	5.5	5.9	5.7
Leaf area ratio (dm^2/g dry wt.)					
Bush	1.40	0.62	0.41	0.29	0.17
Vine	1.19	0.62	0.54	0.39	0.35
Cucumlative net assimilation rate (g dry wt. · m^{-2} leaf area · day^{-1})					
Bush	2.55	3.81	4.34	4.92	7.00[1]
Vine	3.01	3.81	3.34	3.62	3.42

Note: Specific leaf weights are means of five replications, one sample per replication for weeks 4 and 6, four samples per replication for weeks 8, 10, and 12. The LAR and NAR were calculated from specific leaf weights and from dry weight data presented in Table 1.

[1] Value inflated because of leaf senescence on mature plants.

parenchyma is considerably thicker in 'Autumn Pride' (1), and this may contribute to its more efficient production of assimilates.

Productivity of Bush and Vine Strains

Hutchins and Croston (11) compared the productivity of several cultivars of *C. maxima* and their F_1 hybrids at 2.7 × 2.7 m spacing. Fresh weight yields of mature fruit for some of the more common cultivars were 47 t/ha for 'Mammoth Chili', 20.3 for 'Banana', 16.1 for 'Blue Hubbard', and 10 for 'Delicious'. Maximum yields obtained in Vermont by Cummings and Stone (3) and in New Hampshire by Hepler (10) for 'Blue Hubbard' at a 2.4 × 2.4 m spacing (two plants per hill) were about 36 t/ha. None of the above studies reported the use of guard rows around plots nor were dry weight yields determined. Evaluation of crop productivity requires obtaining data on dry as well as fresh mass. For example, percent dry matter is extremely low in 'Mammoth Chili' fruits, and thus the high fresh weight yields reported above (11) for this cultivar are deceiving when comparing productivity.

In all of the above studies, lower than optimum plant densities for maximizing yields were used. In 1980 and 1981 we compared the yield of 'Blue Hubbard' at a near optimum spacing to that of bush 'Autumn Pride' at several plant densities (Table 3). In both years fresh weight yields of 'Autumn Pride' increased with an increase in plant density, with the two highest densities significantly outyielding 'Blue Hubbard'. Also, at the highest plant density the dry weight yield of 'Autumn Pride' was significantly higher than that of 'Blue Hubbard'. Moreover, the fruits of 'Autumn Pride' matured

Table 3. Fresh and dry weight yield of 'Autumn Pride' planted at different densities and of 'Blue Hubbard' planted at a density of 5600 plants/ha

Plants/ha	Spacing	Fresh weight, 1980[1]		Fresh weight, 1981[1]		Dry weight, 1981 (kg/ha)
		t/ha	kg/plant	t/ha	kg/plant	
'Autumn Pride'						
22,000	(0.3 × 1.5 m)	77.0a	3.7a	74.8a	3.5a	5302a
11,000	(0.6 × 1.5 m)	70.3ab	6.6b	61.5b	5.5b	4551bc
7400	(0.9 × 1.5 m)	64.6bcd	9.0c	55.8c	7.7c	4313c
5600	(1.2 × 1.5 m)	61.3cd	10.6d	52.2cd	9.2d	4594bc
'Blue Hubbard'						
5600	(1.2 × 1.5 m)	57.3d	10.4d	50.7d	8.7d	4861b

[1] Values within columns not followed by the same letters are significantly different according to Duncan's multiple range test, 5% level.

more rapidly (95 to 105 days) than those of 'Blue Hubbard' (105 to 115 days), so that higher economic yield was achieved in a shorter time span. It should be noted, however, that at the highest density, the fruits of 'Autumn Pride' were smaller than desirable for marketability and some fruit were misshapen. Because fruit number per plant in 'Autumn Pride' approaches one at plant densities of 1.5 m², the fruit size can be conveniently regulated by choice of plant spacing.

The percent dry matter in fruit was unusually low in the 1981 trials, so that the data by no means represent the ultimate biomass yields of these cultivars in northern latitudes. Considerably higher dry weight economic yields of 'Autumn Pride' were obtained from a 1979 trial in Durham, New Hampshire (6800 kg/ha) and from a 1980 trial in Tully, New York (6700 kg/ha).

Conclusions

Although grown widely throughout temperate latitudes, *C. maxima* can be considered a minor crop in terms of its contribution as a food source. As such, genetic studies and breeding programs to improve this crop have been limited. Our research suggests that the productivity of *C. maxima* could be increased by continued efforts to utilize the bush phenotype. Compared with vine strains, bush strains exhibit a more uniform growth pattern, achieve leaf canopy cover more rapidly at high density planting, and have a higher harvest index. Fruit maturity generally occurs earlier in bush than in vine strains, so that the potential for increased productivity of bush over vine strains will most easily be achieved in regions having relatively short growing seasons. Because vine plants exhibit more indeterminate growth than bush plants, they can be expected to set more fruit and outyield bush plants in a longer growing season, (i.e., > 110 days).

Success in developing bush strains at the University of New Hampshire has been achieved only with large-fruited strains, those that typically bear only one fruit per plant at high density plantings. Although the only release from this program to date has been the pink-skinned 'Autumn Pride', several additional cultivars with eating quality superior to that of 'Autumn Pride' and having a more desirable orange to red-orange skin color appear forthcoming. In addition, we have tested one promising experimental semi-bush F_1 hybrid that exhibited considerable heterosis for growth and productivity (12). Heterosis has been reported previously in *C. maxima* (9, 11), and it is likely that F_1 bush or semi-bush hybrids will continue to be exploited in this species and may eventually dominate that market for large-fruited winter squash.

Developing acceptable small-fruited bush strains of *C. maxima* comparable to the premium cultivar 'Buttercup' is a more difficult task. Small-fruited strains of *C. maxima* tend to set too heavy a fruit load for the size of the plant. As a result, there is more variability in fruit size and fruit dry matter than in vine strains (5) and, therefore, less commercial acceptability. Perhaps this problem can be overcome by introducing vigorous hybrid strains with a bush or semi-bush phenotype. One example of this is the cultivar 'Sweet Mama'.

Literature Cited

1. Broderick, C. E. 1982. Morpho-physiological factors affecting plant productivity in bush and vine forms of winter squash (*Cucurbita maxima* Duch. ex Lam.). Ph.D. dissertation, Univ. New Hampshire, Durham.
2. Chandler, R. F. 1969. Plant morphology and stand geometry in relation to nitrogen. *In* J. D. Eastin, F. A. Haskins, C. Y. Sullivan, and C. H. M. van Bavel, eds., Physiological Aspects of Crop Yield. Amer. Soc. Agron., Madison, WI.
3. Cummings, M. B., and S. C. Stone. 1921. Yield and Quality in Hubbard Squash. Vermont Agric. Exp. Sta. Bull. 222, Burlington.
4. Edmeades, G. O., and T. B. Daynard. 1979. The development of plant to plant variability in maize at different plant densities. Canad. J. Pl. Sci. 59: 561–576.
5. Evans, D., and B. Loy. 1984. Effects of fruit thinning on dry matter content of *Cucurbita maxima* squash. HortScience 19: 586 (Abstract).
6. Evans, L. T., and I. F. Wardlaw. 1976. Aspects of comparative physiology of grain yield in cereals. Advances Agron. 28: 301–359.
7. Gifford, R. M., and L. T. Evans. 1981. Photosynthesis, carbon partitioning and yield. Ann. Rev. Pl. Physiol. 32: 485–509.
8. Good, N. E., and D. H. Bell. 1979. Photosynthesis, plant productivity, and crop yield. *In* P. Carlson, ed., The Biology of Crop Productivity. Academic Press, New York.

9. Hayase, H., and T. Ueda. 1956. *Cucurbita* crosses. IX. Hybrid vigor of reciprocal F_1 crosses in *C. maxima* (in Japanese). Hokkaido Natl. Agric. Sta. Res. Bull. 71: 119–128.

10. Hepler, J. R. 1941. Fertilizer and storage experiments with squash. Proc. Amer. Soc. Hort. Sci. 38: 618–620.

11. Hutchins, A. E., and F. E. Croston. 1941. Productivity of F_1 hybrids in squash *Cucurbita maxima*. Proc. Amer. Soc. Hort. Sci. 39: 332–336.

12. Loy, B., M. Hutton, and B. Carle. 1988. Developmental basis for heterosis in *Cucurbita maxima* winter squash. HortScience 22: 1090 (Abstract).

13. MacGillivray, J. H., G. C. Hanna, and P. A. Minges. 1942. Vitamin, protein, calcium, iron, and caloric yield of vegetables per acre per man-hour. Proc. Amer. Soc. Hort. Sci. 41: 293–297.

14. Shibles, R. M., and C. R. Weber. 1965. Leaf area, solar radiation, and dry matter production by various soybean planting patterns. Crop Sci. 6: 55–59.

15. Wareing, P. F. and J. Patrick. 1975. Source-sink relations and the partition of assimilates in the plant. *In* J. P. Cooper, ed., Photosynthesis and Productivity in Different Environments. Cambridge Univ. Press, Cambridge.

16. Watson, D. J. 1952. The physiological basis of variation in yield. Advances Agron. 4: 101–145.

17. Zack, C. D., and J. B. Loy. 1981. Effect of fruit development on vegetative growth of squash. Canad. J. Pl. Sci. 61: 673–676.

18. Zelitch, I. 1971. Photosynthesis, Photorespiration and Plant Productivity. Academic Press, New York.

Appendix: An Outline Classification of the Cucurbitaceae

Charles Jeffrey

A classification of the Cucurbitaceae, from family to generic rank, including generic synonymy, follows. Brief indications of the size and geographical distribution of each accepted taxon are given.

Familia Cucurbitaceae Juss., Gen. Pl.: 393 (1789) nom. cons.

I. Subfamilia Zanonioideae C. Jeffr., Kew Bull. 15: 345 (1962)
Styles 2–3, free; tendrils bifid, spiralling below and above the point of branching, the bifurcation distal and the branches short; filaments inserted on or about the disk; pollen grains small, tricolporate, striate
18 genera, about 80 species, tropics
  A. Tribus Zanonieae Blume, Bijdr.: 936 (1826)
  Characters and content of the subfamily
    1. Subtribus Fevilleinae Pax in Engl. & Prantl, Nat. Pflanzenfam. IV, 5: 10 (1889)
    Ovary 3-locular; seeds very large, unwinged; fruit operculate
    1 genus, 5 species, Neotropics
      Fevillea L., Sp. Pl. 2: 1013 (1753) and Gen. Pl., ed. 5: 443 (1754)
        Nhandiroba Adans., Fam. Pl. 2: 139, 581 (1763) nom. illegit.
        Anisosperma S. Manso, Enum. Subst. Braz.: 38 (1836)
        Hypanthera S. Manso, Enum. Subst. Braz.: 37 (1836)
        5 species, Neotropics
    2. Subtribus Zanoniinae Pax in Engl. & Prantl, Nat. Pflanzenfam. IV, 5: 12 (1889)
    Ovary 2–3-locular; seeds winged; fruit apically dehiscent; leaves minutely sinuate-dentate to entire

7 genera, 30 species, tropics

Alsomitra (Blume) M. J. Roem., Fam. Nat. Syn. Monogr. 2: 113, 117 (1846)

   Macrozanonia (Cogn.) Cogn., Bull. Herb. Boissier 1: 612 (1893)

   2 species, Malesia

Zanonia L., Sp. Pl. 2: 1028 (1753) and Gen. Pl., ed. 5: 454 (1754)

   Penar-valli Adans., Fam. Pl. 2: 139, 589 (1763) nom. illegit.

   Juppia Merrill, J. Straits Branch Roy. Asiat. Soc. 85: 170 (1922)

   1 species, Indomalesia

Siolmatra Baill., Bull. Mens. Soc. Linn. Paris 1: 458 (1885)

   3 species, South America

Gerrardanthus Harv. ex Hook. f. in Benth. & Hook. f., Gen. Pl. 1: 820, 840 (1867)

   Atheranthera M. T. Mast. in D. Oliver, Fl. Trop. Afr. 2: 519 (1871)

   4 species, tropical Africa

Zygosicyos Humbert, Bull. Soc. Bot. France 91: 166 (1945)

   2 species, Madagascar

Xerosicyos Humbert, Compt. Rend. Hebd. Seances Acad. Sci. 208: 220 (1939)

   3 species, Madagascar

Neoalsomitra Hutch., Ann. Bot. (London), Ser. 2, 6: 97 (1942)

   About 15 species, Indomalesia, China, and Australia

3. Subtribus Gomphogyninae Pax in Engl. & Prantl, Nat. Pflanzenfam. IV, 5: 12 (1889)

Ovary 2–3-locular; seeds winged or unwinged; fruit apically dehiscent; leaves coarsely apiculate-serrate

3 genera, 20 species, eastern Asia, and Indomalesia

Hemsleya Cogn. ex F. Forbes & Hemsl., J. Linn. Soc. Bot. 23: 490 (1888)

   About 15 species, China, and eastern Himalaya

Gomphogyne Griff., Account Bot. Coll. Cantor: 26 (?1841)

   Triceros Griff., Notul. Pl. Asiat. 4: 606 (1854) non Lour. (1790)

   1 species, eastern Himalaya

Gynostemma Blume, Bijdr.: 23 (1825)

   Enkylia Griff., Account. Bot. Coll. Cantor: 26 (?1841)

   Pestalozzia Zoll. & Moritzi in Moritzi, Syst. Verz.: 31 (1846)

   Trirostellum C. P. Wang & Q. Z. Xie, Acta Phytotax. Sin. 19: 481 (1981)

   5 species, eastern Asia, Indomalesia

4. Subtribus Actinostemmatinae C. Jeffr. subtribus nov., typus *Actinostemma* Griff., fructu operculato parvo distincta

Ovary 2–3-locular; seeds winged or unwinged; fruit operculate; leaves sinuate-denticulate

2 genera, 3 species, eastern Asia
  Bolbostemma Franquet, Bull. Mus. Hist. Nat. Paris, Ser. 2, 2: 325 (1930)
    2 species, China
  Actinostemma Griff., Account Bot. Coll. Cantor: 24 (?1841)
    Mitrosicyos Maxim., Mem. Acad. Imp. Sci. St. Petersb. 9: 112 (1859)
    Pomasterion Miq., Ann. Mus. Bot. Lugduno-Batavum 2: 80 (1865)
    1 species, eastern Asia
5. Subtribus Sicydiinae Pax in Engl. & Prantl, Nat. Pflanzenfam. IV, 5: 21 (1889)
Ovary 1-locular; ovule solitary; fruit baccate or samaroid, indehiscent
5 genera, 16 species, tropical America, Africa, and Madagascar
  Sicydium Schlechtend., Linnaea 7: 388 (1832)
    Triceratia A. Rich. in Sagra, Hist. Fis. Cuba: 614 (1845)
    8 species, Neotropics
  Chalema Dieterle, Contr. Univ. Michigan Herb. 14: 71 (1980)
    1 species, Mexico
  Pteropepon (Cogn.) Cogn. in Engl., Pflanzenreich 66 (IV, 275, 1): 260 (1916)
    3 species, South America
  Pseudosicydium Harms, Notizbl. Bot. Gart. Berlin-Dahlem 10: 182 (1927)
    1 species, South America
  Cyclantheropsis Harms, Bot. Jahrb. Syst. 23: 167 (1896)
    3 species, tropical Africa and Madagascar
II. Subfamilia Cucurbitoideae
Styles united into a single column; tendrils usually unbranched or 2–5-fid, spiralling only above the point of branching, the latter proximal and the branches long, very rarely of zanonioid type; filaments inserted on the hypanthium, free from the disk when the latter present; pollen grains various, not striate
100 genera, about 745 species, tropics
  A. Tribus Melothrieae Endl., Gen. Pl.: 936 (1839)
  Flowers comparatively small, with more or less campanulate hypanthium, similar in male and female flowers; anther thecae usually comparatively simple; pollen grains usually tricolporate, rarely 4–6-colporate, or triporate, reticulate
  34 genera, 250 species, tropics
    1. Subtribus Dendrosicyinae C. Jeffr., Kew Bull. 15: 344 (1962)
    Seeds small, tumid or more or less compressed; anther thecae not fringed with hairs; flowers ephemeral
    16 genera, 80 species, America, Africa, Madagascar, Socotra, and Indomalesia

Kedrostis Medic., Philos. Bot. 2: 69 (1791)
  Coniandra Schrad. ex Eckl. & Zeyh., Enum. Pl. Afric. Austral.: 275 (1836)
  Cyrtonema Schrad. ex Eckl. & Zeyh., Enum. Pl. Afric. Austral.: 276 (1836)
  Rhynchocarpa Schrad. ex Endl., Gen. Pl.: 936 (1839)
  Aechmandra Arn., Madras J. Lit. Sci. 12: 49 (1840)
  Bryonopsis Arn., Madras J. Lit. Sci. 12: 49 (1840) non Raf. (1814)
  Pisosperma Sond. in Harv. & Sond., Fl. Cap. 2: 498 (1862)
  Cerasiocarpum Hook. f. in Benth. & Hook. f., Gen. Pl. 1: 832 (1867)
  Toxanthera Hook. f., Hooker's Icon. Pl. 15: 16 (1883)
  ?*Phialocarpus* Deflers, Bull. Soc. Bot. France 42: 304 (1898)
  Gijefa (M. J. Roem.) Kuntze in Post & O. Kuntze, Lex. Gen. Phan.: 248 (1903)
  About 20 species, tropical and southern Africa, Madagascar, southwest Asia, and Indomalesia
Dendrosicyos Balf. f., Proc. Roy. Soc. Edinburgh 11: 513 (1882)
  1 species, Socotra
Corallocarpus Welw. ex Hook. f. in Benth. & Hook. f., Gen. Pl. 1: 831 (1867)
  Calyptrosicyos Keraudren, Compt. Rend. Hebd. Seances Acad. Sci. 248: 3592 (1959)
  17 species, tropical and southern Africa, Madagascar, southwest Asia, and India
Ibervillea Greene, Erythea 3: 75 (1895)
  Maximowiczia Cogn. in A. & C. DC., Monogr. Phan. 3: 726 (1881) non Ruprecht (1856)
  5 species, Mexico and southern United States
Tumamoca Rose, Contr. U.S. Natl. Herb. 16: 21 (1912)
  1 species, Arizona and northern Mexico
Halosicyos Mart. Crovetto, Bol. Soc. Argent. Bot. 2: 84 (1947)
  1 species, Argentina
Ceratosanthes Burm. ex Adans., Fam. Pl. 2: 139, 535 (1763)
  About 5 species, West Indies to Paraguay
Doyerea Grosourdy, Med. Bot. Criollo 1, (2): 338 (1864)
  Anguriopsis J. R. Johnst., Proc. Amer. Acad. Arts 40: 697 (1905)
  1 species, Central America, West Indies, and northern Venezuela and Colombia
Trochomeriopsis Cogn. in A. & C. DC., Monogr. Phan. 3: 661 (1881)
  1 species, Madagascar
Seyrigia Keraudren, Bull. Soc. Bot. France 107: 298 (1961)
  5 species, Madagascar

Dieterlea Lott, Brittonia 38: 407 (1986)
   1 species, Mexico
Cucurbitella Walp., Repert. Bot. Syst. 6: 50 (1846) corr. Walp. Repert. Bot. Syst. 6: 769 (1847)
      Schizostigma Arn., Madras J. Lit. Sci. 12: 50 (1840) non Arn. ex Meissn. (1838)
      Prasopepon Naud., Ann. Sci. Nat. Bot., Ser. 5, 5: 26 (1866)
      About 3 species, South America
Apodanthera Arn., J. Bot. (Hooker) 3: 274 (1841)
   About 15 species, tropical and subtropical America
Guraniopsis Cogn., Bot. Jahrb. Syst. 42: 173 (1908)
   1 species, Peru
Melothrianthus Mart. Crovetto, Notul. Syst. (Paris) 15: 58 (1955)
   1 species, Brazil and Paraguay
Wilbrandia S. Manso, Enum. Subst. Braz.: 30 (1836)
   5 species, South America
2. Subtribus Guraniinae C. Jeffr., Kew Bull. 16: 483 (1963)
Seeds small, subcompressed; anther thecae not fringed with hairs; flowers long-lived
3 genera, 55 species, Neotropics
   Helmontia Cogn., Bull. Soc. Roy. Bot. Belgique 14: 239 (1875)
      2 species, South America
   Psiguria Neck. ex Arn., J. Bot. (Hooker) 3: 274 (1841)
      Anguria Jacq., Enum. Syst. Pl. Carib.: 9, 31 (1760) non Mill. (1754)
      12 species, Neotropics
   Gurania (Schlechtend.) Cogn., Bull. Soc. Roy. Bot. Belgique 14: 239 (1875)
      Dieudonnaea Cogn., Bull. Soc. Roy. Bot. Belgique 14: 239 (1875)
      Ranugia (Schlechtend.) Post & O. Kuntze, Lex. Gen. Phan.: 476 (1903)
      About 40 species, Neotropics
3. Subtribus Cucumerinae Pax in Engl. & Prantl, Nat. Pflanzenfam. IV, 5: 22 (1889)
Seeds small, compressed; anther thecae fringed with hairs; flowers ephemeral
12 genera, 100 species, tropics
   Melancium Naud., Ann. Sci. Nat. Bot., Ser. 4, 16: 175 (1862)
      1 species, eastern and southern Brazil
   Cucumeropsis Naud., Ann. Sci. Nat. Bot., Ser. 5, 5: 30 (1866)
      Cladosicyos Hook. f. in D. Oliver, Fl. Trop. Afr. 2: 534 (1871)
      1 species, western tropical Africa
   Posadaea Cogn., Bull. Acad. Roy. Sci. Belgique, Ser. 2, 20: 476 (1890)

1 species, Neotropics

Melothria L., Sp. Pl. 1: 35 (1753) and Gen. Pl., ed. 5: 21 (1754)

   ?*Alternasemina* S. Manso, Enum. Subst. Braz.: 35 (1836)

   Diclidostigma G. Kunze, Linnaea 17: 576 (1844)

   Allagosperma M. J. Roem., Fam. Nat. Syn. Monogr. 2: 15 (1846)

   Landersia MacFady., Fl. Jamaica 2: 142 (1850)

   12 species, New World

Muellerargia Cogn. in A. & C. DC., Monogr. Phan. 3: 630 (1881)

   2 species, Madagascar and Malesia

Zehneria Endl., Prodr. Fl. Norfolk: 69 (1833)

   Pilogyne Eckl. ex Schrad., Index Sem. Hort. Gott. 1835: 5 (1835)

   Cucurbitula (M. J. Roem.) O. Kuntze ex Post & O. Kuntze, Lex. Gen. Phan.: 152 (1903) non Fuckel (1870)

   About 35 species, Paleotropics

Cucumella Chiov., Fl. Somala 1: 183 (1929)

   7 species, tropical Africa, Madagascar, and India

Cucumis L., Sp. Pl. 2: 1011 (1753) and Gen. Pl., ed. 5: 440 (1754)

   Melo Mill, Gard. Dict. Abridg., ed. 4 (1754)

   About 30 species, Paleotropics, mainly tropical and southern Africa

Oreosyce Hook. f. in D. Oliver. Fl. Trop. Afr. 2: 548 (1871)

   Hymenosicyos Chiov., Ann. Bot. (Rome) 9: 62 (1911)

   2 species, tropical Africa and Madagascar

Myrmecosicyos C. Jeffr., Kew Bull. 15: 357 (1962)

   1 species, Kenya

Mukia Arn., Madras J. Lit. Sci. 12: 50 (1840)

   4 species, Paleotropics

Dicoelospermum C. B. Clarke in Hook. f., Fl. Brit. India 2: 630 (1879) corr. Post & O. Kuntze, Lex. Gen. Phan.: 172 (1903)

   1 species, India

4. Subtribus Trochomeriinae C. Jeffr., Kew Bull. 15: 342 (1962)

Seeds large, tumid, with thick, hard testa

4 genera, 13 species, Africa, Asia, and West Malesia

   Solena Lour., Fl. Cochinch.: 477, 514 (1790) non *Solenia* Pers. ex Fries (1821)

     Karivia Arn., Madras J. Lit. Sci. 12: 50 (1840)

     Juchia M. J. Roem., Fam. Nat. Syn. Monogr. 2: 11 (1846)

     Harlandia Hance ex Walp., Ann. Bot. Syst. 2: 648 (1852)

     1 species, tropical Asia and Malesia

   Trochomeria Hook. f. in Benth. & Hook. f., Gen. Pl. 1: 822 (1867)

     Heterosicyos Welw. ex. Hook. f. in Benth. & Hook. f., Gen. Pl. 1: 822 (1867)

     8 species, tropical and southern Africa

Dactyliandra (Hook. f.) Hook. f. in D. Oliver, Fl. Trop. Afr. 2: 557 (1871)
>2 species, tropical Africa and India
Ctenolepis Hook. f. in Benth. & Hook. f., Gen. Pl. 1: 832 (1867)
>*Ctenopsis* Naud., Ann. Sci. Nat. Bot., Ser. 5, 6: 12 (1866) non De Not. (1847)
>*Blastania* Kotschy & Peyr., Pl. Tinn.: 15 (1867)
>2 species, tropical Africa and India
B. Tribus Schizopeponeae C. Jeffr., Kew Bull. 17: 475 (1964)
Like Melothrieae, but ovules pendulous, solitary in each loculus; fruit 3-valved; and pollen grains small, tricolporate, reticuloid
1 genus, 8 species, eastern Asia
>*Schizopepon* Maxim., Mem. Sav. Etr. Acad. St. Petersb. 9: 110 (1859)
>>8 species, eastern Asia
C. Tribus Joliffieae Schrad., Linnaea 12: 402 (1838)
Flowers comparatively large, with short, shallow hypanthium, especially in female flowers; anther thecae simple to moderately complex; petals with basal ventral scales or sometimes fimbriate; pollen grains tricolporate, reticulate
5 genera, 76 species, Old World
>1. Subtribus Thladianthinae Pax in Engl. & Prantl., Nat. Pflanzenfam. IV, 5: 13 (1889)
Petals with basal ventral scales, not fringed
4 genera, 73 species, Asia, Malesia, and tropical Africa
>*Indofevillea* Chatterjee, Nature 158: 345 (1946) and Kew Bull. 2: 119 (1947)
>>1 species, eastern Himalayas
>*Siraitia* Merrill, Pap. Michigan Acad. Sci. 19: 200 (1934)
>>*Neoluffa* Chakravarty, J. Bombay Nat. Hist. Soc. 50: 895 (1952)
>>*Thladiantha* Bunge subg. *Microlagenaria* C. Jeffr., Kew Bull. 15: 363 (1962)
>>7 species, Southeast Asia and tropical Africa
>*Thladiantha* Bunge, Enum. Pl. China Bor.: 29 (1833)
>>About 20 species, eastern Asia and Indomalesia
>*Momordica* L., Sp. Pl. 2: 1009 (1753) and Gen. Pl., ed. 5: 440 (1754)
>>*Zucca* Comm. ex Juss., Gen. Pl.: 398 (1789)
>>*Muricia* Lour., Fl. Cochinch.: 542, 596 (1790)
>>*Neurosperma* Raf., Amer. Monthly Mag. Crit. Rev. 4: 40 (1818)
>>*Dimorphochlamys* Hook. f. in Benth. & Hook. f., Gen. Pl. 1: 827 (1867)
>>*Raphanocarpus* Hook. f., Hooker's Icon. Pl. 11: 67, t. 1084 (1871)

Raphanistrocarpus (Baill.) Pax in Engl. & Prantl, Nat. Pflanzen-
fam. IV, 5: 22, 25 (1889)
Eulenbergia Pax, Bot. Jahrb. Syst. 39: 654 (1907)
Kedrostis Medic. sect. *Gilgina* Cogn. in Engl., Pflanzenreich 66
(IV. 275, 1): 139, 155 (1916)
Calpidosicyos Harms, Notizbl. Bot. Gart. Berlin-Dahlem 8: 480
(1923)
About 45 species, Paleotropics
2. Subtribus Telfairiinae Pax in Engl. & Prantl, Nat. Pflanzenfam. IV,
5: 22 (1889)
Petals without basal ventral scales, fringed
1 genus, 3 species, tropical Africa
Telfairia Hook., Bot. Mag. 54: t. 2751 and 2752 (1827)
Joliffia Bojer ex Delile, Mem. Soc. Hist. Nat. Paris 3: 318 (1827)
3 species, tropical Africa
D. Tribus Trichosantheae C. Jeffr., Kew Bull. 15: 341 (1962)
Flowers comparatively large, with hypanthium long in both male and
female flowers; anther thecae simple to complex; petals fringed or entire;
pollen grains tricolporate, variously ornamented
10 genera, 75 species, Old World
1. Subtribus Hodgsoniinae C. Jeffr., Kew Bull. 15: 342 (1962)
Petals fringed; ovules 12, in 6 collateral pairs; fruit fleshy, seeds very
large; pollen reticulate
1 genus, 1 species, Indomalesia
Hodgsonia Hook. f. & T. Thoms., Proc. Linn. Soc. London 2: 257
(1854)
1 species, Indomalesia
2. Subtribus Ampelosicyinae C. Jeffr., Kew Bull. 15: 341 (1962)
Petals fringed or not; ovules numerous; fruits fleshy; seeds very large to
small; pollen lirate
2 genera, 23 species, Africa, Madagascar, and Seychelles
Ampelosicyos Thouars, Hist. Veg. Isles Austr. Afr.: 68 (1805)
Delognaea Cogn., Bull. Mens. Soc. Linn. Paris 1: 425 (1884)
3 species, Madagascar
Peponium Engl. in Engl. & Prantl, Nat. Pflanzenfam. Nachtr.: 318
(1897)
Peponia Naud., Ann. Sci. Nat. Bot., Ser. 5, 5: 29 (1866) non Grev.
(1863)
Peponiella O. Kuntze, Rev. Gen. Pl. 3(2): 131 (1898)
About 20 species, tropical and southern Africa, Madagascar, Al-
dabra, and Seychelles
3. Subtribus Trichosanthinae Pax in Engl. & Prantl, Nat. Pflanzenfam.
IV, 5: 31 (1889)

Petals fringed or not; ovules numerous to few; fruit fleshy; seeds rather small; pollen tenuiexinous, with incrassate apertures
3 genera, 45 species, Madagascar, tropical Asia, Malesia, and tropical Australasia

    Gymnopetalum Arn., Madras J. Lit. Sci. 12: 52 (1840)
        Tripodanthera M. J. Roem., Fam. Nat. Syn. Monogr. 2: 48 (1846)
        Scotanthus Naud., Ann. Sci. Nat. Bot., Ser. 4, 16: 172 (1862)
        3 species, Indomalesia
    Trichosanthes L., Sp. Pl. 2: 1008 (1753) and Gen. Pl., ed. 5: 439 (1754)
        Anguina Mill., Fig. Pl. 21 (1755) non Jacq. (1760)
        Cucumeroides Gaertn., Fruct. Sem., Pl. 2: 485 (1791)
        Involucraria Ser., Mem. Soc. Phys. Geneve 3 (1): 27, t. 5 (1825)
        Eopepon Naud., Ann. Sci. Nat. Bot., Ser. 5, 5: 31 (1866)
        Platygonia Naud., Ann. Sci. Nat. Bot., Ser. 5, 5: 33 (1866)
        About 40 species, eastern Asia, Indomalesia, tropical Australia to Fiji; subtropical Eurasia in Miocene and Pliocene
    Tricyclandra Keraudren, Bull. Soc. Bot. France 112: 327 (1966)
        1 species, Madagascar
4. Subtribus Herpetosperminae C. Jeffr., Kew Bull. 15: 342 (1962)
Petals not fringed; ovules numerous; fruits fleshy, or dry and dehiscent into 3 valves; seeds small; pollen suprabaculate, pilate
4 genera, 5 species, northeastern tropical Africa and eastern Himalayas
    Cephalopentandra Chiov., Fl. Somala 1: 187 (1929)
        1 species, northeastern tropical Africa
    Biswarea Cogn., Bull. Soc. Bot. Belgique 21(2): 16 (1882)
        Warea C. B. Clarke, J. Linn. Soc. Bot. 15: 127 (1876) non Nutt. (1834)
        2 species, eastern Himalayas
    Herpetospermum Wall. ex Hook. f., in Benth. & Hook. f., Gen. Pl. 1: 834 (1867)
        Rampinia C. B. Clarke, J. Linn. Soc. Bot. 15: 129 (1876)
        1 species, eastern Himalayas
    Edgaria C. B. Clarke, J. Linn. Soc. Bot. 15: 113 (1876)
        1 species, eastern Himalayas
E. Tribus Benincaseae Ser., Mem. Soc. Phys. Geneve 3(1): 25 (1825)
Flowers comparatively large; hypanthium shallow, especially in female flowers; anther thecae complex; petals entire; pollen grains tricolporate, usually reticulate, rarely spinulose
17 genera, 85 species, all Old World except for three species of *Luffa*
  1. Subtribus Benincasinae (Ser.) C. Jeffr., Kew Bull. 15: 339 (1962)
  Fruit fleshy, not operculate
17 genera, 80 species, Old World

Cogniauxia Baill., Bull. Mens. Soc. Linn. Paris 1: 423 (1884)
  Cogniauxiella Baill., Bull. Mens. Soc. Linn. Paris 1: 424 (1884)
  2 species, tropical Africa
Ruthalicia C. Jeffr., Kew Bull. 15: 360 (1962)
  2 species, western tropical Africa
Lagenaria Ser., Mem. Soc. Phys. Geneve 3(1): 26, t. 2 (1825)
  Adenopus Benth. in Hook., Niger. Fl.: 372 (1849)
  Sphaerosicyos Hook. f. in Benth. & Hook. f., Gen. Pl. 1: 824 (1867)
  6 species, tropical Africa, Madagascar, Indomalesia, and Neotropics
Benincasa Savi, Biblioth. Ital. (Milan) 9: 158 (1818)
  Camolenga Post & O. Kuntze, Lex. Gen. Phan.: 95 (1903) nom. illegit.
  1 species, Southeast Asia and Malesia
Praecitrullus Pang., Bot. Zhurn. SSSR 29: 203 (1944)
  1 species, India
Citrullus Schrad. ex Eckl. & Zeyh., Enum. Pl. Afric. Austral.: 270 (1836) nom. cons.
  Anguria Mill., Gard. Dict. Abridg., ed. 4 (1754) nom. rejic.
  Colocynthis Mill., Gard. Dict. Abridg., ed. 4 (1754) nom. rejic.
  3 species, Old World tropics and subtropics
Acanthosicyos Welw. ex. Hook. f. in Benth. & Hook. f., Gen. Pl. 1: 824 (1867)
  2 species, southern tropical Africa
Eureiandra Hook. f. in Benth. & Hook. f., Gen. Pl. 1: 825 (1867)
  8 species, tropical Africa and Socotra
Bambekea Cogn., Bull. Jard. Bot. Etat 5: 115 (1916)
  1 species, tropical Africa
Nothoalsomitra Telford, Fl. Austral. 8: 388 (1982)
  1 species, Queensland
Coccinia Wight & Arn., Prodr.: 347 (1834)
  Cephalandra Schrad. ex Eckl. & Zeyh., Enum. Pl. Afric. Austral.: 280 (1836)
  Physedra Hook. f. in Benth. & Hook. f., Gen. Pl. 1: 827 (1867)
  Staphylosyce Hook. f. in Benth. & Hook. f., Gen. Pl. 1: 828 (1867)
  About 30 species, Paleotropics, mostly African
Diplocyclos (Endl.) Post & O. Kuntze, Lex. Gen. Phan.: 178 (1903) corr. C. Jeffr., Kew Bull. 15: 354 (1962)
  Ilocania Merrill, Philipp. J. Sci. 13: 65 (1918)
  About 4 species, Paleotropics
Raphidiocystis Hook. f. in Benth. & Hook. f., Gen. Pl. 1: 828 (1867)
  5 species, tropical Africa and Madagascar

Lemurosicyos Keraudren, Bull. Soc. Bot. France 110: 405 (1964)
  1 species, Madagascar
Zombitsia Keraudren, Adansonia, Ser. 2, 3: 167 (1963)
  1 species, Madagascar
Ecballium A. Rich. in Bory de St. Vincent, Dict. Class. Hist. Nat. 6: 19 (1824) nom. cons.
  Elaterium Mill., Gard. Dict. Abridg., ed. 4 (1754) nom. rejic. non Jacq. (1760)
  1 species, Mediterranean region and western and central Asia
Bryonia L. Sp. Pl. 2: 1012 (1753) and Gen. Pl., ed. 5: 442 (1754)
  12 species, Europe, Mediterranean region and western and central Asia
2. Subtribus Luffinae C. Jeffr., Kew Bull. 15: 340 (1962)
Fruit dry, fibrous, operculate
1 genus, 7 species, tropics
  Luffa Mill., Gard. Dict. Abridg., ed. 4 (1754)
    Trevouxia Scop., Introd.: 152 (1777)
    Turia Forssk. ex J. F. Gmel., Syst. Nat., ed. 13, 2: 303 (1791)
    Poppya Neck. ex M. J. Roem., Fam. Nat. Syn. Monogr. 2: 59 (1846)
    7 species, 4 Paleotropics, 3 Neotropics
F. Tribus Cucurbiteae
Flowers comparatively large; hypanthium short, especially in female flowers; anther thecae complex; petals entire; pollen pantoporate, spiny
12 genera, 110 species, all Neotropical except for one Afro-Madagascan species of *Cayaponia*
  Cucurbita L., Sp. Pl. 2: 1010 (1753) and Gen. Pl., ed. 5: 441 (1754)
    Melopepo Mill., Gard. Dict. Abridg., ed. 4 (1754)
    Pepo Mill., Gard. Dict. Abridg., ed. 4 (1754)
    Ozodycus Raf., Atlantic J. 1: 145 (1832)
    ?*Sphenantha* Schrad., Linnaea 12: 416 (1838)
    Mellonia Gasparr., Rendiconti Reale Accad. Sci. Fis. 6: 411 (1847)
    Pileocalyx Gasparr., Ann. Sci. Nat. Bot., Ser. 3, 9: 220 (1848) nom. rejic.
    Tristemon Scheele, Linnaea 21: 586 (1848) non Raf. (1819) nec Raf. (1838) nec Klotsch (1838)
    About 20 species, New World
  Sicana Naud., Ann. Sci. Nat. Bot., Ser. 4, 18: 180 (1863)
    3 species, Neotropics
  Tecunumania Standl. & Steyerm., Field Mus. Nat. Hist., Bot. Ser. 23: 96 (1944)
    1 species, Central America
  Calycophysum Karst. & Triana in Triana, Nuev. Jen. Esp.: 20 (1854)

Edmondia Cogn. in A. & C. DC., Monogr. Phan. 3: 420 (1881) non Cass. (1818)

Bisedmondia Hutch., Gen. Fl. Pl. 2: 398 (1967)

5 species, tropical South America

Peponopsis Naud., Ann. Sci. Nat. Bot., Ser. 4, 12: 88 (1860)

1 species, Mexico

Anacaona A. H. Liogier, Phytologia 47: 190, fig. 10. (1980)

1 species, Hispaniola

Polyclathra Bertol., Novi Comment. Acad. Sci. Inst. Bononiensis 4: 438 (1840) and Fl. Guatimal.: 38 (1840)

Pittiera Cogn., Bull. Soc. Roy. Bot. Belgique 30: 271 (1891)

Roseanthus Cogn., Contr. U.S. Natl. Herb. 3: 577 (1896)

1 species, Central America

Schizocarpum Schrad., Index Sem. Hort. Gott. 1830: 4 (1830)

6 species, Central America

Penelopeia Urb., Feddes Repert. Spec. Nov. Regni Veg. 17: 8 (1921)

1 species, Hispaniola

Cionosicyos Griseb., Fl. Brit. W. I.: 288 (1860) corr. Hook. f. in Benth. & Hook. f., Gen. Pl. 1: 826 (1867)

3 species, Central America and West Indies

Cayaponia S. Manso, Enum. Subst. Braz.: 31 (1836) nom. cons.

Dermophylla S. Manso, Enum. Subst. Braz.: 30 (1836)

Druparia S. Manso, Enum. Subst. Braz.: 35 (1836) non Raf. (1808) nec Clairv. (1811)

Perianthopodus S. Manso, Enum. Subst. Braz.: 28 (1836)

Arkezostis Raf., New Fl. 4: 100 (1838)

Trianosperma (Torr. & A. Gray) Mart., Syst. Mater. Med. Brasil.: 79 (1843)

Cionandra Griseb., Fl. Brit. W. I.: 286 (1860)

Antagonia Griseb., Abh. Konigl. Ges. Wiss. Gottingen 19: 96 (1874)

About 60 species, subtropical and tropical Americas; 1 species, tropical Africa and Madagascar

Selysia Cogn. in A. & C. DC., Monogr. Phan. 3: 735 (1881)

3 species, tropical South America

Abobra Naud., Rev. Hort. 1862: 111 (1862) and Ann. Sci. Nat. Bot., Ser. 4, 16: 196 (1862)

1 species, subtropical South America

G. Tribus Sicyeae Schrad., Linnaea 12: 407 (1838)

Flowers small; anther thecae complex; petals entire; nectaries trichomatous; fruit fleshy, fibrous, dry or woody, operculate, elastically dehiscent or indehiscent; pollen grains 4–10-colporate

19 genera, 140 species, New World, Hawaiian Islands, and Australasia

1. Subtribus Cyclantherinae C. Jeffr. subtribus nov., typus *Cyclanthera*
Schrad., ovulis ascendentibus et pollinis granis 4–8-colporatis punc-
titegillatis distincta
Fruit fibrous or fleshy, operculate or elastically dehiscent; pollen grains
4–8-colporate, punctitegillate
12 genera, 75 species, New World

  Hanburia Seem., Bonplandia 6: 293 (1858)
    Nietoa Seem. ex Shaffn., Naturaleza (Mexico City) 3: 343 (1876)
    2 species, Central America
  Echinopepon Naud., Ann. Sci. Nat. Bot., Ser. 5, 6: 17 (1866)
    About 15 species, New World
  Marah Kellogg, Proc. Calif. Acad. Sci. 1: 38 (1854)
    Megarrhiza Torr. & A. Gray, Rep. Explor. Railr. Pacific 12, 2: 61
    (1860–61)
    7 species, Pacific North America
  Echinocystis Torr. & A. Gray, Fl. N. Amer. 1: 542 (1840) nom.
cons.
    Micrampelis Raf., Med. Repos., Ser. 2, 5: 350 (1808) nom. rejic.
    Hexameria Torr. & A. Gray in Torr., Rep. Pl. New York: 137
    (1839) non R. Br. (1838)
    1 species, North America
  Vaseyanthus Cogn., Zoe 1: 368 (1891)
    Pseudoechinopepon (Cogn.) Cockerell, Bot. Gaz. (Crawfordville)
    24: 378 (1897) nom. illegit.
    1 species, Pacific North America
  Brandegea Cogn., Proc. Calif. Acad. Sci. Ser. 2, 3: 58 (1891)
    1 species, Pacific North America
  Apatzingania Dieterle, Brittonia 26: 131 (1974)
    1 species, Mexico
  Cremastopus P. Wils., Hooker's Icon. Pl. 36: t. 3586 (1962)
    Heterosicyos (S. Wats.) Cockerell, Bot. Gaz. (Crawfordville) 24:
    378 (1897) nom. illegit., non Welw. ex Hook. f. (1867)
    3 species, Mexico
  Elateriopsis Ernst, Flora 56: 257 (1873)
    5 species, tropical Central and South America
  Pseudocyclanthera Mart. Crov., Notul. Syst. (Paris) 15: 56 (1955)
    1 species, subtropical South America
  Cyclanthera Schrad., Index Sem. Hort. Gott. 1831: 2 (1831)
    Discanthera Torr. & A. Gray, Fl. N. Amer. 1: 696 (1840)
    About 30 species, Neotropics
  Rytidostylis Hook. & Arn., Bot. Beechey Voy.: 424 (1840)
    Elaterium Jacq., Enum. Syst. Pl.: 9, 31 (1760) non Mill. (1754)
    9 species, Neotropics

2. Subtribus Sicyinae C. Jeffr. subtribus nov., typus *Sicyos* L., ovario uniloculari, ovulo solitario pendulo et pollinis granis 7–10-colporatis spinulosis distincta

Fruit fleshy, dry or woody, indehiscent, 1-seeded; ovule solitary, pendulous; pollen grains 7–10-colporate, spinulose

7 genera, 65 species, New World, Hawaiian Islands, and Australasia

    Sicyos L., Sp. Pl. 2: 1013 (1753) and Gen. Pl., ed. 5: 443 (1754)

        Sicyoides Mill., Gard. Dict. Abridg., ed. 4 (1754)

        Anomalosicyos Gentry, Bull. Torrey Bot. Club 73: 565 (1946)

        Sicyocaulis Wiggins, Madrono 20: 251 (1970)

        Skottsbergiliana St. John, Pacific Sci. 28: 457 (1975)

        Cladocarpa (St. John) St. John, Bot. Jahrb. Syst. 99: 491 (1978)

        Sarx St. John, Bot. Jahrb. Syst. 99: 491 (1978)

        Sicyocarya (A. Gray) St. John, Bot. Jahrb. Syst. 100: 246 (1978)

        Costarica L. Gomez, Phytologia 53: 97 (1983)

    About 40 species, New World, Hawaiian Islands, and Australasia

    Sicyosperma A. Gray, Smithsonian. Contr. Knowl. 5: 6; Pl. Wright. 2: 62 (1853)

        1 species, southern United States

    Parasicyos Dieterle, Phytologia 32: 289 (1975)

        1 species, Guatemala

    Microsechium Naud., Ann. Sci. Nat. Bot., Ser. 5, 6: 25 (1866)

        1 species, Mexico

    Sechium P. Br., Civ. Nat. Hist. Jamaica: 355 (1756) nom. cons.

        Chocho Adans., Fam. Pl. Pl. 2: 500, 538 (1763) nom. illegit.

        Chayota Jacq., Sel. Stirp. Am. Hist.: 125, t. 245 (1780)

        Frantzia Pitt., Contr. U.S. Natl. Herb. 13: 127 (1910)

        Polakowskia Pitt., Contr. U.S. Natl. Herb. 13: 131 (1910)

        Ahzolia Standl. & Steyerm., Field Mus. Nat. Hist., Bot. Ser. 23: 92 (1944)

        8 species, Central America

    Sechiopsis Naud., Ann. Sci. Nat. Bot., Ser. 5, 6: 23 (1866)

        2 species, Mexico

    Pterosicyos T. Brandegee, Univ. Calif. Publ. Bot. 6: 72 (1914)

        1 species, Central America

H. Subfamilia Cucurbitoideae incertae sedis

    Odosicyos Keraudren, Bull. Soc. Bot. France 127, Lettres Bot. 1980 (5): 518 (1981) and Bull. Soc. Bot. France 129, Lettres Bot. 1982 (2): 149–151 (1982)

        1 species, Madagascar

III. Genera fossilia incertae sedis

Cucurbitites E. W. Berry, John Hopkins Univ. Stud. Geol. 10: 168 (1929)

Tertiary; seed

Cucurbitospermum Chesters, Palaeontographia, Abt. B, Palaophytol. 101: 57, t.21, f. 9–10 (1957)

Miocene, Kenya; Eocene, England; seeds

Index

465